p-Adic Analysis, Arithmetic and Singularities

CONTEMPORARY MATHEMATICS

778

p-Adic Analysis, Arithmetic and Singularities

UIMP-RSME Lluís A. Santaló Summer School
p-Adic Analysis, Arithmetic and Singularities
June 24–28, 2019
Universidad Internacional Menéndez Pelayo
Santander, Spain

Carlos Galindo
Alejandro Melle Hernández
Julio José Moyano-Fernández
Wilson A. Zúñiga-Galindo
Editors

AMS AMERICAN MATHEMATICAL SOCIETY

Real Sociedad Matemática Española

2020 *Mathematics Subject Classification.* Primary 11S40, 14B05, 14E18, 14H20, 14E15, 32P05, 32S45, 82D05, 11S80.

Library of Congress Cataloging-in-Publication Data

Names: Galindo, Carlos, 1962– editor.
Title: p-adic analysis, arithmetic and singularities : UIMP-RSME Lluís A. Santaló Summer School, p-adic Analysis, Arithmetic and Singularities, June 24–28, 2019, Universidad Internacional Menéndez Pelayo, Santander, Spain / Carlos Galindo, Alejandro Melle Hernández, Julio José Moyano-Fernández, Wilson A. Zúñiga-Galindo, editors.
Description: Providence : American Mathematical Society, [2022] | Series: Contemporary mathematics, 0271-4132 ; volume 778 | Includes bibliographical references.
Identifiers: LCCN 2021059076 | ISBN 9781470467791 (paperback) | ISBN 9781470469764 (ebook)
Subjects: LCSH: p-adic analysis–Study and teaching. | AMS: Number theory – Algebraic number theory: local and p-adic fields – Zeta functions and L-functions. | Algebraic geometry – Local theory in algebraic geometry – Singularities in algebraic geometry. | Algebraic geometry – Birational geometry – Arcs and motivic integration. | Algebraic geometry – Curves – Singularities, local rings. | Algebraic geometry – Birational geometry – Global theory and resolution of singularities. | Several complex variables and analytic spaces – Non-Archimedean analysis – Non-Archimedean analysis. | Several complex variables and analytic spaces – Complex singularities – Modifications; resolution of singularities : Lefschetz theorems, topological classification, invariants. | Statistical mechanics, structure of matter – Applications of statistical mechanics to specific types of physical systems – Statistical mechanical studies of gases. | Number theory – Algebraic number theory: local and p-adic fields – Other analytic theory (analogues of beta and gamma functions, p-adic integration, etc.).
Classification: LCC QA241 .P225 2022 | DDC 512.7/4–dc23/eng20220406
LC record available at https://lccn.loc.gov/2021059076

Contemporary Mathematics ISSN: 0271-4132 (print); ISSN: 1098-3627 (online)
DOI: https://doi.org/10.1090/conm/778

---

10 9 8 7 6 5 4 3 2 1   27 26 25 24 23 22

Contents

Preface

The present volume reflects the contents of some of the courses and talks given at the "XX School of Mathematics *Lluís Santaló*: Research Summer School on p-adic Analysis, Arithmetic and Singularities" held at the Magdalena Palace in Santander (Spain), from June 24th to June 28th, 2019, under the support of the UIMP–Universidad Internacional Menéndez Pelayo and the RSME–Real Sociedad Matemática Española. The volume also contains some related contributions.

The School aimed to provide an introduction to a very active research area lying at the intersection of number theory, p-adic analysis, algebraic geometry and singularity theory. The main objects of study here are the *local zeta functions* and some algebraic-combinatorial counterparts called Poincaré series (sometimes Hilbert-Poincaré series), which are in some sense a kind of generating functions associated with suitable ring filtrations. The local zeta functions are ubiquitous objects in mathematics and physics.

There are several types of local zeta functions, for instance p-adic, Archimedean, topological and motivic, among others. In the Archimedean setting, the local zeta functions were introduced in the 50s by Gel'fand and Shilov. The main motivation was that the meromorphic continuation of Archimedean local zeta functions implies the existence of fundamental solutions (i.e. Green functions) for differential operators with constant coefficients. This fact was established, independently, by Atiyah and Bernstein.

In the 60s, Weil studied local zeta functions, in the Archimedean and non-Archimedean settings, in connection with the Poisson-Siegel formula. In the 70s, Igusa developed a uniform theory for local zeta functions and oscillatory integrals, with polynomial phase, in characteristic zero. In the p-adic setting, the local zeta functions are connected with the number of solutions of polynomial congruences mod p^m and with exponential sums mod p^m. Recently Denef and Loeser introduced the motivic zeta functions which constitute a vast generalization of p-adic local zeta functions. There are many intriguing conjectures relating the poles of the local zeta functions with the eigenvalues of the complex monodromy and with the roots of the Bernstein polynomials.

The p-adic strings seem to be related in some interesting ways with ordinary strings. For instance, connections through the adelic relations and through the limit when $p \to 1$. This limit gives rise to a boundary string field theory. Denef and Loeser established that the limit $p \to 1$ of Igusa's local zeta function provides an object, called topological zeta function. By using Denef-Loeser's theory of topological zeta functions, it was established recently that limits $p \to 1$ of tree-level p-adic string amplitudes give rise to certain amplitudes, which are connected with the boundary string field theory mentioned.

The construction of the motivic Igusa zeta functions requires techniques of motivic integration introduced by Kontsevich. On the other hand, Campillo, Delgado and Gusein-Zade, using integration with respect to the Euler characteristic (introduced by Viro), defined the motivic Poincaré series associated with an ideal filtration of a one-dimensional local ring. This Poincaré series is a complete topological invariant of a curve singularity: it coincides (up to a factor in the particular case of an irreducible curve) with the Alexander polynomial, and with the A'Campo zeta function of the singularity.

During the School four courses were offered, including coordinated exercise sessions, and also talks presenting current research problems. The School was aimed at doctoral and postdoctoral students working on themes broadly connected to the issues in the School. According to this spirit, we divide the volume in two parts. The first part contains a series of five lecture notes enlightening the basics of the topics of the School:

- ◇ The article "Archimedean zeta functions and oscillatory integrals" by Edwin León-Cardenal is a short survey about the relation between the Archimedean zeta functions (which were called "complex powers" originally) and oscillatory integrals. The survey provides a historical background, also it presents some central results, including a few key arguments, and it concludes with a challenging question.
- ◇ Julio J. Moyano-Fernández contributes with a survey entitled "Generalized Poincaré series for plane curve singularities", on Poincaré series of multi-index filtrations associated with plane complex curve singularities. The survey starts with an introduction to resolutions of plane curve singularities and their Poincaré series. Then, an interpretation of the Poincaré series as an integral with respect to the Euler characteristic is presented. Finally, the motivic Poincaré series of Campillo, Delgado and Gusein-Zade is introduced.
- ◇ Naud Potemans and Willem Veys give an introduction to p-adic Igusa zeta functions; it contains a careful choice of examples and exercises leading to make it easier to understand the relation of zeta functions with singularity theory, particularly with embedded resolutions. The authors consider, in full generality, the case where the polynomials are defined on fields of characteristic zero and on rings of algebraic integers.
- ◇ In the article "An introduction to p-adic and motivic integration, zeta functions and invariants of singularities", Juan Viu-Sos connects p-adic integration with motivic integration, remarking the main similarities and differences between the two theories. The survey devotes a special attention to the definition of motivic measure in the space of arcs and the formula of change of variables; after this, a relation between the Hodge invariants of birationally equivalent varieties is established.
- ◇ Wilson A. Zúñiga-Galindo contributes with a survey titled "p-adic analysis: a quick introduction", which offers an interesting introduction (with some proofs, exercises and references) to p-adic numbers, p-adic integration and harmonic analysis. This survey offers a fast way to understand the subject assuming basic knowledge in algebra and analysis.

The second part of the volume consists of four research articles dealing with current matters covered by the School:

- ◇ The article "On maximal order poles of generalized topological zeta functions" by Enrique Artal Bartolo and Manuel González Villa shows some examples of topological zeta functions associated to an isolated plane curve singular point and an allowed, in the sense of Némethi and Veys, differential form that have several poles of order two.
- ◇ José Ignacio Cogolludo-Agustín, Tamás László, Jorge Martín-Morales, and András Némethi in "Local invariants of minimal generic curves on rational surfaces" consider the non-easy problem of deciding which analytic invariants are in fact topological. They study the case of reduced curve germs $(C, 0)$ in a normal surface singularity $(X, 0)$. In fact, they recover the delta invariant of C from the topology of the embedding $(C, 0) \subseteq (X, 0)$.
- ◇ In the article "Motivic Poincaré series of cusp surface singularities" by János Nagy and András Némethi, the authors study both the equivariant multivariable analytical and the topological Poincaré series of a germ of curve, as well as their motivic versions. The authors show that they are equal and provide an explicit combinatorial expression for them. The key ingredient is a motivic multivariable series associated with the space of effective Cartier divisors of the reduced exceptional curve.
- ◇ "Non-archimedean electrostatics" by Christopher D. Sinclair discusses the p-adic analogues of 1D log-Coulomb gases and their connections with local zeta functions, which provides an interesting link between number theory and physics.

Finally, we would like to mention that the "XX School of Mathematics *Lluís Santaló*" included a parallel session under the motto *Civil dislocation and mathematical roots*. This consisted of three open talks by Antonio Campillo, Luis Español and Sebastiá Xambó-Descamps on the work and life of Lluís Santaló Sors, the Spanish mathematician who names the Research School.

The sponsors of this School include the UIMP–Universidad Internacional Menéndez Pelayo, the Spanish Ministry called "Ministerio de Ciencia, Innovación y Universidades" at that time, as well as the Royal Spanish Mathematical Society (Real Sociedad Matemática Española–RSME). We thank also the American Mathematical Society (AMS) and RSME for agreeing to publish this volume as one of their common publications. Last but not least we would like to thank all organizers, speakers and participants who contributed to the occasion of this successful summer school.

Carlos Galindo
Alejandro Melle Hernández
Julio J. Moyano-Fernández
Wilson A. Zúñiga-Galindo

Speakers and members of the organizing committee

- ◇ Juan Manuel Burgos, CINVESTAV, Mexico
- ◇ Antonio Campillo López, Universidad de Valladolid, Spain
- ◇ Félix Delgado de la Mata, Universidad de Valladolid, Spain
- ◇ Luis Español González, Universidad de La Rioja, Spain
- ◇ Carlos Galindo Pastor, Universitat Jaume I, Spain
- ◇ Sabir M. Gusein-Zade, Moscow State University, Russia
- ◇ Edwin León-Cardenal, CIMAT, Mexico
- ◇ Francisco Marcellán Español, Universidad Politécnica de Madrid, Spain
- ◇ Alejando Melle Hernández, Universidad Complutense de Madrid, Spain
- ◇ Francisco José Monserrat Delpalillo, Universitat Politècnica de València, Spain
- ◇ Julio José Moyano-Fernández, Universitat Jaume I, Spain
- ◇ Christopher D. Sinclair, University of Oregon, USA
- ◇ Willem Veys, Katholieke Universiteit Leuven, Belgium
- ◇ Sebastiá Xambó-Descamps, Universitat Politècnica de Catalunya, Spain
- ◇ Wilson A. Zúñiga-Galindo, University of Texas Rio Grande Valley, USA, and CINVESTAV, Mexico

List of Participants

Patricio Almirón
Universidad Complutense de Madrid
Spain

Guillem Blanco
Universitat Politècnica de Catalunya
Spain

Juan Manuel Burgos
CINVESTAV
Mexico

Víctor Burgos Guerrero
Universidad de Concepción
Chile

Daniel Camazón Portela
Universidad de Valladolid
Spain

Antonio Campillo López
Universidad de Valladolid
Spain

Félix Delgado de la Mata
Universidad de Valladolid
Spain

Luis Español González
Universidad de La Rioja
Spain

Carlos Galindo Pastor
Universitat Jaume I
Spain

Karol Gryszka
Pedagogical University of Cracow
Poland

Sabir M. Gusein-Zade
Moscow State University
Russia

Carlos Rodrigo Guzmán Durán
Universidad Complutense de Madrid
Spain

Beata Hejmej
Pedagogical University of Cracow
Poland

Zeynep Kısakürek
Heinrich-Heine-Universität Düsseldorf
Germany

Tamás László
MTA Rényi Institute of Mathematics
Hungary

Edwin León-Cardenal
CIMAT
Mexico

Paula M. Lins de Araujo
KU Leuven
Belgium

Francisco Marcellán Español
Universidad Politécnica de Madrid
Spain

Alejando Melle Hernández
Universidad Complutense de Madrid
Spain

Francisco J. Monserrat Delpalillo
Universitat Politècnica de València
Spain

Carlos-Jesús Moreno-Ávila
Universitat Jaume I
Spain

Julio José Moyano-Fernández
Universitat Jaume I
Spain

Naud Potemans
KU Leuven
Belgium

Ivan Proskurnin
Moscow State University
Russia

Otto Romero Germán
CINVESTAV
Mexico

Oswaldo Josué Sevilla Requeno
CIMAT
Mexico

Christopher D. Sinclair
University of Oregon
USA

Tao Song
University of Pennsylvania
USA

María de la Paz Tirado Hernández
Universidad de Sevilla
Spain

Siegfried Van Hille
KU Leuven
Belgium

Willem Veys
KU Leuven
Belgium

Juan Viu-Sos
IMPA
Brazil

Joe Webster
University of Oregon
USA

Brian Andrés Zambrano
CINVESTAV
Mexico

Sebastiá Xambó-Descamps
Universitat Politècnica de Catalunya
Spain

Andoni Zozaya Ursuegui
Euskal Herriko Unibertsitatea
Spain

Wilson A. Zúñiga-Galindo
University of Texas Rio Grande Valley, USA, and CINVESTAV
Mexico

Part I: Surveys

Contemporary Mathematics
Volume **778**, 2022
https://doi.org/10.1090/conm/778/15653

Archimedean zeta functions and oscillatory integrals

Edwin León-Cardenal

ABSTRACT. This note is a short survey of two topics: Archimedean zeta functions and Archimedean oscillatory integrals. We have tried to portray some of the history of the subject and some of its connections with similar devices in mathematics. We present some of the main results of the theory and at the end we discuss some generalizations of the classical objects.

1. Introduction

Probably the first Archimedean zeta function was the Gamma function, introduced by Euler in a letter to Golbach in 1729 with the aim to interpolate the factorial, see [**29**]. If the real part of the complex number s is positive, then the Gamma function is defined via the convergent improper integral

$$\Gamma(s) = \int_0^\infty x^{s-1} e^{-x}\, \mathrm{d}x. \tag{1.1}$$

It is not difficult to show that $\Gamma(s)$ converges absolutely on $\Re(s) > 0$. Moreover the function admits an analytic continuation as a meromorphic function that is holomorphic in the whole complex plane except the non-positive integers, where $\Gamma(s)$ has simple poles.

The Gamma function is just a particular case of the more general definition of an Archimedean zeta function. Take $\mathbb{K} = \mathbb{R}$ or $\mathbb{C}$, and fix an open set U in $\mathbb{K}^n$, by $\mathcal{S}(U)$ we will denote the Schwartz space of U, see Section 2.1. Consider a $\mathbb{K}$-analytic function $f : U \to \mathbb{K}$ and a function $\phi \in \mathcal{S}(U)$. The *local zeta function attached to the pair* (f, ϕ) is the parametric integral

$$Z_\phi(s, f) = \int_{\mathbb{K}^n \smallsetminus f^{-1}(0)} \phi(x)\, |f(x)|_{\mathbb{K}}^s\, |\mathrm{d}x|, \tag{1.2}$$

for $s \in \mathbb{C}$ with $\Re(s) > 0$, where $|\mathrm{d}x|$ is the Haar measure on $\mathbb{K}^n$. For uniformity reasons we will use for $a \in \mathbb{C}$ the convention $|a|_{\mathbb{K}} = \|a\|_{\mathbb{C}}^2$, where $\|a\|_{\mathbb{C}}$ is the standard complex norm.

It is easily seen that $Z_\phi(s, f)$ converges on the half plane $\{s \in \mathbb{C} \mid \Re(s) > 0\}$ and defines a holomorphic function there. Furthermore, $Z_\phi(s, f)$ admits a meromorphic

2020 *Mathematics Subject Classification.* Primary 14G10; Secondary 42B20,11M41,14B05, 14E15, 41A60, 32S05, 32S40.

Key words and phrases. Archimedean zeta functions, Bernstein-Sato polynomials, oscillatory integrals, oscillatory integral operators, multilinear level set operators, D-modules.

The author was partially supported by CONACYT, Grant # 286445.

continuation to the whole complex plane. This was proved by I. N. Bernstein and S. I. Gel'fand in [**17**], then independently by M. Atiyah in [**6**], both proofs make use of Hironaka's theorem on resolution of singularities [**46**]. Later, I. N. Bernstein [**16**] gave a proof by using the following functional equation:

$$P(s, x, \partial/\partial x) \cdot f(x)^{s+1} = b(s) f(x)^s.$$

Here $b(s) \in \mathbb{K}[s]$ and P denotes a polynomial in the 'variables' $s, x_1, \ldots, x_n, \partial/\partial x_1, \ldots, \partial/\partial x_n$ and with coefficients in $\mathbb{K}$. The monic generator of the ideal of functions satisfying the functional equation is called the *Bernstein-Sato polynomial of* f.

In fact, the theory of the Bernstein-Sato polynomials, also known as b-functions, was developed by M. Sato during the 60's at Kyoto under the name of algebraic theory of linear differential systems, as it is mentioned in the translation of his original work in [**92**]. This theory was framed in the theory of prehomogeneous vector spaces, see the comments in [**50**, Chapter 6] and [**40**]. Some years later, the advanced student of I. M. Gel'fand, I.N. Bernstein was thinking about similar questions in Moscow [**14, 15**]. Both works initiated the huge $\mathcal{D}$-module theory which has multiple and fruitful ties with algebraic geometry, singularity theory, topology of varieties, representation theory and (of course) differential equations, among others. See the nice surveys [**1, 40, 98**] and the books [**18, 28**].

The other topic that we treat in this note is *oscillatory integrals*, which are integrals of the form

$$I_\phi(\tau; f) = \int_{\mathbb{R}^n} \exp(\mathrm{i}\tau f(x))\, \phi(x)\, |\mathrm{d}x|,$$

for f real analytic, $\tau \in \mathbb{R}$ and $\phi \in \mathcal{S}(U)$. They can be considered as generalizations of the Fourier transform and can be traced back to the original works of Airy, Stokes, Lipschitz, and Riemann, see the nice historic remarks in the book by E. Stein [**94**, Chapter VIII]. They have played a major role in harmonic analysis, partial differential equations and number theory among others. The function f is called the *phase* and the function ϕ is called the *amplitude*. Here a central question is to know (or, better, estimate) how $I_\phi(\tau; f)$ decays when the real parameter τ tends to ∞. A general principle known as the stationary phase principle or the saddle point method states that the main contribution in the asymptotics of $I_\phi(\tau; f)$ is given by neighbourhoods of the critical points of the phase. Elementary methods can be used to study simple phases, cf. [**94**, Chapter VIII] and [**37**, Section III.4.5], but more complicated phases require more elaborated methods.

The general theorem on asymptotic expansions of $I_\phi(\tau; f)$ (c.f. Theorem 4.2) was proved first by Jeanquartier in [**53**], and then by Malgrange [**75**] and Igusa [**49**]. Again the necessary tool is resolution of singularities, and the conclusion is that the asymptotic behaviour is controlled by the poles of $Z_\phi(s, f)$.

In the proof given by Igusa in [**49**], he initiates the uniform treatment of the subject of local zeta functions over local fields $\mathbb{K}$ of characteristic zero, i.e. $\mathbb{K} = \mathbb{R}$, $\mathbb{C}$, $\mathbb{Q}_p$ or a finite extension of $\mathbb{Q}_p$. Since then, many authors have been working on this theory of local zeta functions, see the book by Igusa [**50**] and the references therein. The theory of p-adic zeta functions (also called Igusa zeta functions) has evolved greatly in the last 30 years, boosted mainly by the definition of the motivic zeta functions of J. Denef and F. Loeser in [**33**]. For more detailed descriptions of the work and legacy of Igusa see the classic survey by J. Denef [**32**] and the more recent survey by D. Meuser [**77**]. It is not our purpose in this note to detail this whole subject, for an introduction to the theory of p-adic zeta functions we recommend [**60**], and also the recent works, including the motivic case, to appear in this Volume [**88, 97**].

The proof by Malgrange of Theorem 4.2 about the asymptotic expansions of $I_\phi(\tau; f)$ brought to the game a collection of new tools of algebraic topological nature. He considered integrals of the form

$$\int_\Gamma \exp(\tau f(x))\, \phi(x)\mathrm{d}x_1 \wedge \cdots \wedge \mathrm{d}x_n,$$

where Γ is a real n-dimensional chain lying in $\mathbb{C}^n$, $f(x)$ is a holomorphic function on $\mathbb{C}^n$ and $\phi \in \mathcal{S}(\mathbb{C}^n)$. Malgrange shows that these integrals also admit asymptotic expansions, but this time related with the monodromy of f at some point of $f^{-1}(0)$. It turns out that this type of integrals is able to recover the real integrals $I_\phi(\tau; f)$, see (4.5), providing a connection between the poles of $Z_\phi(s, f)$ and the eigenvalues of the (complex) monodromy of f at some point of $f^{-1}(0)$. This point of view contributes to enlarge and foster the theory of Archimedean zeta functions.

Finally we want to mention that there have been several generalizations of the classical Archimedean zeta functions $Z_\phi(s, f)$. Most of them use a variant proposed by C. Sabbah in [**90**], where he considers several complex variables $s_1, \dots, s_l$ attached to the same number of functions, see Section 5.1. There have been also some generalizations of the oscillatory integrals $I_\phi(\tau; f)$ in the frame of multilinear (oscillatory) operators, see the work of Phong, Stein, and Sturm in [**86**], and the work of Christ, Li, Tao and Thiele in [**24**].

We now describe the content of the article. In Section 2 we present the basic definitions of the Scwartz space, the Weyl algebra and the Archimedean local zeta functions. We also present in this section our first proof of Theorem 2.2 about the meromorphic continuation of $Z_\phi(s, f)$, by using resolution of singularities. We also present as an application of Theorem 2.2 a proof of the fact that differential operators with constant coefficients admit fundamental solutions. Section 3 is devoted to the presentation of the Bernstein-Sato polynomial and the second proof of the meromorphic continuation of $Z_\phi(s, f)$ in Theorem 3.2. We discuss also here the difference between the sets of candidate poles for the Archimedean zeta functions. Next, in Section 4 we introduce the oscillatory integrals, the Gel'fand Leray form and some interesting fiber integrals. We state in Theorem 4.2 the asymptotic expansion of $I_\phi(\tau; f)$ and relate some of its parameters with the log-canonical threshold of f. We briefly explain in Section 4.1 some of the problems when dealing with oscillatory integrals of real analytic phases. In Section 4.2 we review some of the results of Malgrange about complex oscillatory integrals and their relations with $Z_\phi(s, f)$. Finally in Section 5 we present some generalizations of Archimedean local zeta functions and oscillatory integrals.

We have tried to provide an ample list of references to the classic and current literature about the considered topics. No new proofs are given and only some proofs are presented. Our purpose is not just to expose the classic theory but also to call the attention of the community to some not so well known relations among Archimedean local zeta functions and oscillatory integrals. The most important example of this fact is the almost inexistent connections between real zeta functions and multilinear oscillatory integrals.

2. Archimedean local zeta functions

The first serious study of real and complex zeta functions was carried out in the first volume of the book GENERALIZED FUNCTIONS by I. M. Gel'fand and G. Shilov [**37**]. They begin their study with the following *regularization* technique. Take $s \in \mathbb{C}$ with $\Re(s) > -1$ and let ϕ be a smooth real function with compact support. Note that the integral

$$(x_+^s, \phi) = \int_0^\infty x^s \phi(x)\, \mathrm{d}x,$$

is a holomorphic function on $\Re(s) > -1$. Moreover we can write

$$(x_+^s, \phi) = \int_0^1 x^s \left[\phi(x) - \phi(0)\right] \mathrm{d}x + \frac{\phi(0)}{s+1} + \int_1^\infty x^s \phi(x)\, \mathrm{d}x. \tag{2.1}$$

The first term of the RHS of (2.1) is defined for $\Re(s) > -2$, the second for $s \neq -1$ and the last one for all $s \in \mathbb{C}$. Hence (x_+^s, ϕ) can be continued analytically to $\Re(s) > -2$ and $s \neq -1$. Analogously, (x_+^s, ϕ) can be continued analytically (*regularized*) to the region $\Re(s) > -n - 1$, $s \neq -1, -2, \ldots, -n$, by means of

$$(x_+^s, \phi) = \int_0^1 x^s \left[\phi(x) - \phi(0) - \cdots - \frac{x^{n-1}}{(n-1)!}\phi^{(n-1)}(0)\right] \mathrm{d}x + \sum_{k=1}^{n} \frac{\phi^{(k-1)}(0)}{(k-1)!(s+k)}$$
$$+ \int_1^\infty x^s\, \phi(x)\, \mathrm{d}x.$$

Finally, (x_+^s, ϕ) has an analytic continuation to the whole $\mathbb{C}$ except at the points $s = -1, -2, \ldots$, where it has simple poles.

A complex analogue of (x_+^s, ϕ) is studied in Section 3.6 of the same book. Indeed, a considerable part of this text is devoted to the generalization of (x_+^s, ϕ), replacing x_+ by more elaborated polynomials in several variables over $\mathbb{R}$ or $\mathbb{C}$. During a talk at the ICM of 1954 [**36**], I. M. Gelfand posed the following problem.

PROBLEM. Let $P(x_1, \ldots, x_n)$ be a polynomial and consider the area in which $P > 0$. Let $\phi(x_1, \ldots, x_n)$ be a smooth function with compact support. Then the generalized function

$$(P^s, \phi) = \int_{P>0} P(x_1, \ldots, x_n)^s \phi(x_1, \ldots, x_n)\, |\mathrm{d}x_1 \ldots \mathrm{d}x_n|,$$

is a meromorphic function of s.

In [**37**] the authors had discovered that this problem requires a careful analysis of the singular set of P, Sing_P. While some simple singularities could be handled, the general case appeared to require new ideas. I.M. Gelfand was aware that the solution to this problem would imply the existence of fundamental solutions for differential operators with constant coefficients, see Section 2.2. The new idea that settled the problem definitively was resolution of singularities. By using Hironaka's theorem [**46**], I.N. Bernstein and S.I. Gelfand in [**17**], and independently M. Atiyah in [**6**], reduce the general problem to the monomial case, which can be easily obtained from the example (x_+^s, ϕ). In order to present their proof we introduce some notation.

2.1. Distributions and local zeta functions. Following Igusa [**50**], we denote by

$$D_n = \mathbb{C}[x, \partial/\partial x] = \mathbb{C}[x_1, \ldots, x_n, \partial/\partial x_1, \ldots, \partial/\partial x_n]$$

the n-th Weyl algebra over $\mathbb{C}$, i.e. the ring of differential operators with polynomial coefficients in n variables over $\mathbb{C}$. A smooth function $\phi : \mathbb{R}^n \to \mathbb{C}$, i.e. a function having derivatives of arbitrarily high order, is called a *Schwartz function* if

$$\|P\phi\|_\infty \quad \text{is finite for every} \quad P \in D_n = \mathbb{C}[x, \partial/\partial x],$$

where $\|\phi\|_\infty = \sup_{x\in\mathbb{R}^n} |\phi(x)|$. The set of Schwartz functions is called the *Schwartz space* $\mathcal{S}(\mathbb{R}^n)$. By identifying $\mathbb{C}$ with $\mathbb{R}^2$, we similarly define $\mathcal{S}(\mathbb{C}^n)$. In any case, $\mathcal{S}(\mathbb{K}^n)$ ($\mathbb{K} = \mathbb{R}$ or $\mathbb{C}$), forms a $\mathbb{C}$-vector space. Moreover, $\mathcal{S}(\mathbb{K}^n)$ is a topological vector space whose topology is induced by the family of seminorms

$$\|\phi\|_{i,j} = \sup_{x\in\mathbb{K}^n} |x^i (\partial/\partial x)^j \phi(x)|,$$

for all $i, j \in \mathbb{Z}^n_{\geq 0}$. As a metric space, $\mathcal{S}(\mathbb{K}^n)$ is complete and the continuous linear functionals on it form the space of tempered distributions $\mathcal{S}'(\mathbb{K}^n)$. There are also local versions of these spaces for an open set U of $\mathbb{K}^n$, see [**47**, Sect. 2.1]. It is well known that the space of smooth functions with compact support is a dense subspace of $\mathcal{S}(\mathbb{K}^n)$, see e.g. [**50**, Lemma 5.2.2].

DEFINITION 2.1. The local zeta function of $f(x) \in \mathbb{K}[x_1, \ldots, x_n] \setminus \mathbb{K}$ and $\phi \in \mathcal{S}(\mathbb{K}^n)$ is the element of $\mathcal{S}'(\mathbb{K}^n)$ defined as

$$Z_\phi(s, f) = (f^s, \phi) = \int_{\mathbb{K}^n \setminus f^{-1}(0)} \phi(x)\, |f(x)|^s_{\mathbb{K}}\, |dx|,$$

for $s \in \mathbb{C}$ with $\Re(s) > 0$.

This definition is essentially the same given by Igusa in [**48**]. Since the complex parameter s appears as power in the distribution induced by $|f|_{\mathbb{K}}$, I.M. Gelfand used in [**36**] the terminology *complex power* for $Z_\phi(s, f)$. Note that the definition given in the introduction by (1.2) is slightly more general than the previous one.

We now present our first proof of the meromorphic continuation of $Z_\phi(s, f)$, given originally by [**17**] and [**6**]. We follow closely the presentation of [**50**]. For $f(x) \in \mathbb{K}[x_1, \ldots, x_n] \setminus \mathbb{K}$, we denote by Sing_f the singular locus of f, meaning the set of roots of f in $\mathbb{K}^n$ at which all the partial derivatives simultaneously vanish. A *resolution of singularities* of $f^{-1}(0)$ is a pair (X, π) where X is a smooth manifold and $\pi : X \to \mathbb{K}^n$ is a proper map. In X there is a finite set $\mathcal{E} = \{E\}$ of closed submanifolds of X of codimension 1 with a pair of positive integers (N_E, ν_E) assigned to each E, which are called the numerical data of the resolution. The map π has the following properties: (i) $(f \circ \pi)^{-1}(0) = \cup_{E\in\mathcal{E}} E$, and π induces an isomorphism

$$X \setminus \pi^{-1}\left(\mathrm{Sing}_f\right) \to \mathbb{K}^n \setminus \mathrm{Sing}_f;$$

(ii) at every point p of X if $E_1, \ldots, E_m$ are all the E in $\mathcal{E}$ containing p with local equations $y_1, \ldots, y_m$ around p, then there exist local coordinates of X around p of the form $(y_1, \ldots, y_m, y_{m+1}, \ldots, y_n)$ such that

$$(f \circ \pi)(y) = \varepsilon(y) \prod_{i=1}^m y_i^{N_i}, \quad \pi^*\left(\bigwedge_{1\leq i\leq n} dx_i\right) = \eta(y) \left(\prod_{i=1}^m y_i^{\nu_i - 1}\right) \bigwedge_{1\leq i\leq n} dy_i \tag{2.2}$$

on some neighborhood of p, in which $\varepsilon(y)$, $\eta(y)$ are units of the local ring $\mathcal{O}_p$ of X at p. In particular $\cup_{E\in\mathcal{E}}E$ has normal crossings.

Now consider our integral $Z_\phi(s,f)$ and let $\left|\bigwedge_{1\le i\le n}\mathrm{d}x_i\right|$ denote the measure induced by the differential form $\bigwedge_{1\le i\le n}\mathrm{d}x_i$ on $\mathbb{K}^n$, which agrees with the Haar measure of $\mathbb{K}^n$. Then

$$Z_\phi(s,f)=\int_{\mathbb{K}^n\setminus f^{-1}(0)}\phi(x)\,|f(x)|_{\mathbb{K}}^s\left|\bigwedge_{1\le i\le n}\mathrm{d}x_i\right|.$$

Note that by density, it is enough to make the proof for ϕ a smooth function with compact support. We use π as a change of variables in our integral to get

$$Z_\phi(s,f)=\int_{X\setminus\pi^{-1}(f^{-1}(0))}\phi(\pi(y))\,|f(\pi(y))|_{\mathbb{K}}^s\left|\pi^*\left(\bigwedge_{1\le i\le n}\mathrm{d}x_i\right)(y)\right|.$$

At every point p of $X\setminus\pi^{-1}(f^{-1}(0))$ we can choose a chart U such that (2.2) holds. Moreover we choose a small neighborhood U_p of p over which the above local coordinates are valid and the functions ε and η are locally invertible. Since the map π is proper, we see that $C=\pi^{-1}(\mathrm{Supp}(\phi))$ is also compact and we will consider a partition of the unity associated to C and subordinated to the cover $\{U_p\}$. By this we mean that for every element U_p of the cover, there exists a finite set $\{\rho_{p,k}(y)\}_{k\in K}$ of smooth functions such that for every y in C, only a finite number of $\rho_{p,k}$ are different from zero, and $\sum_{k\in K}\rho_{p,k}(y)=1$. This implies that

$$Z_\phi(s,f)=\sum_{k\in K}\int_{U_p}\left(\phi(\pi(y))\cdot\rho_{p,k}(y)\,|\varepsilon(y)|_{\mathbb{K}}^s\,\eta(y)\right)\prod_{i=1}^m|y_i|_{\mathbb{K}}^{N_is+\nu_i-1}\left|\bigwedge_{1\le i\le n}\mathrm{d}y_i\right|.$$

In every neighborhood U_p the function $\phi(\pi(y))\cdot\rho_{p,k}(y)\,|\varepsilon(y)|_{\mathbb{K}}^s\,\eta(y)$ can be considered as a smooth function with compact support. Therefore, by using the n-dimensional version of the example (x_+^s,ϕ), see e.g. [**4**, Lemma 7.3], one gets the following theorem, stated in Igusa's version [**50**, Theorem 5.4.1].

THEOREM 2.2 ([**6**, Introduction],[**17**, Theorem 1]). *Let $f(x)\in\mathbb{K}[x_1,\dots,x_n]\setminus\mathbb{K}$ and $\phi\in\mathcal{S}(\mathbb{K}^n)$. Assume that (X,π) a resolution of singularities of $f^{-1}(0)$ with total transform $\mathcal{E}=\{E\}$. Let (N_E,ν_E) be the numerical data of (X,π). Then $Z_\phi(s,f)$ admits a meromorphic continuation to the whole complex plane with poles contained in the set*

$$\bigcup_{E\in\mathcal{E}}-\frac{\nu_E+\mathbb{Z}_{\ge0}}{qN_E},\tag{2.3}$$

where $q=1$ if $\mathbb{K}=\mathbb{R}$ and $q=2$ if $\mathbb{K}=\mathbb{C}$. Moreover, the orders of the candidate poles are at most equal to $\min\{n,\mathrm{Card}(\mathcal{E})+1\}$.

2.2. Fundamental solutions of differential operators. The proof by M. Atiyah in [**6**] of Theorem 2.2 contains other interesting consequences of the meromorphic continuation of $Z_\phi(s,f)$. For instance, one may give another proof of the following result of Hörmander and Lojasiewicz, known as the division of distributions problem. Again we follow [**50**, Sect. 5.5].

THEOREM 2.3 ([**6**, Corollary 1]). *If $f(x)\in\mathbb{C}[x_1,\dots,x_n]\setminus\mathbb{C}$, there exists an element T of $\mathcal{S}'(\mathbb{R}^n)$ satisfying $fT=1$.*

The proof is not difficult, and according to Igusa is an elegant one. A nice corollary of Theorem 2.3 is the remarkable fact that there exists fundamental (or elementary) solutions for differential operators with constant coefficients as we shall see below. First we recall that for $P \in D_n = \mathbb{K}[x, \partial/\partial x]$, its *adjoint operator* P^* is the image of the $\mathbb{K}$-involution of D_n given by $x_i^* = x_i$ and $(\partial/\partial x_i)^* = -\partial/\partial x_i$. In particular this implies that $(P_1P_2)^* = P_2^*P_1^*$ and $(P_1^*)^* = P_1$ for every $P_1, P_2 \in D_n$.

This notion of adjoint can be extended up to $\mathcal{S}'(\mathbb{K}^n)$. In particular one has for every $\phi \in \mathcal{S}(\mathbb{R}^n)$ that

$$1^*(\phi) = \phi(0) =: \delta_0(\phi),$$

which means that $1^* = \delta_0$ in $\mathcal{S}'(\mathbb{R}^n)$. The element δ_0 of $\mathcal{S}'(\mathbb{R}^n)$ is known as the *Dirac measure of* $\mathbb{R}^n$ *supported by* 0. Now consider $f(x) \in \mathbb{C}[x_1, \ldots, x_n] \setminus \mathbb{C}$, say $f(x) = \sum_{k \in (\mathbb{Z}_{\geq 0})^n} c_k x^k$ for $c_k \in \mathbb{C}$, and define an element P of $\mathbb{C}[\partial/\partial x_1, \ldots, \partial/\partial x_n]$ (i.e. a *differential operator with constant coefficients*) by

$$P = \sum_k \frac{1}{(2\pi \mathrm{i})^{|k|}} c_k (\partial/\partial x)^k.$$

It is a simple matter to show that for any T in $\mathcal{S}'(\mathbb{K}^n)$ one has $(fT)^* = PT^*$.

COROLLARY 2.4 ([**6**, Corollary 3]). *If P is a differential operator with constant coefficients, then there exists an elementary solution for P, i.e., an element S of $\mathcal{S}'(\mathbb{R}^n)$ satisfying $PS = \delta_0$.*

PROOF. Assume that

$$P = \sum_k \frac{1}{(2\pi \mathrm{i})^{|k|}} c_k (\partial/\partial x)^k,$$

for some $c_k \in \mathbb{C}$. Put $f(x) = \sum_{k \in \mathbb{Z}_{\geq 0}} c_k x^k$ in $\mathbb{C}[x_1, \ldots, x_n] \setminus \mathbb{C}$, then by Theorem 2.3, there is an element T of $\mathcal{S}'(\mathbb{R}^n)$ satisfying $fT = 1$. By taking $S = T^*$, we have that S belongs to $\mathcal{S}'(\mathbb{R}^n)$ and moreover $PS = (fT)^* = 1^* = \delta_0$. □

3. The Bernstein-Sato polynomial

So far our approach to Archimedean local zeta functions appeared only related with objects from analysis, we will see in this section how $Z_\phi(s, f)$ is connected with objects of more algebraic nature. The algebraic analysis theory or algebraic theory of linear differential systems began in the 60's with the school of M. Sato at Kyoto, see [**92**] for the translation of his work in the 60's. A similar development was taking place in Moscow, where I.N. Bernstein was working on similar matters, see [**14, 15**]. The whole mathematical construction is known today as $\mathcal{D}$-module theory and it has connection in many areas of mathematics such as algebraic geometry, singularity theory, topology of varieties, representation theory and of course differential equations, among others. For more on the history of the subject and its connections we recommend the surveys [**1, 40, 98**] or the books [**18, 28**].

In this section we will only present the definition of the Bernstein-Sato polynomial and discuss some of the connections with $Z_\phi(s, f)$. Let us start by considering the Weyl algebra $D_n = \mathbb{K}[x, \partial/\partial x]$. Note that $\mathbb{K}(x)$ is a D_n-module and so is $\mathbb{K}[x]$ because it is mapped to itself under operations of D_n. Moreover, we pick another variable s (in addition to $x_1, \ldots, x_n$) and consider the Weyl algebra

$D = \mathbb{K}(s)[x, \partial/\partial x]$. Now we take an element $f(x) \in \mathbb{K}[x] \setminus \{0\}$ and denote by S the multiplicative subset of $\mathbb{K}[x] \setminus \{0\}$ generated by $f(x)$, then it can be shown that

$$S^{-1}(\mathbb{K}(s)[x]) = \mathbb{K}(s)[x_1, \ldots, x_n, 1/f(x)],$$

is a D-module. Igusa presents Bernstein's main Theorem in the following form

THEOREM 3.1 ([**14**, Lemma 1]). *There exists an element P_0 of D_n satisfying*

$$P_0 \cdot f(x) = 1. \tag{3.1}$$

Multiplying both sides of (3.1) by an arbitrary polynomial of $\mathbb{K}[s] \setminus \{0\}$ we get

$$P \cdot f(x) = b(s),$$

for some polynomial P in the 'variables' $s, x_1, \ldots, x_n, \partial/\partial x_1, \ldots, \partial/\partial x_n$ and with coefficients in $\mathbb{K}$. If now we assume that $s \in \mathbb{Z}$, we obtain the following functional equation

$$P(s, x, \partial/\partial x) \cdot f(x)^{s+1} = b(s) f(x)^s, \tag{3.2}$$

which is proved in detail in [**16**]. The set of all the polynomials $b(s)$ of $\mathbb{K}[s]$ satisfying (3.2) forms an ideal, and the unique monic generator of this ideal is called the *Bernstein-Sato polynomial $b_f(s)$ of f*. The meromorphic continuation of the real $Z_\phi(s, f)$ is stated by Bernstein in the following form.

THEOREM 3.2 ([**16**, Theorem 1']). *For a complex number s with $\Re(s) > 0$, consider $Z_\phi(s, f)$, the local zeta function of $f(x) \in \mathbb{R}[x_1, \ldots, x_n] \setminus \mathbb{R}$ and $\phi \in \mathcal{S}(\mathbb{R}^n)$. Suppose that the Bernstein-Sato polynomial $b_f(s)$ of f factors as*

$$b_f(s) = \prod_\lambda (s + \lambda),$$

and denote by $\Gamma(s)$ the Gamma function, see (1.1). Then $\frac{Z_\phi(s,f)}{\prod_\lambda \Gamma(s+\lambda)}$ has a holomorphic continuation to the whole $\mathbb{C}$.

PROOF. In order to avoid complications with the absolute value we first consider the following simplification. Let f_+^s be the function defined by

$$f_+^s(x) = \begin{cases} f(x)^s & \text{for } f(x) \geq 0 \\ 0 & \text{for } f(x) < 0. \end{cases}$$

Then by the functional equation (3.2) one has

$$\begin{aligned} b_f(s)\, Z_\phi(s, f_+) &= \int_{\mathbb{R}^n \setminus f^{-1}(0)} \phi(x)\, b_f(s) f_+^s(x)\, |dx| = \int_{\mathbb{R}^n \setminus f^{-1}(0)} \phi(x)\, P(s)\, f_+^{s+1}(x)\, |dx| \\ &= \int_{\mathbb{R}^n \setminus f^{-1}(0)} P^*(s) \cdot \phi(x)\, f_+^{s+1}(x)\, |dx|. \end{aligned}$$

Here P^* denotes the adjoint operator of P, and one has that $P^*(s) \cdot \phi(x) \in \mathcal{S}(\mathbb{R}^n)$. We can repeat the above process and, after r-times, we get

$$b_f(s + r - 1) \cdots b_f(s)\, Z_\phi(s, f_+) = \int_{\mathbb{R}^n \setminus f^{-1}(0)} \phi_r(x)\, f_+^{s+r}(x)\, |dx|,$$

where $\phi_r(x) = P^*(s + r - 1) \cdots P^*(s) \cdot \phi \in \mathcal{S}(\mathbb{R}^n)$. Now, by a simple calculation one has that

$$\prod_\lambda \Gamma(s + \lambda)\, b_f(s) \cdots b_f(s + r - 1) = \prod_\lambda \Gamma(s + r + \lambda),$$

which in turn implies that for s with $\Re(s)$ sufficiently large

$$\frac{Z_\phi(s,f_+)}{\prod_\lambda \Gamma(s+\lambda)} = \frac{1}{\prod_\lambda \Gamma(s+r+\lambda)} \int_{\mathbb{R}^n \setminus f^{-1}(0)} \phi_r(x)\, f_+^{s+r}(x)\, |\mathrm{d}x|. \tag{3.3}$$

Note that, while the LHS of (3.3) is holomorphic when $\Re(s) > 0$, the RHS is holomorphic on $\Re(s) > -r$. Since $r \in \mathbb{Z}_{\geq 0}$ is arbitrary, one concludes that $\frac{Z_\phi(s,f_+)}{\prod_\lambda \Gamma(s+\lambda)}$ has a holomorphic continuation to $\mathbb{C}$. □

The previous proof can be easily adapted for the complex $Z_\phi(s,f)$, see [**50**, Theorem 5.3.2].

3.1. Poles of $Z_\phi(s,f)$. By Theorem 3.2 the poles of $Z_\phi(s,f)$ are integer translates of the roots of $b_f(s)$. This agrees with Theorem 2.2 since the roots of the Bernstein-Sato polynomial of f are negative rational numbers [**57**]. Moreover one has

THEOREM 3.3 ([**62**, Theorem 1]). *If $f(x) \in \mathbb{C}[x_1,\dots,x_n] \setminus \mathbb{C}$ and (X,π) is a resolution of singularities of $f^{-1}(0)$, with the notation of Section 2.1, every root of $b_f(s)$ is of the form*

$$\lambda = -\frac{\nu_E + k}{N_E},$$

for some $E \in \mathcal{E}$ and some integer $k \geq 0$.

In most cases, Theorem 3.3 gives only a big list of candidate roots for $b_f(s)$ and the problem of the actual determination of the true roots is largely open. Of course, the same problem translates to the candidate poles of $Z_\phi(s,f)$, since in general, Theorem 2.2 gives a very big list of candidate poles and the determination of the true poles is hard. For instance it is known that, when f is a reduced plane curve singularity or an isolated quasi homogeneous singularity, the set of poles of the complex $Z_\phi(s,f)$ coincides exactly with the set $\{\lambda - r \mid \lambda \text{ is a root of } b_f(s), r \in \mathbb{Z}_{\geq 0}\}$, see [**70**]. See also [**19**] for some recent results about the poles of the complex $Z_\phi(s,f)$ in the case of curves.

It is also known that the roots of the Bernstein-Sato polynomial are deeply connected with the geometry of f. Malgrange [**76**] showed that, if λ is a root of $b_f(s)$ then $\exp(2\pi \mathrm{i}\lambda)$ is an eigenvalue of the local monodromy of f at some point of $f^{-1}(0)$, and all eigenvalues are obtained in this way, see [**7**]. We discuss some of these results in Section 4.2.

We have started this section mentioning the work by M. Sato in the 60's. Part of his interest in the theory of $\mathcal{D}$-modules was motivated by another mathematical theory, namely prehomogeneous vector spaces. If G is a connected reductive algebraic subgroup of $GL_n(\mathbb{C})$ acting transitively on the complement of an absolutely irreducible hypersurface $f^{-1}(0)$ in $\mathbb{C}^n$, then $(G,\mathbb{C}^n)$ is called a *prehomogeneous vector space*. In this case one may show that $f(x)$ is a homogeneous polynomial and also a *relative invariant* of G, that is,

$$f(g\cdot x) = \nu(g) f(x), \quad \text{for all } g \in G,$$

where ν belongs to $\mathrm{Hom}(G,\mathbb{C}^\times)$. See [**92**] for the translation of the original work of M. Sato and [**50**, Chap. 6] for a detailed exposition. This theory of prehomogeneous vector spaces contains Archimedean zeta functions of some group invariants, see e.g.

[**40, 58, 93**]. It may be useful to compute explicitly some zeta functions, as in the following example, given at the end of [**50**, Chap. 6]:

$$\int_{\mathbb{C}^{n^2}} \exp(-2\pi {}^{\mathrm{t}}x\bar{x}) \,|\det(x)|_{\mathbb{C}}^{s}\, |dx| = \frac{1}{(2\pi)^{ns}} \cdot \prod_{r=1}^{n} \frac{\Gamma(s+r)}{\Gamma(r)}.$$

4. Oscillatory integrals

Denote by $\varepsilon(x)$ the standard additive character of $\mathbb{K} = \mathbb{R}$ or $\mathbb{C}$, which is defined as

$$\varepsilon(x) = \begin{cases} \exp(2\pi \mathrm{i} x), & \text{for } x \in \mathbb{R} \\ \exp(4\pi \mathrm{i} \Re(x)), & \text{for } x \in \mathbb{C}. \end{cases}$$

DEFINITION 4.1. If $f(x) \in \mathbb{K}[x_1, \ldots, x_n] \backslash \mathbb{K}$ and $\phi \in \mathcal{S}(\mathbb{K}^n)$, then the *oscillatory integral* attached to (f, ϕ) is defined as

$$I_\phi(\tau; f) = \int_{\mathbb{K}^n \backslash f^{-1}(0)} \phi(x)\, \varepsilon(\tau \cdot f(x))\, |dx|,$$

for $\tau \in \mathbb{K}$.

This type of integrals appears frequently in a great number of situations of physical and mathematical interest, see e.g. [**4**, Chapter 6] and [**94**, Chapter VIII]. More generally f is considered to be a $\mathbb{K}$-analytic function, which is called the *phase* and the function ϕ is called the *amplitude.* One of the main problems in the theory of oscillatory integrals is to study the asymptotic behaviour of $I_\phi(\tau; f)$ when the norm of the parameter τ tends to ∞. The stationary phase principle states that the main contribution in the asymptotics is given by neighbourhoods of the critical points of the phase. Moreover, this asymptotic behaviour is controlled by the poles of $Z_\phi(s, f)$.

It is also known that the integral $I_\phi(\tau; f)$ can be rewritten in terms of one dimensional integrals using the Gel'fand-Leray forms, defined as follows. Assume that f is a $\mathbb{K}$-analytic function, then the Gel'fand-Leray form $\omega_f(x, t)$ is the unique $(n-1)$-form on the level hypersurface $X_t = \{x \mid f(x) = t\}$ $(t \neq 0)$, which satisfies

$$df \wedge \omega_f(x, t) = dx_1 \wedge \cdots \wedge dx_n.$$

If $C_f := \{x \in \mathbb{K}^n \mid \nabla f(x) = 0\}$ denotes *the critical set of* f, and one assumes that $C_f \cap \operatorname{Supp}(\phi) \subset f^{-1}(0)$, then

$$I_\phi(\tau; f) = \int_{\mathbb{K}} \varepsilon(\tau \cdot t) \left(\int_{X_t} \phi(x) \cdot \omega_f(x, t) \right) |dt|. \tag{4.1}$$

THEOREM 4.2. *Consider* $f(x) \in \mathbb{K}[x_1, \ldots, x_n] \setminus \mathbb{K}$ *and* $\phi \in \mathcal{S}(\mathbb{K}^n)$. *If we assume that*

$$C_f \cap \operatorname{Supp}(\phi) \subset f^{-1}(0),$$

then the oscillatory integral $I_\phi(\tau; f)$ *can be expanded in an asymptotic series of the form*

$$\sum_{\alpha} \sum_{k=0}^{n-1} S_{k,\alpha}(\phi) \tau^\alpha (\ln \tau)^k, \quad \text{as } |\tau|_{\mathbb{K}} \to \infty. \tag{4.2}$$

Here the numerical coefficients $S_{k,\alpha}(\phi)$ *belong to* $\mathcal{S}'(\mathbb{K}^n)$, *and the parameter* α *runs through the arithmetic progressions of the form* (2.3), *given by the poles of* $Z_\phi(s, f)$.

The previous statement is a reformulation of [**49**, Theorem 1.6, Ch. III], where three objects are related: the asymptotic expansions of $I_\phi(\tau; f)$ when $|\tau|_{\mathbb{K}} \to \infty$, the asymptotic expansions of the fiber integral

$$\int_{X_t} \phi(x) \cdot \omega_f(x,t), \text{ when } |t|_{\mathbb{K}} \to 0,$$

and the poles of $Z_\phi(s,f)$. For the proof, some functional spaces are defined for each one of the aforementioned objects and then it is shown that these spaces are homeomorphic by using the Fourier and Mellin tranforms. The proof also includes the case of p-adic fields.

Another relevant characteristic of the asymptotic series of $I_\phi(\tau; f)$ is the *oscillation index* β_f of f, which is the leading power of the parameter τ in the expansion (4.2). It turns out that the negative of β_f coincides in some cases with another relevant invariant of the singularity defined by f: the *log canonical threshold* of f. It may be defined in terms of the resolution (X, π) of $f^{-1}(0)$ as

$$\operatorname{lct}(f) := \min_{E \in \mathcal{E}} \frac{\nu_E + 1}{N_E},$$

where we have used the notation of (2.3) in Section 2.1. It is well known that $-\operatorname{lct}(f)$ is exactly the largest pole of $Z_\phi(s,f)$. For properties and more information about $\operatorname{lct}(f)$ and its role in birational geometry see e.g. [**78, 79**].

4.1. Real oscillatory integrals. In the realm of harmonic analysis, the real oscillatory integrals of type $I_\phi(\tau; f)$ have been an essential player for a long time. They have also been the basis for the definition of differential operators and have become a classical subject in harmonic analysis, partial diferential equations and geometry (geometric analysis) of manifolds. For instance, in [**94**] there is the following class of oscillatory integrals of the second kind (being the real $I_\phi(\tau; f)$ the ones of the first kind):

$$T_\tau(f)(x) = \int_{\mathbb{R}^n} \exp(\mathrm{i}\tau f(x,y))\, K(x,y) g(y)\ |\mathrm{d}y|.$$

Here the kernel $K(x,y)$ carries an oscillatory factor and the function g satisfies some smoothness condition. In this case, the important matter are the boundedness properties of $T_\tau(f)(x)$. A key hypothesis in many cases is the analyticity of the phase f, in this case Theorem 4.2 was proved first by Jeanquartier in [**53**], and then by Malgrange [**75**], Igusa [**49**] and Jeanquartier [**54**], among others. See also [**4**, Section 7.3.2].

An important class of real analytic phases for which the asymptotics of $I_\phi(\tau; f)$ and the boundedness of $T_\tau(f)(x)$ (for some T_τ) is better understood is the class of non-degenerate functions with respect to their Newton polyhedron. In his seminal paper [**95**], A. N. Varchenko investigates the oscillatory index of $I_\phi(\tau; f)$ as well as the asymptotic series (4.2) in terms of the geometry of the Newton polyhedron of the phase. Loosely speaking, Varchenko's idea is to attach a Newton polyhedron $\mathcal{NP}(f)$ to the phase function f and then define a non-degeneracy condition with respect to $\mathcal{NP}(f)$. Then one may construct a toric variety associated to the Newton polyhedron, and use the well known toric resolution of singularities to give a list of parameters α in (4.2) (equivalently a list of candidate poles for $Z_\phi(s,f)$). Moreover, this list can be read off from the geometry of $\mathcal{NP}(f)$ and a refinement of its normal fan. For example, Varchenko shows that under some mild conditions

$$\mathrm{lct}(f) = -\beta_f = \frac{1}{t_0},$$

where t_0 is the entry of the point in $\mathbb{R}^n$ given by the intersection of the diagonal of $\mathbb{R}^n$ with the boundary of $\mathcal{NP}(f)$. A more detailed exposition of the original article of Varchenko is presented in [**4**, Chapters 6-8].

It is well known to the specialist that the only data of $\mathcal{NP}(f)$ that give poles of $Z_\phi(s, f)$ are contained in the normal fan of the polyhedron, i.e. the extra rays required for the toric resolution give rise to fake candidate poles. This list of fake poles was removed by Denef and Sargos in [**35**] and by a different method by Aroca, Gómez-Morales and the author in [**5**].

The original ideas of Varchenko have been extensively used and generalized in analysis. In particular they have been used in the study of: more general oscillatory integrals, estimation of the size of sublevel sets, boundedness and decay estimates of operators, among other matters. See for instance the works [**22, 27, 34, 38, 39, 42–45, 51, 55, 56, 81, 85–87, 101**], and the references therein. In particular we should note that Kamimoto and Nose have extended in [**55**] the work of Varchenko for the case when the phase f is smooth and the amplitude has a zero at a critical point of f. They show that the optimal rates of decay, and other related parameters, for some weighted oscillatory integrals can be expressed in terms of the geometry of $\mathcal{NP}(f)$ and a refinement of its normal fan. Similar results are given for the poles of the associated local zeta functions.

4.2. Complex oscillatory integrals. We now consider the case of holomorphic functions f. In some sense, the corresponding counterpart of the oscillatory integrals $I_\phi(\tau; f)$ of real analytic functions are integrals over vanishing cycles, see e.g. [**4, 7, 75, 83, 84**]. We consider two holomorphic functions on $\mathbb{C}^n$, $f(x)$ and $\phi(x)$. The integrals of interest have the form

$$\int_\Gamma \exp(\tau f(x))\, \phi(x) \mathrm{d}x_1 \wedge \cdots \wedge \mathrm{d}x_n, \tag{4.3}$$

where Γ is a real n-dimensional chain lying in $\mathbb{C}^n$, and τ is a large real parameter. These integrals also admit asymptotic expansions, but this time related with the monodromy of f at some point of $f^{-1}(0)$. We only present here the main ideas of Malgrange [**75**], following closely the presentation of [**4**, Chapter 11]. We start by assuming that

$$f : (\mathbb{C}^n, 0) \to (\mathbb{C}, 0)$$

has a critical point x_0 of finite multiplicity, in fact we will assume that x_0 is the origin. We will say that the n-dimensional chain Γ is *admissible* if on its boundary one has $\Re(f(x)) < 0$. This notion admits the following interpretation in terms of the Milnor fibration.

Let B_ϵ be the open ball centered at the origin with radius ϵ and let D_η be the open punctured disk defined by the condition $\{t \in \mathbb{C}\,; |t| < \eta\}$, with $0 < \eta \ll \epsilon \ll 1$. We also put $X = B_\epsilon \cap f^{-1}(D_\eta)$. We recall that the *Milnor fibration* of f at the origin is the smooth locally trivial fibration defined by

$$f|_X : X \to D_\eta. \tag{4.4}$$

We denote by D_η^- the set $\{t \in D_\eta\,; \Re(t) < 0\}$ and by X^- the set $X \cap f^{-1}(D_\eta^-)$. With these definitions, the n-dimensional chain Γ is admissible if $\Gamma \subset X$ and

$\partial(\Gamma) \subset X^-$. Two admissible chains Γ_1, Γ_2 are said to be equivalent if there exists an $(n+1)$-dimensional chain $V \subset X$ such that

$$(\Gamma_1 - \Gamma_2 + \partial V) \subset X^-.$$

The equivalence classes of admissible chains with the operations of addition and multiplication by scalars form a vector space, by definition coinciding with the homology group $H_n(X, X^-)$.

Since $f|_{X^-} : X^- \to D_\eta^-$ is also a trivial fibration, one may show that the homology of X^- is isomorphic to the homology of an arbitrary fiber over D_η^-. This means that for any $t \in D_\eta^-$, there is an isomorphism

$$\partial_t : H_n(X, X^-) \to H_{n-1}(X_t),$$

where X_t is a Milnor fiber (under (4.4)) and $H_{n-1}(X_t)$ is the reduced homology group. Now, for an admissible chain Γ, representing the class $[\Gamma] \in H_n(X, X^-)$, one may contract its boundary in X^- to the fiber X_t, giving a cycle that represents the element $\partial_t[\Gamma] \in H_{n-1}(X_t)$. The family of classes obtained in this way depends continuously on $t \in D_\eta^-$.

THEOREM 4.3 ([**75**, Theorem 7.1]). *Let ω be a holomorphic differential n-form defined on X, and take $[\Gamma] \in H_n(X, X^-)$. Then there exists a small positive number $t_0 \in D_\eta$,such that*

$$\int_\Gamma \exp(\tau f) \cdot \omega \approx \int_0^{t_0} \exp(-\tau t) \left(\int_{\partial_{-t}[\Gamma]} \omega_f(x,t) \right) \mathrm{d}t.$$

The sign $\approx$ means that the integrals are equal up to a negative integer power of the parameter τ when $\tau \to \infty$.

Note the similarity with (4.1). Once we have reached this point, it only rests to state the main Theorem of Malgrange in [**75**].

THEOREM 4.4. *Let ω be a holomorphic differential n-form defined on X, and take $[\Gamma] \in H_n(X, X^-)$. Then the integral $\int_\Gamma \exp(\tau f) \cdot \omega$ can be expanded in an asymptotic series of the form*

$$\sum_{\alpha,k} S_{k,\alpha} \tau^\alpha (\ln \tau)^k, \quad \text{as } \tau \to \infty.$$

Here the parameter α runs through a finite set of arithmetic progressions, depending only on the phase and consisting of negative rational numbers. Moreover, each number α verifies that $\exp(-2\pi \mathrm{i}\alpha)$ is an eigenvalue of the classical monodromy operator of the critical point of the phase. The coefficient $S_{k,\alpha}$ is zero when the monodromy does not have Jordan blocks of dimension greater than or equal to $k+1$ associated with the eigenvalue $\exp(-2\pi \mathrm{i}\alpha)$.

Malgrange also shows that integrals of type (4.3) allow one to recover some real oscillatory integrals, see e.g. [**75**, Section 7] and [**4**, Section 11.2]. If f is a real analytic function and ϕ is supported on a small neighbourhood of the origin, then under some mild conditions on ϕ, there exists an admissible n-chain Γ for which

$$\int_{\mathbb{R}^n} \exp(\mathrm{i}\tau f)\, \phi\, \mathrm{d}x_1 \wedge \cdots \wedge \mathrm{d}x_n \approx \int_\Gamma \exp(\mathrm{i}\tau f) \cdot \phi\, \mathrm{d}x_1 \wedge \cdots \wedge \mathrm{d}x_n. \tag{4.5}$$

From this observation follows the connection between the poles of Archimedean local zeta functions and eigenvalues of the complex monodromy of f at some point of $f^{-1}(0)$.

For some other works related with complex oscillatory integrals or with the ideas of Malgrange in [**75**] see for example [**2–4, 7–13, 21, 25, 26, 30–32, 52, 61, 69–71, 82–84, 91, 99, 100**], and the references therein.

5. Some generalizations

In this last section we want to mention some generalizations of local zeta functions and oscillatory integrals.

5.1. Local zeta functions. In 1987 C. Sabbah introduces in [**90**] the following generalization of $Z_\phi(s, f)$. Let $f_1, \dots, f_l$ be a set of analytic functions defined over a complex (smooth) variety X, and let ϕ be a smooth function with compact support contained in X. Then the *multivariate local zeta function associated to* ϕ and $f = (f_1, \dots, f_l)$ is defined as

$$Z_\phi(s_1, \dots, s_l, f) = \int_X \phi(x)\ |f_1(x)|^{s_1} \cdots |f_l(x)|^{s_l}\ dx,$$

where $s_i \in \mathbb{C}$ with $\Re(s_i) > 0$, for $i = 1, \dots, l$.

This integral converges on the subspace of $\mathbb{C}^l$ defined by $\Re(s_i) > 0$, for $i = 1, \dots, l$. Moreover, the author shows that there is a meromorphic continuation to the whole $\mathbb{C}^l$ with poles contained in the union of some hyperplanes described in terms of some Bernstein polynomial. He shows that there is a finite set $\mathcal{L}$ of linear forms with positive integer coefficients, relatively prime, such that for $i = 1, \dots, l$ the following functional equations holds:

$$\left[\prod_{L \in \mathcal{L}} b_{L,i}(L(s))\right] \cdot f^s = P_k(s, x, \partial_x) f^s \cdot f_k.$$

Here the $b_{L,i}$ denote certain Bernstein polynomials defined previously by the author in [**89**]. The $P_k(s, x, \partial_x)$ are certain differential operators analogous to the ones defined in Section 3. The set $\mathcal{L}$ is called the *set of slopes* and in principle depends on the choice of some filtration of the $\mathcal{D}$-module associated with f.

Apparently at the same time, F. Loeser defines in [**72**] similar zeta functions. He works over a local field $\mathbb{K}$ of characteristic zero. In this case $f_1, \dots, f_l$ are polynomials vanishing at the origin when $\mathbb{K}$ is non-Archimedean, and analytic functions in the Archimedean case. In the p-adic case the set of test functions is given by locally constant functions with compact support, while in the Archimedean case it is the set of smooth functions with compact support. The *multivariate local zeta function* associated to a test function ϕ and $F = (f_1, \dots, f_l)$ is defined as

$$Z_\phi(s_1, \dots, s_l, f) = \int_{\mathbb{K}^n} \phi(x)\ |f_1(x)|_{\mathbb{K}}^{s_1} \cdots |f_l(x)|_{\mathbb{K}}^{s_l}\ |dx|, \tag{5.1}$$

where $s_i \in \mathbb{C}$ with $\Re(s_i) > 0$, for $i = 1, \dots, l$. Actually, his definition is a little more general, since it includes the quasicharacters of $\mathbb{K}$. He uses Hironaka's theorem to show that the poles of the meromorphic continuation to $\mathbb{C}^l$ (as a rational function in the p-adic case) are contained in the union of some hyperplanes described this time in terms of the resolution of $\cup_{i=1}^l f_i^{-1}(0)$. He also shows that the set of slopes is connected with the geometry of the morphism F. More precisely, if one imposes some finiteness condition on F, the set of slopes is contained in the set of normal vectors to the faces of the Newton polyhedron of the discriminant of F at the origin.

Later on, B. Lichtin begins in [**63**] a systematic study of (twisted) multivariate zeta functions essentially of the form (5.1). He was motivated by Igusa's question about the behaviour of the singular series determined by polynomial mappings $P = (P_1, \ldots, P_l) : \mathbb{K}^n \to \mathbb{K}^l$, $(l \leq n)$. He found out that this behaviour depends crucially on the geometric features of the singular locus of P. In the p-adic case his results give also non trivial uniform decay estimates for generalized exponential sums attached to P, [**63–65**]. In the case $\mathbb{K} = \mathbb{R}$ he also found in [**67**] non trivial uniform bounds for some oscillatory integrals of the form

$$\int_{\mathbb{R}^n} \exp(\mathrm{i}\{\tau_1 P_1(x) + \cdots + \tau_l P_l(x)\})\, \phi(x)\, |\mathrm{d}x|,$$

as $|(\tau_1, \ldots, \tau_l)| \to \infty$. For other applications of these techniques see [**66, 68**].

These multivariate zeta functions have been used recently in mathematical physics to study string amplitudes in [**20**] and Log-Coulomb Gases in [**102**]. See the references given there for other uses in p-adic mathematical physics.

Another local zeta function involving several functions was studied in a joint work with W. Veys and W. A. Zúñiga-Galindo [**59**], were we considered Archimedean zeta functions for analytic mappings. If $\mathbb{K} = \mathbb{R}$ or $\mathbb{C}$, let $\boldsymbol{f} = (f_1, \ldots, f_l) : U \to \mathbb{K}^l$ be a $\mathbb{K}$-*analytic mapping* defined on an open U in $\mathbb{K}^n$. Let $\phi : U \to \mathbb{C}$ be a smooth function on U with compact support. Then the local zeta function attached to $(\boldsymbol{f}, \Phi)$ is defined as

$$Z_\phi(s, \boldsymbol{f}) = \int_{\mathbb{K}^n \setminus \boldsymbol{f}^{-1}(0)} \phi(x)\, |\boldsymbol{f}(x)|_{\mathbb{K}}^{s}\, |\mathrm{d}x|, \qquad \text{for } s \in \mathbb{C} \text{ with } \Re(s) > 0.$$

Here $|\boldsymbol{f}(x)|_{\mathbb{K}} = |(f_1(x), \ldots, f_l(x)|_{\mathbb{K}}$ stands for $\sqrt{\sum f_i^2(x)}$ or $\sqrt{\sum |f_i(x)|_{\mathbb{C}}^2}$ depending on whether $\mathbb{K}$ is $\mathbb{R}$ or $\mathbb{C}$, and $|\mathrm{d}x|$ is the Haar measure on $\mathbb{K}^n$. In this case we give a description of the possible poles of $Z_\phi(s, \boldsymbol{f})$ in terms of a log-principalization of the ideal $\mathcal{I}_{\boldsymbol{f}} = \langle f_1, \ldots, f_l \rangle$. We also generalize some of the results of Varchenko presented in Section 4.1, by introducing a new non-degeneracy condition associated to the Newton polyhedron of $\boldsymbol{f}$ (which in general differs from the Minkowski sum of the $\mathcal{NP}(f_i)$).

Very recently W. Veys and W. A. Zúñiga-Galindo extend in [**96**] part of the classical theory of local zeta functions and oscillatory integrals to the case of meromorphic functions defined over a local field $\mathbb{K}$ of characteristic zero. Let $f, g : U \to \mathbb{K}$ be two $\mathbb{K}$-analytic functions defined on an open set $U \subseteq \mathbb{K}^n$, such that f/g is not constant. The *local zeta function attached to a test function* ϕ *and* f/g is defined as

$$Z_\phi(s, f/g) = \int_{U \setminus D_{\mathbb{K}}} \phi(x) \left| \frac{f(x)}{g(x)} \right|_{\mathbb{K}}^{s} |\mathrm{d}x|,$$

where s is a complex number and $D_{\mathbb{K}} = f^{-1}(0) \cup g^{-1}(0)$. In this case a resolution of singularities of $D_{\mathbb{K}}$ is required even to show that $Z_\phi(s, f/g)$ converges in some open strip (α, β), but it turns out that α and β are independent of the chosen resolution. The authors also consider oscillatory integrals, with the notation of Definition 4.1, they have the form

$$\int_{\mathbb{K}^n \setminus D_{\mathbb{K}}} \phi(x)\, \varepsilon\!\left(\tau \cdot \frac{f(x)}{g(x)}\right) |\mathrm{d}x|.$$

They show that the analogue of Theorem 4.2 implies two different asymptotic expansions: one when the norm of the parameter tends to infinity, and another one when the norm of the parameter tends to zero. The first asymptotic expansion is controlled by some poles of the local zeta functions associated to the meromorphic functions $f/g - c$, for certain special values c, while the second expansion is controlled by other poles of $Z_\phi(s, f/g)$.

5.2. Oscillatory integrals. Some generalizations of oscillatory integrals have been considered under the name of multilinear oscillatory operators, see e.g. [**23, 24, 39, 41, 73, 74, 80, 86**] and the references given there. We discuss just a couple of examples to see the kind of problems of interest in the area.

In [**86**] Phong, Stein, and Sturm investigated the multilinear oscillatory integral operator

$$T_D(\tau; g_1, \ldots, g_n) = \int_D \exp(i\tau \cdot f(x_1, \ldots, x_n))\, g_1(x_1) \cdots g_n(x_n)\, |dx_1 \wedge \cdots \wedge dx_n|,$$

where τ is a real parameter, the phase function f is a polynomial and D is a subset of the unit ball in $\mathbb{R}^n$. In addition, the functions g_i belong to the functional space $\mathbf{L}^{p_i}[0, 1]$. One of their main results is the following *decay estimate* for $n \geq 2$,

$$|T_D(\tau; g_1, \ldots, g_n)| \leq C\, |\tau|^{-1/\alpha}\, [\ln(2 + |\tau|)]^{n-1/2} \prod_{i=1}^{n} \|g_i\|_{p_i},$$

where α is some constant depending on a certain Newton polyhedron of f and C is a constant that does not depend on τ.

A more general multilinear oscillatory integral operator was proposed by Christ, Li, Tao and Thiele in [**24**] by

$$I(\tau; g_1, \ldots, g_m) = \int_{\mathbb{R}^n} \exp(i\tau \cdot f(x)) \prod_{j=1}^{m} g_j(\pi_j(x))\, \phi(x)\, |dx|.$$

Here τ is a real parameter, ϕ is a smooth function with compact support and $f : \mathbb{R}^n \to \mathbb{R}$ is a measurable function. Moreover, each π_j is a projection from $\mathbb{R}^n$ to some $\mathbb{R}^{k_j}$, with $1 \leq k_j \leq n - 1$, and the functions $g_j : \mathbb{R}^{k_j} \to \mathbb{C}$ are locally integrable with respect to the measure of $\mathbb{R}^{k_j}$.

The integral $I(\tau; g_1, \ldots, g_m)$ is well defined if all the g_j belong to the functional space $\mathbf{L}^\infty$, and in this case

$$|I(\tau; g_1, \ldots, g_m)| \leq C \prod_j \|g_j\|_\infty.$$

The authors show that if f is a polynomial of bounded degree and the k_j's are $n-1$ or 1 according to some conditions, then it is possible to characterize the decay rates of $I(\tau; g_1, \ldots, g_m)$. Some other cases are considered in [**80**] and [**39**], but the general case is largely open.

Finally we want to comment that to the best of our knowledge, there are no well established connections between these multilinear oscillatory integrals and local zeta functions. In seems also that there are no 'multilinear' analogues of the complex oscillatory integrals of section 4. To establish such connections/analogues may be of some interest for analysts or algebraic geometers.

Acknowledgements

The author wishes to express his gratitude to the referee for several helpful comments and suggestions. The author also wishes to thank to Professor B. Lichtin for drawing the author's attention to his work on multivariate local zeta functions and oscillatory integrals [**63**–**68**].

References

[1] J. Àlvarez Montaner, J. Jeffries, L. Núñez-Betancourt, *Bernstein-Sato Polynomials in Commutative Algebra.* In: Peeva I. (eds) Commutative Algebra. Springer, Cham., 2021. https://doi.org/10.1007/978-3-030-89694-2_1

[2] Jørgen Ellegaard Andersen and William Elbæk Petersen, *Asymptotic expansions of the Witten-Reshetikhin-Turaev invariants of mapping tori I*, Trans. Amer. Math. Soc. **372** (2019), no. 8, 5713–5745, DOI 10.1090/tran/7740. MR4014292

[3] Yann Angeli, *Identités de Bernstein explicites et singularités des intégrales de Riesz généralisées* (French, with French summary), J. Lie Theory **15** (2005), no. 1, 279–297. MR2115243

[4] V. I. Arnol′d, S. M. Guseĭn-Zade, and A. N. Varchenko, *Singularities of differentiable maps. Vol. II*, Monographs in Mathematics, vol. 83, Birkhäuser Boston, Inc., Boston, MA, 1988. Monodromy and asymptotics of integrals; Translated from the Russian by Hugh Porteous; Translation revised by the authors and James Montaldi, DOI 10.1007/978-1-4612-3940-6. MR966191

[5] Fuensanta Aroca, Mirna Gómez-Morales, and Edwin León-Cardenal, *On Archimedean zeta functions and Newton polyhedra*, J. Math. Anal. Appl. **473** (2019), no. 2, 1215–1233, DOI 10.1016/j.jmaa.2019.01.017. MR3912873

[6] M. F. Atiyah, *Resolution of singularities and division of distributions*, Comm. Pure Appl. Math. **23** (1970), 145–150, DOI 10.1002/cpa.3160230202. MR256156

[7] D. Barlet, *Contribution effective de la monodromie aux développements asymptotiques* (French), Ann. Sci. École Norm. Sup. (4) **17** (1984), no. 2, 293–315. MR760679

[8] Daniel Barlet, *Monodromie et pôles du prolongement méromorphe de* $\int_X |f|^{2\lambda}\square$ (French, with English summary), Bull. Soc. Math. France **114** (1986), no. 3, 247–269. MR878239

[9] D. Barlet, *Développement asymptotique des fonctions obtenues par intégration sur les fibres* (French), Invent. Math. **68** (1982), no. 1, 129–174, DOI 10.1007/BF01394271. MR666639

[10] Daniel Barlet, *Real canonical cycle and asymptotics of oscillating integrals*, Nagoya Math. J. **171** (2003), 187–196, DOI 10.1017/S0027763000025563. MR2002018

[11] Daniel Barlet and Ahmed Jeddi, *The distribution* $\int_A f^s\square$ *and the real asymptotic spectrum*, Real and complex singularities, Lecture Notes in Pure and Appl. Math., vol. 232, Dekker, New York, 2003, pp. 137–149. MR2075061

[12] D. Barlet and H.-M. Maire, *Asymptotic expansion of complex integrals via Mellin transform*, J. Funct. Anal. **83** (1989), no. 2, 233–257, DOI 10.1016/0022-1236(89)90020-7. MR995749

[13] D. Barlet and H.-M. Maire, *Asymptotique des intégrales-fibres* (French, with English and French summaries), Ann. Inst. Fourier (Grenoble) **43** (1993), no. 5, 1267–1299. MR1275199

[14] Joseph N. Bernšteĭn, *Feasibility of the analytic continuation* $f_+{}^\lambda$ *for certain polynomials* f, Funkcional. Anal. i Priložen **2** (1968), no. 1, 92–93, English version in Funct. Anal. Its Appl. 2, 85–87 (1968). MR0230125

[15] I. N. Bernšteĭn, *Modules over a ring of differential operators. An investigation of the fundamental solutions of equations with constant coefficients* (Russian), Funkcional. Anal. i Priložen. **5** (1971), no. 2, 1–16. MR0290097

[16] I. N. Bernšteĭn, *Analytic continuation of generalized functions with respect to a parameter*, Funkcional. Anal. i Priložen. **6** (1972), no. 4, 26–40. MR0320735

[17] I. N. Bernšteĭn and S. I. Gel′fand, *Meromorphy of the function* P^λ (Russian), Funkcional. Anal. i Priložen. **3** (1969), no. 1, 84–85. MR0247457

[18] J.-E. Björk, *Rings of differential operators*, North-Holland Mathematical Library, vol. 21, North-Holland Publishing Co., Amsterdam-New York, 1979. MR549189

[19] Guillem Blanco, *Poles of the complex zeta function of a plane curve*, Adv. Math. **350** (2019), 396–439, DOI 10.1016/j.aim.2019.04.048. MR3947649

[20] M. Bocardo-Gaspar, Willem Veys, and W. A. Zúñiga-Galindo, *Meromorphic continuation of Koba-Nielsen string amplitudes*, J. High Energy Phys. **9** (2020), 138, 43, DOI 10.1007/jhep09(2020)138. MR4190286

[21] Jean-Paul Brasselet and Marcos Sebastiani, *Brieskorn and the monodromy*, J. Singul. **18** (2018), 84–104, DOI 10.5427/jsing.2018.18f. MR3899535

[22] Anthony Carbery, Stephen Wainger, and James Wright, *Singular integrals and the Newton diagram*, Collect. Math. **Vol. Extra** (2006), 171–194. MR2264209

[23] Wengu Chen and Shanzhen Lu, *Weighted inequalities for multilinear oscillatory singular integrals*, Hokkaido Math. J. **26** (1997), no. 1, 163–175, DOI 10.14492/hokmj/1351257812. MR1432545

[24] Michael Christ, Xiaochun Li, Terence Tao, and Christoph Thiele, *On multilinear oscillatory integrals, nonsingular and singular*, Duke Math. J. **130** (2005), no. 2, 321–351, DOI 10.1215/00127094-8229909. MR2181092

[25] Raf Cluckers, Georges Comte, Daniel J. Miller, Jean-Philippe Rolin, and Tamara Servi, *Integration of oscillatory and subanalytic functions*, Duke Math. J. **167** (2018), no. 7, 1239–1309, DOI 10.1215/00127094-2017-0056. MR3799699

[26] Yves Colin de Verdière, *Spectrum of the Laplace operator and periodic geodesics: thirty years after* (English, with English and French summaries), Ann. Inst. Fourier (Grenoble) **57** (2007), no. 7, 2429–2463. Festival Yves Colin de Verdière. MR2394548

[27] Tristan C. Collins, Allan Greenleaf, and Malabika Pramanik, *A multi-dimensional resolution of singularities with applications to analysis*, Amer. J. Math. **135** (2013), no. 5, 1179–1252, DOI 10.1353/ajm.2013.0042. MR3117305

[28] S. C. Coutinho, *A primer of algebraic D-modules*, London Mathematical Society Student Texts, vol. 33, Cambridge University Press, Cambridge, 1995, DOI 10.1017/CBO9780511623653. MR1356713

[29] Philip J. Davis, *Leonhard Euler's integral: A historical profile of the gamma function*, Amer. Math. Monthly **66** (1959), 849–869, DOI 10.2307/2309786. MR106810

[30] E. Delabaere, *Addendum to the hyperasymptotics for multidimensional Laplace integrals*, Analyzable functions and applications, Contemp. Math., vol. 373, Amer. Math. Soc., Providence, RI, 2005, pp. 177–190, DOI 10.1090/conm/373/06919. MR2130830

[31] E. Delabaere and C. J. Howls, *Global asymptotics for multiple integrals with boundaries*, Duke Math. J. **112** (2002), no. 2, 199–264, DOI 10.1215/S0012-9074-02-11221-6. MR1894360

[32] Jan Denef, *Report on Igusa's local zeta function*, Astérisque **201-203** (1991), Exp. No. 741, 359–386 (1992). Séminaire Bourbaki, Vol. 1990/91. MR1157848

[33] Jan Denef and François Loeser, *Motivic Igusa zeta functions*, J. Algebraic Geom. **7** (1998), no. 3, 505–537. MR1618144

[34] Jan Denef, Johannes Nicaise, and Patrick Sargos, *Oscillating integrals and Newton polyhedra*, J. Anal. Math. **95** (2005), 147–172, DOI 10.1007/BF02791501. MR2145563

[35] Jan Denef and Patrick Sargos, *Polyèdre de Newton et distribution f_+^s. I* (French), J. Analyse Math. **53** (1989), 201–218, DOI 10.1007/BF02793414. MR1014986

[36] I. Gelfand, *Some aspects of functional analysis and algebra*, Proceedings of the International Congress of Mathematicians, Amsterdam, 1954, Vol. 1, Erven P. Noordhoff N. V., Groningen; North-Holland Publishing Co., Amsterdam, 1957, pp. 253–276. MR0095423

[37] Israel M. Gel'fand and Georgi E. Shilov, *Generalized functions. Vol. 1*, AMS Chelsea Publishing, Providence, RI, 2016, Properties and operations, Translated from the 1958 Russian original [MR0097715] by Eugene Saletan, Reprint of the 1964 English translation [MR0166596].

[38] Maxim Gilula, *Some oscillatory integral estimates via real analysis*, Math. Z. **289** (2018), no. 1-2, 377–403, DOI 10.1007/s00209-017-1956-2. MR3803795

[39] Maxim Gilula, Philip T. Gressman, and Lechao Xiao, *Higher decay inequalities for multilinear oscillatory integrals*, Math. Res. Lett. **25** (2018), no. 3, 819–842, DOI 10.4310/MRL.2018.v25.n3.a5. MR3847336

[40] Michel Granger, *Bernstein-Sato polynomials and functional equations*, Algebraic approach to differential equations, World Sci. Publ., Hackensack, NJ, 2010, pp. 225–291, DOI 10.1142/9789814273244_0006. MR2766095

[41] Michael Greenblatt, *Simply nondegenerate multilinear oscillatory integral operators with smooth phase*, Math. Res. Lett. **15** (2008), no. 4, 653–660, DOI 10.4310/MRL.2008.v15.n4.a5. MR2424903

[42] Michael Greenblatt, *Oscillatory integral decay, sublevel set growth, and the Newton polyhedron*, Math. Ann. **346** (2010), no. 4, 857–895, DOI 10.1007/s00208-009-0424-7. MR2587095

[43] Michael Greenblatt, *Resolution of singularities, asymptotic expansions of integrals and related phenomena*, J. Anal. Math. **111** (2010), 221–245, DOI 10.1007/s11854-010-0016-1. MR2747065

[44] Michael Greenblatt, *L^p Sobolev regularity of averaging operators over hypersurfaces and the Newton polyhedron*, J. Funct. Anal. **276** (2019), no. 5, 1510–1527, DOI 10.1016/j.jfa.2018.05.014. MR3912783

[45] Allan Greenleaf, Malabika Pramanik, and Wan Tang, *Oscillatory integral operators with homogeneous polynomial phases in several variables*, J. Funct. Anal. **244** (2007), no. 2, 444–487, DOI 10.1016/j.jfa.2006.11.005. MR2297031

[46] Heisuke Hironaka, *Resolution of singularities of an algebraic variety over a field of characteristic zero. I, II*, Ann. of Math. (2) **79** (1964), 109–203; ibid. (2) **79** (1964), 205–326, DOI 10.2307/1970547. MR0199184

[47] Lars Hörmander, *The analysis of linear partial differential operators. I*, Classics in Mathematics, Springer-Verlag, Berlin, 2003. Distribution theory and Fourier analysis; Reprint of the second (1990) edition [Springer, Berlin; MR1065993 (91m:35001a)], DOI 10.1007/978-3-642-61497-2. MR1996773

[48] Jun-ichi Igusa, *Complex powers and asymptotic expansions. I. Functions of certain types*, J. Reine Angew. Math. **268(269)** (1974), 110–130, DOI 10.1515/crll.1974.268-269.110. MR347753

[49] Jun-ichi Igusa, *Forms of higher degree*, Tata Institute of Fundamental Research Lectures on Mathematics and Physics, vol. 59, Tata Institute of Fundamental Research, Bombay; by the Narosa Publishing House, New Delhi, 1978. MR546292

[50] Jun-ichi Igusa, *An introduction to the theory of local zeta functions*, AMS/IP Studies in Advanced Mathematics, vol. 14, American Mathematical Society, Providence, RI; International Press, Cambridge, MA, 2000, DOI 10.1090/amsip/014. MR1743467

[51] Isroil A. Ikromov, *Estimates for Fourier transform of measures supported on singular hypersurfaces*, Turkish J. Math. **31** (2007), no. 1, 1–21. MR2311601

[52] Ph. Jacobs, *The distribution $|f|^\lambda$, oscillating integrals and principal value integrals*, J. Anal. Math. **81** (2000), 343–372, DOI 10.1007/BF02788996. MR1785288

[53] Pierre Jeanquartier, *Développement asymptotique de la distribution de Dirac attachée à une fonction analytique* (French), C. R. Acad. Sci. Paris Sér. A-B **201** (1970), A1159–A1161. MR420695

[54] ———, *Transformation de Mellin et développements asymptotiques*, Enseign. Math. (2) **25** (1979), no. 3-4, 285–308 (1980).

[55] Joe Kamimoto and Toshihiro Nose, *Newton polyhedra and weighted oscillatory integrals with smooth phases*, Trans. Amer. Math. Soc. **368** (2016), no. 8, 5301–5361, DOI 10.1090/tran/6528. MR3458382

[56] V. N. Karpushkin, *Uniform estimates for an oscillatory integral and volume with the phase of A. N. Varchenko* (Russian, with Russian summary), Mat. Zametki **72** (2002), no. 5, 688–692, DOI 10.1023/A:1021452805001; English transl., Math. Notes **72** (2002), no. 5-6, 636–640. MR1963131

[57] Masaki Kashiwara, *B-functions and holonomic systems. Rationality of roots of B-functions*, Invent. Math. **38** (1976/77), no. 1, 33–53, DOI 10.1007/BF01390168. MR430304

[58] Tatsuo Kimura, *The b-functions and holonomy diagrams of irreducible regular prehomogeneous vector spaces*, Nagoya Math. J. **85** (1982), 1–80. MR648417

[59] E. León-Cardenal, Willem Veys, and W. A. Zúñiga-Galindo, *Poles of Archimedean zeta functions for analytic mappings*, J. Lond. Math. Soc. (2) **87** (2013), no. 1, 1–21, DOI 10.1112/jlms/jds031. MR3022704

[60] Edwin León-Cardenal and W. A. Zúñiga-Galindo, *An introduction to the theory of local zeta functions from scratch* (English, with English and Spanish summaries), Rev. Integr. Temas Mat. **37** (2019), no. 1, 45–76, DOI 10.18273/revint. MR3920742

[61] G. V. Liakhovetski and R. B. Paris, *Asymptotic expansions of Laplace-type integrals. III*, J. Comput. Appl. Math. **132** (2001), no. 2, 409–429, DOI 10.1016/S0377-0427(00)00443-X. MR1840638

[62] B. Lichtin, *Poles of $|f(z,w)|^{2s}$ and roots of the b-function*, Ark. Mat. **27** (1989), no. 2, 283–304, DOI 10.1007/BF02386377. MR1022282

[63] Ben Lichtin, *On a question of Igusa: towards a theory of several variable asymptotic expansions. I*, Compositio Math. **120** (2000), no. 1, 25–82, DOI 10.1023/A:1001743909924. MR1738214

[64] Ben Lichtin, *Uniform bounds for some exponential sums* (mod p^r) *in two variables*, Proceedings of the Session in Analytic Number Theory and Diophantine Equations, Bonner Math. Schriften, vol. 360, Univ. Bonn, Bonn, 2003, pp. 63. MR2075633

[65] Ben Lichtin, *On a question of Igusa. II. Uniform asymptotic bounds for Fourier transforms in several variables*, Compos. Math. **141** (2005), no. 1, 192–226, DOI 10.1112/S0010437X04000922. MR2099776

[66] Ben Lichtin, *On a question of Igusa. III. A generalized Poisson formula for pairs of polynomials*, J. Reine Angew. Math. **611** (2007), 189–219, DOI 10.1515/CRELLE.2007.079. MR2361089

[67] Ben Lichtin, *Uniform bounds for two variable real oscillatory integrals and singularities of mappings*, J. Reine Angew. Math. **611** (2007), 1–73, DOI 10.1515/CRELLE.2007.073. MR2360603

[68] Ben Lichtin, *Applications of multivariate asymptotics. I. Boundedness of a maximal operator on $L^p(\mathbb{R}^3)$*, Forum Math. **21** (2009), no. 5, 759–795, DOI 10.1515/FORUM.2009.037. MR2560390

[69] J.-M. Lion and J.-P. Rolin, *Intégration des fonctions sous-analytiques et volumes des sous-ensembles sous-analytiques* (French, with English and French summaries), Ann. Inst. Fourier (Grenoble) **48** (1998), no. 3, 755–767. MR1644093

[70] François Loeser, *Quelques conséquences locales de la théorie de Hodge* (French), Ann. Inst. Fourier (Grenoble) **35** (1985), no. 1, 75–92. MR781779

[71] F. Loeser, *Évaluation d'intégrales et théorie de Hodge* (French), Hodge theory (Sant Cugat, 1985), Lecture Notes in Math., vol. 1246, Springer, Berlin, 1987, pp. 125–142, DOI 10.1007/BFb0077535. MR894048

[72] F. Loeser, *Fonctions zêta locales d'Igusa à plusieurs variables, intégration dans les fibres, et discriminants* (French), Ann. Sci. École Norm. Sup. (4) **22** (1989), no. 3, 435–471. MR1011989

[73] Shan Zhen Lu and Gui Ping Tao, (L^p, L^q) *estimates for multilinear oscillatory singular integrals with smooth phases*, Acta Math. Sin. (Engl. Ser.) **19** (2003), no. 4, 645–654, DOI 10.1007/s10114-003-0258-2. MR2023359

[74] Shanzhen Lu and Huoxiong Wu, *A class of multilinear oscillatory singular integrals related to block spaces*, Tohoku Math. J. (2) **56** (2004), no. 3, 299–315. MR2075767

[75] Bernard Malgrange, *Intégrales asymptotiques et monodromie* (French), Ann. Sci. École Norm. Sup. (4) **7** (1974), 405–430 (1975). MR372243

[76] ———, *Polynômes de Bernstein-Sato et cohomologie évanescente*, Analysis and topology on singular spaces, II, III (Luminy, 1981), Astérisque, vol. 101, Soc. Math. France, Paris, 1983, pp. 243–267.

[77] Diane Meuser, *A survey of Igusa's local zeta function*, Amer. J. Math. **138** (2016), no. 1, 149–179, DOI 10.1353/ajm.2016.0006. MR3462883

[78] Mircea Mustaţă, *Invariants of singularities via integration*, Notes of the 2006 AIM Workshop 'Numerical invariants of singularities and of higher-dimensional algebraic varieties'. Available at `www.aimath.org/WWN/singularvariety/mot1.pdf`.

[79] Mircea Mustaţă, *IMPANGA lecture notes on log canonical thresholds*, Contributions to algebraic geometry, EMS Ser. Congr. Rep., Eur. Math. Soc., Zürich, 2012, pp. 407–442, DOI 10.4171/114-1/16. Notes by Tomasz Szemberg. MR2976952

[80] Aleksandra Niepla, Kevin O'Neill, and Zhen Zeng, *Decay rate of multilinear oscillatory integral operators in $\mathbb{R}^2$*, Proc. Amer. Math. Soc. **148** (2020), no. 4, 1689–1695, DOI 10.1090/proc/14857. MR4069206

[81] Toshihisa Okada and Kiyoshi Takeuchi, *Meromorphic continuations of local zeta functions and their applications to oscillating integrals*, Tohoku Math. J. (2) **65** (2013), no. 2, 159–178, DOI 10.2748/tmj/1372182720. MR3079283

[82] V. P. Palamodov, *Asymptotic expansions of integrals in complex and real domains* (Russian), Mat. Sb. (N.S.) **127(169)** (1985), no. 2, 209–238, 287. MR792439

[83] Robin Pemantle, *Analytic combinatorics in d variables: an overview*, Algorithmic probability and combinatorics, Contemp. Math., vol. 520, Amer. Math. Soc., Providence, RI, 2010, pp. 195–220, DOI 10.1090/conm/520/10260. MR2681861

[84] Robin Pemantle and Mark C. Wilson, *Asymptotic expansions of oscillatory integrals with complex phase*, Algorithmic probability and combinatorics, Contemp. Math., vol. 520, Amer. Math. Soc., Providence, RI, 2010, pp. 221–240, DOI 10.1090/conm/520/10261. MR2681862

[85] D. H. Phong, E. M. Stein, and J. A. Sturm, *On the growth and stability of real-analytic functions*, Amer. J. Math. **121** (1999), no. 3, 519–554. MR1738409

[86] D. H. Phong, E. M. Stein, and Jacob Sturm, *Multilinear level set operators, oscillatory integral operators, and Newton polyhedra*, Math. Ann. **319** (2001), no. 3, 573–596, DOI 10.1007/PL00004450. MR1819885

[87] D. H. Phong and Jacob Sturm, *Algebraic estimates, stability of local zeta functions, and uniform estimates for distribution functions*, Ann. of Math. (2) **152** (2000), no. 1, 277–329, DOI 10.2307/2661384. MR1792297

[88] Naud Potemans and Willem Veys, *Introduction to p-adic igusa zeta functions*, p-Adic Analysis, Arithmetic and Singularities (Carlos Galindo, Alejandro Melle Hernández, Julio José Moyano-Fernández, and Wilson A. Zúñiga-Galindo, eds.), Contemporary Mathematics, vol. 778, Amer. Math. Soc., Providence, RI, 2022, to appear.

[89] C. Sabbah, *Proximité évanescente. I. La structure polaire d'un $\mathcal{D}$-module* (French, with English summary), Compositio Math. **62** (1987), no. 3, 283–328. MR901394

[90] C. Sabbah, *Proximité évanescente. II. Équations fonctionnelles pour plusieurs fonctions analytiques* (French, with English summary), Compositio Math. **64** (1987), no. 2, 213–241. MR916482

[91] ———, *Some properties and applications of Brieskorn lattices*, J. Singul. **18** (2018), 238–247.

[92] Mikio Sato, *Theory of prehomogeneous vector spaces (algebraic part)—the English translation of Sato's lecture from Shintani's note*, Nagoya Math. J. **120** (1990), 1–34, DOI 10.1017/S0027763000003214. Notes by Takuro Shintani; Translated from the Japanese by Masakazu Muro. MR1086566

[93] Mikio Sato and Takuro Shintani, *On zeta functions associated with prehomogeneous vector spaces*, Ann. of Math. (2) **100** (1974), 131–170, DOI 10.2307/1970844. MR344230

[94] Elias M. Stein, *Harmonic analysis: real-variable methods, orthogonality, and oscillatory integrals*, Princeton Mathematical Series, vol. 43, Princeton University Press, Princeton, NJ, 1993. With the assistance of Timothy S. Murphy; Monographs in Harmonic Analysis, III. MR1232192

[95] Alexander N. Varčenko, *Newton polyhedra and estimates of oscillatory integrals*, Funkcional. Anal. i Priložen. **10** (1976), no. 3, 13–38.

[96] Willem Veys and W. A. Zúñiga-Galindo, *Zeta functions and oscillatory integrals for meromorphic functions*, Adv. Math. **311** (2017), 295–337, DOI 10.1016/j.aim.2017.02.022. MR3628217

[97] Juan Viu-Sos, *An introduction to p-adic and motivic integration, zeta functions and invariants of singularities*, p-Adic Analysis, Arithmetic and Singularities (Carlos Galindo, Alejandro Melle Hernández, Julio José Moyano-Fernández, and Wilson A. Zúñiga-Galindo, eds.), Contemporary Mathematics, vol. 778, Amer. Math. Soc., Providence, RI, 2022, to appear.

[98] Uli Walther, *Survey on the D-module f^s*, Commutative algebra and noncommutative algebraic geometry. Vol. I, Math. Sci. Res. Inst. Publ., vol. 67, Cambridge Univ. Press, New York, 2015, pp. 391–430. With an appendix by Anton Leykin. MR3525478

[99] Andrzej Weber, *Residue forms on singular hypersurfaces*, Michigan Math. J. **53** (2005), no. 3, 553–569, DOI 10.1307/mmj/1133894165. MR2207207

[100] R. Wong and J. P. McClure, *On a method of asymptotic evaluation of multiple integrals*, Math. Comp. **37** (1981), no. 156, 509–521, DOI 10.2307/2007443. MR628712

[101] Lechao Xiao, *Endpoint estimates for one-dimensional oscillatory integral operators*, Adv. Math. **316** (2017), 255–291, DOI 10.1016/j.aim.2017.06.007. MR3672907

[102] Wilson A. Zúñiga-Galindo, Brian A. Zambrano-Luna, and Edwin León-Cardenal, *Graphs, local zeta functions, log-coulomb gases, and phase transitions at finite temperature*, Journal of Mathematical Physics **63** (2022), no. 1, Paper Number 013506. `https://doi.org/10.1063/5.0070683`

CONACYT–Centro de Investigación en Matemáticas (CIMAT). Unidad Zacatecas , Parque Quantum Ciudad del Conocimiento. Av. Lasec, Andador Galileo Galilei, Manzana 3 Lote 7, C.P. 98160, Zacatecas, ZAC, Mexico

Email address: edwin.leon@cimat.mx

Contemporary Mathematics
Volume **778**, 2022
https://doi.org/10.1090/conm/778/15654

Generalized Poincaré series for plane curve singularities

Julio José Moyano-Fernández

Dedicated to Félix Delgado and Carlos Galindo on the occasion of their 60th birthday and to Sabir Gusein-Zade on the occasion of his 70th birthday

Abstract. The purpose of these notes is to give an introduction to the notion of Poincaré series attached to multi-index filtrations associated with (plane) complex curve singularities. This was introduced by Campillo, Delgado and Kiyek in 1994 in a purely algebraic-combinatorial viewpoint and further developed in a series of papers by Campillo, Delgado and Gusein-Zade from 1999 on; in particular, they found an alternative definition as an integral with respect to the Euler characteristic, which allowed them to generalize the concept to a setting of motivic nature in 2011. In this review we will summarize the basic facts and results mainly about the *generalized Poincaré series*. No claim of novelty —beyond the exposition self— is done.

Contents

1. Introduction

This survey contains introductory material extracted from the books of Brieskorn and Knörrer [**BK86**], Casas-Alvero [**Cs00**], Greuel, Lossen and Shustin [**GLS07**], as well as Wall [**W04**] in Sections 1 and 2. Furthermore, Sections 3, 5, 6 and 7 are based on the work of Campillo, Delgado and Gusein-Zade, mainly on [**CDG03b**]

2020 *Mathematics Subject Classification.* Primary 14H20; Secondary 32S99.

Key words and phrases. Curve singularities; Poincaré series; generalized Poincaré series; embedded resolution.

These notes contain the material explained in the "Workshop on singularities: semigroups, topology and valuations" that took place in the University Complutense of Madrid in October 28th–31st, 2019. Partially supported by MCIN/AEI/10.13039/501100011033/FEDER (ERDF) "A way to make Europe", grants PGC2018-096446-B-C22 and RED2018-102583-T, as well as by Universitat Jaume I, grant UJI-B2021-02.

and [**CDG07**]. Section 4 is a summary of the survey [**G10**] by Gusein-Zade. The reader is referred to those publications for further details.

We are interested in the study of complex *plane curve singularities*, i.e. zero sets given by one equation in two variables with coefficients in $\mathbb{C}$, as those in Figure 1.

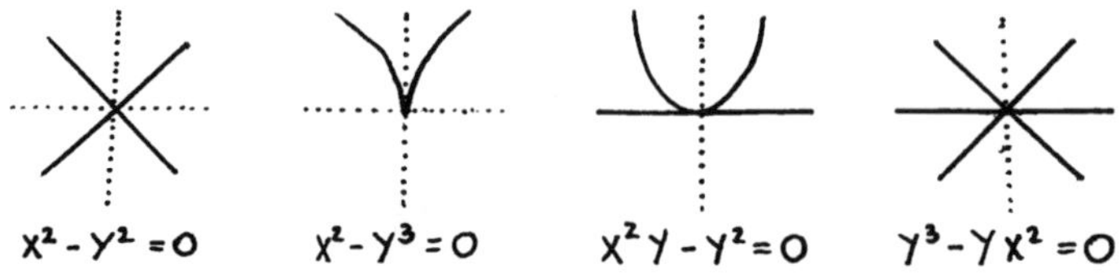

FIGURE 1. Plane curve singularities.

But where do these equations belong to? The polynomials above are supposed to be elements in $\mathbb{C}[x, y]$. The pictures show just the set of *real* solutions but this contains already the most remarkable geometric properties of the original set of *complex* solutions in a small neighbourhood of $(0, 0)$, e.g.

- ◇ the number of irreducible components;
- ◇ the intersection behaviour of these components (transversal or tangencial).

In all these four examples, the *irreducible components* (called "local branches", or simply "branches") of the singularity are given by polynomial equations. But this does not need to be the case; take for example $f := x^2 - y^2(1 + y)$.

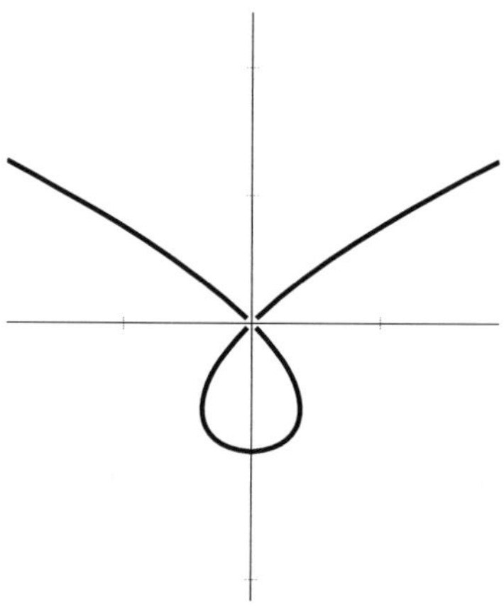

FIGURE 2. A node of equation $x^2 - y^2(1 + y) = 0$.

This is irreducible as a polynomial, i.e. as an element of $\mathbb{C}[x, y]$, and even at the localization $\mathbb{C}[x, y]_{(x,y)}$, but it is reducible as an element of the power series ring $\mathbb{C}\{x, y\}$; indeed

$$f = (x - y\sqrt{1 + y})(x + y\sqrt{1 + y}).$$

This fact corresponds to the intuition developed when looking at the picture, see Figure 2: in a small neighbourhood around $(0, 0)$ we find two irreducible components, but the curve is irreducible in a bigger neighbourhood.

However, the behaviour of singularities such as $x^2 - y^2 = 0$ and $f = 0$ in a small neighbourhood are at a first glance the same: indeed, the factor rings

$$\mathbb{C}\{x, y\}/(x^2 - y^2) \quad \text{and} \quad \mathbb{C}\{x, y\}/(f)$$

are isomorphic (via $x \mapsto x$, $y \mapsto y\sqrt{1+y}$). This is referred as *to have the same analytic type.*

So, one purpose of the singularists is to be able to classify plane curve singularities with respect to some interesting equivalence relation such as "to have the same analytic type". To this end, it is very useful to find *invariants*, i.e. algebraic, easy-to-handle properties that do not change for certain families; the most familiar invariants are given by numbers, polynomials or power series.

There is also a topological classification of plane curve singularities, of a discrete nature, which classifies singularities according to their topological type. This can be described in two particularly remarkable ways:

- ◇ via the characteristic exponents of the Puiseux parametrizations of the local branches and their mutual intersection multiplicities;
- ◇ via the sequence of infinitely near points in the minimal embedded resolution and their multiplicities.

Both descriptions can be used to express the main topological numerical invariants (like the Milnor number, the δ-invariant, etc.) and the relations between them.

Classification plays a central role in Mathematics, as Brieskorn and Knörrer [**BK86**], page 323, pointed out:

> One of the most natural questions in the investigation of a particular class of mathematical objects is always the problem of classification of these objects. One tries to understand the structure of the objects under investigation as far as possible and, on the basis of this understanding, to divide the totality of objects into classes in a suitable way. The finer this classification is, the better.

In these pages we want to describe one of those topological invariants: the Poincaré series associated with a plane curve singularity by means of a multi-index filtration on the ring of germs of functions of two variables defined by the orders of a function on the irreducible components of the curve.

We will start with a review of the main technical tools used in the study of curve singularities.

2. Technical tools

The two most powerful technical tools in the treatment of plane curve singularities are

(1) the parametrization of local branches;
(2) the embedded resolution of singularities by a sequence of blowing-ups.

We cannot spend here so much time for the details on these, but we will present them as briefly as possible.

The first issue to address is the finding of a suitable algebraic framework to work with singularities —real pictures are obviously not enough! As we have said, we are interested in small neighbourhoods of (singular) points at complex plane curves given by holomorphic, i.e. analytic functions. According to the complex analysis, this can be done by using analytic spaces. More precisely, if $U \in \mathbb{C}^2$ is an open subset and f is a holomorphic function defined on U, then

$$C = \{(x, y) \in U : f(x, y) = 0\}$$

is an analytic subspace of U. Of course, we are interested in an arbitrary small neighbourhood of a point $p \in C$, i.e. in the *germ* of C at p; this will be denoted by (C, p). Formally, a germ at p is an equivalence class of holomorphic functions on neighbourhoods of p so that two such functions are said to be equivalent if their restrictions to an adequate neighbourhood of p coincide. The set of all germs of holomorphic functions at p form a ring $\mathcal{O}_{\mathbb{C}^2,p}$, called the ring of germs of homomorphic functions at p. Without loss of generality one may assume $p = (0, 0) = \mathbf{0}$.

The ring of germs of holomorphic functions at $\mathbf{0}$ is isomorphic to the ring of convergent power series $\mathbb{C}\{x, y\}$ via Taylor expansions at $\mathbf{0}$: this is how power series come into the picture. This is a local ring with maximal ideal $\mathfrak{m} = \mathfrak{m}_{C,\mathbf{0}}$. Recall that for the power series ring, the maximal ideal is just (x, y).

The ideal (f) in $\mathcal{O}_{\mathbb{C}^2,\mathbf{0}}$ generated by f allows us to define the local ring $\mathcal{O}_{C,\mathbf{0}} = \mathcal{O}_{\mathbb{C}^2,\mathbf{0}}/(f)$ of $(C, \mathbf{0})$. In order to deal with power series also in the local ring, we need to assume that $\mathcal{O}_{C,\mathbf{0}}$ has no nilpotent elements, i.e. that $(C, \mathbf{0})$ is reduced. Now it should be clear: a plane curve singularity will be a germ of an analytic space $(C, \mathbf{0})$ together with the local ring $\mathcal{O}_{C,\mathbf{0}}$. For the sake of simplicity we will often say "C has the property E in $\mathbf{0}$" or even "f has the property E', meaning actually that "$(C, \mathbf{0})$ has the property E".

The reduceness of a germ plays also a role in the definition of an invariant called the multiplicity: for a reduced germ $(C, \mathbf{0})$ the *multiplicity* $\mathrm{m}(C, \mathbf{0})$ of C at $\mathbf{0}$ is defined to be the number of intersection points of a suitable small representative of C with a general line of $\mathbb{C}^2$ in a small neighbourhood of the origin (but the line should not go through the origin!). Viewing the germ as a power series, this is the minimum order occurring in the Taylor expansion of f at $\mathbf{0}$ i.e., the natural number e such that $f \in \mathfrak{m}^e \setminus \mathfrak{m}^{e+1}$. In other words, if

$$f(x, y) = \sum_{i,j} a_{ij} x^i y^j,$$

then $\mathrm{m}(C, \mathbf{0}) = \min\{i + j : a_{ij} \neq 0\}$.

This role interchange between power series and germs allows us to use the "zero set notation" and write $(C, \mathbf{0}) := V(f) \subset (\mathbb{C}^2, \mathbf{0})$ for $f \in \mathfrak{m} \subset \mathbb{C}\{x, y\}$. Thus, if $f = f_1^{n_1} \cdots f_r^{n_r}$ is the decomposition into irreducible factors of $f \in \mathbb{C}\{x, y\}$, then $V(f) = V(f_1) \cup \ldots \cup V(f_r)$ and we call $(C_i, \mathbf{0}) = V(f_i)$ a *local branch* of $(C, \mathbf{0})$, which is called *reduced* if $n_i = 1$, for all $i = 1, \ldots, r$. As mentioned above, we will consider reduced germs. If $r = 1$ then f is called *irreducible*.

As one may read on page 331 in [**BK86**], "admittedly, algebraic geometers are not very glad to use the ring $\mathbb{C}\{x, y\}$. Among other reasons, this is because they want to consider zero sets of polynomials whose coefficients are not real or complex, but in any field k. In an arbitrary field it is no longer meaningful to speak of convergent power series". The solution to satisfy these wishes is to consider the ring of formal power series $k[\![x, y]\!]$. Of course $\mathbb{C}\{x, y\} \subset \mathbb{C}[\![x, y]\!]$.

Some authors like Casas-Alvero [**Cs00**] consider germs of curves on smooth complex surfaces (connected two-dimensional complex analytic manifolds); but these curves are locally flat in the sense that they lie on surfaces which are locally isomorphic to $\mathbb{C}^2$. Therefore this makes the local study not more general than that of the germs of curves at a point of $\mathbb{C}^2$, although we may make this assumption to fit better the literature.

2.1. Parametrizations of local branches. A parametrization of a branch $f = f(x, y)$ at $\mathbf{0}$ is given by power series $x(t), y(t) \in \mathbb{C}\{t\}$ such that

(1) $f(x(t), y(t)) = 0$ in $\mathbb{C}\{t\}$;
(2) if $\tilde{x}(t)$, $\tilde{y}(t)$ satisfy $f(\tilde{x}(t), \tilde{y}(t)) = 0$, then there is a unique unit $u \in \mathbb{C}\{t\}$ such that $\tilde{x}(t) = x(u \cdot t)$ and $\tilde{y}(t) = y(u \cdot t)$.

One can get parametrizations by using the Newton-Puiseux algorithm (see e.g. [**Cs00**] or [**F94**]) or the Hamburger-Noether expansions (in [**C80**] or [**CF02**]).

EXAMPLE 2.1. For the cusp $f = x^2 - y^3$ an easy computation shows that the power series $x(t) = t^3$ and $y(t) = t^2$ give a parametrization of f at $\mathbf{0}$, while $u(t) = t^6, v(t) = t^4$ do not —the condition (2) fails. Notice that the map $t \mapsto (x(t), y(t))$ is in fact the normalization map.

An immediate application of parametrizations appears in the study of the *intersection multiplicity* of two local branches, defined as the total order of one curve on the parametrizations of the local branches of the other curve:

$$[f, g] = [f, g]_{\mathbf{0}} = \operatorname{ord}_t\Big(g(x(t), y(t))\Big) = \sup\{m \in \mathbb{N} : t^m \text{ divides } g(x(t), y(t))\},$$

with $(x(t), y(t))$ a parametrization of f at $\mathbf{0}$. This is an additive invariant for the irreducible components in the sense that, if $f = f_1^{n_1} \cdots f_r^{n_r}$, then

$$[f, g] = \sum_{i=1}^{r} n_i [f_i, g].$$

In fact, the intersection multiplicity has a bilinear, symmetric behaviour and is finite if and only if the germs do not share branches.

Two germs are said to be *transversal* if their intersection multiplicity is equal to 1.

EXAMPLE 2.2. Consider the plane curve singularities $f = x^2 - y^3$ and $g = x^2 - y^2(1+y)$. Observe that $g = g_1 g_2$ with $g_1 = x - y\sqrt{1+y}$ and $g_2 = x + y\sqrt{1+y}$, and that, as a power series,

$$\sqrt{1+y} = 1 + \frac{y}{2} - \frac{y^2}{8} + \frac{y^3}{16} - \frac{5y^4}{128} + \cdots$$

If we consider the parametrization of f in Example 2.1, then we compute

$$\begin{aligned}
[f, g] =& [f, g_1] + [f, g_2] = \\
=& \operatorname{ord}_t\big(g_1(x(t), y(t))\big) + \operatorname{ord}_t\big(g_2(x(t), y(t))\big) \\
=& \operatorname{ord}_t\big(t^3 - t^2\sqrt{1+t^2}\big) + \operatorname{ord}_t\big(t^3 + t^2\sqrt{1+t^2}\big) \\
=& \operatorname{ord}_t\big(t^3 - t^2 - \frac{t^4}{2} + \frac{t^6}{8} + \cdots\big) + \operatorname{ord}_t\big(t^3 + t^2 + \frac{t^4}{2} - \frac{t^6}{8} + \cdots\big) \\
=& 2 + 2 = 4.
\end{aligned}$$

2.2. Blowing-up. The main tool in the investigations on plane curve singularities is the *blowing-up* of a point p in a smooth complex surface S. This is a *local process*: a point $p \in S$ (called the center of the blowing-up) is replaced with a $\mathbb{P}^1$, and the "view" of S from p in direction $[a : b]$ is attached to a point $[a : b] \in \mathbb{P}^1$. As a result, curves on S meeting at p get separated, i.e., they get smooth curves with different tangents at p, or, at least, their intersection multiplicity decreases: "blowing-ups simplify curve singularities".

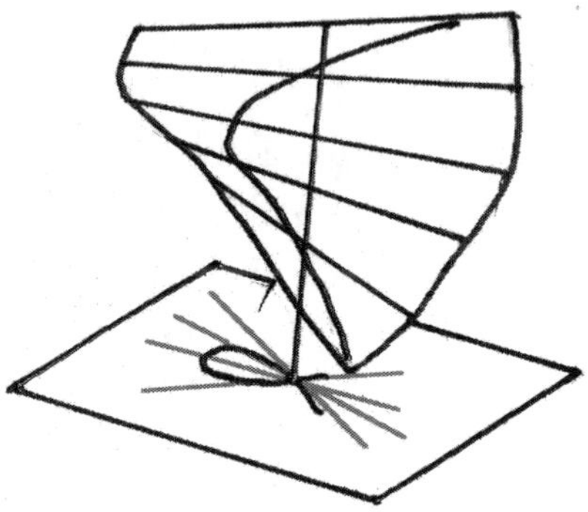

FIGURE 3. Blowing-up a node.

Let us think for a start of the case $S = \mathbb{C}^2$ and $p = \mathbf{0}$. We identify $\mathbb{P}^1$ with the set of lines $L \subset \mathbb{C}^2$ through $\mathbf{0}$ and define the following subspace of $\mathbb{C}^2 \times \mathbb{P}^1$:

$$\begin{aligned} \mathrm{Bl}_{\mathbf{0}}\mathbb{C}^2 =& \{(z, L) \in \mathbb{C}^2 \times \mathbb{P}^1 : z \in L\} \\ =& \{(x, y; s : t) \in \mathbb{C}^2 \times \mathbb{P}^1 : tx - sy = 0\}, \end{aligned}$$

if $\mathbb{P}^1$ is endowed with fixed homogeneous coordinates $[s : t]$, and (x, y) are analytic coordinates in an open neighbourhood U of $\mathbf{0}$.

DEFINITION 2.3. The projection $\pi : \mathrm{Bl}_{\mathbf{0}}\mathbb{C}^2 \to \mathbb{C}^2$ given by $(z, L) \mapsto z$ is called the *blowing-up* of $\mathbf{0} \in \mathbb{C}^2$.

Blowing-up is a purely local process; hence we can generalize the blowing-up of $\mathbf{0} \in \mathbb{C}^2$ to define the blowing-up of a point p in an arbitrary smooth complex surface S. There exists an open neighbourhood of p, say $U \subset S$, which is isomorphic to an open neighbourhood of $\mathbf{0} \in \mathbb{C}^2$. Choosing local coordinates with center p we can apply the above construction and define the blowing-up of U, say $\pi : \mathrm{Bl}_p U \to U \subset S$.

The curve $E := \pi^{-1}(p) = \{p\} \times \mathbb{P}^1$ is called the *exceptional divisor*, or also the first infinitesimal neighbourhood of p; it is isomorphic to $\mathbb{P}^1$. Moreover, the restriction of π outside E yields an isomorphism

$$\mathrm{Bl}_p U \setminus E \to U \setminus \{p\}.$$

In performing blowing-ups with local equations, one has to take into account where is the equation valid, so to speak, i.e. in which chart we are playing the game. Indeed, $\mathbb{P}^1$ is covered by the affine subsets

$$\mathbb{A}^1_s := \{[s : t] \in \mathbb{P}^1 : s \neq 0\} \quad \wedge \quad \mathbb{A}^1_t := \{[s : t] \in \mathbb{P}^1 : t \neq 0\}$$

with affine coordinates defined by

$$y' = \frac{t}{s} \quad \text{resp.} \quad x' = \frac{s}{t}.$$

In this way, $U \times \mathbb{P}^1 = (U \times \mathbb{A}^1_s) \cup (U \times \mathbb{A}^1_t)$. Set

$$\tilde{U}_x := \mathrm{Bl}_p U \cap (U \times \mathbb{A}^1_s)$$
$$\tilde{U}_y := \mathrm{Bl}_p U \cap (U \times \mathbb{A}^1_t)$$

Using local coordinates we have:

(1) $\tilde{U}_x := \{(x, y; y') \in U \times \mathbb{A}^1_s : y = xy'\}$.
Here (x, y') are taken as analytic coordinates in $\tilde{U}_x$.
The morphism π can be described as $\pi : \tilde{U}_x \to U$, $(x, y') \mapsto (x, xy')$.
The exceptional divisor E has local equation $x = 0$ in $\tilde{U}_x$.

(2) $\tilde{U}_y := \{(x, y; x') \in U \times \mathbb{A}^1_t : x = x'y\}$.
Here (x', y) are taken as analytic coordinates in $\tilde{U}_y$.
The morphism π can be described as $\pi : \tilde{U}_y \to U$, $(x', y) \mapsto (x'y, y)$.
The equation of E in $\tilde{U}_y$ is given by $y = 0$.

For the sake of simplicity we say that we may perform two types of changes of coordinates:

◇ **1-change**, corresponding to $\tilde{U}_x$ (called "chart 1" from now on), and given by

$$x = x'$$
$$y = x'y'$$

where (x', y') are local analytic coordinates in $\tilde{U}_x$ and $x' = 0$ is a local equation of E. This covers all the points $[s : t]$ of E except for $[0 : 1]$.

◇ **2-change**, corresponding to $\tilde{U}_y$ (called "chart 2" in the sequel), and given by

$$x = x'y'$$
$$y = y'$$

where (x', y') are local analytic coordinates in $\tilde{U}_y$ and $y' = 0$ is a local equation of E. This covers all the points $[s : t]$ of E except for $[1 : 0]$.

Remark 2.4. After blowing-up, if we want to consider local equations around a point $p = [\beta : \alpha]$ of the exceptional divisor E, we have to perform an additional "translation". In such a case we can include this "translation" in the change of variables as follows:

◇ 1-change ("chart 1")

$$x = x'$$
$$y = x'(y' + \alpha)$$

where $(x', y') = (x, \frac{y}{x-\alpha})$ is a regular system of parameters of the local ring at the point of coordinates $[1 : \alpha]$ in E.

◇ 2-change ("chart 2")

$$x = (x' + \beta)y'$$
$$y = y'$$

where $(x', y') = (\frac{x}{y-\beta}, y)$ is a regular system of parameters of the local ring at the point of homogeneous coordinates $[\beta : 1]$ in E.

We conclude the subsection by observing that one can also consider *global* blowing-ups. Patching together Bl_pU and $S\setminus\{p\}$ by means of the induced morphism

$$\mathrm{Bl}_pU \setminus E \cong U \setminus \{p\}$$

we obtain a surface Bl_pS and a morphism

$$\mathrm{Bl}_pS \to S$$

extending π and whose restriction gives rise to an isomorphism

$$\mathrm{Bl}_pS \setminus E \cong S \setminus \{p\}.$$

DEFINITION 2.5. The couple given by Bl_pS and this (restricted) morphism $\mathrm{Bl}_pS \to S$ is called the blowing-up of S with center at p.

REMARK 2.6. The blowing-up of S at p does not depend on the open subset U and the local coordinates x, y in the sense that different choices for those lead to isomorphic blowing-ups.

2.3. Transforms. Let (C, p) be a germ of curve C at p with local equation $\{f = 0\}$ in an open neighbourhood U of p, and consider the blowing-up $\pi : \mathrm{Bl}_pS \to S$.

DEFINITION 2.7. The *total transform* of C at Bl_pS is the curve C^* (or π^*C) defined by the pull-back of C by π; it has hence an equation given by $\pi^*(f) = f \circ \pi$.

Locally:

- ◇ the defining equation in $\tilde{U}_x$ is $f^* = f(x', x'y') = 0$;
- ◇ the defining equation in $\tilde{U}_y$ is $f^* = f(x'y', y') = 0$.

If we consider consecutive blowing-ups, the total transforms are defined by iteration accordingly.

On the other hand, if $f(x, y) = 0$ is a local equation for C at $p \in S$, we may write

$$f = f_m + f_{m+1} + \cdots + f_i + \cdots$$

where $m = \mathrm{m}(C, p)$ is the multiplicity of C at p.

Now, the equation of the total transform of C in $\tilde{U}_x$ is

$$\begin{aligned} f^* &= f_m(x', x'y') + f_{m+1}(x', x'y') + \cdots \\ &= (x')^m \Big(f_m(1, y') + x' f_{m+1}(1, y') + \cdots \Big) \end{aligned}$$

This works analogously in $\tilde{U}_y$.

The function $\tilde{f} = \frac{f^*}{(x')^m}$ defines a curve in $\tilde{U}_x$ with finitely many intersection points with E, given by the zeroes of $f_m(1, y')$. Analogously, the function $\tilde{f} = \frac{f^*}{(y')^m}$ defines a curve in $\tilde{U}_y$ whose finitely many intersection points with E are given by the zeroes of $f_m(x', 1)$. Patching these two curves together we obtain a curve $\tilde{C}$ (also denoted by $C^{(1)}$) in Bl_pU which is called the *strict transform* of C.

For the relation between the total and the strict transforms it is customary to use the divisorial additive notation, since a (Cartier) divisor D in a complex space is a subspace locally defined by one equation $f = 0$. So we may denote by mD the divisor given by $f^m = 0$, or $D + D'$ the divisor given by $fg = 0$ if D' is another divisor given by $g = 0$, etc. In our situation here, we will write

$$C^* = \tilde{C} + \mathrm{m}(C, p)E.$$

If we consider successive blowing-ups, the strict transforms are defined by iteration in the obvious manner.

2.4. Infinitely near points and clusters. Let p be a point of S. The exceptional divisor E_p associated with the blowing-up of S with center p is called the *first infinitesimal neighbourhood* of p. By induction we define the points in the ith infinitesimal neighbourhood of p, for $i > 0$, as the points in the first infinitesimal neighbourhood of some point in the $(i-1)$th infinitesimal neighbourhood. The points which are in the ith infinitesimal neighbourhood of p for some $i > 0$ are called *infinitely near points* to p. The points in S are called *proper points.*

NOTATION. If a point q is infinitely near to p we write $q \geq p$. This binary relation is really an ordering.

Any infinitely near point q has a unique *immediate predecessor* which is the point p such that q lies in its first infinitesimal neighbourhood E_p.

Consider the finite sequence of blowing ups

$$X = X_s \xrightarrow{\pi_s} X_{s-1} \xrightarrow{\pi_{s-1}} \cdots \longrightarrow X_3 \xrightarrow{\pi_3} X_2 \xrightarrow{\pi_2} X_1 \xrightarrow{\pi_1} X_0 = S, \tag{2.1}$$

where π_i is the blowing up of a closed point $p_{i-1} \in X_{i-1}$, for $i = 1, \ldots, s$. Then X is achieved by blowing up successively those points p_i for $i = 0, \ldots, s-1$.

Set $\pi := \pi_s \circ \pi_{s-1} \circ \cdots \circ \pi_2 \circ \pi_1$.

Therefore the successive centers of blowing-ups in the sequence (2.1) yield a finite set of proper or infinitely near points $\{p_i\}_{1 \leq i \leq s}$. Any set of points like that is called a *cluster* over S. We write $\mathcal{K} = \{p_i\}_{1 \leq i \leq n}$.

We write $E_{i,i}$ for the exceptional divisor of π_i as divisor of X_i, and let $E_{i,j}$ resp. $E^*_{i,j}$ denote the strict transform resp. the total transform of $E_{i,i}$ in X_j by the morphism $X_j \longrightarrow X_i$, for $j > i$. Moreover, let E_i resp. E^*_i denote the strict resp. total transform $E_{i,s}$ resp. $E^*_{i,s}$ by the morphism $X \longrightarrow X_i$.

EXAMPLE 2.8. Consider the plane curve singularity $(C, \mathbf{0})$ given by $f(x, y) = y^8 - x^{11}$. We blow up at $p_0 = \mathbf{0}$ via a morphism π_1 so that we perform the coordinates change $x = x_1, y = x_1 y_1$ corresponding to the chart 1:

$$f(x, y) = f(x_1, x_1 y_1) = x_1^8 (y_1^8 - x_1^3).$$

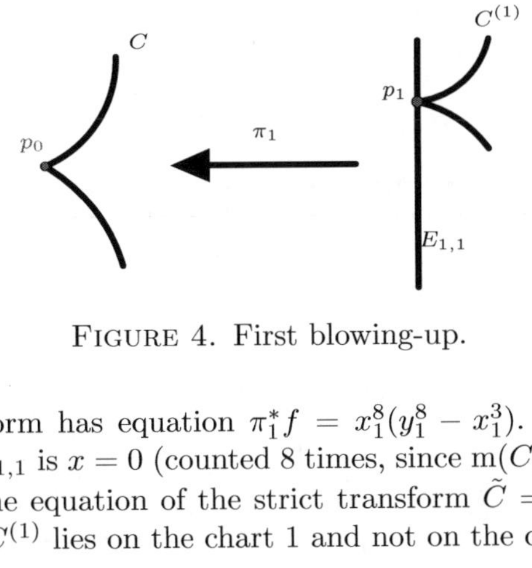

FIGURE 4. First blowing-up.

The total transform has equation $\pi_1^* f = x_1^8 (y_1^8 - x_1^3)$. The equation of the exceptional divisor $E_{1,1}$ is $x = 0$ (counted 8 times, since $\mathrm{m}(C, p_0) = 8$). The second factor $f_1(x_1, y_1)$ is the equation of the strict transform $\tilde{C} = C^{(1)}$. Note that the singular point p_1 of $C^{(1)}$ lies on the chart 1 and not on the chart 2; see Figure 4.

We blow up at p_1, so we apply the change $x_1 = x_2y_2$, $y_1 = y_2$:

$$f_1(x_1, y_1) = f_1(x_2y_2, y_2) = y_2^3(y_2^5 - x_2^3).$$

The total transform C^* of $C^{(1)}$ consists of the exceptional divisor $E_{2,2}$ given by $y_2 = 0$ counted thrice, and the strict transform $C^{(2)}$ given by $f_2(x_2, y_2) = y_2^5 - x_2^3$. The strict transform $E_{1,2}$ of $E_{1,1}$ is given by the local equation $x_2 = 0$. This is represented in Figure 5.

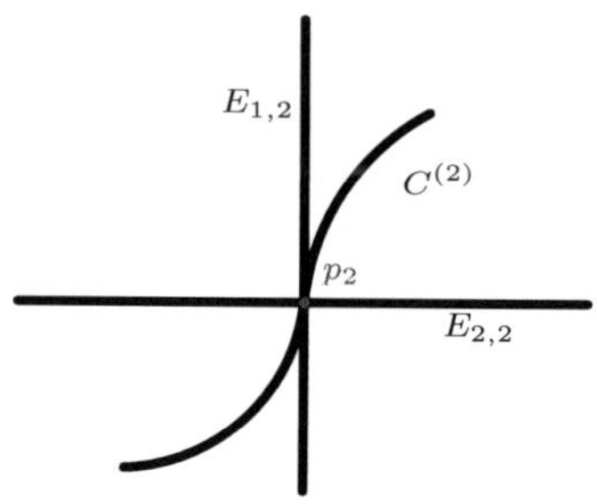

FIGURE 5. Second blowing-up.

We blow up at the singular point p_2 of f_2 by changing the coordinates $x_2 = x_3y_3$, $y_2 = y_3$:

$$f_2(x_2, y_2) = f_2(x_3y_3, y_3) = y_3^3(y_3^2 - x_3^3).$$

The strict transform $E_{1,3}$ of $E_{1,2}$ is given by $x_3 = 0$, whereas the strict transform $E_{2,3}$ of $E_{2,2}$ does not meet the domain of the chosen chart. The exceptional divisor $E_{3,3}$ is given locally by the equation $y_3 = 0$, and the strict transform $C^{(3)}$ of C is given by $f_3(x_3, y_3) = y_3^2 - x_3^3$. This is depicted in Figure 6.

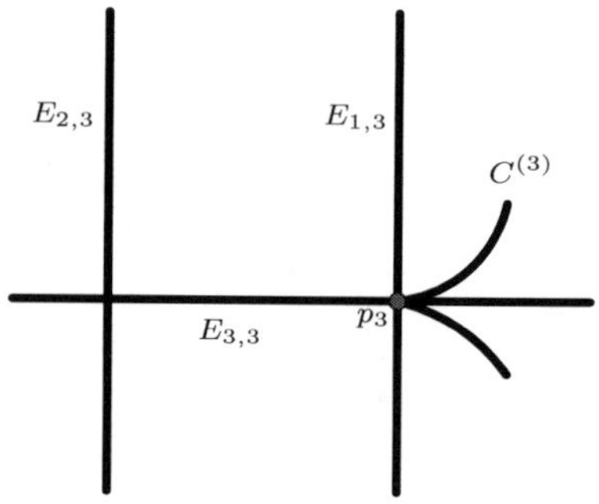

FIGURE 6. Third blowing-up.

Now we blow up at the singular point p_3 of f_3 by doing $x_3 = x_4$, $y_3 = x_4y_4$:

$$f_3(x_3, y_3) = f_3(x_4, x_4y_4) = x_4^2(y_4^2 - x_4).$$

The strict transforms $E_{1,4}$ resp. $E_{2,4}$ of $E_{1,3}$ resp. $E_{2,3}$ do not appear. The strict transform $E_{3,4}$ of $E_{3,3}$ via π_4 is given by the local equation $y_4 = 0$, and the exceptional divisor $E_{4,4}$ created by π_4 is given by $x_4 = 0$. Moreover, the strict transform $C^{(4)}$ of C is the smooth curve $y_4^2 - x_4 = 0$; see Figure 7.

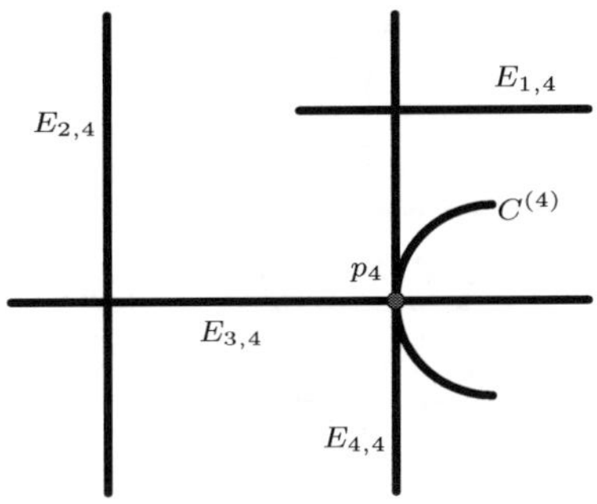

FIGURE 7. Fourth blowing-up.

For some purposes[1] further blowing-ups are required. If we blow up at the point p_5, which is the tangency point between the strict transforms $C^{(4)}$ and $E_{4,4}$, and also lies on $E_{3,4}$, we make $x_4 = x_5 y_5$ and $y_4 = y_5$ to obtain

$$f_4(x_4, y_4) = f_4(x_5 y_5, y_5) = y_5(x_5 - y_5).$$

The appearing strict transforms are $E_{4,5}$ given by $x_5 = 0$, $E_{5,5}$ given by $y_5 = 0$ and $C^{(5)}$ given by $f_5(x_5, y_5) = x_5 - y_5 = 0$, see Figure 8.

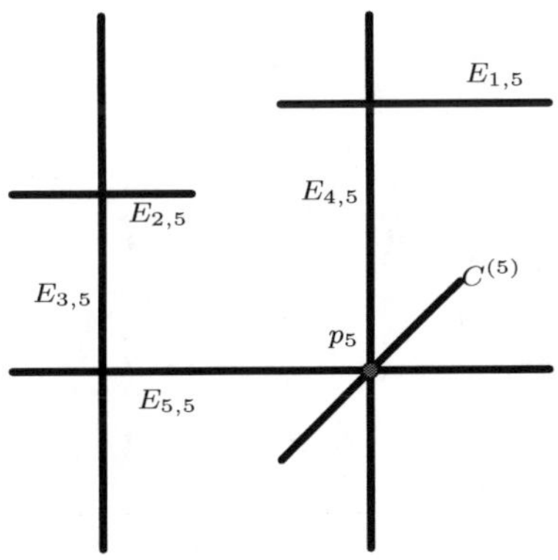

FIGURE 8. Fifth blowing-up.

Any two of these are transverse, but all three go through a single point p_5, so we must blow up once more at p_5: if $x_5 = x_6 y_6$ and $y_5 = y_6$, then

$$f_5(x_5, y_5) = f_5(x_6 y_6, y_6) = y_6(x_6 - 1).$$

[1]For instance, in order to desingularize the curve, see Theorem 2.21.

This blowing-up π_6 produces an exceptional divisor $E_{6,6} = E_6$ meeting $E_{4,6} = E_4$, $E_{5,6} = E_5$ and $C^{(6)}$ transversally at different points, as Figure 9 shows.

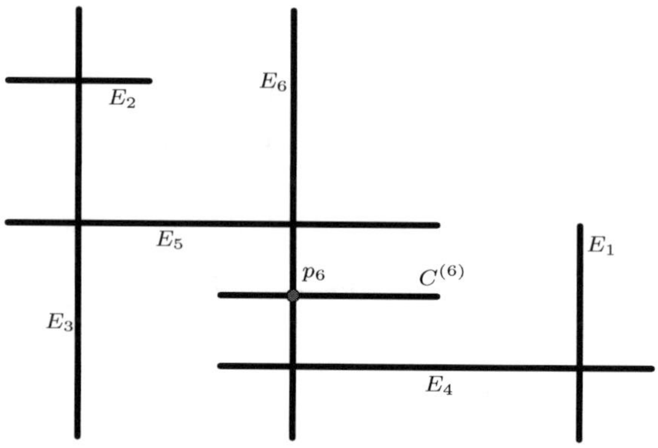

FIGURE 9. Sixth blowing-up.

The blowing up $\pi = \pi_6 \circ \pi_5 \circ \pi_4 \circ \pi_3 \circ \pi_2 \circ \pi_1$ gives rise to the cluster of infinitely near points

$$\mathcal{K} = \{p_0, p_1, p_2, p_3, p_4, p_5\}.$$

The information about how the different components of the exceptional divisor of π intersect each other may be encoded in a graph called the *dual graph*, whose vertices correspond to the irreducible components of the exceptional divisor in X, and two vertices are joined by an edge if and only if they meet. It is customary to add the strict transform of the curve in the representation by means of an arrow. In this example the dual graph is

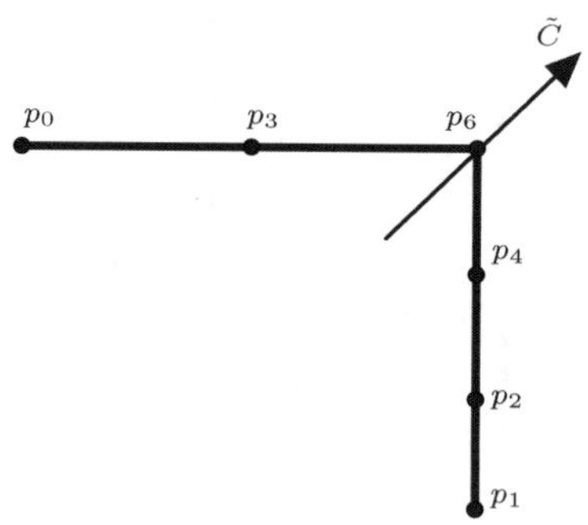

FIGURE 10. Dual graph of $y^8 - x^{11}$.

2.5. Proximity. A crucial notion regarding the relative position of infinitely near points is the proximity:

DEFINITION 2.9. A point q is said to be *proximate* to a point p, and we write $q \to p$, if $q \neq p$ and q belongs to the strict transform of the divisor E_p.

Observe that:

(1) If p is the immediate predecessor of q, then $q \to p$.
(2) Any point q can be proximate to *at most* two other points.

These two remarks can be easily displayed as in Figure 11.

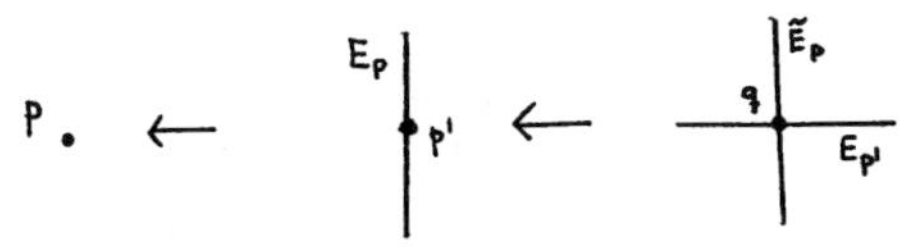

FIGURE 11. $p' \to p$ and $q \to p$; moreover p resp. p' is the immediate predecessor of p' resp. q.

DEFINITION 2.10. An infinitely near point is called *satellite* if it belongs to the intersection of the strict transforms of two exceptional divisors; otherwise the point is called *free.*

Let $\mathcal{K} = \{p_0, \dots, p_{n-1}\}$ be a cluster over S. The proximity relations between the points of $\mathcal{K}$ may be encoded in an $n \times n$-matrix

$$P_{\mathcal{K}} = (p_{ij})$$

which is called the *proximity matrix* of $\mathcal{K}$, where the entries[2] of $P_{\mathcal{K}}$ are defined as

$$p_{ij} = \begin{cases} 1, & \text{if } i = j; \\ -1, & \text{if } p_j \to p_i; \\ 0, & \text{otherwise.} \end{cases}$$

EXAMPLE 2.11. If we come back to Example 2.8, the proximity matrix associated to the cluster $\mathcal{K} = \{p_0, p_1, p_2, p_3, p_4, p_5\}$ is

$$P_{\mathcal{K}} = \begin{pmatrix} 1 & -1 & -1 & -1 & 0 & 0 \\ 0 & 1 & -1 & 0 & 0 & 0 \\ 0 & 0 & 1 & -1 & -1 & 0 \\ 0 & 0 & 0 & 1 & -1 & -1 \\ 0 & 0 & 0 & 0 & 1 & -1 \\ 0 & 0 & 0 & 0 & 0 & 1 \end{pmatrix}.$$

Observe that p_1, p_2 and p_3 are proximate to p_0, the point p_2 is proximate to p_1, the points p_3, p_4 are proximate to p_2, and p_4, p_5 are proximate to p_3, and p_5 is proximate to p_4. This information can be also encoded in a graph called the *proximity graph* whose vertices are the points of the cluster and p_j is joined to p_i by and edge if and only if $p_j \to p_i$, for any $i, j \in \{1, 2, \dots, \sharp\mathcal{K}\}$ with $i < j$. In our example we have the proximity graph depicted in Figure 12.

The entries of the proximity matrix have a nice interpretation in terms of linear algebra. Let $\mathbb{E}$ be the subgroup of 1-cycles of X of the form $\sum_{i=1}^{s} n_i E_i$, for $n_i \in \mathbb{Z}$ (i.e. the free $\mathbb{Z}$-module generated by the divisors E_i). The set $\{E_i\}$ builds a basis

[2]We have committed the crime of using the same letter for both the points in the cluster and the entries of the proximity matrix. We are sorry for that.

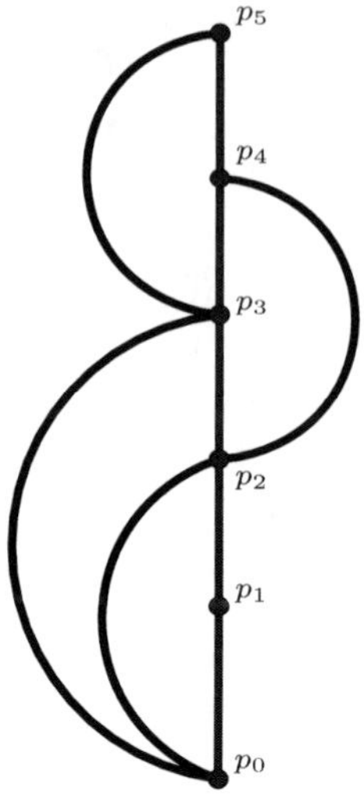

FIGURE 12. Proximity graph for $y^8 - x^{11}$.

of $\mathbb{E}$, and moreover:

LEMMA 2.12. *If E_j^* stands for the total transform of $E_{i,j}$ by the morphism $X \to X_i$, then*

$$E_j = \sum_i p_{ij} E_i^*,$$

where p_{ij} are the entries of the proximity matrix $P_{\mathcal{K}}$. As a consequence, $\{E_j^*\}$ build also a basis of $\mathbb{E}$ and the proximity matrix $P_{\mathcal{K}}$ is the matrix of the basis change from $\{E_i\}$ to $\{E_i^*\}$.

PROOF. The point p_j is proximate to p_i if and only if $p_j \in E_{ij}$, i.e. if and only if $E_{i,j} \cap E_{j,j} \neq \emptyset$. Since the multiplicity of E_{ij} at p_j must be 1, one has

$$\begin{aligned} \pi_j^* E_{i,j-1} &= E_{i,j} + E_{j,j}, \text{ if } p_j \to p_i \\ \pi_j^* E_{i,j-1} &= E_{i,j}, \text{ otherwise} \end{aligned}$$

and then

$$\pi_j^* E_{i,j-1} = E_{i,j} - p_{ij} E_{j,j}.$$

Write $\widehat{E}_{i,j-1}^*$ for the total transform $(\pi_{j-2}^* \circ \cdots \circ \pi_{i+1}^*)(E_i)$ of E_i in X_{j-1}. We will show by induction on $j - i - 1$ that

$$\widehat{E}_{i,j-1}^* + p_{i,i+1}\widehat{E}_{i+1,j-1}^* + \cdots + p_{i,j-1}\widehat{E}_{j-1,j-1}^* = E_{i,j-1} \text{ for } i < j-1.$$

For $j = i + 2$ this is obvious. If it holds for particular values of j and i, then, applying π_j^* to both sides we get

$$\widehat{E}_{ij}^* + p_{i,i+1}\widehat{E}_{i+1,j}^* + \cdots + p_{i,j-1}\widehat{E}_{j,j}^* = \pi_j^* E_{i,j-1}.$$

Hence the result follows for i and j, because

$$p_{ji}\widehat{E}^*_{j,j} + \pi_j^* E_{i,j-1} = E_{ij}.$$

Taking $j-1=k$ in the equation just obtained gives

$$E_i = E_i^* + p_{i,i+1}E_{i+1}^* + \cdots + p_{ik}E_k^*.$$

Since $p_{ii}=1$ and $p_{ij}=0$ for $i>j$, we are done. □

One can also consider the matrix of the basis change from $\{E_i^*\}$ to $\{E_i\}$, say

$$Q_{\mathcal{K}} = (q_{ij}) = P_{\mathcal{K}}^{-1},$$

such that $E_i^* = \sum_j q_{ij}E_j$. This matrix encodes in particular the multiplicities at the infinitely near points p_i of the strict transform of the curve C on X_i: In the obvious manner one may define the multiplicity $m_i(C) := \mathrm{m}(C^{(i)}, p_i)$ at an infinitely near point p_i of the strict transform $C^{(i)}$ of C on X_i, and one has the relation

$$m_i(C) = \sum_{p_j \to p_i} m_j(C). \tag{2.2}$$

In this context we remind a formula relating these multiplicities with the intersection multiplicity of two germs (see e.g. [**Cs00**], Theorem 3.3.1):

THEOREM 2.13 (Noether formula). *Let (C,p) resp. (D,p) two germs of curve with local equations f resp. g. The intersection multiplicity $[f,g]_p$ is finite if and only if the corresponding germs share finitely many points infinitely near to p, and in such a case*

$$[f,g]_p = \sum_q \mathrm{m}(C,q)\mathrm{m}(D,q),$$

where the sum runs over all points q equal or infinitely near to p that lie on C and D.

We turn back to the matrix $Q_{\mathcal{K}}$ to show an interpretation of its last column.

LEMMA 2.14. *For every $i=1,\dots,s$ we have*

$$q_{i,s} = m_{i-1}(C).$$

PROOF. Since the points which are proximate to p_s are not under consideration, equality (2.2) can be rewritten as

$$\sum_{i=1}^{s} p_{ij}m_{i-1}(C) = 0$$

for all $j=1,\dots,s-1$, which is equal to

$$\sum_{i=1}^{s} p_{ij}m_{i-1}(C) = \delta_{j,s}$$

for all $j=1,\dots,s$, where $\delta_{i,j}$ stands for the Kronecker delta. Hence

$$\sum_{i=1}^{s}\sum_{j=1}^{s} q_{ji}p_{ij}m_{i-1}(C) = \sum_{j=1}^{s} q_{ij}\delta_{js}$$

and so $m_{i-1}(C) = q_{i,s}$. □

We can give however a suitable interpretation of all entries of the matrix $Q_{\mathcal{K}}$ through the concept of curvette. Let p_k be an infinitely near point in the sequence of blowing-ups π. Choose a smooth curve $\widetilde{C}_k$ crossing E_k transversally in X_{k+1} at a point not lying on any E_i with $i < k$. We project $\widetilde{C}_k$ to X_0 via π to obtain a possibly singular curve γ_k, which is called a *curvette* at p_k. If we blow up now the singular point of γ_k and successive points in order to desingularize γ_k, then we obtain again the same sequence of infinitely near points p_i for $i \leq k$. Analogous arguments as those in the proof of Lemma 2.14 show that

$$m_{i-1}(\gamma_k) = q_{i,k}$$

for $1 \leq i, k \leq s$.

EXAMPLE 2.15. The inverse of the proximity matrix in Example 2.11 is

$$Q_{\mathcal{K}} = P_{\mathcal{K}}^{-1} = \begin{pmatrix} 1 & 1 & 2 & 3 & 5 & 8 \\ 0 & 1 & 1 & 1 & 2 & 3 \\ 0 & 0 & 1 & 1 & 2 & 3 \\ 0 & 0 & 0 & 1 & 1 & 2 \\ 0 & 0 & 0 & 0 & 1 & 1 \\ 0 & 0 & 0 & 0 & 0 & 1 \end{pmatrix}.$$

These matrices play a role in the theory of intersection one may develop on X, as we will see in a moment. The standard references here are the book of Fulton [**Fu98**] and Chapter V of Hartshorne [**H77**]. First of all, it is possible to define the intersection of two divisors (or 1-cycles) $A, B \in \mathrm{Div}(X)$ on X by means of a symmetric, bilinear intersection form $\deg_X : \mathrm{Div}(X) \times \mathrm{Div}(X) \to \mathbb{Z}$, $(A, B) \mapsto \deg_X(A \cdot B) = (A \cdot B)$. If the cycles A resp. B correspond locally at x with "squarefree" equations f resp. g without common factors, then

$$(A \cdot B) = \sum_{p \in X} [f, g]_p.$$

If we restrict ourselves to the case $A = E_i^*$ and $B = E_j^*$, it is an important result in intersection theory that

$$\deg_X(E_i^*, E_j^*) = -\delta_{ij}, \tag{2.3}$$

where δ_{ij} is the Kronecker delta (it follows from the projection formula (see [**L02**], Theorem 9.2.12, p. 398; see also [**CR94**], Proposition 1.10). Therefore the matrix of the intersection form in the basis $\{E_i^*\}$ is just $-I$ (minus the identity). By Lemma 2.12, the matrix of the intersection form in the basis $\{E_i\}$ is

$$N_{\mathcal{K}} = -P_{\mathcal{K}} \cdot P_{\mathcal{K}}^t,$$

where $P_{\mathcal{K}}^t$ stands for the transpose of the matrix $P_{\mathcal{K}}$. The matrix $N_{\mathcal{K}}$ is called the *intersection matrix* with respect to $\{E_i\}$ associated to the cluster $\mathcal{K}$. This matrix encodes the behaviour of the intersection of the transforms of the exceptional divisor in the following manner:

LEMMA 2.16. *If $N_{\mathcal{K}} = (n_{ij})_{1 \leq i,j \leq s}$ is the intersection matrix associated to a cluster $\mathcal{K}$, then*

$$n_{ij} = \begin{cases} -1 - \sharp\{p_k : p_k \to p_{i-1}\} & \text{if } i = j \\ 1 & \text{if } i \neq j \ \wedge \ \sharp(E_i \cap E_j) = 1 \\ 0 & \text{otherwise} \end{cases}$$

EXAMPLE 2.17. The intersection matrix associated to the cluster

$$\mathcal{K} = \{p_0, p_1, p_2, p_3, p_4, p_5\}$$

of Examples 2.8, 2.11 and 2.15 is therefore

$$N_{\mathcal{K}} = -P_{\mathcal{K}} \cdot P_{\mathcal{K}}^t = \begin{pmatrix} -4 & 0 & 0 & 1 & 0 & 0 \\ 0 & -2 & 1 & 0 & 0 & 0 \\ 0 & 1 & -3 & 0 & 1 & 0 \\ 1 & 0 & 0 & -3 & 0 & 1 \\ 0 & 0 & 1 & 0 & -2 & 1 \\ 0 & 0 & 0 & 1 & 1 & -1 \end{pmatrix}.$$

Curvettes are also involved in the interpretation of the inverse of the intersection matrix $N_{\mathcal{K}}$. Notice that the matrix $N_{\mathcal{K}}$ is invertible, since $\det(N_{\mathcal{K}}) = (-1)^s$, where s is the number of irreducible components of the exceptional divisor in X. Indeed, for $M_{\mathcal{K}} := -N_{\mathcal{K}}^{-1}$ we have:

LEMMA 2.18. *The intersection number between the curvettes γ_i and γ_j is the (i, j) entry of the matrix*

$$M_{\mathcal{K}} = Q_{\mathcal{K}}^t \cdot Q_{\mathcal{K}}.$$

PROOF. This follows from $-N_{\mathcal{K}}^{-1} = (P_{\mathcal{K}} \cdot P_{\mathcal{K}}^t)^{-1} = Q_{\mathcal{K}}^t \cdot Q_{\mathcal{K}}$ together with an application of the Noether formula (2.13) to the curvettes. □

EXAMPLE 2.19. Taking into account the cluster $\mathcal{K}$ of the previous Examples 2.8, 2.11, 2.15 and 2.17, the matrix $M_{\mathcal{K}}$ is

$$M_{\mathcal{K}} = \begin{pmatrix} 1 & 1 & 2 & 3 & 5 & 8 \\ 1 & 2 & 3 & 4 & 7 & 11 \\ 2 & 3 & 6 & 8 & 14 & 22 \\ 3 & 4 & 8 & 12 & 20 & 32 \\ 5 & 7 & 14 & 20 & 35 & 55 \\ 8 & 11 & 22 & 32 & 55 & 88 \end{pmatrix}.$$

2.6. Resolution. Let $\mathcal{N}_p$ denote the set $\{q : q \geq p\} \cup \{p\}$, i.e. the set of all points which are infinitely near to p as well as p itself. Moreover, given a cluster $\mathcal{K}$, the composition of all blowing-ups centered at the points of $\mathcal{K}$ gives rise to a morphism denoted by

$$\pi_{\mathcal{K}} : S_{\mathcal{K}} \to S.$$

DEFINITION 2.20. Let (C, p) be a germ of plane curve. A point $q \in \mathcal{N}_p$ is said to be *singular* if either

- ⋄ $\mathrm{m}(C, q) \geq 2$, or
- ⋄ q is satellite, or
- ⋄ the strict transform $\tilde{C}$ of C at q is tangent to the exceptional divisor.

We conclude this section reminding the concepts of resolution, embedded resolution and minimal embedded resolution.

A *resolution* of a singular germ of plane curve C is a morphism $\pi_{\mathcal{K}} : S_{\mathcal{K}} \to S$ given by the composition of blowing-ups centered at the points of a cluster $\mathcal{K}$ such that the strict transform $\tilde{C}$ of C in $S_{\mathcal{K}}$ is smooth. If, in addition, both $\tilde{C}$ and the union of the strict transforms of the exceptional divisors meet transversally at all their common points, then $\pi_{\mathcal{K}}$ is called an *embedded resolution*. An embedded resolution is said to be *minimal* if it is minimal in the number of point blowing-ups which are necessary to reach the embedded resolution.

Now, if we denote by $S(C)$ the (possible infinite) cluster of infinitely near singular points of C, we have:

THEOREM 2.21 (Desingularization theorem). *Let C be a singular and reduced germ of plane curve. Then the cluster $S(C)$ is finite and the composition of blowing-ups $\pi_{S(C)} : S_{S(C)} \to S$ is a minimal embedded resolution of the plane curve singularity.*

EXAMPLE 2.22. The sequence of blowing-ups $\pi_4 \circ \pi_3 \circ \pi_2 \circ \pi_1$ in Example 2.8 is a resolution of the singular germ C given by $f(x, y) = y^8 - x^{11}$. If we consider moreover the blowing-ups π_5 and π_6, then we build up the morphism π which is an embedded resolution of C, also minimal.

After this brief summary of the main tools for the investigation of curve singularities, we are ready to present the main object of study in the particular case of an irreducible plane curve singularity: the Poincaré series of a local branch of a plane curve.

3. Poincaré series of a local branch

Let C be a germ of an irreducible, reduced plane curve at the origin in $\mathbb{C}^2$, say $(C, \mathbf{0}) \subset (\mathbb{C}^2, \mathbf{0})$ given by an equation $f = 0$. One may look at a parametrization of C

$$\varphi : (C, \mathbf{0}) \to (\mathbb{C}^2, \mathbf{0})$$

with $\mathrm{im}\varphi = C$ and $\mathbb{C}^2 \setminus \{\mathbf{0}\} \cong C \setminus 0$. This map defines a *curve (discrete) valuation* on the ring of (germs of holomorphic) functions $\mathcal{O}_{\mathbb{C}^2,\mathbf{0}}$

$$v : \mathcal{O}_{\mathbb{C}^2,\mathbf{0}} \to \mathbb{Z}_{\geq 0} \cup \{+\infty\}$$

in the sense that, for any $z \in \mathcal{O}_{\mathbb{C}^2,\mathbf{0}}$ we have

$$z \circ \varphi(t) = a(t)t^{v(z)} + \text{terms of higher degree.}$$

The usual definition of valuation requires the consideration of the fraction field of the ring. Remind that (discrete) valuation means that

(1) $v(\lambda z) = v(z)$ for every $\lambda \in \mathbb{C}^*$;
(2) $v(z_1 + z_2) \geq \min\{v(z_1), v(z_2)\}$ for every z_1, z_2;
(3) $v(z_1 z_2) = v(z_1) + v(z_2)$ for every z_1, z_2.

If we consider just (1) and (2) above, the map v is said to be an order function. In any case, a valuation v as above allows us to define ideals on the ring $\mathcal{O}_{\mathbb{C}^2,\mathbf{0}}$ of the form

$$J(w) = \{z \in \mathcal{O}_{\mathbb{C}^2,\mathbf{0}} : v(z) \geq w\} \text{ for } w \in \mathbb{Z}.$$

These ideals define a decreasing filtration

$$\mathcal{O}_{\mathbb{C}^2,\mathbf{0}} = J(0) \supset J(1) \supset J(2) \supset \dots$$

To this filtration we can attach a series

$$P_C(t) = \sum_{w=0}^{\infty} \dim_{\mathbb{C}} \Big(J(w)/J(w+1)\Big) t^w,$$

where $J(w)/J(w+1)$ is a $\mathbb{C}$-vector space of dimension 0 or 1. This series is called the *Poincaré series* attached to the filtration $\{J(w)\}$. Indeed, if we consider the *value semigroup* defined by the curve valuation associated to C, i.e. the numerical semigroup

$$H(C) = \{v(z) : z \in \mathcal{O}_{\mathbb{C}^2,\mathbf{0}}\},$$

then we can be more precise:

LEMMA 3.1. *For every $w \in \mathbb{Z}$,*

$$\dim_{\mathbb{C}} J(w)/J(w+1) = \begin{cases} 0, & \text{if } w \notin H(C); \\ 1, & \text{if } w \in H(C). \end{cases}$$

This lemma implies that $P_C(t)$ is the generating function of the value semigroup $H(C)$, i.e.

$$P_C(t) = \sum_{w \in H(C)} t^w.$$

Moreover, since $H(C)$ is a numerical semigroup, it is clear that

$$P_C(t) = \frac{Q(t)}{1-t}$$

for $Q(t) \in \mathbb{Z}[t]$ with $Q(0) = 1$; i.e., $P_C(t)$ is a rational function.

The value semigroup $H(C)$ is therefore finitely generated, and one can construct a minimal system of generators $\overline{\beta}_0, \overline{\beta}_1, \ldots, \overline{\beta}_h$ of $H(C)$ in terms of an embedded resolution of C (see e.g. [**C80**]). As we have seen, the value semigroup determines the Poincaré series; we can even give a closed formula for $P_C(t)$ in terms of the minimal set of generators of $H(C)$:

THEOREM 3.2. *Let $H(C) = \langle \overline{\beta}_0, \overline{\beta}_1, \ldots, \overline{\beta}_h \rangle$ be the value semigroup of the germ C of plane curve minimally generated by $\overline{\beta}_i$, $i = 0, \ldots, h$. Set $e_i = \gcd(\overline{\beta}_0, \ldots, \overline{\beta}_i)$ for $i = 0, \ldots, h$ and $N_i = \frac{e_{i-1}}{e_i}$ for $i = 1, \ldots, h$. Then*

$$P_C(t) = \frac{\prod_{i=1}^{h}(1 - t^{N_i \overline{\beta}_i})}{\prod_{i=0}^{h}(1 - t^{\overline{\beta}_i})}.$$

PROOF. First we observe that

(1) $N_i \overline{\beta}_i < \overline{\beta}_{i+1}$
(2) $N_i \overline{\beta}_i \in \langle \overline{\beta}_0, \ldots, \overline{\beta}_{i-1} \rangle$
(3) For every $w \in H(C)$ there exists a unique writing of w as

$$w = \lambda_0 \overline{\beta}_0 + \sum_{i=1}^{h} \lambda_i \overline{\beta}_i$$

with $\lambda_0 \geq 0$ and $0 \leq \lambda_i < N_i$ for every $i = 1, \ldots, h$.

(The proof of these properties is left to the reader). By the preceding facts it is now clear that

$$\begin{aligned} P_C(t) &= \sum_{w\in H(C)} t^w \\ &= \sum_{\substack{\lambda_0\geq 0\\ 0\leq\lambda_i<N_i\\ i=1,\dots,h}} t^{\lambda_0\overline{\beta}_0+\lambda_1\overline{\beta}_1+\cdots+\lambda_h\overline{\beta}_h} \\ &= \sum_{\lambda_0\geq 0} t^{\lambda_0\overline{\beta}_0}\cdot \sum_{0\leq\lambda_1<N_1} t^{\lambda_1\overline{\beta}_1}\cdots \sum_{0\leq\lambda_h<N_h} t^{\lambda_h\overline{\beta}_h} \\ &= \frac{1}{1-t^{\overline{\beta}_0}}\cdot\frac{1-t^{N_1\overline{\beta}_1}}{1-t^{\overline{\beta}_1}}\cdots\frac{1-t^{N_g\overline{\beta}_h}}{1-t^{\overline{\beta}_h}}, \end{aligned}$$

as desired. □

The computations in Theorem 3.2 are richer than what one might think: they are closely related to an embedded resolution of C. Let

$$\pi : X = X_s \to X_{s-1} \to \cdots \to X_1 \to X_0 = \mathbb{C}^2$$

be an embedded resolution of C, with $\pi = \pi_s \circ \pi_{s-1} \circ \cdots \circ \pi_1$ and $\pi_i : X_i \to X_{i-1}$ for $i = 1, \dots, s$; moreover, assume that C has equation $f = 0$, for $f \in \mathcal{O}_{\mathbb{C}^2,\mathbf{0}}$.

Let the exceptional divisor $\mathcal{E}$ of π be the union of finitely many components $E_i \cong \mathbb{P}^1$. Moreover, let $\overset{\circ}{E}_i$ be the nonsingular part of E_i, i.e. E_i minus the intersection points with all other components of the total transform $(f\circ\pi)^{-1}(\mathbf{0})$ of the curve C.

Now consider the lifting $f \circ \pi$ of f to the space X_s; let m_i be the multiplicity of this lifting along the component E_i of $\mathcal{E}$. The multiplicities m_i are connected to the minimal generators $\overline{\beta}_i$ of the value semigroup $H(C)$ as follows:

LEMMA 3.3. *The exceptional divisor $\mathcal{E}$ of π in X has three types of irreducible components, namely*

- ◇ *those with 1 singular point, say E_{a_i};*
- ◇ *those with 3 singular points, say E_{b_i};*
- ◇ *those with 2 singular points, say E_{c_i}.*

Then we have that

(i) $\chi(\overset{\circ}{E}_{a_i}) = \chi(E_{a_i}) - 1 = 1;$

(ii) $\chi(\overset{\circ}{E}_{b_i}) = \chi(E_{b_i}) - 3 = -1;$

(iii) $\chi(\overset{\circ}{E}_{c_i}) = \chi(E_{c_i}) - 2 = 0.$

Moreover

- ◇ $m_{a_i} = \overline{\beta}_i$ *for* $i = 1, \dots, h;$
- ◇ $m_{b_i} = N_i\overline{\beta}_i$ *for* $i = 0, \dots, h.$

The proof of Lemma 3.3 is not straightforward; it can be found e.g. in [**W04**], Theorem 8.5.2. In fact, for $E_\sigma \in \mathcal{E}$, we have already seen other interpretation of the numbers m_σ (even for reducible germs): If $L_\sigma \subset (\mathbb{C}^2, \mathbf{0})$ is a curvette at E_σ of equation $g_\sigma = 0$, i.e., a plane branch such that its strict transform is smooth and transversal to the component E_σ at a smooth point, then $m_\sigma = v(g_\sigma)$. So the proof of Lemma 3.3 requires the explicit construction of curvettes cutting the irreducible components of $\mathcal{E}$ in a specific way.

By Lemma 3.3 we deduce:

THEOREM 3.4.

$$P_C(t) = \prod_\sigma (1 - t^{m_\sigma})^{-\chi(\overset{\circ}{E}_\sigma)},$$

where σ indexes the set of irreducible components of $\mathcal{E} \subset X_s$.

PROOF. Application of Theorem 3.2 and Lemma 3.3 shows that

$$\begin{aligned} P_C(t) =& \frac{(1-t^{m_{b_1}})\cdots(1-t^{m_{b_h}})}{(1-t^{m_{a_0}})\cdots(1-t^{m_{a_h}})} \\ =&(1-t^{m_{b_1}})^{-(-1)}\cdots(1-t^{m_{b_h}})^{-(-1)}\cdot(1-t^{m_{a_0}})^{-1}\cdots(1-t^{m_{a_h}})^{-1} \\ =&(1-t^{m_{b_1}})^{-\chi(\overset{\circ}{E}_{b_1})}\cdots(1-t^{m_{b_h}})^{-\chi(\overset{\circ}{E}_{b_h})}\cdot(1-t^{m_{a_0}})^{-\chi(\overset{\circ}{E}_{a_0})}\cdots(1-t^{m_{a_h}})^{-\chi(\overset{\circ}{E}_{a_h})} \\ =&\prod_\sigma (1-t^{m_\sigma})^{-\chi(\overset{\circ}{E}_\sigma)}. \end{aligned}$$

□

Theorem 3.4 yields a formula in terms of an embedded resolution of C which allows us to connect the Poincaré series with the *monodromy zeta function* of the germ C by means of a result due to A' Campo [**AC75**]. This "coincidence" encouraged Campillo, Delgado and Gusein-Zade to further investigate the Poincaré series and generalize the result for reducible germs. Moreover, thanks to the Eisenbud-Neumann formula[3] [**EN85**] the Poincaré series $P_C(t)$ is proven to coincide with the Alexander polynomial $\Delta_C(t)$ (divided by a factor $1-t$) of the knot $C \cap S^3_\varepsilon \subset S^3_\varepsilon$, where S^3_ε stands for a sphere of small radius ε with centre $\mathbf{0} \in \mathbb{C}^2$. Then Theorem 3.4 means that the resolution allows us to determine the Poincaré series. Moreover: conversely, it is also true that the Poincaré series determines the resolution (e.g. the dual graph), that is, the Poincaré series is therefore a complete invariant of the resolution, since the Alexander polynomial is so, as Yamamoto [**Y84**] proved. These results generalize to reducible germs.

EXAMPLE 3.5. Consider the germ given by $f = y^8 - x^{11}$. A parametrization (of Puiseux type) of f is given by

$$x(t) = t^{11}, y(t) = t^8.$$

The value semigroup of C is minimally generated by

$$\overline{\beta}_0 = 8, \overline{\beta}_1 = 11.$$

According to the previous notation we have $e_0 = 8, e_1 = \gcd(\overline{\beta}_0, \overline{\beta}_1) = 1$, hence $N_1 = e_0/e_1 = 8$ and so

$$P_C(t) = \frac{(1-t^{N_1\overline{\beta}_1})}{(1-t^{\overline{\beta}_0})(1-t^{\overline{\beta}_1})} = \frac{(1-t^{88})}{(1-t^8)(1-t^{11})}.$$

[3]The result of Eisenbud and Neumann is valid for arbitrary r; in fact for $r > 1$, the Alexander polynomial and the Poincaré series —to be defined in the following sections— coincide.

EXAMPLE 3.6. Consider the germ given by $f = y^4 - 2x^3y^2 - 4x^5y + x^6 - y^7$. A parametrization (of Puiseux type) of f is given by

$$x(t) = t^4, y(t) = t^6 + t^7.$$

The value semigroup of C is minimally generated by

$$\overline{\beta}_0 = 4, \overline{\beta}_1 = 6, \overline{\beta}_2 = 13.$$

According to the previous notation $e_0 = 4, e_1 = 2, e_2 = 1$, hence $N_1 = e_0/e_1 = 2$ and $N_2 = e_1/e_2 = 2$. Therefore

$$P_C(t) = \frac{(1 - t^{N_1\overline{\beta}_1})(1 - t^{N_2\overline{\beta}_2})}{(1 - t^{\overline{\beta}_0})(1 - t^{\overline{\beta}_1})(1 - t^{\overline{\beta}_2})} = \frac{(1 - t^{12})(1 - t^{26})}{(1 - t^4)(1 - t^6)(1 - t^{13})}.$$

REMARK 3.7. The exponents $N_i\overline{\beta}_i$ are also endowed with a combinatorial interpretation, see Campillo and Marijuán [**CM91**], and Kunz and Herzog [**KH71**] for further details.

4. Integration with respect to the Euler characteristic

Thanks to an idea due to O. Viro [**V88**], it is possible to define an integral taking as a measure the Euler characteristic, but one has to care about the definition of the Euler characteristic. The usual Euler characteristic of a topological space X, say e.g. a CW-complex, defined as the alternating sums of the Betti numbers of X

$$\chi(X) = \sum_{k \geq 0} (-1)^k \dim H^k(X; \mathbb{R}),$$

is not additive if X is not compact: it is easily seen that $0 = \chi(S^1) \neq \chi(\{\text{pt}\}) + \chi(S^1 \setminus \{\text{pt}\}) = 1 + 1$.

We have instead to consider the Euler characteristic defined by cohomology groups with compact support,

$$\chi(X) = \sum_{k \geq 0} (-1)^k \dim H^k_c(X; \mathbb{R})$$

(the cohomology groups $H^k_c(X; \mathbb{R})$ with compact support are nothing but the Borel-Moore cohomology groups, and also the cohomology groups $H^k(X^*, *; \mathbb{R})$ of the one-point compactification X^* of the space X modulo the added point $*$; if X is compact, X^* is just the disjoint union of X with a point $*$). This Euler characteristic behaves additive for spaces which are locally closed unions of cells in a finite CW-complex. These sets do not form an algebra, but the Euler characteristic with compact support extends to an additive function on the algebra of sets consisting of unions of cells in finite CW-complexes.

Observe that we recover the additivity in the previous example for this definition of the Euler characteristic, since now $\chi(S^1) = 0$, $\chi(\{\text{pt}\}) = 1$ and $\chi(S^1 \setminus \{\text{pt}\}) = -1$.

We would like to point out here that the Euler characteristic is multiplicative as well, in the following sense: if $p : E \to B$ is a locally trivial —in the topological sense— fibre bundle with fibre F, then $\chi(E) = \chi(B)\chi(F)$.

However, we consider the Euler characteristic as a function on the algebra of constructible sets (i.e. finite disjoint unions of locally closed subspaces; this type of sets is convenient: they admit cell decompositions, form an algebra of sets —the algebra generated by quasi-projective varieties, i.e. differences of two projective spaces— and are stable under taking images, which in particular makes available the Fubini theorem). In fact one should take care when leaving complex quasi-projective varieties, otherwise the usual Euler characteristic coincides with the compactified one. We cannot explain in detail this here; further explanations can be found in [**G10**].

A function ψ on a constructible set X with values in an abelian group G is said to be *constructible* if for any element $a \in G$ the corresponding level set $\psi^{-1}(a)$ is constructible. Since a sum of constructible functions is constructible, and a constructible function takes just finitely many values, we define:

DEFINITION 4.1. The *integral* of a constructible function ψ with respect to the Euler characteristic over the set X is

$$\int_X \psi d\chi = \sum_{a\in G} \chi(\{\psi^{-1}(a)\})a.$$

This integral satisfies the usual properties of integrals, in particular the Fubini theorem.

DEFINITION 4.2. A map $f : X \to Y$ between constructible sets X, Y is *constructible* if one can write

$$X = \bigcup_{i\in I} X_i, \quad Y = \bigcup_{j\in J} Y_j$$

where the unions are disjoint, I and J are finite, X_i and Y_j are quasi-projective and for any $i \in I$ there exists $j \in J$ with $f(X_i) \subset Y_j$ and the restriction of f to X_i is an algebraic map $X_i \to Y_j$.

THEOREM 4.3 (Fubini). *Let $f : X \to Y$ be a constructible map and let $\psi : X \to G$ be a constructible function on the set X with values in an abelian group G. Then the function $\hat{\psi} : Y \to G$ defined as*

$$\hat{\psi}(y) = \int_{f^{-1}(y)} \psi d\chi$$

for $y \in Y$ over the fibres $f^{-1}(y)$ of f is constructible and

$$\int_X \psi d\chi = \int_Y \Big(\int_{f^{-1}(y)} \psi d\chi\Big) d\chi.$$

As a consequence

$$\chi(X) = \int_Y \chi(f^{-1}(y)) d\chi.$$

It is also possible to define the integral with respect to the Euler characteristic over infinite-dimensional sets, not only over constructible sets. We are interested in the ring $\mathcal{O}_{V,\mathbf{0}}$ of germs of complex analytic functions on the germ $(V, \mathbf{0})$ of a singular complex analytic space (not necessarily of dimension two); more specifically, over its projectivization $\mathbb{P}\mathcal{O}_{V,\mathbf{0}}$.

Let $\mathfrak{m} = \mathfrak{m}_{V,\mathbf{0}}$ be the maximal ideal of the ring $\mathcal{O}_{V,\mathbf{0}}$. Let p be a non-negative integer and let

$$J^p_{V,\mathbf{0}} := \mathcal{O}_{V,\mathbf{0}}/\mathfrak{m}^{p+1}$$

be the space of p-jets over $\mathcal{O}_{V,\mathbf{0}}$, which is a finite-dimensional $\mathbb{C}$-vector space of dimension $d(p)$. For instance, if $V = \mathbb{C}^n$, then $d(p) = \binom{p+n}{n}$.

Let us consider its projectivisation $\mathbb{P}J^p_{V,\mathbf{0}}$ and let us adjoin one point to this (that is, $\mathbb{P}^*J^p_{V,\mathbf{0}} := \mathbb{P}J^p_{V,\mathbf{0}} \cup \{*\}$ with $*$ representing the added point) in order to have a well-defined map $\pi_p : \mathbb{P}\mathcal{O}_{V,\mathbf{0}} \to \mathbb{P}^*J^p_{V,\mathbf{0}}$. Notice that for $k \geq \ell$ the map $\pi_{k,\ell} : \mathbb{P}J^k_{V,\mathbf{0}} \to \mathbb{P}J^\ell_{V,\mathbf{0}}$ is a locally trivial fibration over $\mathbb{P}J^\ell_{V,\mathbf{0}} \subset \mathbb{P}^*J^\ell_{V,\mathbf{0}}$, and the fibre is the complex affine space $\mathbb{C}^{d(k)-d(\ell)}$.

DEFINITION 4.4. A subset $X \subset \mathbb{P}\mathcal{O}_{V,\mathbf{0}}$ is said to be cylindric if there exists a constructible subset $Y \subset \mathbb{P}J^p_{V,\mathbf{0}} \subset \mathbb{P}^*J^p_{V,\mathbf{0}}$ such that $X = \pi_p^{-1}(Y)$. The Euler characteristic $\chi(X)$ of X is defined to be the Euler characteristic $\chi(Y)$ of the set Y.

Of course, one has to check that this Euler characteristic is well-defined: suppose you choose another $p' > p$ such that $X = \pi_p^{-1}(Y) = \pi_{p'}^{-1}(Y')$ for some constructible set $Y' \subset \mathbb{P}J^{p'}_{V,\mathbf{0}}$; we want to show the equality $\chi(Y) = \chi(Y')$. Since Y' is a locally trivial fibration over Y whose fibre is $\mathbb{C}^{d(p)-d(p')}$ and has therefore Euler characteristic 1, then $\chi(Y) = \chi(Y')$, as desired.

Let $\psi : \mathbb{P}\mathcal{O}_{V,\mathbf{0}} \to G$ be a function with values in an abelian group G. The function ψ is called cylindric if the level set $\psi^{-1}(a) \subset \mathbb{P}\mathcal{O}_{V,\mathbf{0}}$ is cylindric for each $a \in G$, $a \neq 0$.

The integral of a cylindric function ψ over the space $\mathbb{P}\mathcal{O}_{V,\mathbf{0}}$ with respect to the Euler characteristic is

$$\int_{\mathbb{P}\mathcal{O}_{V,\mathbf{0}}} \psi d\chi := \sum_{a \in G \setminus \{0\}} \chi\left(\psi^{-1}(a)\right) \cdot a,$$

if this sum makes sense in G. In such a case, the function ψ is said to be integrable.

5. Poincaré series as an integral with respect to the Euler characteristic

For a given topological space X one can define an integral over X with respect to an additive invariant of a function $f : X \to G$ with values in an abelian group G, whenever $f^{-1}(a)$ is "measurable" for every $a \in G$. In the previous section we have considered the topological Euler characteristic as additive invariant, and the condition to define an integral à la Viro is that $f^{-1}(a)$ should have Euler characteristic for any $a \in G$.

If we focus on the case of plane curves[4], we consider a germ $(C, \mathbf{0}) \subset (\mathbb{C}^2, \mathbf{0})$ of a reduced (not necessarily irreducible any more) analytic curve with $C = C_1 \cup \ldots \cup C_r$ its decomposition into irreducible components. Let $\varphi_i : (\mathbb{C}, \mathbf{0}) \to (\mathbb{C}^2, \mathbf{0})$ be the parametrisation of the component C_i of C (i.e., a germ of analytic map such that $\mathrm{Im}\varphi_i = C_i$ and φ_i is an isomorphism between $\mathbb{C}$ and C_i in a neighbourhood of the origin). For a germ $z \in \mathcal{O}_{\mathbb{C}^2,\mathbf{0}}$ let $v_i = v_i(z)$ and $a_i = a_i(z)$ be the power of the leading term and its corresponding coefficient in the power series decomposition of the germ $z \circ \varphi_i : (\mathbb{C}, \mathbf{0}) \to \mathbb{C}$, namely

$$z \circ \varphi_i(\tau) = a_i \cdot \tau^{v_i} + \text{terms of higher degree}$$

with $a_i \neq 0$. Notice that if $z \circ \varphi_i(\tau) \equiv 0$, then $v_i(z)$ is assumed to be ∞ and $a_i(z)$ is not defined. Let $\underline{v}(z) = (v_1(z), \ldots, v_r(z))$. The valuations v_i define a multi-index filtration on the ring $\mathcal{O}_{\mathbb{C}^2,\mathbf{0}}$: this is a family of subspaces $\{J(\underline{v})\}$ of $\mathcal{O}_{\mathbb{C}^2,\mathbf{0}}$ labelled

[4]The curves are assumed to be plane, but this restriction is not necessary in Sections 5 and 6.

by the index $\underline{v} = (v_1, \ldots, v_r) \in \mathbb{Z}^r$, which is determined by the valuations (order functions) $v_i : \mathcal{O}_{\mathbb{C}^2,\mathbf{0}} \to \mathbb{Z} \cup \{+\infty\}$, for $i = 1, \ldots, r$, with

$$J(\underline{v}) := \{z \in \mathcal{O}_{\mathbb{C}^2,\mathbf{0}} : v_i(z) \geq v_i \text{ for } i = 1, \ldots, r\}.$$

The ideals in this multi-index filtration are related to the value semigroup associated to C

$$H(C) = \{\underline{v}(z) : z \in \mathcal{O}_{\mathbb{C}^2,\mathbf{0}} \text{ with } v_i(z) < \infty \text{ for all } i = 1, \ldots, r\}$$

in the following way:

LEMMA 5.1. *The element $\underline{v} \in \mathbb{Z}^r$ belongs to $H(C)$ if and only if $J(\underline{v})/J(\underline{v}+e_i) \neq \{0\}$ for every $i = 1, \ldots, r$.*

Here e_i denotes the unit vector $(0, \ldots, 0, \underset{i}{1}, 0, \ldots, 0)$ in $\mathbb{Z}^r$. We will write $\underline{1} := e_1 + \cdots + e_r$.

Anyway we can consider the factor ideal $J(\underline{v})/J(\underline{v}+\underline{1})$ which is in fact a complex vector space of finite dimension and define the Laurent series

$$L(\underline{t}) := \sum_{\underline{v} \in \mathbb{Z}^r} \dim_{\mathbb{C}} (J(\underline{v})/J(\underline{v}+\underline{1})) \cdot \underline{t}^{\underline{v}}, \tag{5.1}$$

where $\underline{t}$ denotes the vector $(t_1, \ldots, t_r)$ and $\underline{t}^{\underline{v}} := t_1^{v_1} \cdots t_r^{v_r}$.

This is an element of the $\mathbb{Z}[t_1, \ldots, t_r]$-module $\mathbb{Z}[\![t_1, \ldots, t_r, t_1^{-1}, \ldots, t_r^{-1}]\!]$ (or, if you want, even a module over $\mathbb{Z}[t_1, \ldots, t_r, t_1^{-1}, \ldots, t_r^{-1}]$, but not a ring, although the polynomial ring $\mathbb{Z}[t_1, \ldots, t_r]$ can be in a natural way considered to be embedded into it). Hence we cannot multiply those series each other, but just multiply by polynomials.

One can imagine this in a picture representing the monomials: there the Laurent series $L(\underline{t})$ is infinitely long in all directions (given by the unit vectors $\{e_1, \ldots, e_r\}$) since $d(\underline{v}) := \dim_{\mathbb{C}} (J(\underline{v})/J(\underline{v}+\underline{1}))$ can be positive for $\underline{v}$ with (some) negative components v_i as well. It is easy to see that, along each line in the lattice $\mathbb{Z}^r$ parallel to a coordinate one, the coefficients $d(\underline{v})$ stabilize in each direction; that is, if v_i' and v_i'' are negative, or if v_i' and v_i'' are positive and large enough, then $d(v_1, \ldots, v_i', \ldots v_r) = d(v_1, \ldots, v_i'', \ldots v_r)$. This means that multiplication by $(t_1 - 1) \cdots (t_r - 1)$ concentrates all appearing monomials into the first quadrant, therefore

$$P'(\underline{t}) := \prod_{i=1}^{r} (t_i - 1) \cdot L(\underline{t}).$$

is a usual power series in $\underline{t}$. By adjusting by means of a suitable correcting term we ended up with a functional relation which indeed defines the *Poincaré series of the multi-index filtration* $\{J(\underline{v})\}$, namely

$$P(\underline{t}) = \frac{P'(\underline{t})}{t_1 \cdots t_r - 1}. \tag{5.2}$$

This definition can be reformulated in order to define $P(\underline{t})$ as an integral. We will explain how.

First of all, let us restrict ourselves to the finite-dimensional, irreducible case for clarity, and then we will proceed with the general, actual case.

If we consider for a moment a finite-dimensional vector space, say T, instead of the ring $\mathcal{O}_{C,0}$, this is easy to explain: if we have the filtration

$$T = J(0) \supset J(1) \supset \cdots$$

(let us consider here a mere one-index filtration for the sake of a better understanding), then we may consider the valuation map $T \to \mathbb{Z} \cup \{+\infty\}$ which extends to a map $v : \mathbb{P}T \to \mathbb{Z} \cup \{+\infty\}$, and the preimage of an integer k is given by

$$v^{-1}(k) = \mathbb{P}J(k) \setminus \mathbb{P}J(k+1).$$

Since $\chi(\mathbb{P}^s) = s + 1$, it is clear that

$$\begin{aligned}\chi(v^{-1}(k)) =& \chi(\mathbb{P}J(k) \setminus \mathbb{P}J(k+1)) \\ =& \chi(\mathbb{P}J(k)) - \chi(\mathbb{P}J(k+1)) \\ =& \dim_{\mathbb{C}} J(k) - \dim_{\mathbb{C}} J(k+1) \\ =& \dim_{\mathbb{C}} J(k)/J(k+1),\end{aligned}$$

hence

$$\int_{\mathbb{P}T} t^v d\chi = P_T(t) \in \mathbb{Z}[\![t]\!]$$

In our context, however, we do not have finite-dimensional vector spaces, but the projective space $\mathbb{P}\mathcal{O}_{C,\mathbf{0}}$. We have to try to emulate these arguments in a clever way, in particular we have to put into the scene some finite-dimensional vector spaces without loosing the essentials of $\mathbb{P}\mathcal{O}_{C,\mathbf{0}}$, so to speak. For that reason we need to consider the jet spaces introduced before, as we have seen. They are on the background of the proof that is the key result here:

PROPOSITION 5.2. *For every $\underline{k} \in \mathbb{N}^r$, the subset*

$$\{z \in \mathbb{P}\mathcal{O}_{C,\mathbf{0}} : \underline{v}(z) = \underline{k}\}$$

is cylindric, and the function $\underline{t}^{\underline{v}(z)} := t_1^{v_1(z)} \cdots t_r^{v_r(z)} : \mathbb{P}\mathcal{O}_{C,\mathbf{0}} \to \mathbb{Z}[\![t_1, \ldots, t_r]\!]$ is cylindric as well.

This leads to the (usual) definition of the Poincaré series associated to the germ C:

DEFINITION 5.3.

$$P(\underline{t}) := \int_{\mathbb{P}\mathcal{O}_{C,\mathbf{0}}} \underline{t}^{\underline{v}(z)} d\chi.$$

With this definition, eq. (5.2) is a statement:

PROPOSITION 5.4.

$$(t_1 \cdots t_r - 1) \cdot P(\underline{t}) = L(\underline{t}) \cdot \prod_{k=1}^{r} (t_k - 1).$$

PROOF. For a subset $I \subset I_0 = \{1, 2, \ldots, r\}$, let $\sharp(I)$ be the cardinality of the set I and let $\underline{1}_I$ be the element of $\mathbb{Z}^r_{\geq 0}$ with the ith component equal to 1 resp. 0

if $i \in I$ resp. $i \notin I$; in particular $\underline{1}_{I_0} = \underline{1}$. We have:

$$\begin{aligned}
P(\underline{t}) &= \int_{\mathbb{P}\mathcal{O}_{C,\mathbf{0}}} \underline{t}^{\underline{v}(z)} d\chi = \sum_{\underline{v}\in\mathbb{Z}^r} \chi\Big(\mathbb{P}J(\underline{v}) \setminus \bigcup_{i=1}^{r} \mathbb{P}J(\underline{v}+\underline{1}_{\{i\}})\}\Big) \cdot \underline{t}^{\underline{v}} \\
&= \sum_{\underline{v}\in\mathbb{Z}^r} \chi\Big((\mathbb{P}J(\underline{v}) \setminus \mathbb{P}J(\underline{v}+\underline{1})) \setminus \bigcup_{i=1}^{r} (\mathbb{P}J(\underline{v}+\underline{1}_{\{i\}}) \setminus \mathbb{P}J(\underline{v}+\underline{1}))\Big) \cdot \underline{t}^{\underline{v}} \\
&= \sum_{\underline{v}\in\mathbb{Z}^r} \Big(\chi(\mathbb{P}J(\underline{v}) \setminus \mathbb{P}J(\underline{v}+\underline{1})) - \sum_{\substack{I\subset I_0\\ I\neq\emptyset}} (-1)^{\sharp(I)-1} \chi(\mathbb{P}J(\underline{v}+\underline{1}_I) \setminus \mathbb{P}J(\underline{v}+\underline{1}))\Big) \cdot \underline{t}^{\underline{v}} \\
&= \sum_{\underline{v}\in\mathbb{Z}^r} \Big(\sum_{I\subset I_0} (-1)^{\sharp(I)} \chi(\mathbb{P}J(\underline{v}+\underline{1}_I) \setminus \mathbb{P}J(\underline{v}+\underline{1}))\Big) \cdot \underline{t}^{\underline{v}} \\
&= \sum_{\underline{v}\in\mathbb{Z}^r} \Big(\sum_{I\subset I_0} (-1)^{\sharp(I)} \dim_{\mathbb{C}}(J(\underline{v}+\underline{1}_I)/J(\underline{v}+\underline{1}))\Big) \cdot \underline{t}^{\underline{v}}.
\end{aligned}$$

Multiplication by $t_1 \cdots t_r - 1$ yields

$$\begin{aligned}
&(t_1 \cdots t_r - 1) \cdot P(\underline{t}) = \\
&= \sum_{\underline{v}\in\mathbb{Z}^r} \Big(\sum_{I\subset I_0} (-1)^{\sharp I} \left[\dim_{\mathbb{C}} (J(\underline{v}+\underline{1}_I-\underline{1})/J(\underline{v})) - \dim_{\mathbb{C}}(J(\underline{v}+\underline{1}_I)/J(\underline{v}+\underline{1}))\right]\Big) \cdot \underline{t}^{\underline{v}} \\
&= \sum_{\underline{v}\in\mathbb{Z}^r} \Big(\sum_{I\subset I_0} (-1)^{\sharp I} [\dim_{\mathbb{C}}(J(\underline{v}+\underline{1}_I-\underline{1})/J(\underline{v}+\underline{1}_I)) - \dim_{\mathbb{C}}(J(\underline{v})/J(\underline{v}+\underline{1}))]\Big) \cdot \underline{t}^{\underline{v}} \\
&= \sum_{\underline{v}\in\mathbb{Z}^r} \Big(\sum_{I\subset I_0} (-1)^{\sharp I} \dim_{\mathbb{C}} J(\underline{v}+\underline{1}_I-\underline{1})/J(\underline{v}+\underline{1}_I)\Big) \cdot \underline{t}^{\underline{v}} \\
&= \Big(\sum_{\underline{v}\in\mathbb{Z}^r} \dim_{\mathbb{C}}(J(\underline{v})/J(\underline{v}+\underline{1})) \cdot \underline{t}^{\underline{v}}\Big) \cdot \prod_{k=1}^{r} (t_k - 1) \\
&= L(\underline{t}) \cdot \prod_{k=1}^{r} (t_k - 1),
\end{aligned}$$

as desired. □

REMARK 5.5. In our exposition we have circumvented the definition of the Poincaré series via the extended semigroup of a germ. This semigroup goes back to Campillo, Delgado and Gusein-Zade [**CDG98**], and the construction of the Poincaré series in this way can be found e.g. in [**CDG03a**].

REMARK 5.6. As mentioned in Section 3, the Poincaré series of a complex plane curve singularity $C = \bigcup_{i=1}^{r} C_i$ for $r > 1$ equals the Alexander polynomial of the link $C \cap S^3_\epsilon$ for a small enough sphere S^3_ϵ: this is a result by Campillo, Delgado and Gusein-Zade [**CDG03b**]. This implies in particular that the Poincaré series of a plane curve singularity with $r > 1$ branches is in fact a polynomial. This can be also deduced by showing that the polynomial $P'(\underline{t})$ is divisible by $\underline{t} - 1$.

6. Generalized Poincaré series

There are other versions for the Euler characteristic as that presented in the previous section. One of them is a universal additive invariant called *generalized* Euler characteristic. Its definition requires a previous notion. Let $\nu_{\mathbb{C}}$ be the set of quasi-projective complex varieties. The ring $K_0(\nu_{\mathbb{C}})$ generated by classes $[X]$ of such varieties subject to the relations:

(i) if $X_1 \cong X_2$ then $[X_1] = [X_2]$;
(ii) if Y is Zariski-closed in X, then $[X] = [Y] + [X \setminus Y]$

(the multiplication is defined by the fibred product) is called the *Grothendieck ring* of quasi-projective varieties.

This is a kind of obscure object, very little is known about it. For instance, in general is not known whether it is a domain. The classes $[X]$ are difficult to compute in practice; theoretically they are categorical in the sense that you can do a universal construction for them. The class of the complex affine line plays a relevant role and thus has a distinguised notation: $\mathbb{L} := [\mathbb{A}^1_{\mathbb{C}}]$ (the letter $\mathbb{L}$ seems to honour the American mathematician S. Lefschetz (1884–1972)). We know that the class $\mathbb{L}$ is not equal to zero in the ring $K_0(\nu_{\mathbb{C}})$. For convenience, we can consider the localisation $K_0(\nu_{\mathbb{C}})_{(\mathbb{L})}$ of the Grothendieck ring $K_0(\nu_{\mathbb{C}})$ by the class $\mathbb{L}$. The morphisms $\mathbb{Z}[X] \to K_0(\nu_{\mathbb{C}})$ which sends X to $\mathbb{L}$ and $\mathbb{Z}[X]_{(X)} \to K_0(\nu_{\mathbb{C}})_{(\mathbb{L})}$ are inclusions. For a deeper discussion of the Grothendieck ring of varieties we refer the reader to Chapter 7 in [**Mu11**].

Viro's construction can be applied also to this with some minor adjustments. The class $[X]$ enjoys from nice additive properties, but this does not lead directly to a well-defined generalized Euler characteristic. We need an extra factor, so that the generalized Euler characteristic $\chi_g(X)$ of a cylindric subset X is the element $[Y] \cdot \mathbb{L}^{-d(p)}$ in the ring $K_0(\nu_{\mathbb{C}})_{(\mathbb{L})}$, where $Y = \pi_p^{-1}(X)$ is a constructible subset of $\mathbb{P}\mathcal{O}_{V,\mathbf{0}}$, as in the previous subsection.

Now note that $\chi_g(X)$ is well-defined, because if $X = \pi_{p'}^{-1}(Y')$, $Y' \subset \mathbb{P}J^{p'}_{V,0}$ and $p > p'$, then Y is a locally trivial fibration over Y' with the fibre $\mathbb{C}^{d(p)-d(p')}$ and therefore $[Y] = [Y'] \cdot \mathbb{L}^{d(p)-d(p')}$.

DEFINITION 6.1. The generalized Poincaré series $P_g(\underline{t})$ of a filtration $\{J(\underline{v})\}$ defined by $\underline{v}(z) = (v_1(z), \ldots, v_r(z))$ is the integral

$$P_g(\underline{t}) := \int_{\mathbb{P}\mathcal{O}_{C,\mathbf{0}}} \underline{t}^{\underline{v}(z)} d\chi_g \quad \text{“}\in\text{”} \quad K_0(\nu_{\mathbb{C}})_{(\mathbb{L})}[\![t_1, \ldots, t_r]\!]$$

over the projectivisation $\mathbb{P}\mathcal{O}_{C,\mathbf{0}}$ of the ring $\mathcal{O}_{C,\mathbf{0}}$ with respect to the generalized Euler characteristic $\chi_g(\cdot)$.

Observe that we have written "$\in$" in Definition 6.1. Let us explain where does $P_g(\underline{t})$ belong to. By definition of the integral, the subset of $\mathbb{P}\mathcal{O}_{C,\mathbf{0}}$ where $\underline{t}^{\underline{v}(z)}$ is equal to $\underline{t}^{\underline{k}}$ is

$$\{z \in \mathbb{P}\mathcal{O}_{C,\mathbf{0}} : \underline{v}(z) = \underline{k}\}.$$

This is of course equal to

$$\left\{z \in \mathbb{P}\mathcal{O}_{C,\mathbf{0}} : \underline{v}(z) \geq \underline{k}\right\} \setminus \left\{ z \in \mathbb{P}\mathcal{O}_{C,\mathbf{0}} : \begin{array}{rl} \underline{v}(z) \geq & (k_1+1, k_2, k_3, \ldots, k_r) \\ \underline{v}(z) \geq & (k_1, k_2+1, k_3, \ldots, k_r) \\ & \vdots \\ \underline{v}(z) \geq & (k_1, k_2, \ldots, k_{r-1}, k_r+1) \end{array} \right\}.$$

This is a difference of two finite-dimensional vector spaces, what leads to the fact that the coefficients of the series $P_g(\underline{t})$ are polynomials in $q := \mathbb{L}^{-1}$, and we may write $P_g(\underline{t}) \in \mathbb{Z}[q][\![t_1, \dots, t_r]\!]$.

For this integral we can get a proposition in the fashion of Proposition 5.4. Let us see why.

Let us take again the (usual) Euler characteristic $\chi(\cdot)$. Let A resp. A' be a subspace of $\mathcal{O}_{C,\mathbf{0}}$ of finite codimension a resp. a' with $a' > a$. We have

$$\dim_{\mathbb{C}}(A/A') = \chi(\mathbb{P}A \setminus \mathbb{P}A') \tag{6.1}$$

$$= \chi(\mathbb{P}(A/A')). \tag{6.2}$$

These equalities do not hold for the generalized Euler characteristic. Nevertheless we can take Equation (6.1) as definition of a sort of "generalized (or motivic) dimension" just by taking (6.1) with $\chi_g(\cdot)$ instead of the usual $\chi(\cdot)$. Hence using the cellular decomposition of a projective space we get

$$\begin{aligned} \chi_g(\mathbb{P}A \setminus \mathbb{P}A') &= q^a + q^{a+1} + \ldots + q^{a'-1} \\ &= q^{a+1} \cdot \frac{1 - q^{a'-a}}{1-q}. \end{aligned} \tag{6.3}$$

If we set $\ell(\underline{v}) := \dim_{\mathbb{C}}(\mathcal{O}_{C,\mathbf{0}}/J(\underline{v}))$ and put $A = J(\underline{v})$, $A' = J(\underline{v} + \underline{1})$, $a = \ell(\underline{v})$ and $a' = \ell(\underline{v} + \underline{1})$ in eq. (6.3), then we can define a series

$$L_g(\underline{t}) := \sum_{\underline{v} \in \mathbb{Z}^r} q^{\ell(\underline{v})+1} \cdot \frac{1 - q^{\ell(\underline{v}+1) - \ell(\underline{v})}}{1-q} \cdot \underline{t}^{\underline{v}}$$

which is a "motivic version" of the series $L(\underline{t})$ given by the formula (5.1). In the spirit of Proposition 5.4 one can also see (cf. [**CDG07**], Proposition 2) that

PROPOSITION 6.2.

$$(t_1 \cdots t_r - 1) \cdot P_g(\underline{t}) = L_g(\underline{t}) \cdot \prod_{k=1}^{r} (t_k - 1).$$

PROOF. We have

$$\begin{aligned} P_g(\underline{t}) = \int_{\mathbb{P}\mathcal{O}_{C,\mathbf{0}}} \underline{t}^{\underline{v}(z)} d\chi_g &= \sum_{\underline{v} \in \mathbb{Z}^r} \chi_g(\{z \in \mathbb{P}\mathcal{O}_{C,\mathbf{0}} : \underline{v}(z) = \underline{v}\}) \cdot \underline{t}^{\underline{v}} \\ &= \sum_{\underline{v} \in \mathbb{Z}^r} \left[\mathbb{P}J(\underline{v}) \setminus \bigcup_{i=1}^{r} \mathbb{P}J(\underline{v} + \underline{1}_{\{i\}})\right] \cdot \underline{t}^{\underline{v}} \\ &= \sum_{\underline{v} \in \mathbb{Z}^r} \Big(\sum_{I \subset I_0} (-1)^{\sharp(I)} [\mathbb{P}J(\underline{v} + \underline{1}_I) \setminus \mathbb{P}J(\underline{v} + \underline{1})] \Big) \cdot \underline{t}^{\underline{v}} . \end{aligned}$$

Hence

$$\begin{aligned}
(t_1\cdots t_r-1)\cdot P_g(\underline{t}) &= \sum_{\underline{v}\in\mathbb{Z}^r}\sum_{I\subset I_0}(-1)^{\sharp I}\left[\mathbb{P}J(\underline{v}+\underline{1}_I-\underline{1})\setminus\mathbb{P}J(\underline{v})\right]\cdot\underline{t}^{\underline{v}} - \\
&- \sum_{\underline{v}\in\mathbb{Z}^r}\sum_{I\subset I_0}(-1)^{\sharp I}\left[\mathbb{P}J(\underline{v}+\underline{1}_I)\setminus\mathbb{P}J(\underline{v}+\underline{1})\right]\cdot\underline{t}^{\underline{v}} \\
&= \sum_{\underline{v}\in\mathbb{Z}^r}\sum_{I\subset I_0}(-1)^{\sharp I}\left[\mathbb{P}J(\underline{v}+\underline{1}_I-\underline{1})\setminus\mathbb{P}J(\underline{v}+\underline{1}_I)\right]\cdot\underline{t}^{\underline{v}} \\
&= \sum_{\underline{v}\in\mathbb{Z}^r}\sum_{I\subset I_0}(-1)^{\sharp I}q^{\ell(\underline{v}-\underline{1}+\underline{1}_I)}\cdot\frac{1-q^{\ell(\underline{v}-\underline{1}_I)-\ell(\underline{v}-\underline{1}+\underline{1}_I)}}{1-q}\cdot\underline{t}^{\underline{v}} \\
&= \Big(\sum_{\underline{v}\in\mathbb{Z}^r}q^{\ell(\underline{v})}\cdot\frac{1-q^{\ell(\underline{v}+\underline{1})-\ell(\underline{v})}}{1-q}\cdot\underline{t}^{\underline{v}}\Big)\cdot\prod_{k=1}^{r}(t_k-1) \\
&= L_g(\underline{t})\prod_{k=1}^{r}(t_k-1).
\end{aligned}$$

□

Remark 6.3. On the other hand, the equality (6.2) does not hold when we replace $\chi(\cdot)$ with $\chi_g(\cdot)$:

$$\chi_g\left(\mathbb{P}A\setminus\mathbb{P}A'\right)\neq\chi_g\left(\mathbb{P}\left(A/A'\right)\right).$$

If we take $A=J(\underline{v})$, $A'=J(\underline{v}+\underline{1})$, $a=\ell(\underline{v})$ and $a'=\ell(\underline{v}+\underline{1})$ we get

$$\chi_g\left(\mathbb{P}\left(J(\underline{v})/J(\underline{v}+\underline{1})\right)\right)=\frac{q^{\ell(\underline{v})+1}}{q^{\ell(\underline{v}+\underline{1})}}\cdot\frac{1-q^{\ell(\underline{v}+\underline{1})-\ell(\underline{v})}}{1-q},$$

and therefore we can define a series

$$\widehat{L}_g(\underline{t}):=\sum_{\underline{v}\in\mathbb{Z}^r}\frac{q^{\ell(\underline{v})+1}}{q^{\ell(\underline{v}+\underline{1})}}\cdot\frac{1-q^{\ell(\underline{v}+\underline{1})-\ell(\underline{v})}}{1-q}\underline{t}^{\underline{v}}.$$

Hence we may also define a Poincaré series

$$\widehat{P}_g(\underline{t}):=\frac{\widehat{L}_g(\underline{t})\cdot\prod_{i=1}^{r}(t_i-1)}{t_1\cdots t_r-1}.$$

This Poincaré series does not coincide with $P_g(\underline{t})$, and has not led (yet) to any remarkable result.

7. Generalized Poincaré series in terms of an embedded resolution

For the sake of simplicity, we will denote the local ring $\mathcal{O}_{C,\mathbf{0}}$ by R in the sequel. In the previous section, we have introduced the generalized Poincaré series of the multi-index filtration given by the ideals $J(\underline{v})$ as the integral

$$P_g(\underline{t}):=\int_{\mathbb{P}R}\underline{t}^{\underline{v}(z)}d\chi_g,$$

where the monomial $\underline{t}^{\underline{v}(z)}:=t_1^{v_1(z)}\cdots t_r^{v_r(z)}$ is considered as a function on $\mathbb{P}R$ with values in $\mathbb{Z}[\![t_1,\dots,t_r]\!]$ (the vector $\underline{t}^{\underline{v}(z)}$ is supposed to be $\underline{0}$ as soon as at least one of the $v_i(z)$ is $+\infty$).

We can blow up a point of C and even continue blowing up points in order to get an embedded resolution after a finite number of steps

$$X = X_s \xrightarrow{\pi_s} X_{s-1} \xrightarrow{\pi_{s-1}} \cdots \longrightarrow X_3 \xrightarrow{\pi_3} X_2 \xrightarrow{\pi_2} X_1 \xrightarrow{\pi_1} X_0 = \mathbb{C}^2.$$

Set $\pi = \pi_s \circ \pi_{s-1} \circ \cdots \circ \pi_1$. Recall that the exceptional divisor E of π is connected, has finitely many different irreducible components, say $E_1, \ldots, E_s$ (each of them isomorphic to $\mathbb{P}^1$), every two components meet transversally at one point, and no three components meet at a point.

For every $i = 1, \ldots, s$, let $\overset{\circ}{E}_i$ be the component E_i of the exceptional divisor of π minus the intersection points with all other components of the total transform of the curve. In the same way we can consider the set $\overset{\bullet}{E}_i$, which is just the component E_i of the exceptional divisor minus the intersection points with other components E_j with $j \neq i$. Note that both $\overset{\circ}{E}_i$ and $\overset{\bullet}{E}_i$ are quasiprojective varieties.

The goal in this section is to describe the integral defining the generalized Poincaré series in terms of an embedded resolution of curves. First of all, we will distinguish three different kinds of points on the exceptional divisor E, namely the intersection points among components of E, the intersection points between components of E and the components of the strict transform of the curve, and the points in $\overset{\circ}{E}_i$. Considering separately the set of the previous first two kinds of points, as well as the symmetric product of the smooth components $\overset{\circ}{E}_i$, we will build in Subsection 7.1 a semigroup Y and a surjection

$$\mathrm{Init} : \mathbb{P}R^* \to Y$$

from the projectivisation of $R \setminus \{0\}$ to Y (this will be in fact a semigroup homomorphism). The preimage of each point of Y under the map Init will give us an affine space whose codimension $F(\underline{n})$ in $\mathbb{P}R$ will be determined in Subsection 7.2; this let us show finally an explicit formula for the generalized Poincaré series in terms of an embedded resolution in Subsection 7.3

7.1. Definition of the map Init. Consider the set

$$R^* := \{z \in R : v_j(z) < \infty,\ j = 1, \ldots, r\}.$$

For $z \in R^*$, let Γ_z be the strict transform of the curve given by z.

For every $i \in \{1, \ldots, s\}$, the intersection $E_i \cap \Gamma_z$ has finite cardinality, because E_i and Γ_z have no common components.

Let $I_0 := \{\sigma := (i_1, i_2) \in \{1, 2, \ldots, s\} \times \{1, 2, \ldots, s\} : i_1 < i_2,\ E_{i_1} \cap E_{i_2} \neq \emptyset\}$. For every $\sigma = (i_1, i_2) \in I_0$, let $P_\sigma := E_{i_1} \cap E_{i_2}$ and $i_1(\sigma) := i_1, i_2(\sigma) := i_2$.

Let $J_0 := \{1, 2, \ldots, r\}$. For $j \in J_0$, let $P_j := E_{i_1(j)} \cap \widetilde{C}_j$, where $\widetilde{C}_j$ is the strict transform of π of the component C_j of the curve C, and $E_{i_1(j)}$ is the component of the exceptional divisor of π which intersects $\widetilde{C}_j$.

We can distinguish three sorts of points belonging to each irreducible component E_i of the exceptional divisor of π, namely: points of kind P_σ for $\sigma \in I_0$, points of kind P_j for $j \in J_0$, and smooth points of the exceptional divisor.

Points of kind P_σ. They are the intersection points $E_{i_1} \cap E_{i_2}$ when $\sigma = (i_1, i_2)$, for $i_1, i_2 \in \{1, 2, \ldots, s\}$ and $i_1 < i_2$. Take such a point P_σ, for $\sigma \in I_0$. Choose local coordinates x_σ, y_σ at P_σ, so that the components of the total transform of the curve are the coordinate lines $E_{i_1(\sigma)} := \{y_\sigma = 0\}, E_{i_2(\sigma)} := \{x_\sigma = 0\}$. We will denote the local ring at P_σ by R_σ.

The quotient ring $R_\sigma/x_\sigma R_\sigma$ (resp. $R_\sigma/y_\sigma R_\sigma$) is a discrete valuation ring with associated valuation ω_{x_σ} (resp. ω_{y_σ}). We can consider the canonical maps

$$\varphi_{x_\sigma} : R_\sigma \longrightarrow R_\sigma/x_\sigma R_\sigma$$

$$\varphi_{y_\sigma} : R_\sigma \longrightarrow R_\sigma/y_\sigma R_\sigma.$$

For any $z \in R^*$, denoting γ_z the equation at P_σ of the strict transform Γ_z of z on X (which must be different from 0), for $\sigma \in I(z) := \{\sigma \in I_0 \mid P_\sigma \in \Gamma_z\}$, we define

$$n'_\sigma(z) := \omega_{x_\sigma}(\gamma_z \bmod x_\sigma R_\sigma)$$

$$n''_\sigma(z) := \omega_{y_\sigma}(\gamma_z \bmod y_\sigma R_\sigma).$$

Note that $n'_\sigma(z) \neq \infty$ and $n''_\sigma(z) \neq \infty$, since $\gamma_z \notin (x_\sigma)$ and $\gamma_z \notin (y_\sigma)$. Note also that $\varphi_{x_\sigma}(y_\sigma)$ resp. $\varphi_{y_\sigma}(x_\sigma)$ is a uniformizing parameter of the discrete valuation ring $R_\sigma/x_\sigma R_\sigma$ resp. $R_\sigma/y_\sigma R_\sigma$. Therefore we have

$$\varphi_{x_\sigma}(\gamma_z) = \alpha_\sigma \cdot \varphi_{x_\sigma}(y_\sigma)^{n'_\sigma(z)}$$

$$\varphi_{y_\sigma}(\gamma_z) = \beta_\sigma \cdot \varphi_{y_\sigma}(x_\sigma)^{n''_\sigma(z)},$$

where α_σ is a unit of $R_\sigma/x_\sigma R_\sigma$ and β_σ is a unit of $R_\sigma/y_\sigma R_\sigma$. We can also consider the image a_σ of α_σ and the image b_σ of β_σ in the residue field $\mathbb{C}$ of R_σ, which are both different from 0. The elements α_σ, β_σ, a_σ and b_σ depend only on the choice of the equations x_σ and y_σ and also of the equation γ_z of the strict transform Γ_z of z. However it is easy to check that the ratio $\frac{a_\sigma}{b_\sigma} \in \mathbb{C}^*$ is independent of the generator of the principal ideal $\gamma_z S_{P_\sigma}$, where S_{P_σ} stands for the local ring at P_σ. We set

$$\lambda_\sigma(z) = \frac{a_\sigma}{b_\sigma}.$$

Points of kind P_j. They are the points in the intersection between E_i and the strict transform of the j-th irreducible component of the curve. Let us take a point P_j, for $j \in J_0$. We may choose local coordinates x_j, y_j in a neighbourhood of P_j in such a way that the components of the total transform $\pi^{-1}(C)$ of the curve C are $E_{i_1(j)} = \{y_j = 0\}$ and $\widetilde{C}_j = \{x_j = 0\}$, and the set $\{x_j, y_j\}$ is a regular system of parameters of the local ring $R_j := \mathcal{O}_{X,P_j}$. If we consider an element $z \in R^*$, its strict transform Γ_z on X, and a local equation of Γ_z given by γ_z at P_j, then, in a similar way as we have done for the points of kind P_σ, for every element j in $J(z) := \{j \in J_0 \mid P_j \in \Gamma_z\}$, we may define the natural numbers

$$\widetilde{n}'_j(z) = \omega_{x_j}(\gamma_z \bmod x_j R_j) \neq \infty$$

$$\widetilde{n}''_j(z) = \omega_{y_j}(\gamma_z \bmod y_j R_j) \neq \infty,$$

as well as two elements $\alpha_j, \beta_j \in R_j$, whose images a_j, b_j in the residue field $\mathbb{C}$ of R_j are different from 0 and uniquely determined modulo x_j and y_j. We write

$$\mu_j(z) = \frac{a_j}{b_j} \in \mathbb{C}^*.$$

Points of kind "smooth". They are the points belonging to $\overset{\circ}{E}_i$, for every $i = 1, \ldots, s$. Hence, if we take $z \in R^*$ and Γ_z the strict transform of the curve given by z, then we can count the number of intersection points between Γ_z and $\overset{\circ}{E}_i$ with multiplicities.

We want to define a space Y and a map between $\mathbb{P}R^*$ and Y so that the preimage of any point of Y is an affine space of finite codimension, and such a map is a locally trivial fibration over each connected component of Y. This will allow us to apply certain integration rules (the Fubini theorem) to the map in order to make the link with the definition of the generalized Poincaré series.

We need the following definition: For any (topological) space $\mathfrak{X}$, and $m \in \mathbb{N}$, we define the *m-th symmetric power* of $\mathfrak{X}$ to be $\mathcal{S}^m\mathfrak{X} := \mathfrak{X}^m/S_m$, where S_m is the group of permutations of m elements.

Consider the index sets I_0 and J_0. For $I \subset I_0$ and $J \subset J_0$, let be the set

$$\begin{aligned}\mathcal{N}_{I,J} \quad &:= \quad \{\underline{n} := (n_i, n'_\sigma, n''_\sigma, \widetilde{n}'_j, \widetilde{n}''_j) : n_i \geq 0,\ 1 \leq i \leq s;\ n'_\sigma > 0, \\ &\qquad n''_\sigma > 0,\ \sigma \in I;\ \widetilde{n}'_j > 0,\ \widetilde{n}''_j > 0,\ j \in J\}.\end{aligned}$$

For each $\underline{n} \in \mathcal{N}_{I,J}$, we define

$$Y_{\underline{n}} := \prod_{i=1}^{s} \mathcal{S}^{n_i} \overset{\circ}{E}_i \times \prod_{\sigma \in I} \mathbb{C}^*_\sigma \times \prod_{j \in J} \mathbb{C}^*_j,$$

with $\mathcal{S}^{n_i} \overset{\circ}{E}_i$ being the n_i-th symmetric power of $\overset{\circ}{E}_i$, $\mathbb{C}^*_\sigma = \mathbb{C}^*$ the residue field at P_σ minus 0 corresponding to the element σ of I, and $\mathbb{C}^*_j = \mathbb{C}^*$ the residue field minus 0 at P_j corresponding to the element j of J.

Define the set

$$Y := \bigcup_{\substack{I \subset I_0 \\ J \subset J_0}} \bigcup_{\underline{n} \in \mathcal{N}_{I,J}} Y_{\underline{n}} \tag{7.1}$$

as the union of the connected components $Y_{\underline{n}}$. One may prove that Y is a semigroup (cf. [**CDG07**], page 203).

For $R^* = \{z \in R : v_j(z) < \infty,\ j = 1, \ldots, r\}$, we consider the quotient $R^*/\sim$, where "$\sim$" denotes the following equivalence relation: for two elements $a, b \in R^*$, we say that $a \sim b$ if there exists an element $u \in \mathbb{C}$ such that $a = ub$. Now we define the projectivization of R^* as

$$\mathbb{P}R^* := R^*/\sim .$$

The set $\mathbb{P}R^*$ can be endowed with a structure of semigroup just considering the multiplication of functions. Define a semigroup homomorphism

$$\text{Init} : \quad \mathbb{P}R^* \quad \longrightarrow \quad Y = \bigcup_{\substack{I \subset I_0 \\ J \subset J_0}} \bigcup_{\underline{n} \in \mathcal{N}_{I,J}} Y_{\underline{n}}$$

between $\mathbb{P}R^*$ and Y, in which the image $\text{Init}(z)$ of any element $z \in \mathbb{P}R^*$ is an element of the connected component $Y_{\underline{n}}$ of Y corresponding to $I(z), J(z)$ and with

$$\underline{n} = \left(n_i(z), n'_\sigma(z), n''_\sigma(z), \widetilde{n}'_j(z), \widetilde{n}''_j(z)\right),$$

which is defined in every factor (connected component) of $Y_{\underline{n}}$ as follows:

- in $\mathcal{S}^{n_i} \overset{\circ}{E}_i$, $1 \leq i \leq s$, $\text{Init}(z)$ is represented by the set of intersection points between Γ_z and $\overset{\circ}{E}_i$ counted with multiplicities;
- in k^*_σ, $\sigma \in I(z)$, $\text{Init}(z)$ is represented by the quotient $\frac{a_\sigma(z)}{b_\sigma(z)}$;
- in k^*_j, $j \in J(z)$, $\text{Init}(z)$ is represented by the quotient $\frac{a_j(z)}{b_j(z)}$.

LEMMA 7.1. *The map Init is surjective.*

PROOF. Let us fix $I \subset I_0$, $J \subset J_0$, $\underline{n} \in \mathcal{N}_{I,J}$. Consider an element

$$y_{\underline{n}} = \left((\{P_{i,j}\}_{j=1}^{n_i};\ i = 1, \ldots, s), (\lambda_\sigma; \sigma \in I), (\mu_j; j \in J)\right) \in Y_{\underline{n}}.$$

We will construct $z \in R^*$ such that $\mathrm{Init}(z) = y_{\underline{n}}$ step by step, in fact it suffices to show the surjectivity in each factor of the space $Y_{\underline{n}}$.

For a point $p \in \overset{\circ}{E}_i$, there exists an irreducible element $z_p \in R$ such that the strict transform of $z_p R$ in X is smooth and transversal to the exceptional divisor E at the point $p \in \overset{\circ}{E}_i$. If S_p denotes the local ring of X at the point $p \in E$, x_p is a local equation for E_i at p, and γ_p is a local equation at p of the lifting of z_p, then $\{x_p, \gamma_p\}$ is a parameter system of S_p . If we take

$$(\{P_{ij}\}_{j=1}^{n_i};\ i = 1, \dots, s) \in \prod_{i=1}^{s} \mathcal{S}^{n_i} \overset{\circ}{E}_i,$$

then it is clear that $z^{(1)} := \prod_{i,j} z_{p_{ij}} \in R$ defines an element of R such that $I(z^{(1)}) = J(z^{(1)}) = \emptyset$, and moreover $n_i(z^{(1)}) = n_i$ for $i = 1, \dots, s$.

Let be now $\sigma \in I$, $n'_\sigma, n''_\sigma > 0$ and $\lambda_\sigma \in k^*_\sigma$. Let us consider the discrete valuation ω_{x_σ} (resp. ω_{y_σ}) associated with the ring $R_\sigma / x_\sigma R_\sigma$ (resp. $R_\sigma / y_\sigma R_\sigma$) (Recall that $\{x_\sigma, y_\sigma\}$ are the equations of the components of the exceptional divisor at $P_\sigma = E_{i_1(\sigma)} \cap E_{i_2(\sigma)}$). Let $u_\sigma \in R_\sigma$ be a unit such that $u_\sigma + \mathfrak{m}_\sigma = \lambda_\sigma \in \mathbb{C}^*_\sigma$ and let us take $\gamma_1^\sigma = x_\sigma + u_\sigma \cdot y_\sigma^{n'_\sigma - 1}$, $\gamma_2^\sigma = y_\sigma + x_\sigma^{n''_\sigma - 1} \in R_\sigma$. Since $\{y_\sigma, \gamma_1^\sigma\}$ is a regular system of parameters, we have that $\gamma_1^\sigma R_\sigma \cap R$ is a prime ideal of height 1, therefore it is principal and generated by $z_1^\sigma \in R$ whose strict transform in R_σ is just $\gamma_1^\sigma R_\sigma$. In the same manner, we can argue with γ_2^σ to obtain an irreducible element $z_2^\sigma \in R$ whose strict transform in R_σ is just $\gamma_2^\sigma R_\sigma$. Write $z^\sigma = z_1^\sigma \cdot z_2^\sigma$ and $\gamma^\sigma = \gamma_1^\sigma \cdot \gamma_2^\sigma$; notice that γ^σ is the strict transform of z^σ. Then we have

$$\varphi_{x_\sigma}(\gamma^\sigma) = \varphi_{x_\sigma}(u_\sigma) \cdot \varphi_{x_\sigma}(y_\sigma)^{n'_\sigma}$$

$$\varphi_{y_\sigma}(\gamma^\sigma) = \varphi_{y_\sigma}(x_\sigma)^{n''_\sigma},$$

and since $\varphi_{x_\sigma}(u_\sigma)$ is a unit, one has that $n'_\sigma(z^\sigma) = n'_\sigma$ and $n''_\sigma(z^\sigma) = n''_\sigma$. Moreover $a_\sigma = \lambda_\sigma$, $b_\sigma = 1$, and therefore $\frac{a_\sigma}{b_\sigma} = \lambda_\sigma \in \mathbb{C}^*_\sigma$.

Thus let be $z^{(2)} := \prod_{\sigma \in I} z^\sigma \in R$. From the previous construction it follows that $n_i(z^{(2)}) = 0$ for all $i = 1, \dots, s$; $J(z^{(2)}) = \emptyset$; $I(z^{(2)}) = I$ and moreover, for every $\sigma \in I$ one has $n'_\sigma(z^{(2)}) = n'_\sigma$, $n''_\sigma(z^{(2)}) = n''_\sigma$ and $\lambda_\sigma(z^{(2)}) = \lambda_\sigma \in \mathbb{C}^*_\sigma$.

For every $j \in J$ (and consequently for J) one can proceed analogously as in the previous case in order to get an element $z^{(3)} \in R$ such that $n_i(z^{(3)}) = 0$ for all $i = 1, \dots, s$; $I(g^{(3)}) = \emptyset$; $J(z^{(3)}) = J$ and $\widetilde{n}'_j(z^{(3)}) = \widetilde{n}'_j$, $\widetilde{n}''_j(z^{(3)}) = \widetilde{n}''_j$ and $\mu_j(z^{(3)}) = \mu_j \in \mathbb{C}^*_j$.

As a consequence we obtain an element $z = z^{(1)} \cdot z^{(2)} \cdot z^{(3)}$ satisfying that

$$\mathrm{Init}(z) = \big((\{P_{i,j}\}_{j=1}^{n_i};\ i = 1, \dots, s), (\lambda_\sigma; \sigma \in I), (\mu_j; j \in J)\big) \in Y_{\underline{n}}.$$

□

For every $i = 1, \dots, s$, the local ring $\mathcal{O}_{X,E_i}$ of X along E_i is a discrete valuation ring, with associated discrete valuation w_i: these are called *divisorial* valuations.

PROPOSITION 7.2. *For z and z' in R^*, we have that* $\mathrm{Init}(z) = \mathrm{Init}(z')$ *if and only if $z = \zeta z' + \xi$, where $\zeta \in \mathbb{C} \setminus \{0\}$, $w_i(\xi) > w_i(z) = w_i(z')$ for every $i = 1, \dots, s$, and $v_j(\xi) > v_j(z) = v_j(z')$ for every $j = 1, \dots, r$.*

PROOF. Let $z, z' \in R^*$ be such that $\mathrm{Init}(z) = \mathrm{Init}(z')$, and consider the liftings $\widetilde{z} := z \circ \pi$ and $\widetilde{z'} := z' \circ \pi$ to X; define the divisors

$$\mathrm{div}(\widetilde{z} \circ \iota_i) = \sum_{P \in E_i} (\Gamma_z \cdot E_i)_P \cdot P, \quad \text{and} \quad \mathrm{div}(\widetilde{z'} \circ \iota_i) = \sum_{P \in E_i} (\Gamma_{z'} \cdot E_i)_P \cdot P,$$

where Γ_z resp. $\Gamma_{z'}$ stands for the strict transform of z resp. z' by π, and $\iota_i : E_i \to X$ is the inclusion map. The condition $\mathrm{Init}(z) = \mathrm{Init}(z')$ means, in particular, that $\mathrm{div}(\widetilde{z} \circ \iota_i) = \mathrm{div}(\widetilde{z'} \circ \iota_i)$, and then the restriction of the meromorphic function $\psi := \widetilde{z}/\widetilde{z'}$ to E_i is a regular function on E_i (since the poles and zeros cancel out each other), therefore the restriction $\psi_i := (\widetilde{z} \circ \iota_i)/(\widetilde{z'} \circ \iota_i)$ of ψ to E_i is a constant function, say $\zeta_i \in \mathbb{C}^*$. Similarly, since the irreducible components $\widetilde{C}_j$ of the strict transform of the curve C are smooth, we can also show that the restriction of ψ to each $\widetilde{C}_j$ is a regular function on $\widetilde{C}_j$, for every $j = 1, \ldots, r$, and therefore constant.

We will see in a moment that there exists a constant $\zeta \in \mathbb{C} \setminus \{0\}$ such that $\zeta_i = \zeta$ for all $i = 1, \ldots, s$.

OBSERVATION. Given a variety X, a point $p \in X$ and $\mathcal{O}_{X,p}$ the local ring of X at p with maximal ideal $\mathfrak{m}_{X,p}$, we consider a regular function f on X. We define the *value* $f(p)$ of f in p as the image of f in $\mathcal{O}_{X,p}/\mathfrak{m}_{X,p}$.

Let P_σ, $\sigma = (i_1, i_2)$ be the intersection point between E_{i_1} and E_{i_2}, for $i_1 < i_2$. Let $R_\sigma := \mathcal{O}_{X,P_\sigma}$ be the local ring of X at P_σ, and consider the local equation γ_z resp. $\gamma_{z'}$ of Γ_z resp. $\Gamma_{z'}$ at P_σ, and the local equations $\{y_\sigma = 0\}$ resp. $\{x_\sigma = 0\}$ of E_{i_1} resp. E_{i_2}. Consider the canonical homomorphisms

$$\begin{array}{rccc} \varphi_{x_\sigma} : & R_\sigma & \longrightarrow & R_\sigma / x_\sigma R_\sigma \\ & \gamma_z & \mapsto & \gamma_z \bmod x_\sigma \end{array}$$

and

$$\begin{array}{rccc} \varphi_{y_\sigma} : & R_\sigma & \longrightarrow & R_\sigma / y_\sigma R_\sigma \\ & \gamma_z & \mapsto & \gamma_z \bmod y_\sigma. \end{array}$$

By the assumption $\mathrm{Init}(z) = \mathrm{Init}(z')$, it follows that

$$\deg_X (\Gamma_z \cdot E_i) = \deg_X (\Gamma_{z'} \cdot E_i) = n_\sigma$$
$$\deg_X (\Gamma_z \cdot E_j) = \deg_X (\Gamma_{z'} \cdot E_j) = m_\sigma.$$

Then there exist $\lambda_\sigma, \lambda'_\sigma, \mu_\sigma, \mu'_\sigma$ units of R_σ, so that

$$\gamma_z - \lambda_\sigma y_\sigma^{n_\sigma}, \gamma_{z'} - \lambda'_\sigma y_\sigma^{n_\sigma} \in x_\sigma R_\sigma$$
$$\gamma_z - \mu_\sigma x_\sigma^{m_\sigma}, \gamma_{z'} - \mu'_\sigma x_\sigma^{m_\sigma} \in y_\sigma R_\sigma,$$

therefore

$$\varphi_{x_\sigma}(\gamma_z) = \varphi_{x_\sigma}(\lambda_\sigma) \cdot \varphi_{x_\sigma}(y_\sigma)^{n_\sigma}$$
$$\varphi_{x_\sigma}(\gamma_{z'}) = \varphi_{x_\sigma}(\lambda'_\sigma) \cdot \varphi_{x_\sigma}(y_\sigma)^{n_\sigma}$$
$$\varphi_{y_\sigma}(\gamma_z) = \varphi_{y_\sigma}(\mu_\sigma) \cdot \varphi_{y_\sigma}(x_\sigma)^{m_\sigma}$$
$$\varphi_{y_\sigma}(\gamma_{z'}) = \varphi_{y_\sigma}(\mu'_\sigma) \cdot \varphi_{y_\sigma}(x_\sigma)^{m_\sigma}.$$

Thus the value of ψ_{i_1} at the point P_σ is just $\frac{\varphi_{y_\sigma}(\mu_\sigma)}{\varphi_{y_\sigma}(\mu'_\sigma)} = \frac{b_\sigma}{b'_\sigma} = \zeta_{i_1}$, and the value of ψ_{i_2} at P_σ is $\frac{\varphi_{x_\sigma}(\lambda_\sigma)}{\varphi_{x_\sigma}(\lambda'_\sigma)} = \frac{a_\sigma}{a'_\sigma} = \zeta_{i_2}$. Since $\mathrm{Init}(z) = \mathrm{Init}(z')$, we have

$$\frac{\varphi_{x_\sigma}(\lambda_\sigma)}{\varphi_{y_\sigma}(\mu_\sigma)} = \frac{\varphi_{x_\sigma}(\lambda'_\sigma)}{\varphi_{y_\sigma}(\mu'_\sigma)} \Longleftrightarrow \frac{\varphi_{y_\sigma}(\mu_\sigma)}{\varphi_{y_\sigma}(\mu'_\sigma)} = \frac{\varphi_{x_\sigma}(\lambda_\sigma)}{\varphi_{x_\sigma}(\lambda'_\sigma)} \Longleftrightarrow \zeta_{i_1} = \zeta_{i_2}.$$

Then all constant values $\zeta_1, \ldots, \zeta_s$ coincide, so they are all equal say to $\zeta \in \mathbb{C}^*$.

This means, for the divisorial valuations w_i we have

$$0 < w_i\left(\frac{z}{z'} - \zeta\right) = w_i\left(\frac{z-\zeta z'}{z'}\right) = w_i(z-\zeta z') - w_i(z'),$$

hence $w_i(z-\zeta z') > w_i(z) = w_i(z')$ for all $i = 1, \dots, s$.

For the valuations $v_1, \dots, v_r$, an analogous reasoning applies.

Conversely, if $z = \zeta z' + \xi$ with $\zeta \in \mathbb{C} \setminus \{0\}$, $w_i(z-\zeta z') > w_i(z) = w_i(z')$ for $i = 1, \dots, s$ and $v_j(z-\zeta z') > v_j(z) = v_j(z')$ for all $j = 1, \dots, r$, then the function ψ is regular on every component E_i of E for $i = 1, \dots, s$, and on every component $\widetilde{C}_j$ of the strict transform of the curve, therefore $\psi|_E \equiv \zeta$. Hence $\operatorname{div}(\widetilde{z' \circ \iota_i}) = \operatorname{div}(z \circ \iota_i)$ for $i = 1, \dots, s$ and the intersection points of the strict transforms Γ_z and $\Gamma_{z'}$ with each component E_i of E coincide (counting multiplicities). For $j = 1, \dots, r$ we also have

$$\left(\Gamma_z \cdot \widetilde{C}_j\right)_{P_j} = v_j(z) - w_{i_1(j)}(z) = v_j(z') - w_{i_1(j)}(z') = \left(\Gamma_{z'} \cdot \widetilde{C}_j\right)_{P_j}.$$

Moreover, for points of kind P_σ (the same holds for points of kind P_j) and following previous notations, the value of $\widetilde{z} \circ \iota_i / \widetilde{z'} \circ \iota_i$ at P_σ is $\zeta = \frac{\varphi_{x_\sigma}(\lambda_\sigma)}{\varphi_{x_\sigma}(\lambda'_\sigma)} = \frac{a_\sigma}{a'_\sigma}$, and at the same time $\zeta = \frac{b_\sigma}{b'_\sigma}$, then $\frac{a_\sigma}{a'_\sigma} = \frac{b_\sigma}{b'_\sigma}$. □

COROLLARY 7.3. *For any point $\varsigma \in Y_{\underline{n}} \subset Y$, the preimage* $\operatorname{Init}^{-1}(\varsigma)$ *consists of an affine space given by a point z plus the ideal $I_{\underline{n}}$, where*

$$I_{\underline{n}} = \{z \in R^* : w_i(z) > w_i(\underline{n}),\ i = 1, \dots, s;\ v_j(z) > v_j(\underline{n}),\ j = 1, \dots, r\}.$$

This space has finite codimension $F(\underline{n})$ in $\mathbb{P}R$. Furthermore, over each connected component $Y_{\underline{n}}$ of Y, the map Init *is a locally trivial fibration.*

7.2. Computation of the codimension $\mathbf{F}(\underline{\mathbf{n}})$. Consider an element $z \in R^*$ and let w_i be the divisorial valuation corresponding to the component E_i of π, for all $i = 1, \dots, s$. Set $\underline{w}(z) := (w_1(z), \dots, w_s(z))$; recall that the value $w_i(z)$ is the multiplicity of the lifting $z \circ \pi$ of z along E_i, for all $i = 1, \dots, s$. Let $\mathcal{D}(z)$ be the total transform by π of the curve given by $z = 0$; i.e. $\mathcal{D}(z)$ is the divisor in X

$$\mathcal{D}(z) = \sum_{i=1}^{s} w_i(z) E_i + \Gamma_z,$$

where Γ_z is the strict transform of the curve defined by $z = 0$. We have that

$$0 = \deg_X(\mathcal{D}(z) \cdot E_i) = \deg_X(\Gamma_z \cdot E_i) + \sum_{j=1}^{s} w_j(z) \deg_X(E_i \cdot E_j)$$

and so $\widehat{n}_i(z) := \deg_X(\Gamma_z \cdot E_i) = -\sum_{j=1}^{s} w_j(z) n_{ij}$, where $N = (n_{ij})$ is the intersection matrix. Thus

$$\underline{w}(z) := (w_1(z), \dots, w_s(z)) = (\widehat{n}_1(z), \dots, \widehat{n}_s(z))M,$$

which provides the precise relation between $\underline{w}(z)$ and $\widehat{\underline{n}}(z) := (\widehat{n}_1(z), \dots, (\widehat{n}_s(z))$.

Let us fix $\widehat{\underline{n}} = (\widehat{n}_1, \dots, \widehat{n}_s) \in \mathbb{Z}^s_{\geq 0}$ and the corresponding divisor $A = \sum_{i=1}^{s} w_i E_i$, where $\underline{w} = (w_1, \dots, w_s) = \widehat{\underline{n}} M$. Let $J^D(\underline{w})$ be the divisorial ideal of R defined by

$$J^D(\underline{w}) = \{z \in R^* : w_i(z) \geq w_i,\ i = 1, \dots, s\},$$

and $h^D(\underline{w}) = \dim_{\mathbb{C}} R/J^D(\underline{w})$.

The following result is known as the Hoskin-Deligne formula:

PROPOSITION 7.4. *Let* $\mathbb{K} = \sum_{i=1}^{s} E_i^*$ *be the canonical divisor on* X *and let* $A = \sum_{i=1}^{s} \alpha_i E_i^*$ *be the expression of the divisor* A *on the basis* $\{E_i^*\}$. *Then*

$$h^D(\underline{w}) = \frac{1}{2}\sum_{i=1}^{s} \alpha_i(\alpha_i + 1).$$

PROOF. The genus formula gives us that

$$h^0(\mathcal{O}_A) = h^D(\underline{w}) = -\frac{\deg_X(A \cdot A) + \deg_X(A \cdot \mathbb{K})}{2}.$$

Eq. (2.3) allows us to deduce

$$\deg_X(A \cdot A) = -\sum_{i=1}^{s} \alpha_i^2$$

$$\deg_X(A \cdot \mathbb{K}) = -\sum_{i=1}^{s} \alpha_i.$$

Then the codimension of the ideal $J^D(\underline{w})$ is equal to $\frac{1}{2}\sum_{i=1}^{s} \alpha_i(\alpha_i + 1)$. □

The point is now to compute the intersection degrees $\deg_X(A \cdot A)$ as well as $\deg_X(A \cdot \mathbb{K})$ in terms of the intersection matrix and $\widehat{n}_1, \dots, \widehat{n}_s$.

Let $\overset{\bullet}{\nu}_i$ (resp. $\overset{\circ}{\nu}_i$) be the number of components E_j of E intersecting E_i for $j \neq i$ (resp. this number plus the number of strict transforms of the curve C intersecting E_i). Notice that $\overset{\bullet}{\nu}_i = \deg_X(E_i \cdot (\sum_{j \neq i} E_j)) = \sum_{j \neq i} n_{ij}$.

LEMMA 7.5. *For* $i = 1, \dots, s$, *write* $\varepsilon_i = 2 - \overset{\bullet}{\nu}_i$. *Then we have*

$$\deg_X(A \cdot \mathbb{K}) = \underline{\widehat{n}} \cdot \underline{1}^t - \underline{\widehat{n}} \cdot M \cdot \underline{\varepsilon}^t.$$

PROOF. First of all we claim that, for every $i = 1, \dots, s$,

$$\deg_X(E_i \cdot E_i) = -2 - \deg_X(\mathbb{K} \cdot E_i). \tag{7.2}$$

In fact, let $P = (p_{ij})_{1 \le i,j \le s}$ be the proximity matrix of π. From the definitions of intersection matrix and the divisor $\mathbb{K}$, we have that $\deg_X(E_i \cdot E_i) = -\sum_{j=1}^{s} p_{ij}^2$ and $\deg_X(\mathbb{K} \cdot E_i) = -\sum_{j=1}^{s} p_{ij}$, that is,

$$\deg_X(E_i \cdot E_i) + \deg_X(\mathbb{K} \cdot E_i) = -\sum_{j=1}^{s} p_{ij}(p_{ij} + 1).$$

For $j = 1, \dots, s$, it is $p_{ij}(p_{ij} + 1) = 0$ if and only if $p_{ij} = 0$ or $p_{ij} = -1$, which occurs whenever $i \neq j$. Then, for every $i = 1, \dots, s$ we have

$$\deg_X(E_i \cdot E_i) + \deg_X(\mathbb{K} \cdot E_i) = -(p_{ii}^2 + p_{ii}) = -(1 + 1) = -2$$

and eq. (7.2) has been shown. Then we can apply it to get, for every $i = 1, \dots, s$,

$$\sum_{j=1}^{s} p_{ij} = -\deg_X(\mathbb{K} \cdot E_i) = 2 + \deg_X(E_i \cdot E_i).$$

On the other hand, the definition of the intersection matrix N implies that

$$N \cdot \begin{pmatrix} 1 \\ \vdots \\ 1 \end{pmatrix} = \begin{pmatrix} \deg_X(E_1 \cdot E_1) + \overset{\bullet}{\nu}_1 \\ \vdots \\ \deg_X(E_s \cdot E_s) + \overset{\bullet}{\nu}_s \end{pmatrix}.$$

If we write $\underline{1} := (1, 1, \ldots, 1)$ and $\underline{\varepsilon} := (\varepsilon_1, \ldots, \varepsilon_s)$, with $\varepsilon_i := 2 - \overset{\bullet}{\nu}_i$, for $i = 1, \ldots, s$, then

$$N \cdot \underline{1}^t + \underline{\varepsilon}^t = P \cdot \underline{1}^t.$$

Hence

$$\begin{aligned} -\deg_X(A \cdot \mathbb{K}) &= (\alpha_1, \ldots, \alpha_s) \cdot (1, \ldots, 1)^t \\ &= \underline{w} \cdot P \cdot (1, \ldots, 1)^t \\ &= \underline{\widehat{n}} \cdot M \cdot P \cdot \underline{1}^t \\ &= \underline{\widehat{n}} \cdot M \cdot (N \cdot \underline{1}^t + \underline{\varepsilon}^t) \\ &= -\underline{\widehat{n}} \cdot \underline{1}^t + \underline{\widehat{n}} \cdot M \cdot \underline{\varepsilon}^t, \end{aligned}$$

as desired. □

REMARK 7.6. Let β_i be the number of components from $\{E_j : j \neq i\}$ such that $E_i \cap E_j \neq \emptyset$. Then $\overset{\bullet}{\nu}_i = \beta_i$ and so $\varepsilon_i = 2 - \beta_i$, i.e. it is the Euler characteristic of the space $\overset{\bullet}{E}_i = E_i \setminus \bigcup_{j \neq i} E_j$.

LEMMA 7.7. *We have the equality*

$$-\deg_X(A \cdot A) = \underline{\widehat{n}} \cdot M \cdot \underline{\widehat{n}}^t.$$

PROOF. If N stands for the intersection matrix, $M = -N^{-1}$, $\underline{w} = \underline{\widehat{n}} \cdot M$, and observing that M is a symmetric matrix, we have

$$\begin{aligned} -\deg_X(A \cdot A) &= (\alpha_1, \ldots, \alpha_s) \cdot (\alpha_1, \ldots, \alpha_s)^t \\ &= (\underline{w} \cdot P) \cdot (\underline{w} \cdot P)^t \\ &= -\underline{w} \cdot N \cdot \underline{w}^t \\ &= -\underline{\widehat{n}} \cdot M \cdot N \cdot (\underline{\widehat{n}} \cdot M)^t \\ &= \underline{\widehat{n}} \cdot M^t \cdot \underline{\widehat{n}}^t \\ &= \underline{\widehat{n}} \cdot M \cdot \underline{\widehat{n}}^t. \end{aligned}$$

□

COROLLARY 7.8. *The codimension of the ideal $J^D(\underline{w})$ is*

$$h^D(\underline{w}) = \frac{1}{2}\left(\sum_{i,i'=1}^{s} m_{ii'}\widehat{n}_i\widehat{n}_{i'} + \sum_{i=1}^{s} \widehat{n}_i \cdot \left(\sum_{i'=1}^{s} m_{ii'}(2 - \overset{\bullet}{\nu}_{i'}) - 1\right)\right).$$

Let $I \subset I_0$, $J \subset J_0$, $\underline{n} \in \mathcal{N}_{I,J}$ and $\underline{y} \in Y_{\underline{n}}$. Recall that $\mathrm{Init}^{-1}(\underline{y})$ is an affine space in $\mathbb{P}R$ of finite codimension $F(\underline{n})$. Let $\underline{n} = (n_i, n'_\sigma, n''_\sigma, \tilde{n}'_j, \tilde{n}''_j)$ with $i \in \{1, \ldots, s\}$, $\sigma \in I$, $j \in J$. Considering $\underline{\widehat{n}} \in \mathbb{Z}^s_{\geq 0}$ the element with entries

$$\widehat{n}_i := n_i + \sum_{\substack{\sigma \in I \\ i_1(\sigma) = i}} n'_\sigma + \sum_{\substack{\sigma \in I \\ i_2(\sigma) = i}} n''_\sigma + \sum_{\substack{j \in J \\ i_1(j) = i}} \tilde{n}'_j,$$

we may define

$$\begin{aligned} \underline{w}(\underline{n}) &:= \widehat{\underline{n}} \cdot M \\ v_j(\underline{n}) &:= w_{i_1(j)}(\underline{n}) + \widetilde{n}_j'' \end{aligned}$$

for $\underline{n} \in \mathcal{N}_{I,J}$ and $j = 1, \ldots, r$. Note that $v_j(\underline{n})$ is nothing but the order of a function g such that $\mathrm{Init}(g) \in Y_{\underline{n}}$ on the component C_j of the curve, $w_i(\underline{n})$ is the multiplicity along the component E_i of the exceptional divisor of the lifting of a function z such that $\mathrm{Init}(z) \in Y_{\underline{n}}$, and $\widetilde{n}_j'' = \widetilde{n}_j''(z)$.

PROPOSITION 7.9. *We have that*

$$\begin{aligned} F(\underline{n}) &= \frac{1}{2}\left(\sum_{i,i'=1}^{s} m_{ii'}\widehat{n}_i\widehat{n}_{i'} + \sum_{i=1}^{s}\widehat{n}_i \cdot \left(\sum_{i'=1}^{s} m_{ii'}(2 - \overset{\bullet}{\nu}_{i'}) + 1\right)\right) \\ &\quad + \sum_{j \in J} \widetilde{n}_j''. \end{aligned}$$

PROOF. By Corollary 7.3, the codimension $F(\underline{n})$ is equal to the codimension in R of the ideal

$$I_{\underline{n}} = \{z \in R^* : w_i(z) > w_i(\underline{n}),\ j = 1, \ldots, s;\ v_j(z) > v_j(\underline{n}),\ j = 1, \ldots, r\}$$

minus 1. Making additional blowing-ups at the intersection points of the strict transforms of the curve C with the exceptional divisor, we can reduce our problem to the case $J = \emptyset$.

Then we have to compute $h^D(\underline{w}(\underline{n}) + \underline{1}) - 1$, where, for $\underline{w} = (w_1, \ldots, w_s) \in \mathbb{Z}^s$,

$$h^D(\underline{w}) := \dim_{k_R} R/J^D(\underline{w}),$$

and $J^D(\underline{w}) := \{z \in R^* : w_i(z) \geq w_i,\ i = 1, \ldots, s\}$. By Proposition 7.4, Lemma 7.5 and Lemma 7.7, we get

$$\begin{aligned} h^D(\underline{w}(\underline{n})) &= -\frac{1}{2}\left(\deg_X(A \cdot A) + \deg_X(A \cdot \mathbb{K})\right) \\ &= \frac{1}{2}\left(\widehat{\underline{n}} \cdot M \cdot \widehat{\underline{n}}^t - \widehat{\underline{n}} \cdot \underline{1}^t + \widehat{\underline{n}} \cdot M \cdot \underline{\varepsilon}^t\right). \end{aligned}$$

In general we cannot apply the above formula to compute $h^D(\underline{w} + \underline{1})$ because $\underline{w} + \underline{1}$ must not be of the form $\widehat{n} \cdot M$ (i.e. it does not belong to the semigroup defined by the considered divisorial valuations). Of course, if this would be the case, the computation would follow in the same way. So we have to look for an alternative reasoning.

The map Init induces a fibration

$$\mathbb{P}J^D(\underline{w}(\underline{n})) \supset Z \longrightarrow \prod_{i=1}^{s} \mathcal{S}^{\widehat{n}_i}\overset{\bullet}{E}_i = W,$$

where $Z = \mathrm{Init}^{-1}(W)$ is an open subset of $\mathbb{P}J^D(\underline{w}(\underline{n}))$. The fibre is the set of elements in Z whose image via the map Init coincides (i.e., the map Init applied to a non-zero element $z \in \mathbb{P}R$ with $I(z) = \emptyset$); this set is by Proposition 7.2 equal to $J^D(\underline{w}(\underline{n}) + \underline{1})$. Then

$$h^D(\underline{w}(\underline{n}) + \underline{1}) = 1 + h^D(\underline{w}(\underline{n})) + \dim_{k_R}\left(\prod_{i=1}^{s} \mathcal{S}^{\widehat{n}_i}\overset{\bullet}{E}_i\right),$$

therefore

$$\begin{aligned} F(\underline{n}) &= h^D(\underline{w}(\underline{n})) + \dim_{\mathbb{C}} \left(\prod_{i=1}^{s} \mathcal{S}^{\widehat{n}_i} \overset{\bullet}{E}_i \right) \\ &= h^D(\underline{w}(\underline{n})) + \sum_{i=1}^{s} \widehat{n}_i , \end{aligned}$$

and the formula follows straightforward. □

7.3. An explicit formula. The generalized Euler characteristic satisfies the Fubini rule (see Theorem 4.3; also in [**V88**], §3, pp. 128–129) so we can apply this formula to the map *Init* to get

$$P_g(\underline{t}; \mathbb{L}) = \int_Y \mathbb{L}^{-F(\underline{n})} \underline{t}^{\underline{v}(\underline{n})} d\chi_g = \sum_{\substack{I \subset I_0 \\ J \subset J_0}} \sum_{\underline{n} \in \mathcal{N}_{I,J}} \mathbb{L}^{-F(\underline{n})} [Y_{\underline{n}}] \cdot \underline{t}^{\underline{v}(\underline{n})}. \tag{7.3}$$

We recall that

$$Y_{\underline{n}} = \prod_{i=1}^{s} \mathcal{S}^{n_i} \overset{\circ}{E}_i \times \prod_{\sigma \in I} \mathbb{C}_{\sigma}^* \times \prod_{j \in J} \mathbb{C}_j^*.$$

Then the class of $Y_{\underline{n}}$ in the Grothendieck ring is

$$[Y_{\underline{n}}] = \prod_{i=1}^{s} [\mathcal{S}^{n_i} \overset{\circ}{E}_i] \cdot \prod_{\sigma \in I} [\mathbb{C}_{\sigma}^*] \cdot \prod_{j \in J} [\mathbb{C}_j^*],$$

where $[\mathbb{C}_{\sigma}^*] = [\mathbb{C}_{\sigma}] - 1 = \mathbb{L} - 1$ and $[\mathbb{C}_j^*] = [\mathbb{C}_j] - 1 = \mathbb{L} - 1$; the computation of the class $[\mathcal{S}^{n_i} \overset{\circ}{E}_i]$ is shown in Lemma 7.10:

LEMMA 7.10. *Preserving notations as above,*

$$[\mathcal{S}^{n_i} \overset{\circ}{E}_i] = \mathbb{L}^{n_i} \sum_{\ell=0}^{\min\{n_i, \overset{\circ}{\nu}_i - 1\}} (-1)^{\ell} \binom{\overset{\circ}{\nu}_i - 1}{\ell} \mathbb{L}^{-\ell}.$$

PROOF. Since $[\{p\}] = 1$, we have

$$[\overset{\circ}{E}_i] = [E_i] - \overset{\circ}{\nu}_i = \mathbb{L} + 1 - \overset{\circ}{\nu}_i.$$

On the other hand, Theorem 1 of [**GLM04**, page 51] shows that

$$\begin{aligned} \sum_{\ell=0}^{\infty} [\mathcal{S}^{\ell} \overset{\circ}{E}_i] t^{\ell} &= (1-t)^{-[\overset{\circ}{E}_i]} \\ &= (1-t)^{-(\mathbb{L}+1-\overset{\circ}{\nu}_i)} \\ &= (1-t)^{-\mathbb{L}} (1-t)^{\overset{\circ}{\nu}_i - 1} \\ &= \sum_{\ell=0}^{\infty} \mathbb{L}^{\ell} t^{\ell} \cdot (1-t)^{\overset{\circ}{\nu}_i - 1}, \end{aligned}$$

therefore

$$
\begin{aligned}
[\mathcal{S}^{n_i}\overset{\circ}{E}_i] &= \sum_{\ell=0}^{\min\{n_i,\overset{\circ}{\nu}_i-1\}} (-1)^\ell \mathbb{L}^{n_i-\ell}\binom{\overset{\circ}{\nu}_i-1}{\ell} \\
&= \mathbb{L}^{n_i}\sum_{\ell=0}^{\min\{n_i,\overset{\circ}{\nu}_i-1\}} (-1)^\ell \binom{\overset{\circ}{\nu}_i-1}{\ell}\mathbb{L}^{-\ell}.
\end{aligned}
$$

□

Lemma 7.10 and eq. (7.3) allow us to finish with the description of the generalized Poincaré series in terms of an embedded resolution:

THEOREM 7.11 (Campillo, Delgado, Gusein-Zade).

$$
\begin{aligned}
P_g(t_1,\ldots,t_r;\mathbb{L}) &= \sum_{\substack{I\subset I_0\\ J\subset J_0}}\sum_{\underline{n}\in\mathcal{N}_{I,J}} \mathbb{L}^{\sharp(I)+\sharp(J)+\sum_{i=1}^s n_i - F(\underline{n})}\left(1-\mathbb{L}^{-1}\right)^{\sharp(I)+\sharp(J)} \\
&\times \prod_{i=1}^s\left(\sum_{\ell=0}^{\min\{n_i,\overset{\circ}{\nu}_i-1\}}(-1)^\ell\binom{\overset{\circ}{\nu}_i-1}{\ell}\mathbb{L}^{-\ell}\right)\underline{t}^{\underline{v}(\underline{n})}.
\end{aligned}
$$

Notice that $F(\underline{n})$ has been already computed in Proposition 7.9 in terms of the intersection matrix.

REMARK 7.12. The substitution $\mathbb{L}=1$ gives the connection with the "classical" Poincaré series $P(\underline{t})$, namely

$$
P_g(\underline{t};1) = \sum_{\substack{I\subset I_0\\ J\subset J_0}}\sum_{\underline{n}\in\mathcal{N}_{I,J}} \underline{t}^{\underline{v}(\underline{n})} = \int_Y \underline{t}^{\underline{v}(\underline{n})}d\chi_g = P(\underline{t}).
$$

Theorem 7.11 means that, for plane curve singularities, the generalized Poincaré series does not say more than the classical Poincaré series: two plane curve singularities with identical Poincaré series are topologically equivalent, thus they have equivalent embedded resolutions and, thanks to Theorem 7.11, they have the same generalized Poincaré series. For non-plane curve singularities this is no longer the case: see the example in [**CDG07**, p. 200].

An algorithmic treatment of the generalized Poincaré series was developed by Gorsky in [**G11**]. In order to see how difficult to compute is the generalized Poincaré series, we present the easiest possible example (which is already complicated enough!): a smooth germ.

EXAMPLE 7.13. We compute the generalized Poincaré series of a smooth germ C. In order to obtain an embedded resolution, we have to blow up once, so that we obtain one component for the exceptional divisor E and the strict transform $\tilde{C}$ of C meets E transversally, with multiplicity 1. In this case we have $r=s=1$, $I_0=\emptyset$, $J_0=\{1\}$, $\chi(\overset{\circ}{E})=2-1=1$, and therefore $1-\chi(\overset{\circ}{E})=0$, and $\chi(\overset{\bullet}{E})=2-0=2$. So the formula in Theorem 7.11 reduces to

$$
P_g(t;q) = \sum_{J\subset J_0}\sum_{\underline{n}\in\mathcal{N}_{\emptyset,J}} q^{F(\underline{n})-\sharp(J)-n_1}(1-q)^{\sharp(J)}t^{v(\underline{n})}.
$$

The first summation yields just two summands, depending on whether $J=\emptyset$ or $J=\{1\}$. If $J=\emptyset$, then $\tilde{n}_1=n_1$, $m_{11}=1$, $v(\underline{n})=n_1$, and

$$
F(\underline{n}) = \frac{1}{2}\Big(m_{11}\tilde{n}_1\tilde{n}_1+\tilde{n}_1(2m_{11}+1)\Big) = \frac{1}{2}(n_1^2+3n_1).
$$

Hence the second summation is equal to

$$\sum_{n_1=0}^{\infty} q^{F(\underline{n})-n_1} t^{v(\underline{n})} = \sum_{n_1=0}^{\infty} q^{\frac{n_1^2+3n_1}{2}-n_1} t^{n_1} = \sum_{n_1=0}^{\infty} q^{\frac{n_1^2+3n_1}{2}} q^{-n_1} t^{n_1}.$$

On the other hand, for the summand of the first summation corresponding to $J=\{1\}$ we have $\mathcal{N}_{\emptyset,\{1\}} = \{\underline{n} = (n_1, \tilde{n}_1', \tilde{n}_1'') : n_1 \geq 0, \tilde{n}_1' > 0, \tilde{n}_1'' > 0\}$, $m_{11} = 1$, $\tilde{n}_1 = n_1 + \tilde{n}_1'$ and $v(\underline{n}) = \tilde{n}_1 + \tilde{n}_1'' = n_+ \tilde{n}_1' + \tilde{n}_1''$. Moreover,

$$F(\underline{n}) = \frac{1}{2}\Big(\tilde{n}_1^2 + 3\tilde{n}_1\Big) + \tilde{n}_1''.$$

This leads to

$$\begin{aligned}
\sum_{\underline{n}\in\mathcal{N}_{\emptyset,\{1\}}} q^{F(\underline{n})-n_1-1}(1-q)t^{v(\underline{n})} &= \sum_{\underline{n}\in\mathcal{N}_{\emptyset,\{1\}}} q^{\frac{\tilde{n}_1^2+3\tilde{n}_1}{2}+\tilde{n}_1''} q^{-n_1-1}(1-q)t^{\tilde{n}_1}t^{\tilde{n}_1''} \\
&= \sum_{\tilde{n}_1=1}^{\infty} q^{\frac{1}{2}(\tilde{n}_1^2+3\tilde{n}_1)} t^{\tilde{n}_1} \cdot \sum_{n_1=0}^{\tilde{n}_1-1} q^{-n_1-1}(1-q) \cdot \sum_{\tilde{n}_1''=1}^{\infty} q^{\tilde{n}_1''} t^{\tilde{n}_1''} \\
&= \sum_{\tilde{n}_1=1}^{\infty} q^{\frac{(\tilde{n}_1^2+3\tilde{n}_1)}{2}} t^{\tilde{n}_1} \cdot (q^{-\tilde{n}_1} - 1) \cdot \frac{qt}{1-qt}.
\end{aligned}$$

If we sum the two expressions for $J=\emptyset$ and $J=\{1\}$, we get

$$\begin{aligned}
P_g(t) =& 1 + \sum_{n=1}^{\infty} q^{\frac{n^2+3n}{2}} \big(q^{-n} + \frac{qt}{1-qt}(q^{-n}-1)\big)t^n \\
=& 1 + \sum_{n=1}^{\infty} q^{\frac{n^2+3n}{2}} \big(\frac{q^{-n}(1-qt) + qt(q^{-n}-1)}{1-qt}\big)t^n \\
=& 1 + \frac{1}{1-qt} \sum_{n=1}^{\infty} q^{\frac{n^2+3n}{2}} (q^{-n} - qt)t^n \\
=& 1 + \frac{1}{1-qt}\Big(\sum_{n=1}^{\infty} q^{\frac{n^2+3n}{2}} q^{-\frac{2n}{2}} t^n - \sum_{n=1}^{\infty} q^{\frac{n^2+3n}{2}} qt^{n+1}\Big) \\
=& 1 + \frac{1}{1-qt}\Big(\sum_{n=1}^{\infty} q^{\frac{n^2+n}{2}} t^n - \sum_{n=1}^{\infty} q^{\frac{n^2+3n+2}{2}} t^{n+1}\Big) \\
=& 1 + \frac{1}{1-qt}\Big(\sum_{n=1}^{\infty} q^{\frac{(n+1)n}{2}} t^n - \sum_{n=1}^{\infty} q^{\frac{(n+2)(n+1)}{2}} t^{n+1}\Big) \\
=& 1 + \frac{1}{1-qt}\Big(\sum_{n=1}^{\infty} q^{\frac{(n+1)n}{2}} t^n - \sum_{h=2}^{\infty} q^{\frac{(h+1)h}{2}} t^h\Big) \\
=& 1 + \frac{1}{1-qt} tq^{\frac{(1+1)\cdot 1}{2}} \\
=& 1 + \frac{tq}{1-tq} \\
=& \frac{1}{1-qt},
\end{aligned}$$

as expected.

Some of the ideas contained in this survey can be generalized to curves defined over other fields different from $\mathbb{C}$. The reader is referred to the papers by Delgado and Moyano-Fernández [**DM09**], Moyano-Fernández and Zúñiga-Galindo [**MZ10**], and Moyano-Fernández [**M11a**], [**M11b**] and [**M13**] for further reading.

References

[AC75] Norbert A'Campo, *La fonction zêta d'une monodromie* (French), Comment. Math. Helv. **50** (1975), 233–248, DOI 10.1007/BF02565748. MR371889

[BK86] E. Brieskorn and H. Knörrer, *Plane curves*, Birkhäuser, Basel, Boston, Stuttgart, 1986.

[C80] Antonio Campillo, *Algebroid curves in positive characteristic*, Lecture Notes in Mathematics, vol. 813, Springer, Berlin, 1980. MR584440

[CDG98] S. M. Guseĭn-Zade, F. Del′gado, and A. Kampil′o, *Extended semigroup of a plane curve singularity* (Russian), Tr. Mat. Inst. Steklova **221** (1998), 149–167; English transl., Proc. Steklov Inst. Math. **2(221)** (1998), 139–156. MR1683692

[CDG99a] S. M. Guseĭn-Zade, F. Del′gado, and A. Kampil′o, *On the monodromy of a plane curve singularity and the Poincaré series of its ring of functions* (Russian), Funktsional. Anal. i Prilozhen. **33** (1999), no. 1, 66–68, DOI 10.1007/BF02465144; English transl., Funct. Anal. Appl. **33** (1999), no. 1, 56–57. MR1711890

[CDG99b] S. M. Guseĭn-Zade, F. Del′gado, and A. Kampil′o, *The Alexander polynomial of a plane curve singularity, and the ring of functions on the curve* (Russian), Uspekhi Mat. Nauk **54** (1999), no. 3(327), 157–158, DOI 10.1070/rm1999v054n03ABEH000160; English transl., Russian Math. Surveys **54** (1999), no. 3, 634–635. MR1728649

[CDG00] S. M. Guseĭn-Zade, F. Del′gado, and A. Kampil′o, *Integration with respect to the Euler characteristic over a function space, and the Alexander polynomial of a plane curve singularity* (Russian), Uspekhi Mat. Nauk **55** (2000), no. 6(336), 127–128, DOI 10.1070/rm2000v055n06ABEH000338; English transl., Russian Math. Surveys **55** (2000), no. 6, 1148–1149. MR1840363

[CDG03a] A. Campillo, F. Delgado, and S. M. Gusein-Zade, *The Alexander polynomial of a plane curve singularity and integrals with respect to the Euler characteristic*, Internat. J. Math. **14** (2003), no. 1, 47–54, DOI 10.1142/S0129167X03001703. MR1955509

[CDG03b] A. Campillo, F. Delgado, and S. M. Gusein-Zade, *The Alexander polynomial of a plane curve singularity via the ring of functions on it*, Duke Math. J. **117** (2003), no. 1, 125–156, DOI 10.1215/S0012-7094-03-11712-3. MR1962784

[CDG07] A. Campillo, F. Delgado, and S. M. Gusein-Zade, *Multi-index filtrations and generalized Poincaré series*, Monatsh. Math. **150** (2007), no. 3, 193–209, DOI 10.1007/s00605-006-0437-1. MR2308548

[CDK94] A. Campillo, F. Delgado, and K. Kiyek, *Gorenstein property and symmetry for one-dimensional local Cohen-Macaulay rings*, Manuscripta Math. **83** (1994), no. 3-4, 405–423, DOI 10.1007/BF02567623. MR1277539

[CF02] A. Campillo and J. I. Farrán, *Symbolic Hamburger-Noether expressions of plane curves and applications to AG codes*, Math. Comp. **71** (2002), no. 240, 1759–1780, DOI 10.1090/S0025-5718-01-01390-4. MR1933054

[CM91] A. Campillo and C. Marijuan, *Higher order relations for a numerical semigroup*, Sém. Théor. Nombres Bordeaux (2) **3** (1991), no. 2, 249–260. MR1149797

[CR94] A. Campillo and A.-J. Reguera, *Combinatorial aspects of sequences of point blowing ups*, Manuscripta Math. **84** (1994), no. 1, 29–46, DOI 10.1007/BF02567441. MR1283325

[Cs00] Eduardo Casas-Alvero, *Singularities of plane curves*, London Mathematical Society Lecture Note Series, vol. 276, Cambridge University Press, Cambridge, 2000, DOI 10.1017/CBO9780511569326. MR1782072

[DM09] Félix Delgado de la Mata and Julio José Moyano Fernández, *On the relation between the generalized Poincaré series and the Stöhr zeta function*, Proc. Amer. Math. Soc. **137** (2009), no. 1, 51–59, DOI 10.1090/S0002-9939-08-09469-0. MR2439424

[DL99] Jan Denef and François Loeser, *Germs of arcs on singular algebraic varieties and motivic integration*, Invent. Math. **135** (1999), no. 1, 201–232, DOI 10.1007/s002220050284. MR1664700

[EN85] David Eisenbud and Walter Neumann, *Three-dimensional link theory and invariants of plane curve singularities*, Annals of Mathematics Studies, vol. 110, Princeton University Press, Princeton, NJ, 1985. MR817982

[F94] G. Fischer, *Ebene algebraische Kurven*, Vieweg, Braunschweig/Wiesbaden, 1994.

[Fu98] William Fulton, *Intersection theory*, 2nd ed., Ergebnisse der Mathematik und ihrer Grenzgebiete. 3. Folge. A Series of Modern Surveys in Mathematics [Results in Mathematics and Related Areas. 3rd Series. A Series of Modern Surveys in Mathematics], vol. 2, Springer-Verlag, Berlin, 1998, DOI 10.1007/978-1-4612-1700-8. MR1644323

[G11] E. Gorsky, *Combinatorial computation of the motivic Poincaré series*, J. Singul. **3** (2011), 48–82, DOI 10.5427/jsing.2011.3d. MR2785859

[GLS07] G.-M. Greuel, C. Lossen, and E. Shustin, *Introduction to singularities and deformations*, Springer Monographs in Mathematics, Springer, Berlin, 2007. MR2290112

[G10] S. M. Guseĭn-Zade, *Integration with respect to the Euler characteristic and its applications* (Russian, with Russian summary), Uspekhi Mat. Nauk **65** (2010), no. 3(393), 5–42, DOI 10.1070/RM2010v065n03ABEH004679; English transl., Russian Math. Surveys **65** (2010), no. 3, 399–432. MR2682719

[GLM04] S. M. Gusein-Zade, I. Luengo, and A. Melle-Hernández, *A power structure over the Grothendieck ring of varieties*, Math. Res. Lett. **11** (2004), no. 1, 49–57, DOI 10.4310/MRL.2004.v11.n1.a6. MR2046199

[H77] Robin Hartshorne, *Algebraic geometry*, Graduate Texts in Mathematics, No. 52, Springer-Verlag, New York-Heidelberg, 1977. MR0463157

[KV04] K. Kiyek and J. L. Vicente, *Resolution of curve and surface singularities: In characteristic zero*, Algebra and Applications, vol. 4, Kluwer Academic Publishers, Dordrecht, 2004, DOI 10.1007/978-1-4020-2029-2. MR2106959

[KH71] J. Herzog and E. Kunz, *Die Wertehalbgruppe eines lokalen Rings der Dimension* 1 (German), S.-B. Heidelberger Akad. Wiss. Math. Natur. Kl. (1971), 27–67. MR0291149

[L02] Qing Liu, *Algebraic geometry and arithmetic curves*, Oxford Graduate Texts in Mathematics, vol. 6, Oxford University Press, Oxford, 2002. Translated from the French by Reinie Erné; Oxford Science Publications. MR1917232

[M11a] Julio José Moyano-Fernández, *Generalised Poincaré series and embedded resolution of curves*, Monatsh. Math. **164** (2011), no. 2, 201–224, DOI 10.1007/s00605-010-0259-z. MR2837115

[M11b] Julio José Moyano-Fernández, *Curvettes and clusters of infinitely near points*, Rev. Mat. Complut. **24** (2011), no. 2, 439–463, DOI 10.1007/s13163-010-0048-1. MR2806354

[M13] Julio José Moyano-Fernández, *Poincaré series for curve singularities and its behaviour under projections*, J. Pure Appl. Algebra **219** (2015), no. 6, 2449–2462, DOI 10.1016/j.jpaa.2014.09.009. MR3299740

[MZ10] J. J. Moyano-Fernández and W. A. Zúñiga-Galindo, *Motivic zeta functions for curve singularities*, Nagoya Math. J. **198** (2010), 47–75, DOI 10.1215/00277630-2009-007. MR2666577

[Mu11] M. Mustaţă, *Zeta functions in algebraic geometry*, Lecture notes for a course in Winter 2011. Available in `http://www-personal.umich.edu/~mmustata/zeta_book.pdf`

[V88] O. Ya. Viro, *Some integral calculus based on Euler characteristic*, Topology and geometry—Rohlin Seminar, Lecture Notes in Math., vol. 1346, Springer, Berlin, 1988, pp. 127–138, DOI 10.1007/BFb0082775. MR970076

[W04] C. T. C. Wall, *Singular points of plane curves*, London Mathematical Society Student Texts, vol. 63, Cambridge University Press, Cambridge, 2004, DOI 10.1017/CBO9780511617560. MR2107253

[Y84] Makoto Yamamoto, *Classification of isolated algebraic singularities by their Alexander polynomials*, Topology **23** (1984), no. 3, 277–287, DOI 10.1016/0040-9383(84)90011-9. MR770564

[Z86] Oscar Zariski, *Le problème des modules pour les branches planes* (French), 2nd ed., Hermann, Paris, 1986. MR861277

Universitat Jaume I, Campus de Riu Sec, Departamento de Matemáticas & Institut Universitari de Matemàtiques i Aplicacions de Castelló, 12071 Castellón de la Plana, Spain

Email address: moyano@uji.es

Contemporary Mathematics
Volume **778**, 2022
https://doi.org/10.1090/conm/778/15655

Introduction to p-adic Igusa zeta functions

Naud Potemans and Willem Veys

ABSTRACT. These notes constitute the content of the series of lectures *Introduction to local zeta functions* by the second author, in the Lluís Santaló Research Summer School 2019: p-Adic Analysis, Arithmetic and Singularities, June 24-28, at the Palacio de la Magdalena in Santander. We want to thank the organizers of the school for their excellent work.

The lectures were intended as an elementary introduction to p-adic Igusa zeta functions and related topics. We hope that these notes reflect that goal. Our text is complementary to the one of León-Cardenal and Zúñiga-Galindo [Rev. Integr. Temas Mat., 37, 2019, 45–76]. The notes of Nicaise [MSJ Mem., 21, 2010, 141–166] are an introduction to p-adic and motivic zeta functions. Substantial survey articles are the 'old' Bourbaki report of Denef [Séminaire Bourbaki, 1990/91, 1992, 359–386] and the more recent paper of Meuser [Amer. J. Math., 138, 2016, 149–179].

CONTENTS

1. Polynomial congruences

1.1. Setting and examples. Consider the classical problem of counting the number of solutions of polynomial congruences. More specifically, fix a polynomial $f \in \mathbb{Z}[x_1, \ldots, x_n]$. Given a positive integer m, we are interested in the number of solutions of f mod m, that is, in the cardinality of the set $\{a \in (\mathbb{Z}/m\mathbb{Z})^n \mid f(a) \equiv$

2020 *Mathematics Subject Classification.* Primary 11S40, 11S80, 14E15; Secondary 11M41, 14G10, 14G20, 14H20.

The authors were partially supported by the Research Foundation Flanders (FWO) project G.0792.18.

$0 \bmod m\} = \{a \in (\mathbb{Z}/m\mathbb{Z})^n \mid f(a) = 0 \text{ in } \mathbb{Z}/m\mathbb{Z}\}$. By the Chinese Remainder Theorem, it suffices to investigate the cases where m is a prime power.

THEOREM 1.1 (Chinese Remainder Theorem). *Let $m = \prod_{j=1}^{r} p_j^{k_j}$ be the decomposition of m as product of primes (where all p_j are different).*

(1) *The map*

$$\frac{\mathbb{Z}}{m\mathbb{Z}} \xrightarrow{\cong} \prod_{j=1}^{r} \frac{\mathbb{Z}}{p_j^{k_j}\mathbb{Z}} : x \bmod m \mapsto (x \bmod p_1^{k_1}, \ldots, x \bmod p_r^{k_r})$$

is an isomorphism of rings.

(2) *Under the induced isomorphism*

$$\left(\frac{\mathbb{Z}}{m\mathbb{Z}}\right)^n \xrightarrow{\cong} \prod_{j=1}^{r} \left(\frac{\mathbb{Z}}{p_j^{k_j}\mathbb{Z}}\right)^n$$

the set $\{a \in (\mathbb{Z}/m\mathbb{Z})^n \mid f(a) \equiv 0 \bmod m\}$ is mapped to $\prod_{j=1}^{r}\{a_j \in (\mathbb{Z}/p_j^{k_j}\mathbb{Z})^n \mid f(a_j) \equiv 0 \bmod p_j^{k_j}\}$.

Given a prime number p, define

$$M_i = M_i(p) := \sharp\left\{a \in \left(\frac{\mathbb{Z}}{p^i\mathbb{Z}}\right)^n \mid f(a) \equiv 0 \bmod p^i\right\}$$

for $i \in \mathbb{Z}_{\geq 0}$ (where $M_0 = 1$). We want to study these values, especially how they vary with i.

From now on, we assume p to be a fixed prime number. All values of M_i are with respect to this prime number p.

EXAMPLE 1.2. Let $f = y - x^2 \in \mathbb{Z}[x, y]$. It is not hard to see that $M_i = p^i$ for all $i \geq 0$.

EXAMPLE 1.3. Let $f = xy \in \mathbb{Z}[x, y]$. We claim that $M_i = (i+1)p^i - ip^{i-1}$. To see this, we consider the following table which contains all solutions exactly once.[1]

x	**y**	number of solutions
$x = 0$	y is free	$1 \cdot p^i$
$x \neq 0$ but $p^{i-1} \mid x$	$p \mid y$	$(p-1) \cdot p^{i-1}$
$p^{i-1} \nmid x$ but $p^{i-2} \mid x$	$p^2 \mid y$	$(p^2 - p) \cdot p^{i-2}$
$\vdots$	$\vdots$	$\vdots$
$p^2 \nmid x$ but $p \mid x$	$p^{i-1} \mid y$	$(p^{i-1} - p^{i-2}) \cdot p$
$p \nmid x$	$y = 0$	$(p^i - p^{i-1}) \cdot 1$

Summing all values in the last column yields the stated result.

[1] Read the table as follows. We make a case distinction for x. Next, we count for each value of x the values of y for which $x \cdot y \equiv 0 \bmod p^i$.

EXAMPLE 1.4. Let $f = y^2 - x^3 \in \mathbb{Z}[x,y]$. It is not hard to see that $M_1 = p$ (consider the parametrization $t \mapsto (t^2, t^3)$, valid over any field). The next values, less 'uniform in i' than in the previous examples, are as follows.

$$
\begin{array}{lll}
M_2 = p(2p-1) & M_6 = p^5(p^2+p-1) & M_8 = p^7(2p^2-1) \\
M_3 = p^2(2p-1) & M_7 = p^6(p^2+p-1) & M_9 = p^8(2p^2-1) \\
M_4 = p^3(2p-1) & & M_{10} = p^9(2p^2-1) \\
M_5 = p^4(2p-1) & & M_{11} = p^{10}(2p^2-1)
\end{array}
$$

EXERCISE 1.5. Let $f = x^2+y^2 \in \mathbb{Z}[x,y]$. Compute the value of M_i for all $i \geq 0$. Hint: make a distinction between the cases $p = 2$, $p \equiv 1 \bmod 4$ and $p \equiv 3 \bmod 4$. In which cases is -1 a square mod p?

REMARK 1.6. In general, it is hard to compute the M_i by hand. Moreover, the examples above are somewhat misleading in the sense that their $M_i(p)$ for fixed i are 'uniform' in p, even of polynomial form. Examples as $f = y^2 - x(x-1)(x-\lambda)$ with $\lambda \neq 0, 1$ and $p \neq 2$ (elliptic curves) already exhibit a more complicated behaviour: here M_1 is of the form $p - 2\sqrt{p}\cos\theta(p)$, where $\theta(p)$ depends in a complicated way on p.

QUESTION 1.7. Is there in general some structure in the values M_i for varying i? Does it suffice to know finitely many such values in order to know all of them?

A partial answer (in easy situations) to Question 1.7 is given by Hensel's Lemma.

LEMMA 1.8 (Hensel's Lemma). *Let $f \in \mathbb{Z}[x_1, \ldots, x_n]$.*

(1) *Let $a \in \mathbb{Z}^n$ such that $f(a) \equiv 0 \bmod p$. If there exists $j \in \{1, \ldots, n\}$ such that $\frac{\partial f}{\partial x_j}(a) \not\equiv 0 \bmod p$, then*

$$\sharp \left\{ x \in \left(\frac{\mathbb{Z}}{p^i\mathbb{Z}}\right)^n \mid f(x) \equiv 0 \bmod p^i \text{ and } x \equiv a \bmod p \right\} = p^{(i-1)(n-1)}.$$

(2) *Hence, if all $a \in \mathbb{Z}^n$ for which $f(a) \equiv 0 \bmod p$ have this property, then $M_i = M_1 \cdot p^{(i-1)(n-1)}$. In particular, all M_i are then determined by M_1.*

REMARK 1.9. Example 1.2 is an illustration of Hensel's Lemma.

1.2. Poincaré series. A typical way to encode and study a countable set of numbers is via their generating series.

DEFINITION 1.10. The *Poincaré series* $P_p(f,T)$ is defined as (a slight adaptation of) the generating series of the values M_i, that is,

$$P(T) = P_p(f,T) := \sum_{i \geq 0} M_i (p^{-n}T)^i = \sum_{i \geq 0} \frac{M_i}{p^{ni}} T^i.$$

The factors p^{ni} are the cardinalities of the sets $(\mathbb{Z}/p^i\mathbb{Z})^n$. So, more precisely, the series $P(T)$ is the generating series of the 'counting measure' of the sets $\{a \in (\mathbb{Z}/p^i\mathbb{Z})^n \mid f(a) \equiv 0 \bmod p^i\}$. This interpretation will pop up naturally in the link with the Igusa zeta function.

EXAMPLE 1.11. Recall Examples 1.2, 1.3 and 1.4.

(1) For $f = y - x^2$, we have

$$P(T) = \sum_{i\geq 0} \frac{M_i}{p^{2i}} T^i = \sum_{i\geq 0} \frac{1}{p^i} T^i = \frac{p}{p-T}.$$

More generally, it is easy to compute the Poincaré series when we can apply Hensel's Lemma. Indeed, in that case $M_i = M_1 \cdot p^{(i-1)(n-1)}$ and we obtain

$$P(T) = \sum_{i\geq 0} M_1 \cdot p^{(i-1)(n-1)} (p^{-n}T)^i = M_1 \cdot \sum_{i\geq 0} p^{-i-n+1} T^i = \frac{M_1}{p^{n-1}} \cdot \frac{p}{p-T}.$$

(2) For $f = xy$, one computes $P(T) = \frac{p^2-T}{(p-T)^2}$.

(3) For $f = y^2 - x^3$, we claim that $P(T) = \frac{p^6+(p^4-p^3)T^2-T^6}{(p-T)(p^5-T^6)}$. This will be discussed later (in Exercise 4.12).

CONJECTURE 1.12 (Borewicz-Šafarevič, 1966 [**BS**]). *The Poincaré series $P(T)$ is a rational function in T.*

1.3. Relation with p-adics. Igusa proved the conjecture above in 1975 [**Ig1**]. His strategy was relating $P(T)$ to a p-adic integral and to use change of variables (geometrically given by a resolution of singularities) to study this integral. As a consequence, Question 1.7 has a positive answer!

REMARK 1.13. The rings $\mathbb{Z}/p^i\mathbb{Z}$ in the polynomial congruences introduced above have a natural link with the ring of p-adic integers $\mathbb{Z}_p$ (and its fraction field $\mathbb{Q}_p$); indeed, $\mathbb{Z}_p = \varprojlim \mathbb{Z}/p^i\mathbb{Z}$ via the natural projections

$$\frac{\mathbb{Z}}{p^{i+1}\mathbb{Z}} \to \frac{\mathbb{Z}}{p^i\mathbb{Z}} : a_0 + a_1 p + \cdots + a_{i-1}p^{i-1} + a_i p^i \mapsto a_0 + a_1 p + \cdots + a_{i-1}p^{i-1},$$

where all $a_\ell \in \{0, \ldots, p-1\}$, inducing also the projections

$$\mathbb{Z}_p \xrightarrow{\pi} \frac{\mathbb{Z}_p}{p^i\mathbb{Z}_p} \cong \frac{\mathbb{Z}}{p^i\mathbb{Z}} \quad \text{and} \quad \mathbb{Z}_p^n \xrightarrow{\pi_n} \left(\frac{\mathbb{Z}_p}{p^i\mathbb{Z}_p}\right)^n \cong \left(\frac{\mathbb{Z}}{p^i\mathbb{Z}}\right)^n.$$

NOTATION 1.14. In the subsequent sections we denote by $\mathrm{ord}_p(\cdot)$ and $|\cdot|_p = p^{-\mathrm{ord}_p(\cdot)}$ the p-order and the standard p-adic norm on $\mathbb{Q}_p$, respectively. That is, writing $z \in \mathbb{Q}_p, z \neq 0$, as $z = p^k u$ with $k \in \mathbb{Z}$ and u a unit in $\mathbb{Z}_p$, we have $\mathrm{ord}_p(z) = k$ and $|z|_p = p^{-\mathrm{ord}_p(z)} = p^{-k}$. Furthermore, $\mathrm{ord}_p(0) = +\infty$ and $|0|_p = 0$.

Also, $|dx| = |dx_1 dx_2 \ldots dx_n|$ denotes the Haar measure on $\mathbb{Q}_p^n$, normalized such that $|dx|(\mathbb{Z}_p^n) = 1$.

REMARK 1.15. Consider $C \subseteq (\mathbb{Z}_p/p^i\mathbb{Z}_p)^n \cong (\mathbb{Z}/p^i\mathbb{Z})^n$. The inverse image $\pi_n^{-1}(C)$ lives in $\mathbb{Z}_p^n$ and is called a *cylindrical set* or *cylinder*. Then

$$|dx|(\pi_n^{-1}(C)) = \frac{\sharp C}{p^{in}}.$$

Indeed, taking for each $\bar{a} \in C$ a fixed representative $a \in \mathbb{Z}_p^n$, the set $\pi_n^{-1}(C)$ is the disjoint union of the sets $a + (p^i\mathbb{Z}_p)^n, \bar{a} \in C$, all with measure p^{-in}.

2. p-adic Igusa (local) zeta functions

2.1. Definition and examples.

DEFINITION 2.1. Consider $f \in \mathbb{Q}_p[x_1, \ldots, x_n]$ and $s \in \mathbb{C}$ with $\Re(s) > 0$. The *Igusa zeta function* of f is defined as

$$Z(s) = Z_p(f; s) := \int_{\mathbb{Z}_p^n} |f(x)|_p^s \, |dx|.$$

More generally, one can integrate over 'balls' $c + (p^e\mathbb{Z}_p)^n$ or $c + (\prod_{i=1}^n p^{e_i}\mathbb{Z}_p)$ for some $c \in \mathbb{Q}_p^n$, or even consider

$$\int_{\mathbb{Q}_p^n} \varphi(x)|f(x)|_p^s \, |dx|,$$

where φ is a *test function*, i.e., a locally constant function with compact support. Observe that $Z(s)$ is in fact this last integral with φ equal to the characteristic function of $\mathbb{Z}_p^n$.

REMARK 2.2. Write $s = a + bi$ with $a, b \in \mathbb{R}$. Then

$$Z(s) = \int_{\mathbb{Z}_p^n} |f(x)|_p^a \, |f(x)|_p^{bi} \, |dx|.$$

Observe that $|f(x)|_p^{bi}$ has modulus 1. Moreover, since f is a continuous function and $\mathbb{Z}_p^n$ is compact, $|f(x)|_p^a$ is bounded for $a > 0$. Hence, $Z(s)$ is indeed well-defined if $\Re(s) > 0$.

At this point, it is also not difficult to show that $Z(s)$ is a holomorphic function in the domain $\Re(s) > 0$. We will prove a stronger result in Theorem 3.7.

REMARK 2.3. The terminology in the literature is sometimes confusing. One uses the adjectives *p-adic*, *Igusa* and/or *local* in all possible combinations to indicate the zeta function $Z(s)$. Here 'local' refers to the fact that $\mathbb{Q}_p$ is a local field.

However, when it is clear that one works in the p-adic setting, the term local zeta function could also mean the integral $\int_{(p\mathbb{Z}_p)^n} |f(x)|_p^s \, |dx|$ (or, more generally, $\int_{(p^e\mathbb{Z}_p)^n} |f(x)|_p^s \, |dx|$ with $e \geq 1$), where now 'local' refers to the fact that the integration domain is a smaller neighbourhood of the origin.

REMARK 2.4. In the sequel we will also consider such integrals for some $s \in \mathbb{C}$ with $\Re(s) \leq 0$. Strictly speaking, we then integrate over $\{x \in \mathbb{Z}_p^n \mid f(x) \neq 0\}$. But, since the excluded set $\{x \in \mathbb{Z}_p^n \mid f(x) = 0\}$ has measure zero, we keep notation as before.

The following example shows that this integral might be ill-defined if $\Re(s) \leq 0$.

EXAMPLE 2.5. Let $f = x \in \mathbb{Q}_p[x]$ and $s = -1 \in \mathbb{C}$. In order to compute $\int_{\mathbb{Z}_p} |x|^{-1}|dx|$, we consider the partition

$$\begin{aligned}\mathbb{Z}_p \setminus \{0\} &= \{x \in \mathbb{Q}_p \mid \exists l \geq 0 \text{ such that } |x|_p = p^{-l}\} \\ &= \bigsqcup_{l \geq 0} \{x \in \mathbb{Q}_p \mid |x|_p = p^{-l}\}.\end{aligned}$$

Denote $\{x \in \mathbb{Q}_p \mid |x|_p = p^{-l}\}$ by A_l and observe that

$$|dx|(A_l) = |dx| \left(\{x \in \mathbb{Q}_p \mid |x|_p \leq p^{-l}\} \setminus \{x \in \mathbb{Q}_p \mid |x|_p \leq p^{-l-1}\}\right) = p^{-l} - p^{-l-1}.$$

Therefore, since $\{0\}$ has measure zero, we have

$$\begin{aligned}\int_{\mathbb{Z}_p} |x|_p^{-1}|dx| &= \sum_{l\geq 0}\int_{A_l} |x|_p^{-1}|dx| = \sum_{l\geq 0}\int_{A_l} p^l\,|dx| = \sum_{l\geq 0} p^l \int_{A_l} |dx| \\ &= \sum_{l\geq 0} p^l(p^{-l}-p^{-l-1}) = \sum_{l\geq 0}(1-p^{-1}) = +\infty.\end{aligned}$$

EXERCISE 2.6. On the other hand, $Z(s)$ can be well defined in a larger half-plane than $\Re(s) > 0$. Verify the following basic examples; they will be important later on.

(1) $\int_{\mathbb{Z}_p} |x|_p^s\,|dx| = \frac{1-p^{-1}}{1-p^{-(s+1)}}$ when $\Re(s) > -1$.
(2) More generally, $\int_{p^e\mathbb{Z}_p} |x|_p^s\,|dx| = \frac{(1-p^{-1})p^{-e(s+1)}}{1-p^{-(s+1)}}$ when $\Re(s) > -1$.
(3) And still more generally (for positive integers N and ν),

$$\int_{p^e\mathbb{Z}_p} |x|_p^{Ns+\nu-1}|dx| = \frac{(1-p^{-1})p^{-e(Ns+\nu)}}{1-p^{-(Ns+\nu)}}$$

when $\Re(s) > -\frac{\nu}{N}$.

REMARK 2.7. We have

$$\begin{aligned}Z(s) = Z_p(f;s) &= \int_{\mathbb{Z}_p^n} |f(x)|_p^s\,|dx| = \int_{\mathbb{Z}_p^n} \left(p^{-\mathrm{ord}_p(f(x))}\right)^s |dx| \\ &= \int_{\mathbb{Z}_p^n} p^{-s\cdot\mathrm{ord}_p(f(x))}|dx| = \int_{\mathbb{Z}_p^n} \left(p^{-s}\right)^{\mathrm{ord}_p(f(x))} |dx|.\end{aligned}$$

Hence, the Igusa zeta function $Z(s)$ is actually a function of p^{-s}. From now on we will write $t = p^{-s}$ and $Z'(t) = Z'(p^{-s}) := Z(s)$. The change of coordinates is given by the map $\psi : \mathbb{C} \to \mathbb{C}^* : s \mapsto t = p^{-s}$.

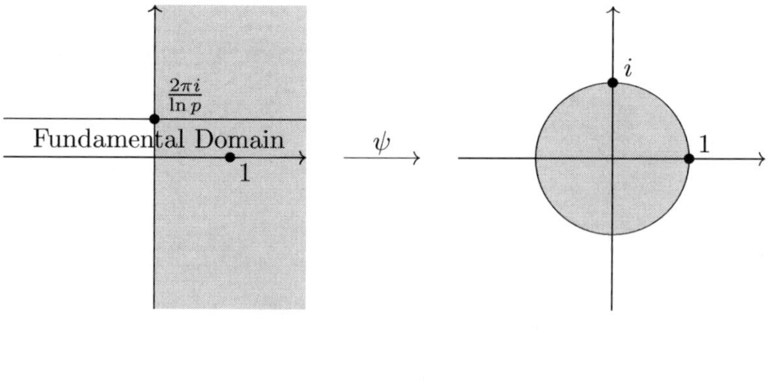

The relations between the parameters s and t are given in the following table.

parameter s	**parameter $\mathbf{t} = \psi(\mathbf{s}) = \mathbf{p}^{-\mathbf{s}}$**
$\Re(s) > 0$	$0 < \lvert\psi(s)\rvert < 1$
$\Re(s) = 0$	$\lvert\psi(s)\rvert = 1$
$\Re(s) < 0$	$\lvert\psi(s)\rvert > 1$
$s - s' \in \frac{2\pi i}{\ln p}\mathbb{Z}$	$\psi(s) = \psi(s')$

2.2. Link with Poincaré series.

PROPOSITION 2.8. *Let $f \in \mathbb{Z}_p[x_1, \dots, x_n]$. Consider its Poincaré series $P(T)$ as given in Definition* 1.10. *Then*

$$Z(s) = \frac{(p^{-s} - 1)P(p^{-s}) + 1}{p^{-s}}$$

or, equivalently,

$$P(t) = \frac{tZ'(t) - 1}{t - 1}.$$

PROOF. Rewrite the Igusa zeta function in the following way:

$$Z(s) = \int_{\mathbb{Z}_p^n} |f(x)|_p^s \, |dx| = \sum_{l \geq 0} \int_{\mathrm{ord}_p f(x) = l} p^{-ls} |dx| = \sum_{l \geq 0} p^{-ls} \int_{\mathrm{ord}_p f(x) = l} |dx|$$

$$= \sum_{l \geq 0} p^{-ls} \Big(|dx|(\{x \in \mathbb{Z}_p^n \mid \mathrm{ord}_p f(x) \geq l\}) - |dx|(\{x \in \mathbb{Z}_p^n \mid \mathrm{ord}_p f(x) \geq l + 1\}) \Big).$$

Recall the projection (see Remark 1.13)

$$\mathbb{Z}_p^n \xrightarrow{\pi_n} \left(\frac{\mathbb{Z}}{p^l \mathbb{Z}} \right)^n : x \mapsto x \bmod p^l$$

and observe that $\{x \in \mathbb{Z}_p^n \mid \mathrm{ord}_p(f(x)) \geq l\} = \pi_n^{-1}(\{x \in (\mathbb{Z}/p^l\mathbb{Z})^n \mid f(x) \equiv 0 \bmod p^l\})$. By Remark 1.15 we see that

$$|dx| \left(\pi_n^{-1} \left(\left\{ x \in \left(\frac{\mathbb{Z}}{p^l \mathbb{Z}} \right)^n \mid f(x) \equiv 0 \bmod p^l \right\} \right) \right) = \frac{M_l}{p^{ln}}.$$

Hence

$$Z(s) = \sum_{l \geq 0} p^{-ls} \left(\frac{M_l}{p^{ln}} - \frac{M_{l+1}}{p^{(l+1)n}} \right) = \sum_{l \geq 0} p^{-ls - ln} M_l - \sum_{l \geq 0} p^{-ls - (l+1)n} M_{l+1}$$

$$= \sum_{l \geq 0} (p^{-s} p^{-n})^l M_l - \sum_{l \geq 1} p^{-(l-1)s - ln} M_l = P(p^{-s}) - p^s \sum_{l \geq 1} (p^{-s} p^{-n})^l M_l$$

$$= P(p^{-s}) - p^s (P(p^{-s}) - 1) = \frac{(p^{-s} - 1)P(p^{-s}) + 1}{p^{-s}}.$$

The other equality follows immediately. □

2.3. Igusa's stationary phase formula. A possible strategy to compute p-adic integrals –which can be practical or (in)famously hard– is Igusa's Stationary Phase Formula. The starting point of this formula is the obvious equality

$$\int_{\mathbb{Z}_p^n} |f(x)|_p^s \,|dx| = \sum_{a \in (\mathbb{Z}/p\mathbb{Z})^n} \int_{a+(p\mathbb{Z}_p)^n} |f(x)|_p^s \,|dx|, \tag{2.1}$$

where we use the same notation for $a \in (\mathbb{Z}/p\mathbb{Z})^n$ and a chosen representative $a \in \mathbb{Z}_p^n$. We denote $\int_{a+(p\mathbb{Z}_p)^n} |f(x)|_p^s \,|dx|$ by I_a. We partition the values of a as follows:

(1) a such that $f(a) \not\equiv 0 \bmod p$,
(2) a such that $f(a) \equiv 0 \bmod p$ and there exists $j \in \{1, \dots, n\}$ such that $\frac{\partial f}{\partial x_j}(a) \not\equiv 0 \bmod p$,
(3) a such that $f(a) \equiv 0 \bmod p$ and for all $j \in \{1, \dots, n\} : \frac{\partial f}{\partial x_j}(a) \equiv 0 \bmod p$.

The first two cases are easy to compute:

(1) $I_a = \int_{a+(p\mathbb{Z}_p)^n} 1^s |dx| = |dx|\,(a + (p\mathbb{Z}_p)^n) = p^{-n}$,
(2)
$$\begin{aligned} I_a &= \int_{a+(p\mathbb{Z}_p)^n} |f(x)|_p^s \,|dx| \\ &= \sum_{l \geq 0} p^{-ls} |dx|\,(\{x \in a + (p\mathbb{Z}_p)^n \mid \mathrm{ord}_p f(x) = l\}) \\ &= \sum_{l \geq 1} p^{-ls} |dx|\,(\{x \in a + (p\mathbb{Z}_p)^n \mid \mathrm{ord}_p f(x) = l\}) \\ &= \sum_{l \geq 1} p^{-ls} \left(\frac{M_l}{p^{ln}} - \frac{M_{l+1}}{p^{(l+1)n}} \right) \\ &= \sum_{l \geq 1} p^{-ls} \left(\frac{p^{(n-1)(l-1)}}{p^{ln}} - \frac{p^{(n-1)l}}{p^{(l+1)n}} \right) \\ &= (p-1)p^{-n} \sum_{l \geq 1} \left(p^{-s-1}\right)^l \\ &= (p-1)p^{-n} \frac{p^{-s-1}}{1 - p^{-s-1}}. \end{aligned}$$

For the fourth equality, we used a result that is explained in the proof of Proposition 2.8, and for the fifth one, we used Hensel's Lemma (Lemma 1.8).

THEOREM 2.9 (Igusa's Stationary Phase Formula, 1994 [**Ig3**]). *Define*

$$V := \left\{ a \in \left(\frac{\mathbb{Z}}{p\mathbb{Z}} \right)^n \mid f(a) \equiv 0 \bmod p \right\},$$

$$S := \left\{ a \in V \mid \frac{\partial f}{\partial x_j}(a) \equiv 0 \bmod p \text{ for all } j \in \{1, \dots, n\} \right\}.$$

Then

$$Z(s) = \frac{p^n - \sharp V}{p^n} + (\sharp V - \sharp S) \frac{p-1}{p^n} \frac{p^{-s-1}}{1 - p^{-s-1}} + \sum_{a \in S} \int_{a+(p\mathbb{Z}_p)^n} |f(x)|_p^s \,|dx|.$$

PROOF. Use equation (2.1) and split the sum according to the three cases discussed above. Next, use the previous computations for the cases (1) and (2). This yields the formula. □

EXAMPLE 2.10. We compute the integral $I := \int_{\mathbb{Z}_p^3} |xy-z^2|_p^s\,|dxdydz|$ (for $p \neq 2$) using Theorem 2.9. First, remark that $\sharp V = p^2$. Indeed, if $x = 0$, then y is free and $z = 0$. If $x \neq 0$, then $y = x^{-1}z^2$ and hence x is free (non-zero), z is free and y is fixed. Therefore, $\sharp V = 1 \cdot p \cdot 1 + (p-1) \cdot 1 \cdot p = p^2$. Remark that $\sharp S = 1$ since $S = \{(0,0,0)\}$. Using Igusa's Stationary Phase Formula, we have

$$I = \frac{p^3 - p^2}{p^3} + (p^2-1)\frac{p-1}{p^3}\frac{p^{-s-1}}{1-p^{-s-1}} + \int_{(p\mathbb{Z}_p)^3} |xy - z^2|_p^s\,|dxdydz|.$$

Since $xy - z^2$ is a homogeneous polynomial of degree two, we can rewrite the last integral, using the change of coordinates $(x,y,z) = (px', py', pz')$, as

$$\int_{\mathbb{Z}_p^3} |p^2|_p^s\,|x'y' - (z')^2|_p^s\,|p^3|_p\,|dx'dy'dz'| = p^{-2s-3}I,$$

and thus

$$I = \frac{p^3 - p^2}{p^3} + (p^2-1)\frac{p-1}{p^3}\frac{p^{-s-1}}{1-p^{-s-1}} + p^{-2s-3}I.$$

An easy calculation then yields

$$I = \frac{p-1}{p} \cdot \frac{1-p^{-s-3}}{(1-p^{-s-1})(1-p^{-2s-3})}.$$

There are some intriguing open problems concerning this formula.

(1) Prove for arbitrary f that $Z(s)$ is rational in p^{-s} using Igusa's Stationary Phase Formula. Recall that this result was already proven by Igusa using other techniques [**Ig1**], see Theorem 3.7.
(2) Show that one can compute $Z(s)$ explicitly for arbitrary f using finitely many iterations of Igusa's Stationary Phase Formula.

These two problems are probably strongly related, but we are not aware of a clear strategy to attack them. In concrete examples, repeated application of Theorem 2.9 often results in an explicit expression for $Z(s)$, which is a rational function in p^{-s}. Typically such an expression arises in two possible ways. Finitely many iterations either yield immediately a concrete expression, or at some point the original integral pops up again (one or more times) as in Example 2.10, and then rewriting the resulting relation yields a concrete expression.

Can this experimental fact lead to a theorem for general f?

EXERCISE 2.11. Compute in some way $\int_{\mathbb{Z}_p^2} |x^2 + y^2|_p^s\,|dxdy|$ (depending on p, there are three cases). Verify that your findings are consistent with Exercise 1.5, using Proposition 2.8.

3. p-adic manifolds and rationality of the zeta function

3.1. p-adic manifolds. We only state briefly the essential features of p-adic manifolds, and refer for a detailed explanation to [**Ig4**, Section 2.4]. Anyway, this notion is analogous to classical real or complex analytic manifolds, defined via appropriate atlases.

DEFINITION 3.1 (Informal). An n-dimensional $\mathbb{Q}_p$-*analytic manifold* or p-*adic manifold* is a Hausdorff topological space that is locally (analytically) of the form $c + (p^e\mathbb{Z}_p)^n$ for some $c \in \mathbb{Q}_p^n$.

REMARK 3.2. A special and crucial feature in the p-adic case is that any n-dimensional p-adic manifold can be covered by (two by two) *disjoint* balls as above.

We want to mention the following remarkable structure theorem due to Serre (although it will not be used in the sequel).

THEOREM 3.3 (Serre, 1965 [**Se**]). *Let X be a compact n-dimensional p-adic manifold. Then X is isomorphic (as a p-adic manifold) to the disjoint union of r times $\mathbb{Z}_p^n$, for a unique $r \in \{1, \ldots, p-1\}$.*

PART OF THE PROOF. For the existence of such r, observe that X is compact and thus a disjoint union of finitely many sets of the form $c + (p^e\mathbb{Z}_p)^n$ by Remark 3.2. That is, $X = \bigsqcup_{i=1}^k c_i + (p^{e_i}\mathbb{Z}_p)^n$ for some positive k.

Moreover, we have the following isomorphism of p-adic manifolds:

$$c + (p^e\mathbb{Z}_p)^n \xrightarrow{\cong} Z_p^n : c + p^e y \mapsto y.$$

Therefore, X is isomorphic to $\bigsqcup_{i=1}^k \mathbb{Z}_p^n$. Recall now that

$$\mathbb{Z}_p = p\mathbb{Z}_p \sqcup (1 + p\mathbb{Z}_p) \sqcup \cdots \sqcup ((p-1) + p\mathbb{Z}_p) \cong \bigsqcup_{j=1}^{p} \mathbb{Z}_p.$$

Similarly, we have that $\mathbb{Z}_p^n$ is isomorphic (as a p-adic manifold) to $\bigsqcup_{j=1}^p \mathbb{Z}_p^n$. (Indeed, $\mathbb{Z}_p^n \cong \mathbb{Z}_p^{n-1} \times \mathbb{Z}_p \cong \mathbb{Z}_p^{n-1} \times \bigsqcup_{j=1}^p \mathbb{Z}_p \cong \bigsqcup_{j=1}^p \mathbb{Z}_p^n$.)

Hence, if k is larger than $p-1$ in $\bigsqcup_{i=1}^k \mathbb{Z}_p^n$, we can reduce k by $p-1$ using the isomorphism above. Doing this until we obtain $1 \leq k \leq p-1$ yields the existence of r. □

For the proof of uniqueness of r we refer to [**Se**]. Nowadays r is called the (p-adic) Serre invariant of X.

REMARK 3.4. On an n-dimensional p-adic manifold, one associates a measure to a given $\mathbb{Q}_p$-analytic differential form of degree n, see e.g. [**Ig3**, Section 7.4] or [**Se**]. For example, the Haar measure $|dx|$ on $\mathbb{Q}_p^n$ is associated to the differential form $dx_1 \wedge \cdots \wedge dx_n$. In what follows, we will use this correspondence.

3.2. Resolution: analytic version. The statement below on embedded resolution follows essentially from Hironaka's work [**Hi**]. See also [**BEV**, Section 8], [**ENV**, Section 5], [**Wł2**], [**Wł3**], and especially [**DvdD**, Theorem 2.2] for this p-adic setting.

THEOREM 3.5 (Embedded Resolution of Singularities; p-adic analytic version). *Let $f \in \mathbb{Q}_p[x_1, \ldots, x_n] \setminus \mathbb{Q}_p$ and consider its zero locus $\{f = 0\} \subseteq \mathbb{Q}_p^n$. Then there exists an n-dimensional p-adic manifold Y, a proper $\mathbb{Q}_p$-analytic map $h : Y \to \mathbb{Q}_p^n$, and finitely many closed $(n-1)$-dimensional submanifolds $E_i, i \in I$, of Y, equipped with* numerical data $(N_i, \nu_i) \in (\mathbb{Z}_{>0})^2, i \in I$, *satisfying*

(1) *h is an isomorphism of p-adic manifolds outside the inverse image by h of the singular locus of $\{f = 0\}$,*
(2) *$h^{-1}(\{f = 0\}) = \cup_{i\in I} E_i$,*
(3) *for all $b \in Y$ there exist local coordinates $(y_1, \ldots, y_n)$ around b such that, if $b \in E_1, \ldots, E_r$ (relabel if necessary), then $y_i = 0$ is the equation of E_i*

for $1 \leq i \leq r$ *(i.e.,* $h^{-1}(\{f=0\})$ *is a normal crossings divisor), and such that*

$$f \circ h = \varepsilon(y_1, \ldots, y_n) \prod_{i=1}^{r} y_i^{N_i}, \quad \textit{and}$$

$$h^*(dx_1 \wedge \cdots \wedge dx_n) = \eta(y_1, \ldots, y_n) \prod_{i=1}^{r} y_i^{\nu_i - 1} dy_1 \wedge \cdots \wedge dy_n$$

on some neighbourhood of b*, where* $\varepsilon(y_1, \ldots, y_n)$ *and* $\eta(y_1, \ldots, y_n)$ *are invertible.*

Such a map h *is called an embedded resolution of* f*.*

REMARK 3.6. It is possible to construct the map h as a composition of finitely many *admissible* blow-ups. More precisely, each centre of blow-up is a closed submanifold C of codimension at least 2 in the ambient manifold, having normal crossings with the exceptional components created by previous blow-ups. That is, there are local coordinates $z_1, \ldots, z_n$ around each point c of C, such that each previously created exceptional component containing c is given by $z_i = 0$ for some $i \in \{1, \ldots, n\}$, and C is given by $z_{i_1} = \cdots = z_{i_d} = 0$ with $d \geq 2$ and $\{i_1, \ldots, i_d\} \subseteq \{1, \ldots, n\}$.

3.3. Rationality and candidate poles.

THEOREM 3.7 (Igusa, 1974 [**Ig1**]). *Consider* $f \in \mathbb{Q}_p[x_1, \ldots, x_n]$*. Then the following statements hold.*

(1) *The Igusa zeta function* $Z(s)$ *is a rational function in* $p^{-s} = t$*. Therefore, it has a meromorphic continuation to the whole of* $\mathbb{C}$*.*
(2) *Fix an embedded resolution* h *of* f *and use notation as in Theorem 3.5. Then*

$$Z(s) = \frac{Q(p^{-s})}{\prod_{j \in I}(1 - p^{-(N_j s + \nu_j)})},$$

or, equivalently,

$$Z'(t) = \frac{Q(t)}{\prod_{j \in I}(1 - p^{-\nu_j} t^{N_j})},$$

where Q *is a polynomial with rational coefficients.*
(3) *The poles of* $Z(s)$ *or* $Z'(t)$ *have order at most* n*.*

In particular, the poles of $Z(s)$ *are contained in the list*

$$-\frac{\nu_j}{N_j} + \frac{2\pi k}{(\ln p) N_j} i, \; j \in I \textit{ and } k \in \mathbb{Z},$$

or, equivalently, the poles of $Z'(t)$ *are contained in the list*

$$\xi p^{\nu_j / N_j}, \; j \in I \textit{ and } \xi^{N_j} = 1.$$

PROOF. We use a fixed embedded resolution $h : Y \to \mathbb{Q}_p^n$ to perform a classical change of variables computation (for a proof in the p-adic setting, see e.g. [**Ig4**,

Proposition 7.4.1]):

$$\begin{aligned} Z(s) &= \int_{\mathbb{Z}_p^n} |f(x)|_p^s \, |dx| \\ &= \int_{h^{-1}(\mathbb{Z}_p^n)} |(f \circ h)(y)|_p^s \, |\operatorname{Jac} h(y)|_p \, |dy| \\ &= \int_{h^{-1}(\mathbb{Z}_p^n)} |(f \circ h)(y)|_p^s \, |h^* dx|. \end{aligned}$$

Observe that the p-adic manifold $h^{-1}(\mathbb{Z}_p^n)$ is compact because h is proper. Therefore, we can write $h^{-1}(\mathbb{Z}_p^n)$ as a disjoint union of finitely many sufficiently small (open and compact) 'balls' B, that can be considered as neighbourhoods of points b as in Theorem 3.5. We can take those balls of the form $(p^e Z_p)^n$, but for concrete computations it is sometimes useful to allow more flexibility and to take balls B of the form $\prod_{j=1}^n p^{e_j}\mathbb{Z}_p$. We compute the integral over such a fixed ball/neighbourhood:

$$\int_B \left| \varepsilon(y) \prod_{j=1}^r y_j^{N_j} \right|_p^s \left| \eta(y) \prod_{j=1}^r y_j^{\nu_j - 1} \right|_p |dy| = \int_B |\varepsilon(y)|_p^s \, |\eta(y)|_p \left| \prod_{j=1}^r y_j^{N_j s + \nu_j - 1} \right|_p |dy|.$$

We may suppose (by taking the ball small enough) that $|\varepsilon(y)|_p$ and $|\eta(y)|_p$ are constant. Indeed, ε and η (and their norms) are continuous functions and their value at b is non-zero. Hence, since locally the p-adic norms $|\varepsilon(y)|_p$ and $|\eta(y)|_p$ take discrete values (integer powers of p), they must be constant on a sufficiently small neighbourhood. Then we can compute the integral above by separating variables as

$$|\varepsilon(b)|_p^s \, |\eta(b)|_p \prod_{j=1}^r \int_{p^{e_j} Z_p} |y_j|_p^{N_j s + \nu_j - 1} |dy_j| \prod_{j=r+1}^n \int_{p^{e_j} Z_p} |dy_j|.$$

Since $|\varepsilon(b)|$, $|\eta(b)|$ and $\prod_{j=r+1}^n \int_{p^{e_j} Z_p} |dy_j|$ are all constants (integer powers of p), the integral is equal to the product of a constant, an integer power of p^{-s}, and the expression

$$\prod_{j=1}^r \frac{(1 - p^{-1}) p^{-e_j (N_j s + \nu_j)}}{1 - p^{-(N_j s + \nu_j)}},$$

using Exercise 2.6. Summing these contributions over all balls B, and putting everything on a common denominator, we obtain that $Z(s)$ is a rational function in p^{-s} of the described form.

Moreover, it is clear that all contributions have poles of order at most n. □

REMARK 3.8. For a polynomial $f \in \mathbb{R}[x_1, \ldots, x_n]$ and $s \in \mathbb{C}$ with $\Re(s) > 0$, one classically studies integrals

$$\int_{\mathbb{R}^n} \varphi(x) |f(x)|^s \, |dx|,$$

analogous to those in Definition 2.1, where now the test function φ is a C^∞ function with compact support and $|dx|$ the Lebesgue measure on $\mathbb{R}^n$ (and similarly for a complex polynomial). It is again easy to verify that such an integral is a holomorphic function of s in the domain $\Re(s) > 0$. I. Gel'fand asked in 1954 whether there exists a meromorphic continuation of this integral to the whole of

$\mathbb{C}$. Independently, Bernstein and S. Gel'fand [**BG**] in 1969 and Atiyah [**At**] in 1970 proved this using resolution of singularities. Alternatively, Bernstein [**Be**] showed it using Bernstein-Sato polynomials in 1972. Note that both techniques were not yet available in 1954.

REMARK 3.9. The possible poles of $Z(s)$, namely $\{-\frac{\nu_j}{N_j} + \frac{2\pi k}{(\ln p)N_j} i \mid j \in I, k \in \mathbb{Z}\}$, are called the candidate poles (associated to the resolution h). Intriguingly, in general most of the candidate poles turn out to be no poles of the Igusa zeta function (this is why elements of this set are called *candidate* poles). The reason for this behaviour is not yet understood in full generality. When $n = 2$, we will determine the actual poles in Section 6.

In general, it is quite difficult to obtain an explicit formula for concrete examples via the technique in the proof of Theorem 3.7. We sketch an easy case where this is doable.

EXAMPLE 3.10. Consider $Z(s) = \int_{\mathbb{Z}_p^2} |xy(x-y)|_p^s \, |dxdy|$. As embedded resolution $h : Y \to \mathbb{Q}_p^2$ of $xy(x-y)$ we take the blow-up at the origin.

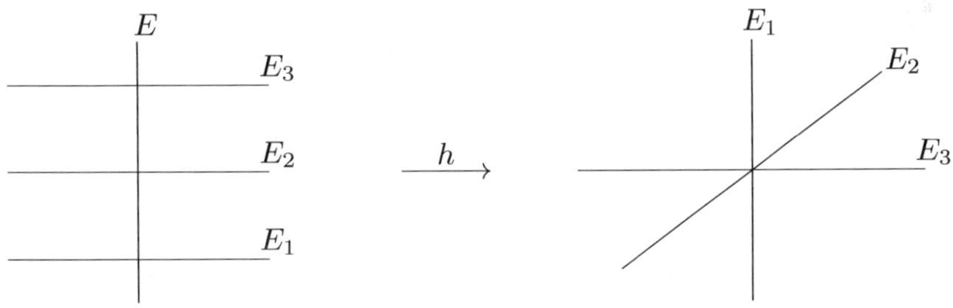

Then Y is covered by two affine charts, both isomorphic to $\mathbb{Q}_p^2$. The restrictions h_1 and h_2 of h to those charts can be described in affine coordinates by

$$h_1 : (x, u) \mapsto (x, y = ux) \qquad \text{and} \qquad h_2 : (y, v) \mapsto (x = vy, y),$$

respectively. The exceptional curve E is a projective line with equation $x = 0$ and $y = 0$ in the first and second chart, respectively.

A crucial point is an adequate description of $h^{-1}(\mathbb{Z}_p^2)$. Writing for example

$$\mathbb{Z}_p^2 = \{(x, y) \mid \text{ord}_p y \geq \text{ord}_p x \geq 0\} \sqcup \{(x, y) \mid 0 \leq \text{ord}_p y < \text{ord}_p x\},$$

we can interpret $h^{-1}(\mathbb{Z}_p^2)$ as the disjoint union of $\mathbb{Z}_p^2$ (with coordinates (x, u) in the first chart), and $\mathbb{Z}_p \times p\mathbb{Z}_p$ (with coordinates (y, v) in the second chart). Then, using change of variables with respect to h, we obtain that

$$Z(s) = \int_{\mathbb{Z}_p^2} |x^3 u(1-u)|_p^s \, |x|_p \, |dxdu| + \int_{\mathbb{Z}_p \times p\mathbb{Z}_p} |y^3 v(v-1)|_p^s \, |y|_p \, |dydv|.$$

To compute the first integral, use separation of variables:

$$\int_{\mathbb{Z}_p^2} |x^3 u(1-u)|_p^s \, |x|_p \, |dxdu| = \int_{\mathbb{Z}_p} |x|_p^{3s+1} \, |dx| \cdot \int_{\mathbb{Z}_p} |u(1-u)|_p^s \, |du|.$$

The first factor can be computed using Exercise 2.6, the second one by doing a straightforward computation or using Theorem 2.9. Similarly, one can compute the second integral. We leave this as an exercise. Finally, we obtain

$$Z(s) = \frac{(p-1)p^{3s}(-1+2p-2p^{s+1}+p^{s+2})}{(p^{s+1}-1)(p^{3s+2}-1)}.$$

Remark that the numerical data associated to the exceptional divisor E and to the strict transforms E_1, E_2 and E_3 are given by $(3,2)$ and $(1,1)$, respectively.

EXERCISE 3.11. Verify the result in Example 3.10 using Igusa's Stationary Phase Formula.

REMARK 3.12. As a corollary of Theorem 3.7, the Igusa zeta function $Z(s)$ always converges on a larger part of $\mathbb{C}$ than $\Re(s) > 0$, namely for all $s \in \mathbb{C}$ such that $\Re(s) > -\min_{i\in I} \frac{\nu_i}{N_i}$. The value $\min_{i\in I} \frac{\nu_i}{N_i}$ is an important invariant, sometimes called the *p-adic log canonical threshold* of f.

When working with polynomials over $\mathbb{C}$ (or any algebraically closed field of characteristic zero) and their embedded resolutions, the analogous minimum is a famous singularity invariant, called the *(geometric) log canonical threshold.* Note that the geometric one can be smaller than the p-adic one, due to singular points that are not defined over $\mathbb{Q}_p$. Take for example the polynomial $f = y^2 - x^3(x^2+1)^5$. Its zero locus over $\mathbb{C}$ contains the three singular points $(0,0)$, $(i,0)$ and $(-i,0)$. It is not difficult to verify that the last two induce the log canonical threshold $\frac{7}{10}$ of f. However, for $p \equiv 3 \bmod 4$, the origin is the only singular point of the zero locus of f over $\mathbb{Q}_p$, inducing the p-adic log canonical threshold $\frac{5}{6}$ of f.

THEOREM 3.13 ([**Ig2**], [**VZ1**], [**VZ2**]). *Let $f \in \mathbb{Q}_p[x_1, \dots, x_n]$. Fix an embedded resolution of f and use notation as in Theorem 3.5. Write $\sigma := \min_{i\in I} \frac{\nu_i}{N_i}$. Then σ does not depend on the chosen resolution and $-\sigma$ is always a pole of $Z(s)$.*

4. Denef's formula

There is an explicit formula, due to Denef, in terms of an algebro-geometric version of embedded resolution. This formula is however only valid for all but finitely many p. We need several preparations to state this result. We first mention such a version of embedded resolution, valid over an arbitrary field K of characteristic zero (not necessarily algebraically closed). In the formulation below all varieties and morphisms are defined over K and such a variety is called irreducible if it is irreducible as K-variety.

4.1. Resolution: algebro-geometric version.

THEOREM 4.1 (Embedded Resolution of Singularities; algebro-geometric version [**Hi**]). *Let K be a field of characteristic 0 and $f \in K[x_1, \dots, x_n] \setminus K$. Then there exists an n-dimensional smooth and irreducible variety Y, a proper birational morphism $h : Y \to \mathbb{A}^n_K$, and finitely many closed $(n-1)$-dimensional smooth and irreducible subvarieties $E_i, i \in I$, of Y, equipped with* numerical data *$(N_i, \nu_i) \in (\mathbb{Z}_{>0})^2, i \in I$, satisfying*

(1) *h is an isomorphism of algebraic varieties outside the inverse image by h of the singular locus of $\{f = 0\}$,*
(2) *$h^{-1}(\{f=0\}) = \cup_{i\in I} E_i$,*

(3) $h^{-1}(\{f=0\})$ *is a normal crossings divisor, and*

$$\operatorname{div}(f\circ h)=h^*\operatorname{div}(f)=\sum_{i\in I}N_iE_i \quad \textit{and} \quad K_{Y/\mathbb{A}^n_K}=\operatorname{div}(\operatorname{Jac}h)=\sum_{i\in I}(\nu_i-1)E_i,$$

where $K_{Y/\mathbb{A}^n_K}=K_Y-h^*K_{\mathbb{A}^n_K}$ *denotes the relative canonical divisor of the map* h*, which in this case is just the canonical divisor* K_Y*, since* $K_{\mathbb{A}^n_K}=0$.

Note that writing the last part in local coordinates is compatible with the last part of Theorem 3.5.

REMARK 4.2. It is again possible to construct the map h as a composition of finitely many *admissible* blow-ups at closed smooth and irreducible centres (all defined over K), i.e., each centre has normal crossings with the previously created exceptional components.

4.2. Reduction mod p**.** Let $f\in\mathbb{Z}[x_1,\dots,x_n]\setminus p\mathbb{Z}[x_1,\dots,x_n]$. Writing f explicitly as $f=\sum_{j_1,\dots,j_n}a_{j_1,\dots,j_n}x_1^{j_1}\dots x_n^{j_n}$, the reduction mod p of f is

$$\bar{f}=f \bmod p:=\sum_{j_1,\dots,j_n}\overline{a_{j_1,\dots,j_n}}x_1^{j_1}\dots x_n^{j_n}\in\mathbb{F}_p[x_1,\dots,x_n],$$

where $\overline{a_{j_1,\dots,j_n}}$ is the class of $a_{j_1,\dots,j_n}\in\mathbb{Z}$ in $\mathbb{Z}/p\mathbb{Z}=\mathbb{F}_p$. Similarly, one defines the reduction mod p of $f\in\mathbb{Z}_p[x_1,\dots,x_n]\setminus p\mathbb{Z}_p[x_1,\dots,x_n]$, also denoted by $\bar{f}$.

More generally, let $f\in\mathbb{Q}_p[x_1,\dots,x_n]$ be a nonzero polynomial and consider $D:=\operatorname{div}(f)$ in $\mathbb{A}^n_{\mathbb{Q}_p}$. Pick the unique $e\in\mathbb{Z}$ such that $p^ef\in\mathbb{Z}_p[x_1,\dots,x_n]\setminus p\mathbb{Z}_p[x_1,\dots,x_n]$. Define the reduction mod p of D as $\overline{D}=D \bmod p:=\operatorname{div}(\overline{p^ef})$, where $\overline{p^ef}$ is the reduction mod p of p^ef as explained above. Hence $\overline{D}$ is a divisor in $\mathbb{A}^n_{\mathbb{F}_p}$, and can be considered as a subscheme of $\mathbb{A}^n_{\mathbb{F}_p}$.

EXAMPLE 4.3. Consider $f=y-p^3x^2\in\mathbb{Z}_p[x,y]\setminus p\mathbb{Z}_p[x,y]$. Then $D:=\operatorname{div}(f)$ is a parabola in $\mathbb{A}^2_{\mathbb{Q}_p}$, but $\overline{D}=\operatorname{div}(\overline{y-p^3x^2})=\operatorname{div}(y)$ is a line in $\mathbb{A}^2_{\mathbb{F}_p}$. The polynomial $f=p^{-2}y-px^2\in\mathbb{Q}_p[x,y]$ induces the same divisors D and $\overline{D}$.

In fact, there is a well-defined notion of reduction mod p for any algebraic variety or scheme Z over $\mathbb{Q}_p$. It is a scheme defined over $\mathbb{F}_p$, denoted by $\overline{Z}$. Similarly, there is a well-defined notion of reduction mod p for any morphism g of $\mathbb{Q}_p$-schemes, being a morphism $\overline{g}$ of the corresponding $\mathbb{F}_p$-schemes. See for example [**De1**] and the references therein.

We now introduce the necessary terminology and notation to state Denef's formula. Consider an embedded resolution $h:Y\to\mathbb{A}^n_{\mathbb{Q}_p}$ of $f\in\mathbb{Q}_p[x_1,\dots,x_n]\setminus\mathbb{Q}_p$, for which we use notation as in Theorem 4.1. Then

$$E_i \hookrightarrow Y \xrightarrow{h} \mathbb{A}^n_{\mathbb{Q}_p} \hookleftarrow D=\operatorname{div}(f)$$

has a reduction mod p:

$$\overline{E}_i \hookrightarrow \overline{Y} \xrightarrow{\overline{h}} \mathbb{A}^n_{\mathbb{F}_p} \hookleftarrow \overline{D}.$$

DEFINITION 4.4 (Denef [**De1**]). We say that h has *good reduction* mod p if

(1) $\overline{Y}$ and all $\overline{E}_i, i\in I$, are smooth,
(2) $\bigcup_{i\in I}\overline{E}_i$ is a normal crossings divisor,
(3) $\overline{E}_i$ and $\overline{E}_j$ have no common components for $i\neq j$.

EXAMPLE 4.5 (Bad reduction). We give examples that show how these conditions might fail.

(1) The zero locus of $g = px + x^2 - y^2$ is smooth, but the zero locus of $\overline{g} = x^2 - y^2 = (x+y)(x-y)$ is not.
(2) The zero loci of $g = x$ and $g' = x + py - y^2$ have normal crossings, but those of $\overline{g} = x$ and $\overline{g'} = x - y^2$ do not.
(3) The zero loci of $g = px + y + p$ and $g' = px + y$ are disjoint, but $\overline{g} = \overline{g'}$.

NOTATION 4.6. For $J \subseteq I$, define $\overline{E}_J = \bigcap_{j\in J} \overline{E}_j$ and $\overset{\circ}{\overline{E}}_J = \overline{E}_J \setminus \bigcup_{k\notin J} \overline{E}_k$. Note that in particular $\overline{E}_\emptyset = \overline{Y}$ and $\overset{\circ}{\overline{E}}_\emptyset = \overline{Y} \setminus \bigcup_{i\in I} \overline{E}_i$.

4.3. The formula.

THEOREM 4.7 (Denef's formula, 1987 [**De1**]).

(1) *Let $f \in \mathbb{Q}_p[x_1, \dots, x_n] \setminus \mathbb{Q}_p$. If $h : Y \to \mathbb{A}^n_{\mathbb{Q}_p}$ is an embedded resolution of f with good reduction mod p, then*
$$Z(s) = \frac{1}{p^n} \sum_{J\subseteq I} \sharp(\overset{\circ}{\overline{E}}_J(\mathbb{F}_p)) \prod_{j\in J} \frac{p-1}{p^{N_j s + \nu_j} - 1}.$$
(2) *Let $f \in \mathbb{Q}[x_1, \dots, x_n] \setminus \mathbb{Q}$. Let $h : Y \to \mathbb{A}^n_{\mathbb{Q}}$ be an embedded resolution of f. For any prime p, we can also consider $f \in \mathbb{Q}_p[x_1, \dots, x_n]$ and view Y and h as defined over $\mathbb{Q}_p$. Then, for all but finitely many p, the resolution h (considered over $\mathbb{Q}_p$) has good reduction mod p.*

Note that for a fixed prime p, it is possible that there exists no embedded resolution of a given $f \in \mathbb{Q}[x_1, \dots, x_n]$ with good reduction mod p.

REMARK 4.8. Traditionally, the factors in the product above are written as $\frac{(p-1)p^{-N_j s - \nu_j}}{1 - p^{-N_j s - \nu_j}}$, in order to view $Z(s)$ explicitly as a function in p^{-s}. The presentation above is often somewhat more efficient.

REMARK 4.9. We define the contribution of E_i to the Igusa zeta function $Z(s)$ to be
$$\frac{1}{p^n} \sum_{i\in J\subseteq I} \sharp(\overset{\circ}{\overline{E}}_J(\mathbb{F}_p)) \prod_{j\in J} \frac{p-1}{p^{N_j s + \nu_j} - 1}.$$
If this contribution has no pole at $s_0 = -\nu_i/N_i$, we say that E_i does not contribute to the candidate pole s_0 of $Z(s)$. Note that in this case $-\nu_i/N_i$ might still be a pole of $Z(s)$; this is however only possible if there exists some other E_j satisfying $s_0 = -\nu_j/N_j = -\nu_i/N_i$ that does contribute to the pole s_0.

EXAMPLE 4.10. Consider $f = xy(x-y)$ as in Example 3.10. The blow-up at the origin h is also an embedded resolution in the sense of Theorem 4.1, and moreover with good reduction mod p for all primes p. Denote the exceptional curve of h by E and the strict transforms in Y of the three components of $\operatorname{div}(f)$ by E_1, E_2, E_3. Then, with the notation of Theorem 4.7, $\overline{E}$ is a projective line, and (for $i = 1, 2, 3$) $\overline{E}_i$ is an affine line and $\overline{E} \cap \overline{E}_i$ a point, all defined over $\mathbb{F}_p$. Hence $\sharp(\overset{\circ}{\overline{E}}_\emptyset(\mathbb{F}_p)) = p^2 - (3p-2)$, $\sharp(\overset{\circ}{\overline{E}}(\mathbb{F}_p)) = (p+1) - 3$ and $\sharp(\overset{\circ}{\overline{E}}_i(\mathbb{F}_p)) = p - 1$ for $i = 1, 2, 3$. Adding everything, Denef's formula yields

$$Z(s) = \frac{1}{p^2}\left(p^2 - 3p + 2 + (p-2)\frac{p-1}{p^{3s+2}-1}\right.$$
$$\left. +3(p-1)\frac{p-1}{p^{s+1}-1} + 3\cdot 1\frac{(p-1)^2}{(p^{3s+2}-1)(p^{s+1}-1)}\right).$$

Verify that this is the same result as in Example 3.10.

EXAMPLE 4.11. Consider $f = x^2 + y^2$ as in Exercises 1.5 and 2.11.

(1) Let $p \equiv 1 \bmod 4$. Take $\alpha \in \mathbb{Z}_p$ such that $\alpha^2 = -1$. Then the zero locus of $f = (x - \alpha y)(x + \alpha y)$ consists of two affine lines E_1 and E_2 over $\mathbb{Q}_p$, intersecting transversely in the origin. Hence we can take the identity as resolution map h. Check that it has good reduction mod p. Compute $Z(s)$ using Denef's formula (there are four terms), and compare with Exercise 2.11.

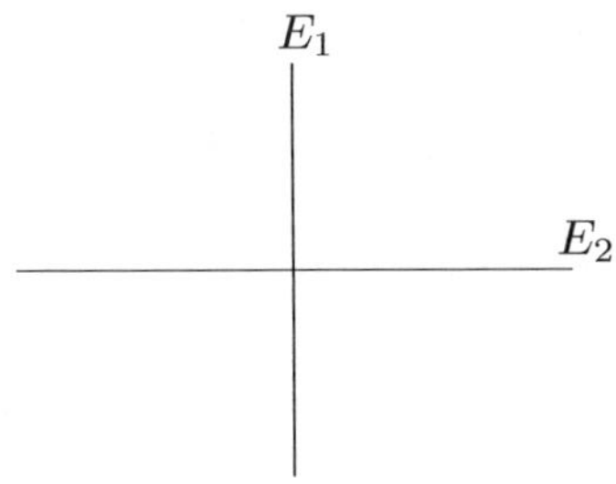

(2) Let $p \equiv 3 \bmod 4$. Then $E_0 = \operatorname{div}(f)$ is irreducible as $\mathbb{Q}_p$-scheme, and its only point with coordinates in $\mathbb{Q}_p$ is the origin. In the figure below, we 'suggest' E_0 by indicating with dashed lines its two components over the quadratic extension of $\mathbb{Q}_p$ in which -1 is a square. Take the blow-up at the origin as resolution map $h : Y \to \mathbb{A}^2_{\mathbb{Q}_p}$. The exceptional curve E is a projective line over $\mathbb{Q}_p$, and the strict transform of E_0 in Y is a 1-dimensional irreducible $\mathbb{Q}_p$-scheme (now without $\mathbb{Q}_p$-points), given in one chart by $1 + u^2 = 0$.

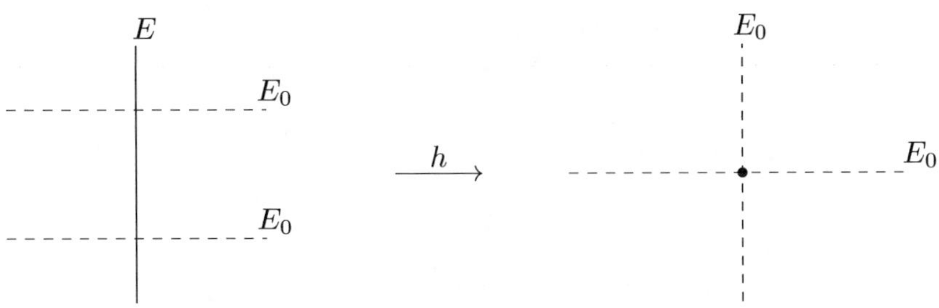

Hence $\overline{E}$ is a projective line over $\mathbb{F}_p$, $\overline{E}_0$ in $\overline{Y}$ is an irreducible $\mathbb{F}_p$-scheme *without* $\mathbb{F}_p$*-points*, still given in one chart by $1+u^2=0$, and $\overline{E}\cap\overline{E}_0$ is a 0-dimensional irreducible $\mathbb{Q}_p$-scheme, given in one chart by $x=1+u^2=0$, also without $\mathbb{F}_p$-points. In particular, h has good reduction mod p. Compute $Z(s)$ using Denef's formula (there are two terms), and compare with Exercise 2.11.

(3) Let $p=2$. Then again $\operatorname{div}(f)$ is irreducible as $\mathbb{Q}_2$-scheme, and the blow-up at the origin is a resolution map h. Verify however that h does not have good reduction mod 2. In this case we cannot apply Denef's formula.

EXERCISE 4.12.

(1) Compute the Igusa zeta function for $f=y^2-x^3$, using Denef's formula. Then use Proposition 2.8 to compute its Poincaré series; you should obtain what was claimed in Example 1.11.
(2) Verify the result of Example 2.10 for $f=xy-z^2$, using Denef's formula.

REMARK 4.13. The relatively easy examples that we considered up to now are somewhat misleading. In general, it is very hard to determine whether a resolution h has good reduction mod p for a given (small) prime p. Denef's formula is typically used when $f\in\mathbb{Q}[x_1,\dots,x_n]$ and one wants to compute $Z_p(f;s)$ or prove properties of $Z_p(f;s)$ for large enough p.

REMARK 4.14. Another technique to compute the Igusa zeta function is by using toric geometry. Denef and Hoornaert [**DH**] give a combinatorial formula to compute $Z(s)$ when $f\in\mathbb{Q}_p[x_1,\dots,x_n]$ is non-degenerate (mod p) with respect to its Newton polyhedron. That formula contains typically less candidate poles than the list given by an embedded resolution.

5. Back to polynomial congruences

Recall the original problem: given $f\in\mathbb{Z}[x_1,\dots,x_n]$, we want to study the values

$$M_i=M_i(p)=\sharp\left\{a\in\left(\frac{\mathbb{Z}}{p^i\mathbb{Z}}\right)^n \mid f(a)\equiv 0 \bmod p^i\right\}.$$

In this section, we discuss how to use the machinery we introduced before to obtain information about the M_i.

5.1. Asymptotic behaviour of the M_i.

We start with an example.

EXAMPLE 5.1. Let $f=y^2-x^3\in\mathbb{Z}[x,y]$. Recall from Example 1.11 or Exercise 4.12 that

$$P(T)=\frac{p^6+(p^4-p^3)T^2-T^6}{(p-T)(p^5-T^6)}=\frac{1+(p^{-2}-p^{-3})T^2-p^{-6}T^6}{(1-p^{-1}T)(1-p^{-5}T^6)}.$$

Using partial fraction decomposition, we know that there exists a constant $q\in\mathbb{Q}$ and a polynomial $Q(T)\in\mathbb{Q}[T]$ (of degree at most 5) such that

$$P(T)=\frac{q}{1-p^{-1}T}+\frac{Q(T)}{1-p^{-5}T^6}.$$

Combining this with the definition of $P(T)$ we obtain

$$\sum_{i\geq 0}M_i(p^{-2}T)^i=P(T)=q\sum_{i\geq 0}(p^{-1}T)^i+Q(T)\sum_{i\geq 0}(p^{-5}T^6)^i.$$

EXERCISE 5.2. Compute q and Q of Example 5.1 explicitly and give explicit formulas for the M_i (depending on the value of $i \bmod 6$). For example, when $i = 6e$ ($e \in \mathbb{Z}_{>0}$), you should obtain $M_{6e} = (p+1)p^{7e-1} - p^{6e-1}$. Conclude that the asymptotic behaviour of the M_i is roughly $p^{\frac{7}{6}i}$.

In general, it is not difficult to conclude the following (using Theorem 3.13 to identify σ), see for instance [**VZ1**, Theorem 2.7] and [**VZ2**, Remark 3.11].

PROPOSITION 5.3. *Let $f \in \mathbb{Z}[x_1, \ldots, x_n]$. Denote by $-\sigma$ the largest real part of a pole of $Z(s)$, i.e., σ is the p-adic log canonical threshold of f. Equivalently, p^σ is the smallest modulus of a pole of $Z'(t)$ or $P(T)$. Then the main contribution to the M_i is determined by σ: roughly, the M_i behave asymptotically as $p^{(n-\sigma)i}$. More precisely: $\limsup_i M_i^{1/i} = p^{n-\sigma}$.*

Moreover, if $-\sigma$ is a pole of $Z(s)$ of order d (where necessarily $d \leq n$), then there exists a positive constant C such that $M_i \leq C i^{d-1} p^{(n-\sigma)i}$ for all i.

Note that for the cusp (see Example 5.1 and Exercise 5.2) we have indeed $\limsup_i M_i^{1/i} = p^{2-5/6} = p^{7/6}$.

REMARK 5.4. An exact expression for the M_i, with 'dominating term' as in Proposition 5.3, involves necessarily data from *all* poles of $P(T)$. Such a complete description of the M_i in terms of the actual poles of $P(T)$ and their orders (and partial fraction decomposition) was given by Segers in full generality in [**Sg2**]. (Note that there is a crucial typo in [**Sg2**]. On the last line of page 4 the numbers involving e must be augmented by 1.) The outcome of Exercise 5.2 is a very special case of this description.

Here we should stress that determining the actual poles is in general a difficult problem. When $n = 2$, this is understood and explained in the next section.

5.2. Divisibility properties of the M_i. There is a remarkable theorem by Segers stating (1) divisibility properties of the M_i by powers of p, (2) a sharp lower bound for the possible real parts of poles of $Z(s)$, and (3) the relation between those two properties. In order to place this result into context, we first mention an 'easy' lower bound for the real parts of poles.

PROPOSITION 5.5. *Let $f \in K[x_1, \ldots, x_n] \setminus K$, where K is any field of characteristic zero and $n \geq 2$. Consider an embedded resolution $h : Y \to \mathbb{A}^n_K$ of f, that is constructed as a finite composition of admissible blow-ups at smooth centres Z_i (always contained in the strict transform of the divisor of f). Each such blow-up induces an exceptional component E_i with numerical data (N_i, ν_i).*

(1) *Let Z_j be contained in $E_{i_1}, \ldots, E_{i_l}$ (and not in the other E_ℓ). Denote by μ_j the multiplicity of the strict transform of the divisor of f along Z_j. Then*

$$N_j = \sum_{k=1}^{l} N_{i_k} + \mu_j,$$

$$\nu_j = \sum_{k=1}^{l} (\nu_{i_k} - 1) + \operatorname{codim}(Z_j),$$

where $\operatorname{codim}(Z_j) = n - \dim(Z_j)$ *is the codimension of Z_j in the ambient space.*

(2) *Assume that each centre Z_j is contained in an exceptional component or in the singular locus of the strict transform. Then all $\frac{\nu_i}{N_i} \leq n-1$.*

PROOF. We explain the argument with local analytic coordinates in the analytic setting when K is $\mathbb{Q}_p$, $\mathbb{R}$ or $\mathbb{C}$. A similar argument can be given with étale coordinates in a general algebro-geometric setting.

(1) Choose local coordinates $z_1, \dots, z_n$ around a general point b of Z_j such that E_{i_k} is given by $\{z_k = 0\}$ for $k = 1, \dots, l$ and Z_j is given by $\{z_1 = \dots = z_d = 0\}$ in a neighbourhood of b. Note that $d = \operatorname{codim}(Z_j) \geq l$. Suppose that, locally around b, the strict transform of f is given by $f_{j-1}(z_1, \dots, z_n)$. With local coordinates $x_1, \dots, x_n$ in the first chart of the blow-up at the center Z_j, this blow-up is described by the change of variables

$$\begin{aligned}&(x_1, x_2, \dots, x_d, x_{d+1}, \dots, x_n) \mapsto \\ &(z_1 = x_1, z_2 = x_1 x_2, \dots, z_d = x_1 x_d, z_{d+1} = x_{d+1}, \dots, z_n = x_n).\end{aligned}$$

Hence the pullback of

$$\varepsilon(z_1, \dots, z_n) z_1^{N_1} z_2^{N_2} \dots z_l^{N_l} f_{j-1}(z_1, \dots, z_n)$$

is

$$\varepsilon'(x_1, \dots, x_n) x_1^{N_1 + \dots + N_l + \mu_j} x_2^{N_2} \dots x_l^{N_l} f_j(x_1, \dots, x_n),$$

with f_j not divisible by x_i for $i = 1, \dots, l$ and $\varepsilon, \varepsilon'$ invertible. Therefore, N_k equals $\sum_{k=1}^{l} N_{i_k} + \mu_j$.

Similarly, the pullback of

$$\eta(z_1, \dots, z_n) z_1^{\nu_1 - 1} z_2^{\nu_2 - 1} \dots z_l^{\nu_l - 1} dz_1 \wedge dz_2 \wedge \dots \wedge dz_n$$

is

$$\eta'(x_1, \dots, x_n) x_1^{\nu_1 - 1 + \nu_2 - 1 + \dots + \nu_l - 1 + d - 1} x_2^{\nu_2 - 1} \dots x_l^{\nu_l - 1} dx_1 \wedge dx_2 \wedge \dots \wedge dx_n,$$

with η, η' invertible. Hence, $\nu_j = \sum_{k=1}^{l} (\nu_{i_k} - 1) + d = \sum_{k=1}^{l} (\nu_{i_k} - 1) + \operatorname{codim}(Z_j)$.

(2) Concerning the first blow-up with centre Z_1 and exceptional component E_1, we know that $(N_1, \nu_1) = (\mu_1, d_1)$ with μ_1 the multiplicity of f at Z_1 and $d_1 = \operatorname{codim}(Z_1)$. By assumption, Z_1 is contained in the singular locus of $\{f = 0\}$, hence $\mu_1 \geq 2$. Consequently, $\frac{\nu_1}{N_1} = \frac{d_1}{\mu_1} \leq \frac{n}{2} \leq n - 1$.

Next, we suppose that Z_j is contained in at least one exceptional component. We use the notation of part (1), where we may assume by induction that $\frac{\nu_{i_k}}{N_{i_k}} \leq n-1$ for $k = 1, \dots, l$. Compute

$$\begin{aligned}\frac{\nu_j}{N_j} &= \frac{\sum_{k=1}^{l} (\nu_{i_k} - 1) + \operatorname{codim}(Z_j)}{\sum_{k=1}^{l} N_{i_k} + \mu_j} \\ &\leq \frac{\sum_{k=1}^{l} (n-1) N_{i_k} - l + n}{\sum_{k=1}^{l} N_{i_k} + \mu_j} \\ &= \frac{(n-1) \sum_{k=1}^{l} N_{i_k}}{\sum_{k=1}^{l} N_{i_k} + \mu_j} - \frac{l - n}{\sum_{k=1}^{l} N_{i_k} + \mu_j} \\ &= n - 1 - \frac{(n-1)\mu_j + l - n}{\sum_{k=1}^{l} N_{i_k} + \mu_j}.\end{aligned}$$

Since $\mu_j \geq 1$ and $l \geq 1$, we conclude that $\frac{\nu_j}{N_j} \leq n - 1$. □

Note that the assumption in Proposition 5.5(2) is always satisfied if no 'stupid' blow-ups are used to construct the resolution h. A stupid example is blowing up at a smooth point of the divisor of f; this yields an exceptional component with numerical data $(N_1, \nu_1) = (1, n)$ and then $\frac{n}{1}$ is not bounded by $n - 1$.

As a corollary of Proposition 5.5, all real parts of poles of zeta functions $Z(s)$ are larger than or equal to $-(n-1)$. The following much stronger result of Segers was in fact not conjectured before, and is proven by quite elementary arguments and techniques. In fact, he first shows the divisibility properties, and derives the bound on the poles from it.

THEOREM 5.6 (Segers, 2006 [**Sg1**]). *Let $f \in \mathbb{Z}[x_1, \ldots, x_n]$, where $n \geq 2$, and fix any prime p. Then*

(1) *M_i is divisible by $p^{\lceil n/2(i-1) \rceil}$ for all i,*
(2) *all real parts of poles of $Z(s)$ are larger than or equal to $-\frac{n}{2}$.*

EXERCISE 5.7. Verify that for $f = x_1^2 + x_2^2 + \cdots + x_n^2$ the lower bound $-\frac{n}{2}$ is an actual pole of the zeta function $Z(s)$.

6. Igusa zeta function for plane curves

For polynomials in two variables, the poles of its Igusa zeta function are well understood.

6.1. Example and non-contribution.

EXAMPLE 6.1. Consider the polynomial $f = y^2 - x^{2019}$ and its minimal embedded resolution h, where E_0 denotes the strict transform.

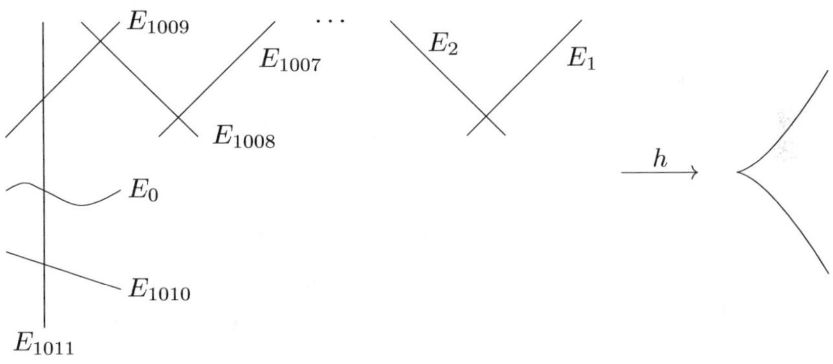

The numerical data associated to this minimal resolution are as follows:

$$(N_i, \nu_i) = \begin{cases} (1,1) & \text{if } i = 0, \\ (2i, i+1) & \text{if } 1 \leq i \leq 1009, \\ (2019, 1011) & \text{if } i = 1010, \\ (4038, 2021) & \text{if } i = 1011. \end{cases}$$

Hence, there are 1012 different real values of candidate poles $-\frac{\nu_i}{N_i}$ of $Z(s)$. However, using Denef's formula, and either a lot of courage or a computer, one computes that

$$Z(s) = \frac{\text{a polynomial in } p^{-s}}{\left(1-p^{-(s+1)}\right)\left(1-p^{-(4038s+2021)}\right)}.$$

REMARK 6.2. Using either the formula for non-degenerate polynomials of Denef and Hoornaert [**DH**], or the formula in [**Ve5**] in terms of the relative log canonical model (which is a certain partial resolution), there are only two candidate real parts of poles, and one obtains this result quite immediately.

Example 6.1 illustrates the general 'geometric determination' of the actual poles of $Z(s)$. Roughly speaking, exceptional components E_i that intersect exactly once or twice other components do not contribute to the poles, and other exceptional components or components of the strict transform do contribute. We explain the main point of the non-contribution argument, in the setting of Denef's formula.

LEMMA 6.3. *Let $f \in \mathbb{Q}_p[x,y] \setminus \mathbb{Q}_p$ and $h : Y \to \mathbb{A}^2_{\mathbb{Q}_p}$ an embedded resolution of f with good reduction mod p. Let E (with numerical data (N,ν)) be an exceptional component of h, intersecting two other components, say E_1 and E_2 (with numerical data (N_1,ν_1) and (N_2,ν_2), respectively). Here we mean 'intersecting over $\mathbb{Q}_p$', i.e., E is a projective line over $\mathbb{Q}_p$ and each of the two intersection points is defined over $\mathbb{Q}_p$. Assume that E induces a candidate pole of order 1, i.e., $\frac{\nu}{N} \neq \frac{\nu_i}{N_i}$ for $i = 1,2$. Then E does not contribute to a pole (in the sense of Remark 4.9).*

PROOF. The contribution of E to the Igusa zeta function is equal to

$$\begin{aligned}
&\frac{1}{p^2}\Bigg(\sharp((\overline{E}\setminus(\overline{E}_1\cup\overline{E}_2))(\mathbb{F}_p))\frac{p-1}{p^{Ns+\nu}-1} + \sharp((\overline{E}\cap\overline{E}_1)(\mathbb{F}_p))\frac{p-1}{p^{Ns+\nu}-1}\cdot\frac{p-1}{p^{N_1s+\nu_1}-1} \\
&\qquad\qquad + \sharp((\overline{E}\cap\overline{E}_2)(\mathbb{F}_p))\frac{p-1}{p^{Ns+\nu}-1}\cdot\frac{p-1}{p^{N_2s+\nu_2}-1}\Bigg) \\
&= \frac{1}{p^2}\left((p-1)\frac{p-1}{p^{Ns+\nu}-1} + \frac{p-1}{p^{Ns+\nu}-1}\cdot\frac{p-1}{p^{N_1s+\nu_1}-1} + \frac{p-1}{p^{Ns+\nu}-1}\cdot\frac{p-1}{p^{N_2s+\nu_2}-1}\right) \\
&= \frac{(p-1)^2}{p^2}\left(\frac{(p^{N_1s+\nu_1}-1)(p^{N_2s+\nu_2}-1)+p^{N_2s+\nu_2}-1+p^{N_1s+\nu_1}-1}{(p^{Ns+\nu}-1)(p^{N_1s+\nu_1}-1)(p^{N_2s+\nu_2}-1)}\right) \\
&= \frac{(p-1)^2}{p^2}\left(\frac{p^{(N_1+N_2)s+\nu_1+\nu_2}-1}{(p^{Ns+\nu}-1)(p^{N_1s+\nu_1}-1)(p^{N_2s+\nu_2}-1)}\right).
\end{aligned}$$

By Lemma 6.5, there exists some $\kappa \in \mathbb{Z}_{>0}$ such that $\kappa N = N_1+N_2$ and $\kappa\nu = \nu_1+\nu_2$. Therefore, the contribution of E is given by

$$\begin{aligned}
&\frac{(p-1)^2}{p^2}\left(\frac{p^{\kappa(Ns+\nu)}-1}{(p^{Ns+\nu}-1)(p^{N_1s+\nu_1}-1)(p^{N_2s+\nu_2}-1)}\right) \\
&= \frac{(p-1)^2}{p^2}\left(\frac{\sum_{j=0}^{\kappa-1}p^{j(Ns+\nu)}}{(p^{N_1s+\nu_1}-1)(p^{N_2s+\nu_2}-1)}\right).
\end{aligned}$$

Hence, E does not contribute to a pole of $Z(s)$. □

REMARK 6.4. When E, defined over $\mathbb{Q}_p$, intersects only once another component, say E_1, then the intersection point is automatically defined over $\mathbb{Q}_p$. Assuming that $\frac{\nu}{N} \neq \frac{\nu_1}{N_1}$, the same proof (putting formally $(N_2,\nu_2) = (0,1)$) yields that E does not contribute to a pole.

The following lemma was first proven for arbitrary plane curves by Loeser [**Lo**] (preceded by some partial results by Strauss, Meuser and Igusa). We present a conceptual proof, which is the starting point of a generalized theory of relations and congruences between numerical data in arbitrary dimension n [**Ve1**], [**Ve2**].

LEMMA 6.5. *Let $h : Y \to \mathbb{A}^2$ be an embedded resolution of singularities of a plane curve C with defining equation g. Write $h^{-1}(C) = E \cup \bigcup_{i=1}^{k} E_i$, where E is a fixed exceptional component. Let E intersect exactly r times other components, say $E_1, \ldots, E_r$. Then there exists some $\kappa \in \mathbb{Z}_{>0}$ such that $\kappa N = \sum_{i=1}^{r} N_i$ and $\kappa\nu = \sum_{i=1}^{r}(\nu_i - 1) + 2$. More precisely, $\kappa = -E^2$ where $E^2 = E \cdot E$ is the self-intersection number of E on Y.*

PROOF. Recall that the Picard group $\mathrm{Pic}(Y)$ of Y is defined as the group of Cartier divisors on Y modulo linear equivalence. In $\mathrm{Pic}(Y)$ we have that

$$NE + \sum_{i=1}^{k} N_i E_i = h^* \mathrm{div}(g) = \mathrm{div}(h^*(g)) = 0$$

and

$$(\nu - 1)E + \sum_{i=1}^{k} (\nu_i - 1)E_i = K_Y - h^* K_{\mathbb{A}^2} = K_Y.$$

We compute the intersection product of both expressions with E (in $\mathrm{Pic}(Y)$). First,

$$0 = E \cdot 0 = E \cdot \left(NE + \sum_{i=1}^{k} N_i E_i \right) = E \cdot NE + E \cdot \left(\sum_{i=1}^{r} N_i E_i \right) = NE^2 + \sum_{i=1}^{r} N_i.$$

Hence, $\kappa N = -NE^2 = \sum_{i=1}^{r} N_i$. Similarly, we compute

$$\begin{aligned} E \cdot K_Y &= E \cdot \left((\nu - 1)E + \sum_{i=1}^{k} (\nu_i - 1)E_i \right) \\ &= E \cdot (\nu - 1)E + E \cdot \left(\sum_{i=1}^{r} (\nu_i - 1)E_i \right) \\ &= (\nu - 1)E^2 + \sum_{i=1}^{r} (\nu_i - 1). \end{aligned}$$

Using the adjunction formula, we obtain

$$-2 = \deg K_E = (K_Y + E) \cdot E = \nu E^2 + \sum_{i=1}^{r} (\nu_i - 1),$$

and hence $\kappa\nu = -\nu E^2 = \sum_{i=1}^{r}(\nu_i - 1) + 2$. □

REMARK 6.6. Observe that Lemma 6.5 agrees (as it should) with the numerical data in Example 6.1.

6.2. Structure of dual graph and general result. Before stating the general theorem on the determination of the poles of $Z(s)$ for $n = 2$, it is useful to know the structure of the dual graph of the minimal embedded resolution of f, decorated with the values $\frac{\nu_i}{N_i}$. This is a geometrical and 'local' result; so we assume now that f is defined over $\mathbb{C}$, and we consider only the germ of f at the origin. (We also exclude the trivial case where this germ has already normal crossings.)

NOTATION 6.7. Let $h : Y \to \mathbb{A}^n_{\mathbb{C}}$ be an embedded resolution of the germ at 0 of $\{f = 0\}$. The dual graph associated to this resolution is a graph that consists of the following data. The vertices of the graph correspond to the irreducible components of $h^{-1}(\{f = 0\})$. The exceptional components are represented by a dot, the (analytically irreducible) components of the strict transform by a circle. Two vertices are connected by an edge precisely when the associated components intersect. It is well known that this dual graph is a tree, where all circles are end vertices.

A vertex with at least three edges is depicted in the following way.

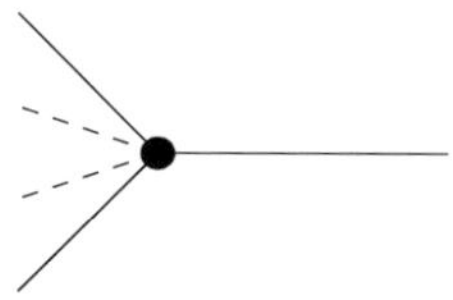

THEOREM 6.8 (Veys, 1995 [**Ve4**]). *Consider the minimal embedded resolution* $h : Y \to \mathbb{A}^n_{\mathbb{C}}$ *of* f *with irreducible components* $E_i, i \in I$, *and associated numerical data* $(N_i, \nu_i), i \in I$. *Denote the locus (with edges) where* $\frac{\nu_i}{N_i}$ *is minimal by* $\mathcal{M}$. *Then* $\mathcal{M}$ *is connected and has one of the following possible forms, where* $r \geq 0$.

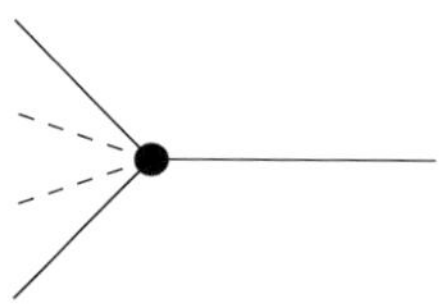

(1)

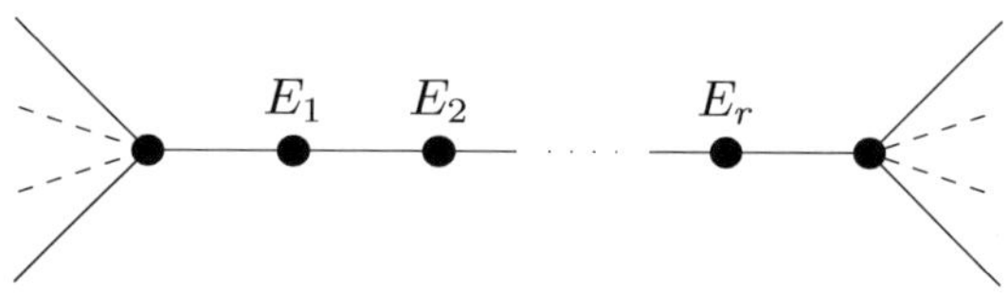

(2)

(3)

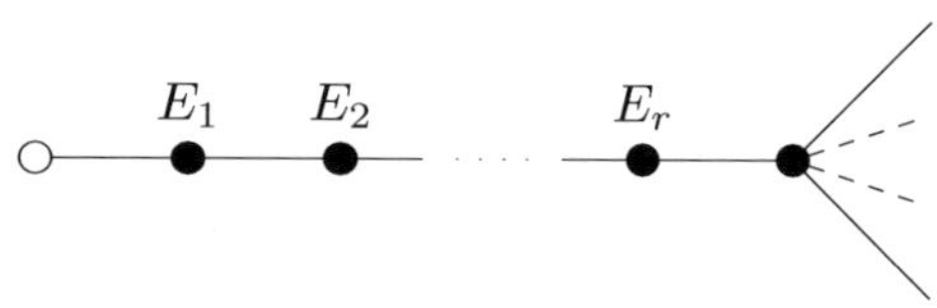

(4)

When f *is reduced, the last two cases (3) and (4), involving a component of the strict transform, cannot occur.*

Furthermore, starting from $\mathcal{M}$, *the values* $\frac{\nu_i}{N_i}$ *strictly increase along each path away from* $\mathcal{M}$.

When $n = 2$, we already knew that poles of $Z(s)$ have order at most 2. Theorem 6.8 implies that there is at most one real part of a pole of order 2; here is the precise statement.

COROLLARY 6.9 (Veys, 1995 [**Ve4**]). *Let* $f \in \mathbb{Q}_p[x, y] \setminus \mathbb{Q}_p$. *Let* $h : Y \to \mathbb{Q}_p^2$ *be an embedded resolution of* f *as in Theorem 3.5. A real number* s_0 *is (the real part* *of) a pole of order 2 of $Z(s)$ if and only if there exist $i, j \in I$ with $s_0 = -\frac{\nu_i}{N_i} = -\frac{\nu_j}{N_j}$ for* intersecting *components E_i and E_j. In that case, s_0 equals minus the p-adic log canonical threshold of f.*

REMARK 6.10. The following generalisation (to higher dimension) was conjectured by the second author in 1999 [**LV**], and proven by Nicaise and Xu in 2016 [**NX**]. Let $f \in \mathbb{Q}_p[x_1, \ldots, x_n]$ and consider its associated Igusa zeta function $Z(s)$. The real part of a pole of order n of $Z(s)$ must be equal to minus the p-adic log canonical threshold of f. In particular, there is at most one possible real part of a pole of order n.

Also, the real part of a pole of order n is of the form $-\frac{1}{k}$ for some positive integer k [**LV**].

Concerning the poles of order 1, 'morally' an exceptional E_i does not contribute if and only if it intersects other components in exactly one or two points defined over $\mathbb{Q}_p$. In order to formulate this precisely, it is more convenient to state it in terms of the local equation of the strict transform before the creation of E_i in the resolution process.

Also, as for the structure of the dual resolution graph, the result is 'local' in the sense that it is formulated for one singular point.

THEOREM 6.11 (Ibadula-Segers, 2012 [**IS**]). *Let $f \in \mathbb{Q}_p[x, y]$. Suppose that the origin is the only singular point of $\{f = 0\}$. Let $h : Y \to \mathbb{Q}_p^2$ be an embedded resolution of f as in Theorem 3.5, constructed as a finite composition of admissible blow-ups. Let $s_0 = -\frac{\nu_i}{N_i}$ for some $i \in I$, such that s_0 is not a pole of order 2 of $Z(s)$.*

(1) *Suppose that $s_0 \neq -\frac{\nu_i}{N_i}$ for all components E_i of the strict transform of f. Then s_0 is a pole of order 1 if and only if there is an exceptional component E_i with $s_0 = -\frac{\nu_i}{N_i}$ satisfying the following. Write the equation of the strict transform of f, at the stage of h before the blow-up with centre C where E_i is created as exceptional component, as a convergent power series in local analytic coordinates around C; then the lowest degree part of this power series is* not *a power of a linear form (over $\mathbb{Q}_p$) or a product of two such powers.*

(2) *Let $s_0 = -\frac{\nu_i}{N_i}$ for a component E_i of the strict transform of f. Then s_0 is always 'locally' a pole of order 1 in the sense that it is a pole of $\int_{(p^k\mathbb{Z}_p)^2} |f(x, y)|_p^s \, |dxdy|$ for some large enough k.*

EXERCISE 6.12. Let $f = y(x^2+y^2)+x^4$ and fix a prime p such that $p \equiv 3 \bmod 4$. The blow-up at the origin is an embedded resolution of f. In the figure below, we sketch the germ of f at the origin, where the dashed lines 'suggest' the tangent directions that are only defined over the quadratic extension of $\mathbb{Q}_p$ in which -1 is a square.

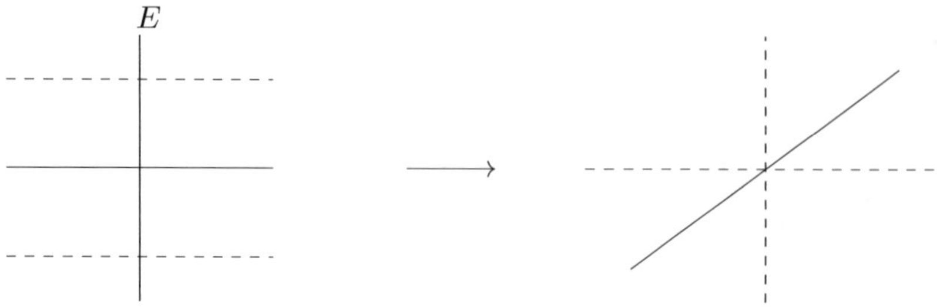

The exceptional curve E has numerical data $(N, \nu) = (3, 2)$. In the context of p-adic manifolds, E intersects the strict transform only in one point. However, geometrically (already over that quadratic extension of $\mathbb{Q}_p$), the exceptional curve E intersects the strict transform three times. This is reflected in the lowest degree part of f, which is over $\mathbb{Q}_p$ the product of a linear form and an irreducible quadratic form. Hence, by Theorem 6.11 we know that $-\frac{3}{2}$ is a pole of order 1 of $Z(s)$.

REMARK 6.13. The determination of non-real poles is less clear. Consider for example $f = x^2 + y^2$ as in Exercises 1.5 and 2.11 and Example 4.11. When $p \equiv 3 \bmod 4$, we have that both -1 and $-1 + \frac{\pi}{\ln p} i$ are poles of order 1 of $Z(s)$. On the other hand, when $p = 2$, we have that -1 is a pole of order 1 of $Z(s)$, but $-1 + \frac{\pi}{\ln 2} i$ is not a pole.

7. Topological and motivic zeta function

The p-adic Igusa zeta function has two remarkable 'spin-offs'.

7.1. Topological zeta function. Recall Denef's formula for the zeta function of $f \in \mathbb{Q}[x_1, \dots, x_n]$ in terms of an embedded resolution h defined over $\mathbb{Q}$, and valid for all primes p for which h has good reduction mod p. Using the notation of Section 4, we have

$$Z_p(f; s) = \frac{1}{p^n} \sum_{J \subseteq I} \sharp(\overset{\circ}{\overline{E}}_J(\mathbb{F}_p)) \prod_{i \in J} \frac{p-1}{p^{N_i s + \nu_i} - 1}.$$

We now indicate a funny heuristic argument to compute the limit of this expression, when p tends to 1 (whatever that means ...). First, using l'Hôpital's rule, $\frac{p-1}{p^{N_i s+\nu_i}-1}$ tends to $\frac{1}{N_i s+\nu_i}$. More challenging, by Grothendieck's trace formula, $\sharp(\overset{\circ}{\overline{E}}_J(\mathbb{F}_p))$ can be written as an alternating sum of traces of the pth power Frobenius operator, acting on adequate ℓ-adic cohomology groups of $\overset{\circ}{\overline{E}}_J$ [**De2**]. The limit when p tends to 1 of this operator is morally the identity operator, and then that alternating sum tends to the alternating sum of the dimensions of the cohomology groups of $\overset{\circ}{\overline{E}}_J$, that is, to an ℓ-adic Euler characteristic. Finally, a comparison theorem motivates to consider the usual topological Euler characteristic $\chi(\overset{\circ}{E}_J)$ of the complex points of $\overset{\circ}{E}_J$ as the limit of $\sharp(\overset{\circ}{\overline{E}}_J(\mathbb{F}_p))$. (It should be clear what we mean by $\overset{\circ}{E}_J$; we make it explicit below.) We conclude that heuristically

$$\lim_{p\to 1} Z_p(f;s) = \sum_{J\subseteq I} \chi(\mathring{E}_j) \prod_{i\in J} \frac{1}{N_i s + \nu_i}.$$

Denef and Loeser were inspired by this heuristic argument to define a new singularity invariant for any $f \in \mathbb{C}[x_1,\ldots,x_n]$.

Consider an embedded resolution $h : Y \to \mathbb{A}^n_{\mathbb{C}}$ of f as in Theorem 4.1 with $K = \mathbb{C}$. Furthermore, for $J \subseteq I$, we denote $E_J = \bigcap_{j\in J} E_j$ and $\mathring{E}_J = E_J \setminus \bigcup_{k\notin J} E_k$. (In particular, $\mathring{E}_\emptyset = Y \setminus \bigcup_{i\in I} E_i$.)

DEFINITION 7.1 (Denef-Loeser, 1991 [**DL1**]). Let $f \in \mathbb{C}[x_1,\ldots,x_n] \setminus \mathbb{C}$ and choose an embedded resolution $h : Y \to \mathbb{A}^n_{\mathbb{C}}$ of f. The *topological zeta function of f* is

$$Z_{top}(s) = Z_{top}(f;s) := \sum_{J\subseteq I} \chi(\mathring{E}_J) \prod_{i\in J} \frac{1}{N_i s + \nu_i}.$$

REMARK 7.2. It is not clear that the topological zeta function is well-defined, i.e., that it is independent of the chosen embedded resolution. Denef and Loeser proved this by turning the heuristic discussion above into an exact argument, using ℓ-adic interpolation. An easier and more geometric approach is to use the following *weak factorisation theorem* (that was not known in 1991). It reduces the problem to showing that the defining expression of $Z_{top}(s)$ is invariant under an admissible blow-up, which is an easy calculation.

THEOREM 7.3 (Weak factorisation theorem, [**AKMW**][**Wł1**]).

(1) *Consider a proper birational map between smooth, irreducible varieties $\phi : Y \dashrightarrow Y'$ and an open set $U \subseteq Y$ such that the restriction of ϕ to U is an isomorphism. Then ϕ can be factored into a sequence of blow-ups and blow-downs with smooth centres disjoint from U.*

(2) *If $Y\setminus U$ and $Y'\setminus\phi(U)$ are normal crossings divisors, then that factorisation can be chosen such that at each step the images or inverse images of these divisors are also normal crossings divisors and such that the centres of blow-up intersect these divisors transversely.*

The topological zeta function is an interesting singularity invariant on its own, and also a useful test case for studying (the poles of) the Igusa zeta function.

QUESTION 7.4. It is a challenging open problem to give an intrinsic definition of the topological zeta function.

7.2. Motivic zeta function. Kontsevich suggested the idea of motivic integration (an analogue of p-adic integration) in a lecture at Orsay in 1995. This theory has been further developed by Denef and Loeser, amongst others. The motivic zeta function is defined as an analogue of the Igusa zeta function, using motivic integration instead of p-adic integration.

DEFINITION 7.5 (Grothendieck ring of k-varieties). Let k be any field. The Grothendieck group of varieties $K_0(\mathrm{Var}_k)$ is the quotient of the free group generated by the symbols $[X]$, where X runs over all varieties over k, by the relations

$$[X] = [Y] \text{ if } X \cong Y,$$
$$[X \setminus Y] + [Y] = [X] \text{ if } Y \text{ is a closed subvariety of } X.$$

Multiplication determined by

$$[X] \cdot [Y] := [X \times_k Y]$$

turns $K_0(\mathrm{Var}_k)$ into a commutative ring, called the *Grothendieck ring of k-varieties*. We denote by $\mathbb{L}$ the class of the affine line $\mathbb{A}^1_k$ and by $(K_0(\mathrm{Var}_k))_{\mathbb{L}}$ the localization of $K_0(\mathrm{Var}_k)$ with respect to $\mathbb{L}$.

The motivic zeta function is defined as an integral over the 'arc space' $k[[t]]^n$, which is the analogue of the space of p-adic integers $\mathbb{Z}_p^n$. It is a power series in the formal variable T, which is sometimes written as $\mathbb{L}^{-s}$ (with a formal s), to stress the analogy with p^{-s}. That arc space carries a natural *motivic measure* $d\mu$, with values in $(K_0(\mathrm{Var}_k))_{\mathbb{L}}$, which is the analogue of the Haar measure on $\mathbb{Z}_p^n$. For a detailed explanation, we refer to [**DL2**] and [**DL3**].

DEFINITION 7.6 (Motivic zeta function). Let k be a field of characteristic zero and $f \in k[x_1, \ldots, x_n]$. The associated *motivic zeta function* is a formal power series over $(K_0(\mathrm{Var}_k))_{\mathbb{L}}$ in the variable $T = \mathbb{L}^{-s}$, given by the motivic integral

$$Z_{mot}(f;s) = Z'_{mot}(f;T) := \int_{k[[t]]^n} (\mathbb{L}^{-s})^{\mathrm{ord}_t f(x)} d\mu.$$

More concretely, define

$$\mathcal{X}_i = \left\{ \gamma \in \left(\frac{k[[t]]}{(t^{i+1})} \right)^n \mid \mathrm{ord}_t f_i(\gamma) = i \right\},$$

where $f_i : (k[[t]]/(t^{i+1}))^n \to k[[t]]/(t^{i+1})$ is the natural extension of $'f : k^n \to k$, and for $\gamma \in k[[t]]/(t^{i+1})$ we denote by $\mathrm{ord}_t(\gamma) \in \{0, 1, \ldots, i, +\infty\}$ the highest power of t that divides γ. Then we have

$$Z_{mot}(f;s) = \frac{1}{\mathbb{L}^n} \sum_{i \geq 0} [\mathcal{X}_i] \mathbb{L}^{-in-is}.$$

In order to see also the analogy of this expression with the Igusa zeta function, verify that, for $f \in \mathbb{Q}_p[x_1, \ldots, x_n]$, we have

$$Z_p(f;s) = \frac{1}{p^n} \sum_{i \geq 0} \sharp \left(\left\{ a \in \left(\frac{\mathbb{Z}_p}{(p^{i+1})} \right)^n \mid \mathrm{ord}_p f(a) = i \right\} \right) p^{-in-is}.$$

The following table compares the notions related to the Igusa zeta function with those related to the motivic zeta function.

Igusa zeta function	**Motivic zeta function**
$f \in \mathbb{Z}[x_1, \dots, x_n]$	$f \in k[x_1, \dots, x_n]$
$M_i = \{$number of solutions of $f = 0$ over ${}^{\mathbb{Z}}/_{p^i\mathbb{Z}}\}$	$\mathcal{M}_i$ is the class of $\{$solutions of $f = 0$ over ${}^{k[t]}/_{(t^i)}\}$ in $K_0(\mathrm{Var}_k)$
$P(T) := \sum_{i \geq 0} M_i (p^{-n}T)^i$	$\mathcal{P}(T) := \sum_{i \geq 0} \mathcal{M}_i (\mathbb{L}^{-n}T)^i \in K_0(\mathrm{Var}_k)_{\mathbb{L}}[[T]]$
$\mathbb{Z}_p = \varprojlim_i {}^{\mathbb{Z}}/_{p^i\mathbb{Z}}$	$k[[t]] = \varprojlim_i {}^{k[t]}/_{(t^i)}$
$Z_p(f; s) := \int_{\mathbb{Z}_p^n} (p^{-s})^{\mathrm{ord}_p f(x)} \lvert dx \rvert$	$Z_{mot}(f; s) := \int_{k[[t]]^n} (\mathbb{L}^{-s})^{\mathrm{ord}_t f(x)} d\mu$
$P(T) \rightsquigarrow Z_p(f; s)$	$\mathcal{P}(T) \rightsquigarrow Z_{mot}(f; s)$

With the last line, we mean that $Z_{mot}(f; s)$ is related to $\mathcal{P}(T)$, similarly as in Proposition 2.8 for the p-adic case.

THEOREM 7.7 (Denef-Loeser [**DL2**]). *Let k be a field of characteristic zero and $f \in k[x_1, \dots, x_n] \setminus k$. Consider an embedded resolution $h : Y \to \mathbb{A}_k^n$ of f as in Theorem 4.1, and put again $\mathring{E}_J = (\bigcap_{j \in J} E_j) \setminus (\bigcup_{k \notin J}) E_k$ for $J \subseteq I$. Then*

$$Z_{mot}(f; s) = \frac{1}{\mathbb{L}^n} \sum_{J \subseteq I} [\mathring{E}_J] \prod_{i \in J} \frac{\mathbb{L} - 1}{\mathbb{L}^{N_i s + \nu_i} - 1}.$$

In particular, this implies that $Z_{mot}(s)$ and $\mathcal{P}(T)$ are rational functions in $T = \mathbb{L}^{-s}$.

REMARK 7.8 ([**DL2**]).

(1) When $k = \mathbb{C}$, the motivic zeta function $Z_{mot}(f; s)$ specialises to the topological zeta function $Z_{top}(f; s)$. This gives a new proof that $Z_{top}(f; s)$ is independent of the chosen resolution.
(2) When $k = \mathbb{Q}$, the motivic zeta function $Z_{mot}(f; s)$ specialises to the Igusa zeta functions $Z_p(f; s)$ for all but finitely many p.

QUESTION 7.9. Take again $f \in \mathbb{Z}[x_1, \dots, x_n]$ and fix a prime number p. Define

$$M_i' := \sharp \left\{ \text{solutions of } f = 0 \text{ in } \left(\frac{\mathbb{F}_p[u]}{(u^i)} \right)^n \right\}.$$

Define $P'(T) := \sum_{i \geq 0} M_i' (p^{-n}T)^i$. It is an open problem whether $P'(T)$ is a rational function in T.

REMARK 7.10. For a given $f \in \mathbb{Z}[x_1, \dots, x_n]$, it is known that $P'(T) = P(T)$ for all large enough p.

8. Miscellaneous

We briefly indicate some generalisations and problems that are not treated here.

8.1. Number fields. Everything discussed before generalises to number fields in the following way. Let F be a number field with ring of integers $\mathcal{O}$, and fix a maximal ideal $\mathfrak{p}$ of $\mathcal{O}$. Instead of counting solutions of $f \in \mathbb{Z}[x_1, \ldots, x_n]$ in $(\mathbb{Z}/p^k\mathbb{Z})^n$, we count solutions of $f \in \mathcal{O}[x_1, \ldots, x_n]$ in $(\mathcal{O}/\mathfrak{p}^k\mathcal{O})^n$.

$$\begin{array}{ccccc} \mathbb{Q} & \hookleftarrow & \mathbb{Z} & & \text{prime } p \\ \downarrow & & \downarrow & & \downarrow \\ \text{number field } F & \hookleftarrow & \text{ring of integers } \mathcal{O} & & \text{maximal ideal } \mathfrak{p} \end{array}$$

Similarly, $\mathbb{Q}_p$ and $\mathbb{Z}_p$ can be replaced by a finite extension K of $\mathbb{Q}_p$ and its valuation ring $\mathcal{R}$.

$$\begin{array}{ccc} \mathbb{Q}_p & \hookleftarrow & \mathbb{Z}_p \\ \downarrow & & \downarrow \\ \text{finite field extension } K & \hookleftarrow & \text{valuation ring } \mathcal{R} \end{array}$$

The corresponding Igusa zeta function in this setting is (the meromorphic continuation of) $\int_{\mathcal{R}^n} |f(x)|_{\mathfrak{p}}^s\,|dx|$.

8.2. Contribution and non-contribution when n ≥ 3. For polynomials in $n = 2$ variables, the determination of the poles of the associated Igusa zeta function $Z(s)$ is well-understood, as discussed in Section 6. However, as soon as n is at least 3, this is an open problem. For some partial results concerning non-contribution of candidate poles, mainly for $n = 3$, we refer to [**Ve3**] and [**Ve6**]. For some contribution results in arbitrary dimension, see [**Ro**].

8.3. Congruences of multiple polynomials. Instead of counting the solutions of a single polynomial, one can generalise this to studying congruences of several polynomials. More precisely, given polynomials $f_1, \ldots, f_r \in \mathbb{Z}[x_1, \ldots, x_n]$ and a fixed prime number p, define $M_i = \{a \in (\mathbb{Z}/p^i\mathbb{Z})^n \mid f_k(a) \equiv 0 \bmod p^i \text{ for all } k = 1, \ldots, r\}$ for $i \in \mathbb{Z}_{\geq 0}$. Once again, the goal is the understand/compute these numbers. In this setting there is also a corresponding Igusa zeta function $Z(s)$, related to the generating Poincaré series of the M_i; it is defined as

$$Z(s) := \int_{\mathbb{Z}_p^n} (\max_i |f_i|_p)^s |dx|,$$

where $\max_i |\cdot|_p$ is the usual norm on $\mathbb{Q}_p^r$. For results in this setting we refer to [**VZ1**].

8.4. Monodromy conjecture. One of the most intriguing conjectures in singularity theory is the *monodromy conjecture*. Given $f \in \mathbb{Q}[x_1, \ldots, x_n]$, it relates the poles of its associated p-adic Igusa zeta functions (for all but finitely many p), which are number theoretic invariants of f, to eigenvalues of local monodromy of f, which are geometric/topological invariants of f, considered as function on $\mathbb{C}^n$. A variant is also considered for the topological and motivic zeta functions. See for example [**Ni**] for an introduction to this topic.

Acknowledgment

We thank the referee for carefully reading these notes and for providing some valuable remarks.

References

[At] M. F. Atiyah, *Resolution of singularities and division of distributions*, Comm. Pure Appl. Math. **23** (1970), 145–150, DOI 10.1002/cpa.3160230202. MR256156

[AKMW] D. Abramovich, K. Karu, K. Matsuki, and J. Włodarczyk, *Torification and factorization of birational maps*, J. Amer. Math. Soc. **15** (2002), no. 3, 531–572, DOI 10.1090/S0894-0347-02-00396-X. MR1896232

[Be] I. N. Bernšteĭn, *Analytic continuation of generalized functions with respect to a parameter*, Funkcional. Anal. i Priložen. **6** (1972), no. 4, 26–40. MR0320735

[BEV] A. M. Bravo, S. Encinas, and O. Villamayor U., *A simplified proof of desingularization and applications*, Rev. Mat. Iberoamericana **21** (2005), no. 2, 349–458, DOI 10.4171/RMI/425. MR2174912

[BG] I. N. Bernšteĭn and S. I. Gel′fand, *Meromorphy of the function* P^λ (Russian), Funkcional. Anal. i Priložen. **3** (1969), no. 1, 84–85. MR0247457

[BS] S. I. Borewicz and I. R. Šafarevič, *Zahlentheorie* (German), Lehrbücher und Monographien aus dem Gebiete der Exakten Wissenschaften, Mathematische Reihe, Band 32, Birkhäuser Verlag, Basel-Stuttgart, 1966. Aus dem Russischen übersetzt von Helmut Koch. MR0195802

[De1] J. Denef, *On the degree of Igusa's local zeta function*, Amer. J. Math. **109** (1987), no. 6, 991–1008, DOI 10.2307/2374583. MR919001

[De2] J. Denef, *Local zeta functions and Euler characteristics*, Duke Math. J. **63** (1991), no. 3, 713–721, DOI 10.1215/S0012-7094-91-06330-1. MR1121152

[De3] J. Denef, *Report on Igusa's local zeta function*, Astérisque **201-203** (1991), Exp. No. 741, 359–386 (1992). Séminaire Bourbaki, Vol. 1990/91. MR1157848

[DvdD] J. Denef and L. van den Dries, *p-adic and real subanalytic sets*, Ann. of Math. (2) **128** (1988), no. 1, 79–138, DOI 10.2307/1971463. MR951508

[DH] J. Denef and K. Hoornaert, *Newton polyhedra and Igusa's local zeta function*, J. Number Theory **89** (2001), no. 1, 31–64, DOI 10.1006/jnth.2000.2606. MR1838703

[DL1] J. Denef and F. Loeser, *Caractéristiques d'Euler-Poincaré, fonctions zêta locales et modifications analytiques* (French), J. Amer. Math. Soc. **5** (1992), no. 4, 705–720, DOI 10.2307/2152708. MR1151541

[DL2] J. Denef and F. Loeser, *Motivic Igusa zeta functions*, J. Algebraic Geom. **7** (1998), no. 3, 505–537. MR1618144

[DL3] J. Denef and F. Loeser, *Germs of arcs on singular algebraic varieties and motivic integration*, Invent. Math. **135** (1999), no. 1, 201–232, DOI 10.1007/s002220050284. MR1664700

[ENV] S. Encinas, A. Nobile, and O. Villamayor U., *On algorithmic equi-resolution and stratification of Hilbert schemes*, Proc. London Math. Soc. (3) **86** (2003), no. 3, 607–648, DOI 10.1112/S0024611502013862. MR1974392

[Hi] H. Hironaka, *Resolution of singularities of an algebraic variety over a field of characteristic zero. I, II*, Ann. of Math. (2) **79** (1964), 109–203; ibid. (2) **79** (1964), 205–326, DOI 10.2307/1970547. MR0199184

[Ig1] J.-i. Igusa, *Complex powers and asymptotic expansions I*, J. Reine Angew. Math. 268/269 (1974), 110–130; II, ibid. 278/279 (1975), 307–321.

[Ig2] J.-i. Igusa, *Some aspects of the arithmetic theory of polynomials*, Discrete groups in geometry and analysis (New Haven, Conn., 1984), Progr. Math., vol. 67, Birkhäuser Boston, Boston, MA, 1987, pp. 20–47, DOI 10.1007/978-1-4899-6664-3_2. MR900822

[Ig3] J.-i. Igusa, *A stationary phase formula for p-adic integrals and its applications*, Algebraic geometry and its applications (West Lafayette, IN, 1990), Springer, New York, 1994, pp. 175–194. MR1272029

[Ig4] J.-i. Igusa, *An introduction to the theory of local zeta functions*, AMS/IP Studies in Advanced Mathematics, vol. 14, American Mathematical Society, Providence, RI; International Press, Cambridge, MA, 2000, DOI 10.1090/amsip/014. MR1743467

[IS] D. Ibadula and D. Segers, *Determination of the real poles of the Igusa zeta function for curves*, Rev. Mat. Complut. **25** (2012), no. 2, 581–597, DOI 10.1007/s13163-011-0073-8. MR2931426

[Lo] F. Loeser, *Fonctions d'Igusa p-adiques et polynômes de Bernstein* (French), Amer. J. Math. **110** (1988), no. 1, 1–21, DOI 10.2307/2374537. MR926736

[LV] A. Laeremans and W. Veys, *On the poles of maximal order of the topological zeta function*, Bull. London Math. Soc. **31** (1999), no. 4, 441–449, DOI 10.1112/S0024609398005566. MR1687560

[LZ] E. León-Cardenal and W. A. Zúñiga-Galindo, *An introduction to the theory of local zeta functions from scratch* (English, with English and Spanish summaries), Rev. Integr. Temas Mat. **37** (2019), no. 1, 45–76, DOI 10.18273/revint. MR3920742

[Me] D. Meuser, *A survey of Igusa's local zeta function*, Amer. J. Math. **138** (2016), no. 1, 149–179, DOI 10.1353/ajm.2016.0006. MR3462883

[Ni] J. Nicaise, *An introduction to p-adic and motivic zeta functions and the monodromy conjecture*, Algebraic and analytic aspects of zeta functions and L-functions, MSJ Mem., vol. 21, Math. Soc. Japan, Tokyo, 2010, pp. 141–166. MR2647606

[NX] J. Nicaise and C. Xu, *Poles of maximal order of motivic zeta functions*, Duke Math. J. **165** (2016), no. 2, 217–243, DOI 10.1215/00127094-3165648. MR3457672

[Ro] B. Rodrigues, *On the geometric determination of the poles of Hodge and motivic zeta functions*, J. Reine Angew. Math. **578** (2005), 129–146, DOI 10.1515/crll.2005.2005.578.129. MR2113892

[Se] J.-P. Serre, *Classification des variétés analytiques p-adiques compactes* (French), Topology **3** (1965), 409–412, DOI 10.1016/0040-9383(65)90005-4. MR179170

[Sg1] D. Segers, *Lower bound for the poles of Igusa's p-adic zeta functions*, Math. Ann. **336** (2006), no. 3, 659–669, DOI 10.1007/s00208-006-0016-8. MR2249763

[Sg2] D. Segers, *The asymptotic behaviour of the number of solutions of polynomial congruences*, An. Ştiinţ. Univ. "Ovidius" Constanţa Ser. Mat. **19** (2011), no. 1, 255–262. MR2785701

[Ve1] W. Veys, *Relations between numerical data of an embedded resolution*, Amer. J. Math. **113** (1991), no. 4, 573–592, DOI 10.2307/2374841. MR1118455

[Ve2] W. Veys, *Congruences for numerical data of an embedded resolution*, Compositio Math. **80** (1991), no. 2, 151–169. MR1132091

[Ve3] W. Veys, *Poles of Igusa's local zeta function and monodromy* (English, with English and French summaries), Bull. Soc. Math. France **121** (1993), no. 4, 545–598. MR1254752

[Ve4] W. Veys, *Determination of the poles of the topological zeta function for curves*, Manuscripta Math. **87** (1995), no. 4, 435–448, DOI 10.1007/BF02570485. MR1344599

[Ve5] W. Veys, *Zeta functions for curves and log canonical models*, Proc. London Math. Soc. (3) **74** (1997), no. 2, 360–378, DOI 10.1112/S0024611597000130. MR1425327

[Ve6] W. Veys, *Vanishing of principal value integrals on surfaces*, J. Reine Angew. Math. **598** (2006), 139–158, DOI 10.1515/CRELLE.2006.072. MR2270570

[VZ1] W. Veys and W. A. Zúñiga-Galindo, *Zeta functions for analytic mappings, log-principalization of ideals, and Newton polyhedra*, Trans. Amer. Math. Soc. **360** (2008), no. 4, 2205–2227, DOI 10.1090/S0002-9947-07-04422-4. MR2366980

[VZ2] W. Veys and W. A. Zúñiga-Galindo, *Zeta functions and oscillatory integrals for meromorphic functions*, Adv. Math. **311** (2017), 295–337, DOI 10.1016/j.aim.2017.02.022. MR3628217

[Wł1] J. Włodarczyk, *Toroidal varieties and the weak factorization theorem*, Invent. Math. **154** (2003), no. 2, 223–331, DOI 10.1007/s00222-003-0305-8. MR2013783

[Wł2] J. Włodarczyk, *Simple Hironaka resolution in characteristic zero*, J. Amer. Math. Soc. **18** (2005), no. 4, 779–822, DOI 10.1090/S0894-0347-05-00493-5. MR2163383

[Wł3] J. Włodarczyk, *Resolution of singularities of analytic spaces*, Proceedings of Gökova Geometry-Topology Conference 2008, Gökova Geometry/Topology Conference (GGT), Gökova, 2009, pp. 31–63. MR2500573

KU Leuven, Departement Wiskunde, Celestijnenlaan 200B, 3001 Leuven, Belgium
Email address: naud.potemans@kuleuven.be

KU Leuven, Departement Wiskunde, Celestijnenlaan 200B, 3001 Leuven, Belgium
Email address: wim.veys@kuleuven.be

Contemporary Mathematics
Volume **778**, 2022
https://doi.org/10.1090/conm/778/15656

An introduction to p-adic and motivic integration, zeta functions and invariants of singularities

Juan Viu-Sos

ABSTRACT. Motivic integration was introduced by Kontsevich to show that birationally equivalent Calabi-Yau manifolds have the same Hodge numbers. To do so, he constructed a certain motivic measure on the arc space of a complex variety, taking values in a completion of the Grothendieck ring of algebraic varieties. Later, Denef and Loeser, together with the works of Looijenga and Batyrev, developed in a series of articles a more complete theory of the subject, with applications in the study of varieties and singularities. In particular, they developed a motivic zeta function, generalizing the usual (p-adic) Igusa zeta function and Denef-Loeser topological zeta function.

These notes are a basic introduction to geometric motivic integration, the precedent p-adic ideas associated with it, and the theory of the above zeta functions related to them. We focus in practical computations and ideas, providing examples and a recent formula obtained by means of partial resolutions.

Introduction

Motivic integration ideas were originally developed by KONTSEVICH [**Kon95**] as a tool to study the relation between additive invariants of varieties, e.g. Betti numbers or Hodge numbers, in the context of Mirror Symmetry in String Theory.

In such a context, the main objects of study are *Calabi-Yau* varieties, i.e. compact, complex algebraic varieties admitting a non-vanishing form of maximal degree. Motivated by the relation between birationally equivalent Calabi-Yau varieties, BATYREV proves the following result.

THEOREM (BATYREV'95, [**Bat99a**]). *Let X and Y be two d-dimensional smooth Calabi-Yau varieties. If X and Y are birationally equivalent, then they have the same Betti numbers, i.e.*

$$b_i(X) = \dim_{\mathbb{C}} H^i(X,\mathbb{C}) = \dim_{\mathbb{C}} H^i(Y,\mathbb{C}) = b_i(Y), \qquad \forall i = 0,\ldots,d.$$

The ingredients of Batyrev's proof are:

- Hironaka's desingularization theorem: creates a common smooth birational model of X and Y.

2020 *Mathematics Subject Classification.* Primary 14B05; Secondary 14E18, 14G10, 11S80, 32S25, 32S45.

Key words and phrases. p-adic integration, motivic integration, zeta functions, Monodromy conjecture, resolution of singularities, quotient singularities, embedded $\mathbb{Q}$-resolutions.

The author was supported by a PNPD/CAPES grant and by a postdoctoral grant #2016/14580-7 by *Fundação de Amparo à Pesquisa do Estado de São Paulo* (FAPESP).

- Reduction mod p^m and Weil's conjectures: strong results proven by DWORK, GROTHENDIECK and DELIGNE about rationality, functional equations and relation with Betti numbers for a zeta function associated with counting points in $X(\mathbb{F}_{p^m})$ for smooth varieties.
- p-adic integration: the p-adic integers encode reductions modulo p^m of algebraic varieties, and p-adic integration provides a way to relate different p-adic volumes (based on counting $X(\mathbb{F}_{p^m})$) by a "change of variables formula".
- Comparison theorem between ℓ-adic cohomology ($\ell \neq p$) and usual Betti numbers.

In a talk made at ORSAY at the end of the same year, KONTSEVICH provided a more direct approach avoiding p-adic integration and Weil's conjectures: he used *arc spaces* $\mathbb{C}[\![t]\!]$ ("t-adic" spaces) instead of p-adic numbers as domain of integration, and constructed an integral with respect to a measure based on a *universal additive invariant of varieties*, i.e. the Grothendieck ring of complex varieties $K_0(\mathrm{Var}_{\mathbb{C}})$. This integration theory comes with a *change of variables formula* relating integrals of different birational equivalent varieties, where the "jacobian" is expressed in terms of the contact order of arcs along divisors. *Motivic integration* was born.

Using these ideas, he generalizes Batyrev's Theorem as a direct consequence of the change of variables formula.

THEOREM (KONTSEVICH, DEC'95). *Under the hypothesis above, X and Y have the same Hodge numbers, i.e.*

$$h^{p,q}(X) = \dim_{\mathbb{C}} H^q(X, \Omega_X^p) = \dim_{\mathbb{C}} H^q(Y, \Omega_Y^p) = h^{p,q}(Y),$$

for any $p, q = 0, \dots, d$.

This theory was quickly developed for arbitrary (in particular singular) algebraic varieties on an algebraic closed field k, char $k = 0$, in a series of articles [**DL98**, **DL99**, **DL02b**, **Loo02**] by DENEF and LOESER, as well as LOOIJENGA. Also, BATYREV [**Bat98**, **Bat99a**] used motivic integration ideas to produce new "stringy" invariants of singularities, generalizing additive invariants "twisted" by numerical values of the relative canonical divisor. Later, CLUCKERS and LOESER [**CL08**] gave a general framework for motivic integration based on model theory.

Focusing on singularities, one of the main applications of this theory is the study of certain zeta functions associated with a polynomial f: the *(p-adic) Igusa zeta function* $Z_{\mathrm{Igusa}}(f;s)$ and the *Denef-Loeser topological zeta function* $Z_{\mathrm{top}}(f;s)$. The function $Z_{\mathrm{Igusa}}(f;s)$ is defined as a p-adic parametric integral of a polynomial f with coefficients in a p-adic field [**Igu74**], and $Z_{\mathrm{top}}(f;s)$ is a rational function constructed in terms of the numerical data associated with an embedded resolution of $V_f = \{f = 0\} \subset \mathbb{C}^d$, with f being now a complex polynomial [**DL92**]. In particular, $Z_{\mathrm{top}}(f;s)$ is obtained as certain limit of $Z_{\mathrm{Igusa}}(f;s)$.

The interesting information is contained in the *poles* of these zeta functions. It is believed that the behavior of the poles is controlled by the topology of the Minor fiber: each pole should correspond to an eigenvalue of the monodromy of the Milnor fiber at some point of V_f. This is known as the *Monodromy conjecture*.

It turns out that both $Z_{\mathrm{Igusa}}(f;s)$ and $Z_{\mathrm{top}}(f;s)$ are specializations of the *motivic zeta function* $Z_{\mathrm{mot}}(f;s)$, defined as a parametric motivic integral [**DL98**]. This gives a more geometric way to understand and manipulate both functions, as well as opens the door to better understand the relation with the topology of the Milnor fiber.

Kontsevich's original theory (sometimes called *"naive"* motivic integration) admits more sophisticated extensions, as the *equivariant* or *monodromic* motivic integration, as well as the *relative* motivic integration with respect to X-varieties. The first one was studied in [**DL99, Loo02**], and they derive the local *motivic Milnor fiber* $\mathcal{S}_{f,0}$, which is an object lying in a ring similar to the one constructed from $K_0(\mathrm{Var}_{\mathbb{C}})$ but also codifying the monodromy action. This object is a "motivic incarnation" of the Milnor fiber $\mathcal{F}_{f,0}$, in the sense that they both have same cohomological invariants, e.g. the Hodge-Steenbrink spectrum, Euler characteristic, etc. In the second case, developed in [**Alu07, dFLNU07**], the authors construct motivic versions of the *Chern-MacPherson classes* for singular varieties using a *relative Grothendieck of X-varieties* $\{V \to X\} \in K_0(\mathrm{Var}_X)$.

Recently, a version of motivic integration in *real algebraic geometry* was developed by several authors [**KP03, Fic05, Cam16, Fic17, Cam17, CFKP19**], with some applications to the study of blow-analytic equivalence, or to Lipschitz inverse map theorems of germs of arc-analytic homeomorphisms.

A first version of these notes was produced as support material of a mini-course about this subject for graduate students which I lectured at ICMC-USP. The notes were completed after a second mini-course at IMPA as part of the *Thematic Program on Singularity Theory 2020.*

The main purpose of these notes is to give a geometric introduction of motivic integration in complex geometry and the related zeta functions for graduated students, focusing on concrete examples and practical computations. One of the main goals is that the reader could feel comfortable once she/he deals with more advanced techniques and papers based on motivic integration and zeta functions. In order to do that, we present a panorama of the basic theory, organized as follows:

- SECTION 1: some prehistory in arithmetic problems about counting solutions modulo p^m, p-adic integration and the Igusa zeta function, Monodromy conjecture and the Denef-Loeser topological zeta function.
- SECTION 2: construction of "naive" motivic integration and its specializations, change of variables formula, proof of Kontsevich's theorem, formulas from normal crossings and embedded resolutions.
- SECTION 3: first applications on singularities, Bartyrev's stringy invariants and the motivic zeta function, techniques using quotient singularities and formulas from embedded $\mathbb{Q}$-resolutions.

We stress the use of resolution of singularities to obtain formulas, detailing one of the classic proofs of such a formula for the motivic integral. In addition, we introduce a recent formula obtained in [**LMVV20**], which computes motivic integrals and zeta functions from "partial resolutions" involving orbifolds and is very practical in order to make computations.

The content of these notes is mainly produced following [**Pop**] (for a complete introduction of the relation between p-adic integration, topology of algebraic varieties, Weil and Igusa zeta functions and the basic theory of motivic integration), the detailed introductions [**Vey06, Cra04**] (for a more general theory on motivic integration and stringy invariants) and also [**Bli11, Nic10**]. More complete and good surveys on the motivic subject can be found in [**Loo02**] and [**DL01**]. Recently, CHAMBERT-LOIR, NICAISE and SEBAG published a book on motivic integration [**CLNS18**] which covers a huge part of the current theory. There is also recent nice introductions to Igusa zeta functions and its relations with oscillatory integrals, to appear in the present Volume [**PV22, LC22**].

Several exercises appear among these notes, the hardest ones are placed at the end of the first and last section.

1. Prehistory: Counting $\mathbb{F}_p$-points, p-adic integration and the Igusa zeta function

The theory of p-adic integration and zeta functions is a rich area which involves analysis over totally disconnected fields, theory of local zeta functions, arithmetic problems, ... and has multiple interactions which real and complex analysis, theory of partial differential equations, algebraic geometry as well as singularity theory. See [**Meu16, LZ19**] for recent surveys on p-adic local zeta functions and related problems, as well as many bibliography references. For an extensive introduction, see IGUSA's book [**Igu00**], and also DENEF's report [**Den91**].

1.1. A problem from number theory. Let $f \in \mathbb{Z}[X_1, \dots, X_d]$ and fix p an arbitrary prime number. We want to investigate the number of solutions of f modulo a power p^m, i.e.

$$m \geq 0: \quad \mathcal{N}_m(f) := \#\{x \in (\mathbb{Z}/p^m\mathbb{Z})^d \mid f(x) \equiv 0 \mod p^m\}$$

with the convention $\mathcal{N}_0(f) = 1$.

EXAMPLE 1.1.

(1) $f_0(x) = x$: we have $x \equiv 0 \mod p^m$ if and only if $x = 0 \in \mathbb{Z}/p^m\mathbb{Z}$, then one has $\mathcal{N}_m(f_0) = 1$ for all $m \geq 0$.

(2) $f_1(x) = x^2$: so $x^2 \equiv 0 \mod p^m$. Studying it for different values of m, we see that:

- $p|x^2 \Leftrightarrow p|x$, then $\mathcal{N}_1(f_1) = 1$.
- $p^2|x^2 \Leftrightarrow p|x$, then $\mathcal{N}_2(f_1) = p$.
- $p^3|x^2 \Leftrightarrow p^2|x$, then $\mathcal{N}_3(f_1) = p$.
- $p^4|x^2 \Leftrightarrow p^2|x$, then $\mathcal{N}_4(f_1) = p^2$.
- $p^5|x^2 \Leftrightarrow p^3|x$, then $\mathcal{N}_5(f_1) = p^2$.
- It is easy to show (EXERCISE): $\mathcal{N}_{2k}(f_1) = \mathcal{N}_{2k+1}(f_1) = p^k$.

(3) $f_2(x, y) = y - x^2$: Fixing an arbitrary $x \in \mathbb{Z}/p^m\mathbb{Z}$, y is uniquely determined by $y \equiv x^2 \mod p^m$. Then, $\mathcal{N}_m(f_2) = p^m$.

(4) $f_3(x, y) = xy$: EXERCISE: $\mathcal{N}_m(f_3) = (m + 1)p^m - mp^{m-1}$.

(5) $f_4(x, y) = y^2 - x^3$: we list the first values. We have that $\mathcal{N}_1(f_4) = p$ and $\mathcal{N}_m(f_4) = p^{m-1}F(p)$ as follows:

- $2 \leq m \leq 5$: $F(p) = 2p - 1$.
- $6 \leq m \leq 7$: $F(p) = p^2 + p - 1$.
- $8 \leq m \leq 11$: $F(p) = 2p^2 - 1$.
- $12 \leq m \leq 13$: $F(p) = p^3 + p^2 - 1$.
- $14 \leq m \leq 17$: $F(p) = 2p^3 - 1$.

etc.

Note that, if we look at the complex sets $V_i = \{f_i = 0\}$, V_2 is smooth, V_3 has a simple nodal singularity and V_4 is an ordinary cusp. In fact, the behavior of $\mathcal{N}_m(f)$ turns out to be "more complicated" precisely when $\{f = 0\} \subset \mathbb{C}^d$ has singularities.

1.2. An arithmetic-geometric approach. Consider the associated *Poincaré power series*

$$Q(f;T) := \sum_{m \geq 0} \mathcal{N}_m(f) T^m \in \mathbb{Z}[\![T]\!], \tag{1.1}$$

codifying the number of solutions modulo p^m. Coming back to the previous examples:

EXAMPLE 1.2.

(1) $Q(f_0;T) = 1 + T + T^2 + \cdots = \frac{1}{1-T}$.

(2) $Q(f_1;T) = 1 + T + pT^2 + pT^3 + \cdots = (1+T)(1+pT^2+p^2T^4+\cdots) = \frac{1+T}{1-pT^2}$.

(3) $Q(f_2;T) = \frac{1}{1-pT}$.

(4) We claim that $Q(f_4;T) = \frac{1-(p-p^2)T^2-p^6T^6}{(1-p^7T^6)(1-pT)}$.

EXERCISE 1.3. Compute $Q(f_3;T)$ and $Q(g;T)$ where $g = x_1^{N_1} \cdots x_d^{N_d}$ and $N_1, \ldots, N_d \geq 1$.

In fact, $\mathcal{N}_m(f)$ has a regular behavior, as it was conjectured by BOREWICZ and SHAFAREVICH and then proved by Igusa.

THEOREM 1.4 (IGUSA,[**Igu74**]). *$Q(f;T)$ is a rational function, i.e. $Q(f;T) \in \mathbb{Q}(T)$.*

We are going to prove the previous theorem at the end of Section 1.6. The main ideas of Igusa's proof are:

- Take $T = p^{-s}$ and express $Q(f, p^{-s})$ in terms of a p-adic integral $\int_{\mathbb{Z}_p} |f|_p^s \, |dx|$ (the *Igusa zeta function*).
- An embedded resolution of singularities of $\{f = 0\} \subset \mathbb{C}^d$.
- A change of variables formula for p-adic integrals.

REMARK 1.5. In fact, $Q(f;T)$ can be computed from an embedded resolution of $\{f = 0\} \subset \mathbb{C}^d$.

1.3. Basics on p-adic numbers.

1.3.1. *Construction.* Fix a prime number p. The p-adics give an analytic way to deal with problems of polynomials in $\mathbb{Z}/p^m\mathbb{Z}$, since any root modulo p^m can be lifted in a p-adic root (Hensel's Lemma).

DEFINITION 1.6. Let $0 \neq x \in \mathbb{Q}$. Consider the unique presentation $x = p^m \cdot \frac{a}{b}$, where $m \in \mathbb{Z}$, a/b irreducible and both $p \nmid a$ and $p \nmid b$. We define in $\mathbb{Q}$:

- The *order* $\mathrm{ord}_p : \mathbb{Q} \to \mathbb{Z} \sqcup \{\infty\}$ by
$$\mathrm{ord}_p(x) := \begin{cases} m & \text{if } x \neq 0 \\ +\infty & \text{if } x = 0 \end{cases}.$$
- The *(p-adic) absolute value:*
$$|x|_p := \begin{cases} p^{-\mathrm{ord}_p(x)} & \text{if } x \neq 0 \\ 0 & \text{if } x = 0 \end{cases}.$$

The idea here is that numbers which are divisible by large powers of p are considered small.

EXERCISE 1.7. Prove that the map $|\cdot|_p : \mathbb{Q} \to \mathbb{R}_{\geq 0}$ is a *non-archimedean absolute value* (or *ultrametric*) on $\mathbb{Q}$, i.e. for any $x, y \in \mathbb{Q}$:

(1) $|x|_p \geq 0$, and $|x|_p = 0$ if and only if $x = 0$.

(2) $|xy|_p = |x|_p \cdot |y|_p$.

(3) $|x+y|_p \leq \max\{\, |x|_p, |y|_p \,\}$.

We define a topology on $\mathbb{Q}$ induced by the *distance* $d(x,y) := |x-y|_p$, for any $x, y \in \mathbb{Q}$.

THEOREM 1.8 (OSTROWSKI, [**Ost16**]). *Let* $\|\cdot\|$ *be a non-trivial absolute value on* $\mathbb{Q}$*. Then* $\|\cdot\|$ *is equivalent either to the usual absolute value* $|\cdot|$*, or to a p-adic one* $|\cdot|_p$.

DEFINITION 1.9. The *field of p-adic numbers* $\mathbb{Q}_p$ is defined as the completion of $(\mathbb{Q}, |\cdot|_p)$, i.e. the set of equivalence classes of Cauchy sequences with respect to $|\cdot|_p$.

REMARK 1.10.

(1) We have an embedding $\mathbb{Q} \hookrightarrow \mathbb{Q}_p$, identifying any $x \in \mathbb{Q}$ with the constant sequence $(x, x, x, x, x, \ldots)$.

(2) As a consequence, $\mathrm{char}\, \mathbb{Q}_p = 0$.

(3) Every $x \in \mathbb{Q}_p$ could be standardly represented by a unique "Laurent series expansion in base p", i.e.
$$x = \sum_{k \geq \mathrm{ord}_p(x)} a_k p^k$$
with $a_{\mathrm{ord}_p(x)} \neq 0$ and $a_k \in \{0, \ldots, p-1\}$ for any k.

EXERCISE 1.11. Prove that the following identities:
$$-1 = (p-1) + (p-1)p + (p-1)p^2 + \cdots \quad \text{and} \quad \frac{1}{1-p} = 1 + p + p^2 + \cdots,$$
hold in $\mathbb{Q}_p$.

1.3.2. *Topology.* As $\mathbb{Q}_p$ is a normed space, we can consider the unit disk space, which is also a subring of $\mathbb{Q}_p$.

DEFINITION 1.12. The *ring of p-adic integers* is defined as
$$\mathbb{Z}_p := \left\{ x \in \mathbb{Q}_p \mid |x|_p \leq 1 \right\},$$

or, equivalently, the set of $x = a_k p^k + a_{k+1}p^{k+1} + \cdots \in \mathbb{Q}_p$ such that $a_k = 0$ for any $k < 0$.

PROPOSITION 1.13.

(1) *$\mathbb{Q}_p$ is a totally disconnected locally compact topological space.*

(2) *$\mathbb{Z}_p$ is open and closed in $\mathbb{Q}_p$, moreover $\mathbb{Z}_p$ is compact.*

(3) *$\mathbb{Z}_p$ is a local ring, with maximal ideal $p\mathbb{Z}_p = \left\{x \in \mathbb{Z}_p \mid |x|_p < 1\right\}$ and residue field*

$$\mathbb{Z}_p/p\mathbb{Z}_p \simeq \mathbb{F}_p.$$

As a consequence, one has a decomposition in equivalence classes:

$$\mathbb{Z}_p = p\mathbb{Z}_p \sqcup (1 + p\mathbb{Z}_p) \sqcup \cdots \sqcup (p - 1 + p\mathbb{Z}_p).$$

(4) *$\mathbb{Q}_p$ has a basis of open and closed neighborhoods given by elements of the form*

$$a + p^m\mathbb{Z}_p = \left\{x \in \mathbb{Q}_p \;\middle|\; |x - a|_p \leq p^{-m}\right\},$$

for any $a \in \mathbb{Q}_p$ and $m \in \mathbb{N}$.

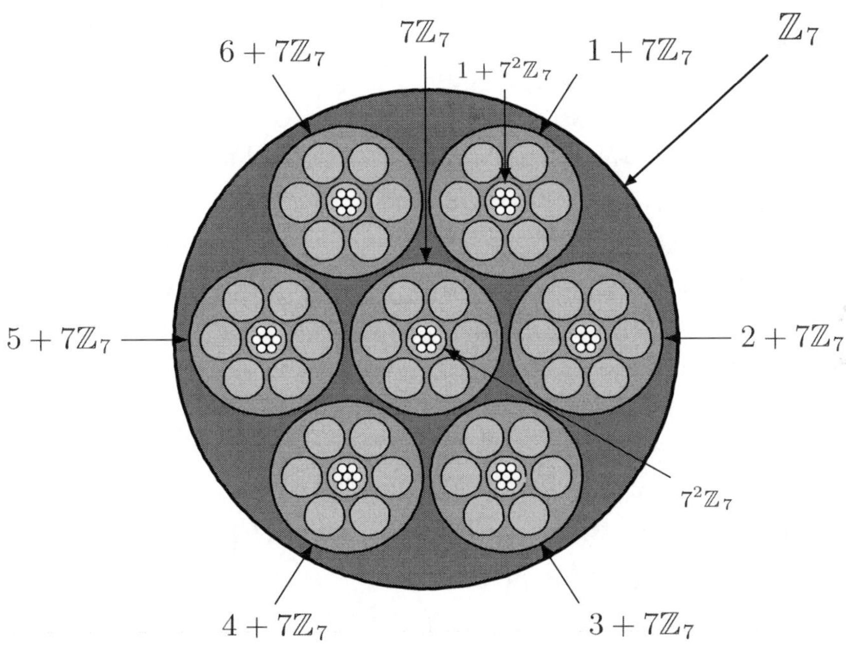

FIGURE 1. Topology of $\mathbb{Z}_7$, as the union of translations of $7^m\mathbb{Z}_7$, for $m \geq 0$.

REMARK 1.14.

(1) There is an algebraic way to construct the p-adic numbers: for any $m, n \in \mathbb{N}$, $m \geq n$, consider the natural projections

$$\pi_n^m : \mathbb{Z}/p^{m+1}\mathbb{Z} \longrightarrow \mathbb{Z}/p^{n+1}\mathbb{Z}$$

given by the natural reduction modulo p^{n+1}. We have thus an *inverse system*

$$\left\{ \left(\mathbb{Z}/p^{m+1}\mathbb{Z}\right)_{m\in\mathbb{N}}, \left(\pi_n^m\right)_{m,n\in\mathbb{N}} \right\}.$$

The p-adic integers are expressed as the *inverse limit*:

$$\mathbb{Z}_p = \varprojlim_m \left(\mathbb{Z}/p^{m+1}\mathbb{Z}\right).$$

Then, $\mathbb{Q}_p$ is simply the field of fractions of $\mathbb{Z}_p$. Moreover, using the representations we deduce that $\mathbb{Q}_p \simeq \left(p^{\mathbb{N}}\right)^{-1}\mathbb{Z}_p^\times$, i.e. the localization of $\mathbb{Z}_p^\times = \left\{x \in \mathbb{Q}_p \mid |x|_p = 1\right\}$, the *units* of $\mathbb{Z}_p$, by the multiplicative system given by non-negative powers of p.

(2) The previous p-adic construction can be extended to any *discrete valuation ring* $(R, \mathfrak{m})$ via completions in the $\mathfrak{m}$-adic topology (see for example [**Pop**]). Then, taking K a finite extension of $\mathbb{Q}_p$ and its corresponding integral closure $\mathcal{O}_K$ of $\mathbb{Z}_p$ in K, all the following theory in this section can be generalized.

1.4. Affine p-adic integration.

1.4.1. *p-adic Haar measure.* We use the basis of neighborhoods described in Proposition 1.13 to define a *Haar measure* in $\mathbb{Q}_p$. See [**Hal50**, Ch. XI] for more details about Haar measures.

DEFINITION 1.15. Let G be a locally compact topological abelian group. A *Haar measure* defined on G is a Borel measure μ on G satisfying:

(1) $\mu(gE) = \mu(E)$, for any $g \in G$ and for any Borel-measurable $E \subset G$.

(2) $\mu(U) > 0$ for any open set $U \subset G$.

(3) $\mu(K) < +\infty$ for any compact $K \subset G$.

PROPOSITION 1.16 ([**Hal50**, Sec. 60, Thm. C]). *Any locally compact topological abelian group G admits a unique Haar measure up to scalars.*

DEFINITION 1.17. The *normalized Haar measure* μ on $\mathbb{Q}_p$ is defined by

$$\mu(a + p^m\mathbb{Z}_p) = \frac{1}{p^m},$$

for any $a \in \mathbb{Q}_p$ and $m \in \mathbb{N}$.

In particular, μ is invariant by translation and $\mu(\mathbb{Z}_p) = 1$.

EXERCISE 1.18. Verify that $\mu(\mathbb{Z}_p^\times) = 1 - p^{-1}$.

1.4.2. *p-adic cylinders.* Consider $\mathbb{Q}_p^d$ with the product topology and the natural projection $\pi_m : \mathbb{Z}_p^d \to \left(\mathbb{Z}/p^{m+1}\mathbb{Z}\right)^d$, $m \geq 0$. We introduce an elementary concept which will play a central role in the construction of the motivic measure.

DEFINITION 1.19. A subset $C \subset \mathbb{Z}_p^d$ is called a *cylinder* if $C = \pi_m^{-1}(\pi_m(C))$, for some $m \geq 0$.

REMARK 1.20. Any $p^{m+1}\mathbb{Z}_p^d = \pi_m^{-1}(0) = \pi_m^{-1}\left(\pi_m(p^{m+1}\mathbb{Z}_p^d)\right)$ is a cylinder.

The measure of a cylinder can be computed using the following "silly" property, considering the cardinal $|\pi_m(C)|$ of the basis $\pi_m(C) \subset \left(\mathbb{Z}/p^{m+1}\mathbb{Z}\right)^d$.

PROPOSITION 1.21. *For any cylinder $C \subset \mathbb{Z}_p^d$, the sequence*

$$\left(\frac{|\pi_m(C)|}{p^{d(m+1)}}\right)_{m\geq 0}$$

is constant for $m \gg 0$, and its limit is equal to $\mu(C)$. Moreover, if we choose $m_0 > 0$ such that $C = \pi_{m_0}^{-1}(\pi_{m_0}(C))$, then for any $m \geq m_0$, we have

$$\frac{|\pi_m(C)|}{p^{d(m+1)}} = \frac{|\pi_{m_0}(C)|}{p^{d(m_0+1)}} = \mu(C).$$

PROOF. Remember that $|\pi_m(C)|$ is a finite set. For $m \geq m_0$, C can be written as

$$C = \bigsqcup_{a\in\pi_m(C)} a + \left(p^{m+1}\mathbb{Z}_p\right)^d$$

By invariance of the Haar measure,

$$\mu(C) = \sum_{a\in\pi_m(C)} \mu\left(\left(p^{m+1}\mathbb{Z}_p\right)^d\right) = |\pi_m(C)| \cdot \frac{1}{p^{d(m+1)}}.$$

Now, considering the projection $\pi_{m_0}^m : \left(\mathbb{Z}/p^{m+1}\mathbb{Z}\right)^d \to \left(\mathbb{Z}/p^{m_0+1}\mathbb{Z}\right)^d$, it is easy to see that $\left|\left(\pi_{m_0}^m\right)^{-1}(a)\right| = p^{d(m-m_0)}$ for any $a \in \left(\mathbb{Z}/p^{m_0+1}\mathbb{Z}\right)^d$. Since $\pi_{m_0} = \pi_{m_0}^m \circ \pi_m$, we have

$$|\pi_m(C)| = \left|\pi_m\left(\pi_{m_0}^{-1}(\pi_{m_0}(C))\right)\right| = \left|\left(\pi_{m_0}^m\right)^{-1}(\pi_{m_0}(C))\right| = |\pi_{m_0}(C)| \cdot p^{d(m-m_0)},$$

and the result holds. □

1.4.3. *Integration over $\mathbb{Z}_p^n$.* Let $F : \mathbb{Q}_p \to \mathbb{C}$ be a measurable function. Assume that the image $\mathrm{Im}(F)$ is a countable subset and take $A \subset \mathbb{Q}_p$ a measurable set. For any $c \in \mathrm{Im}(F)$, consider the level sets of F in A:

$$A_F(c) := \{x \in A \mid F(x) = c\}.$$

Then,

$$\int_A F(x)\mathrm{d}\mu = \sum_{c\in\mathrm{Im}(F)} \int_{A_F(c)} F(x)\mathrm{d}\mu = \sum_{c\in\mathrm{Im}(F)} \mu(A_F(c)) \cdot c.$$

We are interested in functions of the form $F(x) = |f(x)|_p^s$, where $s \in \mathbb{C}$ such that $\mathrm{Re}(s) > 0$. Note that, in this case:

$$\int_A |f(x)|_p^s \,\mathrm{d}\mu = \int_A p^{-\operatorname{ord}_p(f(x))s}\mathrm{d}\mu = \sum_{n\geq 0} \mu\left\{x \in A \mid \operatorname{ord}_p(f(x)) = m\right\} \cdot p^{-ms}.$$

EXAMPLE 1.22. For $N \geq 0$, consider

$$\int_{\mathbb{Z}_p} \left|x^N\right|_p^s \mathrm{d}\mu.$$

Taking $F(x) = \left|x^N\right|_p^s = p^{-\operatorname{ord}_p(x)Ns}$, the image is countable and only depends on the order of x. In fact, for any $m \geq 0$,

$$A_F\left(p^{-mNs}\right) = p^m\mathbb{Z}_p \setminus p^{m+1}\mathbb{Z}_p.$$

which gives a partition of $\mathbb{Z}_p$. Thus,

$$\begin{aligned}\int_{\mathbb{Z}_p} \left|x^N\right|_p^s \mathrm{d}\mu &= \sum_{m\geq 0} \mu\left(p^m\mathbb{Z}_p \setminus p^{m+1}\mathbb{Z}_p\right)\cdot p^{-mNs} = \sum_{m\geq 0}\left(p^{-m} - p^{-(m+1)}\right)\cdot p^{-mNs} \\ &= \sum_{m\geq 0}\left(1-p^{-1}\right)\cdot p^{-m(Ns+1)} = \left(1-p^{-1}\right)\sum_{m\geq 0}\left(p^{-(Ns+1)}\right)^m \\ &= \left(1-p^{-1}\right)\frac{1}{1-p^{-(Ns+1)}} = \frac{p-1}{p-p^{-Ns}}.\end{aligned}$$

EXERCISE 1.23. Prove that $\int_{\mathbb{Z}_p^d} \left|x_1^{N_1}\cdots x_d^{N_d}\right|_p^s \mathrm{d}\mu = \prod_{i=1}^d \frac{p-1}{p-p^{-N_i s}}$.
(<u>HINT</u>: factorize the sums coming from different x_i)

1.5. The Igusa zeta function. We introduce the first of the zeta functions, studied by IGUSA in order to determine the numbers $\mathcal{N}_m(f)$.

DEFINITION 1.24. Let $f \in \mathbb{Z}_p[X_1,\ldots,X_d]$ and let $s \in \mathbb{C}$. The *(local) Igusa zeta function* of f is given by

$$Z_{\text{Igusa}}(f;s) := \int_{\mathbb{Z}_p^d} |f(x)|_p^s \,\mathrm{d}\mu.$$

REMARK 1.25. $Z_{\text{Igusa}}(f;s)$ is holomorphic for any $s \in \{z \in \mathbb{C} \mid \operatorname{Re}(z) > 0\}$.

We can recover the power series $Q(f;T)$ in (1.1) from this zeta function as follows.

PROPOSITION 1.26. *We have the following relation*

$$Z_{\text{Igusa}}(f;s) = Q\left(f, \frac{1}{p^{s+d}}\right)(1-p^s) + p^s,$$

where $Q(f,T)$ is the Poincaré power series defined in (1.1).

Before proving the relation above, we notice the following relation between level sets of $|f|_p$ and the numbers $\mathcal{N}_m(f)$ previously defined.

LEMMA 1.27. *The set $V_m = \left\{x \in \mathbb{Z}_p^d \mid |f(x)|_p \leq p^{-m}\right\}$ is a cylinder in $\mathbb{Z}_p^d$. Moreover, $\mu(V_m) = \mathcal{N}_m(f)\cdot p^{-dm}$.*

PROOF. We can rewrite $V_m = \left\{x \in \mathbb{Z}_p^d \mid \operatorname{ord}_p f(x) \geq m\right\}$. Note that $V_0 = \mathbb{Z}_p^d$ and the formula holds in this case. For any $m \geq 1$, we have $\pi_{m-1}(V_m) = \left\{x \in (\mathbb{Z}/p^m\mathbb{Z})^d \mid f(x) = 0 \mod p^m\right\}$. Moreover, the set $V_m = (\pi_{m-1}^{-1}\circ\pi_{m-1})(V_m)$ is a cylinder, and by Proposition 1.21, it follows

$$\mu(V_m) = \frac{|\pi_{m-1}(V_m)|}{p^{dm}} = \frac{\mathcal{N}_m(f)}{p^{dm}}.$$

□

PROOF OF PROPOSITION 1.26. For any $m \geq 0$, the level sets of $|f(x)|_p^s$ can be expressed as $V_m \setminus V_{m+1}$. Thus

$$\begin{aligned} Z_{\text{Igusa}}(f;s) &= \sum_{m\geq 0} \mu(V_m \setminus V_{m+1}) \cdot p^{-ms} \\ &= \sum_{m\geq 0} \left(\frac{\mathcal{N}_m(f)}{p^{dm}} - \frac{\mathcal{N}_{m+1}(f)}{p^{d(m+1)}} \right) \cdot p^{-ms} \\ &= \sum_{m\geq 0} \frac{\mathcal{N}_m(f)}{p^{m(d+s)}} - \sum_{m\geq 0} \frac{\mathcal{N}_{m+1}(f)}{p^{d(m+1)+ms}} \\ &= \sum_{m\geq 0} \mathcal{N}_m(f) \left(\frac{1}{p^{d+s}} \right)^m - \sum_{m\geq 1} \mathcal{N}_m(f) \left(\frac{1}{p^{m(d+s)-s}} \right) \\ &= Q\left(f, \frac{1}{p^{s+d}}\right) - p^s \left(Q\left(f, \frac{1}{p^{s+d}}\right) - 1 \right). \end{aligned}$$

□

REMARK 1.28. Taking the substitution $T = p^{-s}$ in $Z_{\text{Igusa}}(f;s)$, the above relation can be rewritten as

$$Q(f;p^{-d}T) = \frac{T \cdot Z_{\text{Igusa}}(f;s) - 1}{T - 1}.$$

EXAMPLE 1.29. For $f(x) = x^N$, we obtain

$$Q(f;T) = \frac{pT \cdot \frac{p-1}{p-p^N T^N} - 1}{pT - 1} = \frac{-1 + (p-1)T + p^{N-1}T^N}{(1 - p^{N-1}T^N)(pT - 1)}.$$

EXERCISE 1.30. Obtain $Q(f;T)$ in the same way for $f(x) = x_1^{N_1} \cdots x_d^{N_d}$.

We have a way of computing the series $Q(f;s)$ using p-adic integration! But it should be noticed that the previous integrals are easy to compute since f was a monomial and we have a formula for the norm of a product, but as soon as f involves sums, the computations become harder and complicated. For example, the reader could compute

$$\int_{\mathbb{Z}_p^2} |xy(x-y)|_p^s \,\mathrm{d}\mu.$$

Solution: we can use resolution of singularities to transform f in a function which locally seems like a monomial. Thus, we need to introduce integration in $\mathbb{Q}_p$-analytic manifolds and prove a change of variables formula.

1.6. p-adic integration in manifolds and change of variables formula.

1.6.1. *Integration in $\mathbb{Q}_p$-analytic manifolds.* The notion of manifolds, analytic functions and its integration theory over $\mathbb{Q}_p$ are quite natural. However, this theory has some advantages over the p-adics, since closed balls are both open and compact sets.

DEFINITION 1.31.

(1) For any open $U \subset \mathbb{Q}_p^d$, a function $f : U \to \mathbb{Q}_p$ is called $\mathbb{Q}_p$-*analytic map* if for any $x \in U$, there exists a neighborhood $V \subset U$ such that $f_{|V}$ is given by a convergent power series.

(2) We call $f = (f_1, \dots, f_d) : U \to \mathbb{Q}_p^d$ a $\mathbb{Q}_p$-analytic map if any f_i is $\mathbb{Q}_p$-analytic.

(3) A *$\mathbb{Q}_p$-analytic manifold of dimension* d is a Hausdorff topological space X together with an atlas $(U_i, \varphi_i)_{i\in I}$ in $\mathbb{Q}_p^d$ and such that any change of charts $\varphi_j \circ \varphi_i^{-1}$ is bianalytic, $i, j \in I$.

REMARK 1.32.

(1) A $\mathbb{Q}_p$-analytic manifold is a locally compact, totally disconnected topological space.

(2) Every open $U \subset \mathbb{Q}_p^d$ is a $\mathbb{Q}_p$-analytic manifold. In particular, $U = \mathbb{Z}_p$ is a compact $\mathbb{Q}_p$-analytic manifold.

EXAMPLE 1.33. Consider the *p-adic projective line*

$$\mathbb{P}^1(\mathbb{Q}_p) = \{[u : v] \mid (u, v) \sim \lambda(u', v'),\ \lambda \in \mathbb{Q}_p^\times\}.$$

One can see that $\mathbb{P}^1(\mathbb{Q}_p)$ is covered by two disjoint compact open sets:

$$U = \left\{[u : v] \mid v \neq 0, |u/v|_p \leq 1\right\} \quad \text{and} \quad V = \left\{[u : v] \mid u \neq 0, |v/u|_p < 1\right\}, \tag{1.2}$$

since we have two bianalytic maps $\varphi_U : U \to \mathbb{Z}_p$, $\varphi_U[u : v] = u/v$, and $\varphi_V : V \to p\mathbb{Z}_p$, $\varphi_V[u : v] = v/u$. Note that $p\mathbb{Z}_p$ is homeomorphic to $\mathbb{Z}_p$.

EXAMPLE 1.34. Let $\pi : \mathrm{Bl}_0(\mathbb{Q}_p^2) \to \mathbb{Q}_p^2$ be the *blow-up at the origin* of the $\mathbb{Q}_p$-affine plane, i.e. the 2-dimensional $\mathbb{Q}_p$-analytic manifold

$$\mathrm{Bl}_0(\mathbb{Q}_p^2) := \{((x, y), [u : v]) \in \mathbb{Q}_p^2 \times \mathbb{P}^1(\mathbb{Q}_p) \mid xv - yu = 0\},$$

with π being the natural projection on $\mathbb{Q}_p^2$. By using the (disjoint) charts of $\mathbb{P}^1(\mathbb{Q}_p) = U \sqcup V$ defined in (1.2), we get two disjoint charts of $\mathrm{Bl}_0(\mathbb{Q}_p^2) = \widetilde{U} \sqcup \widetilde{V}$, i.e.

$$\widetilde{U} = \mathrm{Bl}_0(\mathbb{Q}_p^2) \cap (\mathbb{Q}_p^2 \times U) \quad \text{and} \quad \widetilde{V} = \mathrm{Bl}_0(\mathbb{Q}_p^2) \cap (\mathbb{Q}_p^2 \times V).$$

Both charts are bianalytic to $\mathbb{Q}_p^2$ via the maps

$$\begin{array}{ccc} \mathbb{Q}_p^2 & \longrightarrow & \widetilde{U} \\ (s_1, t_1) & \longmapsto & ((s_1, s_1 t_1), [1 : t_1]) \end{array} \quad \text{and} \quad \begin{array}{ccc} \mathbb{Q}_p^2 & \longrightarrow & \widetilde{V} \\ (s_2, t_2) & \longmapsto & ((s_2 t_2, t_2), [s_2 : 1]) \end{array}.$$

Thus, the projection map $\pi : \mathrm{Bl}_0(\mathbb{Q}_p^2) \to \mathbb{Q}_p^2$ is expressed in the affine charts of $\widetilde{U}$ and $\widetilde{V}$ via the morphisms

$$\begin{array}{cccc} \varphi_1 : & \mathbb{Q}_p^2 & \longrightarrow & \mathbb{Q}_p^2 \\ & (s_1, t_1) & \longmapsto & (s_1, s_1 t_1) \end{array} \quad \text{and} \quad \begin{array}{cccc} \varphi_2 : & \widetilde{V} & \longrightarrow & \mathbb{Q}_p^2 \\ & (s_2, t_2) & \longmapsto & (s_2 t_2, t_2) \end{array},$$

respectively. Denote by $E = \pi^{-1}(0)$ the *exceptional divisor*, note that $\mathrm{Bl}_0(\mathbb{Q}_p^2) \setminus E \overset{\pi}{\simeq} \mathbb{Q}_p^2 \setminus O$, i.e. $\varphi_{1|\{s_1\neq 0\}}$ and $\varphi_{2|\{t_2\neq 0\}}$ are bianalytic maps. For a $\mathbb{Q}_p$-analytic submanifold $X \subset \mathbb{Q}_p^2$, we define its *strict transform*, denoted by $\widehat{X}$, as the Zariski closure of $\pi^{-1}(X \setminus 0)$.

Take $Y = \pi^{-1}(\mathbb{Z}_p^2)$. We can express it in two disjoint compact charts $Y = Y_U \cup Y_V$ using $\widetilde{U}$ and $\widetilde{V}$. Looking locally:

$$\varphi_1^{-1}(\mathbb{Z}_p^2) = \left\{(s_1, t_1) \in \mathbb{Q}_p^2 \mid |s_1|_p \leq 1,\ |s_1|_p \cdot |t_1|_p \leq 1\right\},$$

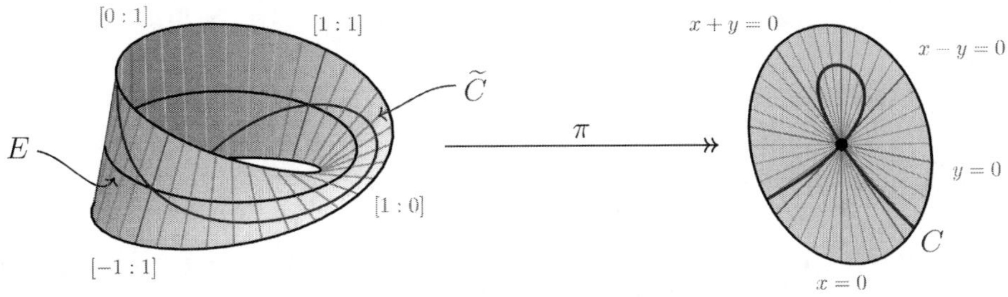

FIGURE 2. Local blow-up of $C = \{x^2+y^2(y-1) = 0\}$ at the origin in $\mathbb{R}^2$.

and analogously for φ_2^{-1}. We obtain a partition of $\mathbb{Z}_p^2$ with respect to the "metric diagonal" $\left\{|x|_p = |y|_p\right\}$ by setting the two compact polydisks:

$$Y_U = \left\{(s_1,t_1) \mid |s_1|_p \leq 1,\ |t_1|_p \leq 1\right\} \quad \text{and} \quad Y_V = \left\{(s_2,t_2) \mid |s_2|_p < 1,\ |t_2|_p \leq 1\right\}.$$

And we have:

$$\pi(Y_U) = \left\{(x,y) \mid |y|_p \leq |x|_p \leq 1\right\} \quad \text{and} \quad \pi(Y_V) = \left\{(x,y) \mid |x|_p < |y|_p \leq 1\right\}.$$

PROPOSITION 1.35. *Every compact $\mathbb{Q}_p$-analytic manifold of dimension d is bi-analytic to a finite disjoint union of copies of $\mathbb{Z}_p^d$ (so, open-compact).*

REMARK 1.36. We can define k-differential forms $\omega \in \Omega^k(X) = \Gamma(\bigwedge^k T^*X)$ on X in the usual way, such that for any open $U \subset X$ with coordinates $x_1,\ldots,x_d$:

$$\omega_{|U} = \sum_{1\leq i_1<\cdots<i_k\leq d} f_{i_1,\ldots,i_k} \mathrm{d}x_{i_1} \wedge \cdots \wedge \mathrm{d}x_{i_k},$$

where the functions $f_{i_1,\ldots,i_k} : U \to \mathbb{Q}_p$ are differentiable. For any top d-degree form, we can define the associated measure $\mu_\omega = |\omega|$ on X using the basis of the product topology, i.e. for any compact-open polycylinder:

$$A \simeq (a_1 + p^{m_1}\mathbb{Z}_p) \times \cdots \times (a_d + p^{m_d}\mathbb{Z}_p) \subset X,$$

such that $\omega_{|A} = f(x) \cdot \mathrm{d}x_1 \wedge \cdots \wedge \mathrm{d}x_d$, we define:

$$\mu_\omega(A) := \int_A |f(x)|_p \,\mathrm{d}\mu,$$

with respect to the usual normalized Haar measure. The above defines a Borel measure over X.

THEOREM 1.37 (LOCAL CHANGE OF VARIABLES). *Let $\varphi : \mathbb{Q}_p^d \to \mathbb{Q}_p^d$ be a $\mathbb{Q}_p$-analytic map. Suppose that $x \in \mathbb{Q}_p^d$ such that* $\det \mathrm{Jac}(\varphi)(x) = \left(\frac{\partial \varphi_i}{\partial x_j}(x)\right) \neq 0$. *Then there exist $U, V \subset \mathbb{Q}_p^d$ open neighborhoods of x and $\varphi(x)$ respectively such that $\varphi_{|U} : U \to V$ is a bianalytic isomorphism and*

$$\int_{\varphi(A)} \mathrm{d}\mu^V = \int_A |\det \mathrm{Jac}(\varphi)(x)|_p \,\mathrm{d}\mu^U,$$

for any measurable $A \subset U$.

COROLLARY 1.38. *Let X be a compact $\mathbb{Q}_p$-analytic manifold, and ω a $\mathbb{Q}_p$-analytic form of top degree on X. Then there exists a globally defined measure μ_ω on X. In particular, for any continuous function $f : X \to \mathbb{C}$, the integral $\int_X f \mathrm{d}\mu_\omega$ is well defined.*

THEOREM 1.39 (CHANGE OF VARIABLES FORMULA). *Let $\varphi : Y \to X$ be a $\mathbb{Q}_p$-analytic map of compact $\mathbb{Q}_p$-analytic manifolds. Assume that there exists $Z \subset Y$ a closed subset of zero measure such that $\varphi_{|Y \setminus Z} : Y \setminus Z \to X \setminus \varphi(Z)$ is bianalytic. If ω is a $\mathbb{Q}_p$-analytic top-degree form on X and $f : X \to \mathbb{Q}_p$ is a $\mathbb{Q}_p$-analytic map, then*

$$\int_X |f|_p^s \,\mathrm{d}\mu_\omega = \int_Y |f \circ \varphi|_p^s \,\mathrm{d}\mu_{\varphi^*\omega}.$$

REMARK 1.40. We will denote by $|\mathrm{d}x|_p = |\mathrm{d}x_1 \wedge \cdots \wedge \mathrm{d}x_d|_p$ the normalized Haar measure in local coordinates.

EXAMPLE 1.41. We can compute the integral

$$\int_{\mathbb{Z}_p^2} \left|x^a y^b (x-y)^c\right|_p^s |\mathrm{d}x \wedge \mathrm{d}y|_p$$

by blowing up the origin $\pi : \mathrm{Bl}_0(\mathbb{Q}_p) \to \mathbb{Q}_p$, in order to get locally monomials on $f(x,y) = x^a y^b (x-y)^c$. We show in Example 1.34 that $Y = \pi^{-1}(\mathbb{Z}_p^2)$ is covered by two disjoint compact polydisks

$$Y_U = \left\{(s_1,t_1) \mid |s_1|_p \le 1,\ |t_1|_p \le 1\right\} \quad \text{and} \quad Y_V = \left\{(s_2,t_2) \mid |s_2|_p < 1,\ |t_2|_p \le 1\right\}.$$

By partitioning $\mathbb{Z}_p^2 = \pi(Y_U) \sqcup \pi(Y_V)$, we have

$$\int_{\mathbb{Z}_p^2} |f|_p^s |\mathrm{d}x \wedge \mathrm{d}y|_p = \underbrace{\int_{Y_U} |\pi^* f|_p^s |\pi^*(\mathrm{d}x \wedge \mathrm{d}y)|_p}_{\mathcal{I}_U} + \underbrace{\int_{Y_V} |\pi^* f|_p^s |\pi^*(\mathrm{d}x \wedge \mathrm{d}y)|_p}_{\mathcal{I}_V}.$$

Since $\mathcal{I}_U$ is computed over the first chart $\pi_{|Y_U} = \varphi_1 : (s_1,t_1) \to (s_1, s_1 t_1)$, we have:

$$\begin{aligned}
\mathcal{I}_U &= \int_{|s_1|_p \le 1,\ |t_1|_p \le 1} \left|s_1^{a+b+c} t_1^b (1-t_1)^c\right|_p^s |s_1 \mathrm{d}s_1 \wedge \mathrm{d}t_1|_p \\
&= \left(\int_{|s_1|_p \le 1} |s_1|_p^{(a+b+c)s+1} |\mathrm{d}s_1|_p\right) \cdot \left(\int_{|t_1|_p \le 1} \left|t_1^b (1-t_1)^c\right|_p^s |\mathrm{d}t_1|_p\right) \\
&= \frac{p-1}{p - p^{-((a+b+c)s+1)}} \int_{|t_1|_p \le 1} \left|t_1^b (1-t_1)^c\right|_p^s |\mathrm{d}t_1|_p.
\end{aligned}$$

We have already seen the first integral in Example 1.22. For the second one, we need to understand how the non-monomial part changes in the poly-disk of integration. Remember that we can decompose $\mathbb{Z}_p = \bigsqcup_{k=0}^{p-1} (k + p\mathbb{Z}_p)$, thus

$$\left\{|t_1|_p \le 1\right\} = \bigsqcup_{k=0}^{p-1} T_k, \quad \text{where } T_k = \left\{|t_1 - k|_p \le \frac{1}{p}\right\},$$

and now

$$\int_{|t_1|_p \le 1} \left|t_1^b (1-t_1)^c\right|_p^s |\mathrm{d}t_1|_p = \sum_{k=0}^{p-1} \int_{T_k} \left|t_1^b (1-t_1)^c\right|_p^s |\mathrm{d}t_1|_p.$$

With the previous decomposition, every integral can be computed as a "monomial integral". Note that, in $T_0 = \left\{|t_1|_p \leq 1/p\right\}$, we show that $|t-1|_p = 1$ (using for example the p-adic representation), and then:

$$\int_{T_0} \left|t_1^b(1-t_1)^c\right|_p^s |\mathrm{d}t_1|_p = \int_{|t_1|_p \leq 1/p} |t_1|_p^{bs} |\mathrm{d}t_1|_p = \frac{(p-1)p^{-bs}}{p-p^{-bs}}.$$

In the same way, in $T_1 = \left\{|t_1 - 1|_p \leq 1/p\right\}$, we have $|t_1|_p = |(t_1-1)+1|_p = 1$ and

$$\int_{T_1} \left|t_1^b(1-t_1)^c\right|_p^s |\mathrm{d}t_1|_p = \int_{|t_1-1|_p \leq 1/p} |t_1 - 1|_p^{cs} |\mathrm{d}t_1|_p = \frac{(p-1)p^{-cs}}{p-p^{-cs}}.$$

For $k \geq 2$, it is easy to see that $\left|t_1^b(t_1-1)\right|_p = 1$ in T_k, then

$$\int_{T_k} \left|t_1^b(1-t_1)^c\right|_p^s |\mathrm{d}t_1|_p = \mu(T_k) = \mu(p\mathbb{Z}_p) = \frac{1}{p}.$$

Regrouping the above integrals, we have:

$$\mathcal{I}_U = \frac{p-1}{p-p^{-((a+b+c)s+1)}} \cdot \left(\frac{(p-1)p^{-bs}}{p-p^{-bs}} + \frac{(p-1)p^{-cs}}{p-p^{-cs}} + \frac{p-2}{p}\right).$$

Following similar arguments, we can compute $\mathcal{I}_V$ only by monomial computations and obtain $\mathcal{I} = \mathcal{I}_U + \mathcal{I}_V$ (EXERCISE).

REMARK 1.42. Geometrically, we are resolving the singularities of $f(x,y) = 0$, i.e. constructing a birational model by blowing-up the origin in $\mathbb{Z}_p^2$ where $\pi^* f$ is locally a monomial. As char $\mathbb{Q}_p = 0$, Hironaka's Theorem on resolution of singularities applies and we can always obtain this model in any dimension. Note there exist explicit algorithmic resolutions of singularities (see VILLAMAYOR's work, e.g. [**Vil89**]).

In the case of $\mathbb{Q}_p$-manifolds, the resolution of singularities can be formulated in terms of (Weil) divisors as follows. Recall that a divisor D is called a *simple normal crossing (SNC) divisor* if $D = \sum_{i=0}^m a_i D_i$ for some codimension one irreducible smooth subvarieties $\{D_i\}_{i=0}^m$, such that at any point $p \in D$ the support of D is locally bianalytic to the union of some coordinate hyperplanes.

THEOREM 1.43 ($\mathbb{Q}_p$-ANALYTIC EMBEDDED RESOLUTION OF SINGULARITIES). *Let $f \in \mathbb{Q}_p[X_1,\ldots,X_d]$ be non-constant. Then there exist a $\mathbb{Q}_p$-analytic manifold Y, $\dim Y = d$, a proper surjective $\mathbb{Q}_p$-analytic map $\pi : Y \to \mathbb{Q}_p^d$ which is an isomorphism outside a set of measure zero, and finitely many submanifolds $E_0,\ldots,E_r \subset Y$, $\operatorname{codim} E_i = 1$, such that:*

(1) *$\sum_{i=0}^r E_i$ is a SNC divisor,*

(2) *$\operatorname{div}(\pi^* f) = \sum_{i=0}^r N_i E_i$, for some $N_0,\ldots,N_r \in \mathbb{Z}_{\geq 0}$,*

(3) *$\operatorname{div}(\operatorname{Jac}(\pi)) = \operatorname{div}(\pi^*(\mathrm{d}x_1 \wedge \cdots \wedge \mathrm{d}x_d)) = \sum_{i=0}^r (\nu_i - 1)E_i$, for some $\nu_0,\ldots,\nu_r \in \mathbb{Z}_{\geq 1}$.*

Moreover, $\pi : Y \to \mathbb{Q}_p^d$ can be constructed as a composition of successive blowing-ups over smooth centers.

REMARK 1.44.

(1) The divisor corresponding to the strict transform of $\operatorname{div}(f)$ by π can be expressed as $\sum_{j\in S_0} N_j E_j$, for some $S_0 \subset \{0,\ldots,r\}$.

(2) In terms of equations, the result above assures that in suitable local coordinates $y = (y_1 \dots, y_d)$ around any point $a \in Y$, we have
$$\pi^* f = y_1^{N_{i_1}} \cdots y_k^{N_{i_k}} \cdot u(y), \quad u(a) \neq 0,$$
and
$$\pi^*(\mathrm{d}x_1 \wedge \cdots \wedge \mathrm{d}x_d) = y_1^{\nu_{i_1}-1} \cdots y_k^{\nu_{i_k}-1} \cdot v(y) \cdot \mathrm{d}y_1 \wedge \cdots \wedge \mathrm{d}y_d, \quad v(a) \neq 0,$$
for some $1 \leq k \leq d$. The sequence $\{(N_i, \nu_i)\}_{i=0}^r$ is called the *numerical data of the resolution* (or *discrepancies*).

EXAMPLE 1.45. Consider the cusp $\mathcal{C} : y^2 - x^3 = 0$. After successive blow-ups at the origin (see FIGURE 3), we obtain three exceptional divisors E_1, E_2, E_3 with
$$\operatorname{div}(\pi^* f) = \widehat{\mathcal{C}} + 2E_1 + 3E_2 + 6E_3 \quad \text{and} \quad \operatorname{div}(\operatorname{Jac} \pi) = E_1 + 2E_2 + 4E_3,$$
where $E_i \simeq \mathbb{P}^1$ and $\widehat{\mathcal{C}} \simeq \mathbb{A}^1$.

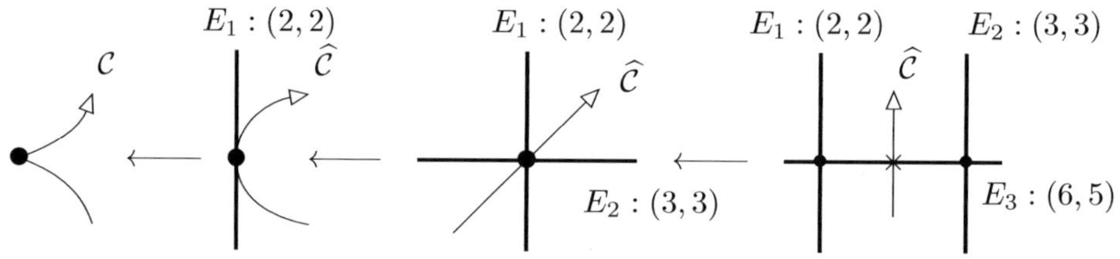

FIGURE 3. Embedded resolution of the cusp by successive blowing-ups.

THEOREM 1.46. *$Z_{\mathrm{Igusa}}(f; s)$ is a rational function on p^{-s}. Moreover, if $\pi : Y \to \mathbb{Q}_p^d$ is an embedded resolution of $\{f = 0\}$ with numerical data $\{(N_i, \nu_i)\}_{i=0}^r$, then*
$$Z_{\mathrm{Igusa}}(f; s) = \frac{P(p^{-s})}{\left(1 - p^{-(N_0 s + \nu_0)}\right) \cdots \left(1 - p^{-(N_r s + \nu_r)}\right)},$$
where $P(T)$ is a polynomial with coefficients in $\mathbb{Z}[1/p]$.

PROOF. We study the restriction of the embedded resolution $\pi : Y \to \mathbb{Z}_p$. Recall that $Y = \pi^{-1}(\mathbb{Z}_p)$ is compact and it can be covered by a finite number of disjoint compact-open charts $\{U_j\}_{j \in I}$, where locally one has
$$\pi^* f = y_1^{N_{i_1}} \cdots y_k^{N_{i_k}} \cdot u(y), \quad u(0) \neq 0,$$
and
$$\pi^*(\mathrm{d}x_1 \wedge \cdots \wedge \mathrm{d}x_d) = y_1^{\nu_{i_1}-1} \cdots y_k^{\nu_{i_k}-1} \cdot v(y) \cdot \mathrm{d}y_1 \wedge \cdots \wedge \mathrm{d}y_d, \quad v(0) \neq 0,$$
for any $y \in U_j$ and some $1 \leq k \leq d$. Also, $\{(N_{i_1}, \nu_{i_1}), \dots, (N_{i_k}, \nu_{i_k})\}$ is a subset of the numerical data of π. Using the previous decomposition and the change of variables, one express $Z_{\mathrm{Igusa}}(f; s)$ as the sum
$$\sum_{j \in I} \underbrace{\int_{U_j} |y_1|_p^{N_{i_1} s + \nu_{i_1} - 1} \cdots |y_k|_p^{N_{i_k} s + \nu_{i_k} - 1} |u(y)|_p^s \, |v(y)|_p \cdot |\mathrm{d}y_1 \wedge \cdots \wedge \mathrm{d}y_d|_p}_{\mathcal{I}_{U_j}}.$$

<u>FACT</u>: $|u(y)|_p$ and $|v(y)|_p$ are finite (since $u(0)v(0) \neq 0$) and locally constant. We can assume that they are constant in U_i, say $|u(y)|_p = p^{-a}$ and $|v(y)|_p = p^{-b}$, for

any $y \in U_j$. Each of the U_j is identified with a polydisk $P_j = \left\{ |y_k|_p \leq p^{-m_k} \mid k = 1, \dots, d \right\}$ for some $m_1, \dots, m_d \in \mathbb{Z}_{\geq 0}$, then

$$\begin{aligned}
\mathcal{I}_{U_j} &= p^{-as-b} \cdot \int_{P_j} |y_1|_p^{N_{i_1}s+\nu_{i_1}-1} \cdots |y_k|_p^{N_{i_k}s+\nu_{i_k}-1} \cdot |\mathrm{d}y_1 \wedge \cdots \wedge \mathrm{d}y_d|_p \\
&= p^{-as-b} \cdot \int_{|y_1|_p \leq p^{-m_1}} |y_1|_p^{N_{i_1}s+\nu_{i_1}-1} \mathrm{d}y_1 \cdot \cdots \cdot \int_{|y_k|_p \leq p^{-m_k}} |y_k|_p^{N_{i_k}s+\nu_{i_k}-1} \mathrm{d}y_d \\
&= p^{-as-b} \cdot \left(\frac{p-1}{p} \right)^d \cdot \frac{p^{-m_1(N_{i_1}s+\nu_{i_1})}}{1-p^{-m_1(N_{i_1}s+\nu_{i_1})}} \cdots \frac{p^{-m_k(N_{i_k}s+\nu_{i_k})}}{1-p^{-m_k(N_{i_k}s+\nu_{i_k})}}.
\end{aligned}$$

It remains to check that the numerator is a polynomial over p^{-s} with coefficients on $\mathbb{Z}[1/p]$, since the number a above could be negative. In fact, we know that $(f \circ \pi)(P_j) \subset \mathbb{Z}_p$ and this implies that $|\pi^* f|_p \leq 1$ on P_j. This is equivalent to

$$1 \geq |y_1|_p^{N_{i_1}} \cdots |y_k|_p^{N_{i_k}} |u(y)|_p = p^{-a-\sum_{\ell=1}^k m_\ell N_{i_\ell}} \iff a + \sum_{\ell=1}^k m_\ell N_{i_\ell} \geq 0.$$

Thus, p^{-s} appears with a non-negative power in the numerator of $\mathcal{I}_{U_j}$. □

COROLLARY 1.47. *The power series $Q(f;T)$ is rational.*

REMARK 1.48.

(1) In the previous theorem, we have deduced that the poles of $Z_{\mathrm{Igusa}}(f;s)$ are of the form

$$s_0 = -\frac{\nu}{N} + \frac{2\pi\sqrt{-1}k}{N} \log p,$$

for $k \in \mathbb{Z}$, where $-\nu/N$ belongs to the set

$$\left\{ -\frac{\nu_1}{N_1}, \dots, -\frac{\nu_r}{N_r} \right\} \subset \mathbb{Q}_{<0},$$

obtained from the numerical data $\{(N_i, \nu_i)\}_{i=0}^r$ of any embedded resolution of $\{f = 0\}$. However, it is a hard problem in general to explicitly determine which are actually poles of $Z_{\mathrm{Igusa}}(f;s)$.

(2) Under some technical mild assumptions over the above $\pi : Y \to \mathbb{Q}_p^d$ (i.e. "π has good reduction modulo p"), DENEF **[Den91]** obtained an expression of $Z_{\mathrm{Igusa}}(f;s)$ from a stratification based on the divisors of the total transform $\{E_i\}_{i=0}^r$, with numerical data $\{(N_i, \nu_i)\}_{i=0}^r$. Here, it suffices to know that the previous condition is satisfied for almost any p. Consider the stratification defined by

$$E_I^\circ := \bigcap_{i \in I} E_i \setminus \bigcup_{j \notin I} E_j,$$

for any $I \subset \{0, \dots, r\}$ (see FIGURE 5). Then, for $p \gg 0$,

$$Z_{\mathrm{Igusa}}(f;s) = p^{-d} \sum_{I \subset \{1,\dots,r\}} |E_I^\circ(\mathbb{F}_p)| \prod_{i \in I} \frac{(p-1)p^{-(N_i s+\nu_i)}}{1-p^{-(N_i s+\nu_i)}}, \tag{1.3}$$

where $|E_I^\circ(\mathbb{F}_p)|$ is the number of $\mathbb{F}_p$-rational points in E_I° reduced modulo p.

(3) Another practical way to compute $Z_{\text{Igusa}}(f;s)$ without involving resolutions of singularities is Igusa's *Stationary Phase Formula*, which can be found in [**LZ19**, Propo. 6.1].

The above expression allows to explore the computation of $Z_{\text{Igusa}}(f;s)$ and the complete determination of its poles form the combinatorics of resolutions, i.e. the possible exceptional divisors and associated numerical data appearing for fixed f. In particular, the determination of poles is involved in the study of an astonishing problem on Singularity Theory, which is discussed in the next section.

EXERCISE 1.49. Using Denef's formula above, compute $Z_{\text{Igusa}}(f_i;s)$ for the polynomials $f_1,\dots,f_4$ appearing in Examples 1.1 and 1.2, and verify the rational expressions describing $Q(f_i;s)$.

REMARK 1.50 (Final remarks).

(1) We can extend the previous notion of measure of cylinders to varieties in the p-adics. Let X be a smooth d-dimensional subvariety of $\mathbb{Z}_p^n$, defined algebraically. We can prove that the sequence

$$\left(\frac{|\pi_m(X)|}{p^{d(m+1)}}\right)_{m\geq 0},$$

is constant for $m \gg 0$, and it is called the *volume* $\mu(X)$ of X.

(2) (OESTERLÉ [**Oes82**]) Now, consider X singular. One defines its volume as

$$\mu(X) := \lim_{\varepsilon\to 0} \mu(X \setminus T_\varepsilon(X_{\text{sing}})) \in \mathbb{R},$$

where $T_\varepsilon(X_{\text{sing}})$ is a tubular neighborhood of radius $\varepsilon > 0$. Then, one has

$$\mu(X) = \lim_{n\to\infty} \frac{|\pi_m(X)|}{p^{d(m+1)}}.$$

(3) One can prove (EXERCISE) that if X is a d-dimensional smooth variety defined over $\mathbb{Z}_p$ and it admits a nowhere vanishing d-form ω, then

$$\int_{X(\mathbb{Z}_p)} \mathrm{d}\mu_\omega = \frac{|X(\mathbb{F}_p)|}{p^d} = \frac{|X(\mathbb{F}_p)|}{|\mathbb{F}_p^d|} \in \mathbb{Z}[1/p].$$

This is in fact one of the main points of Batyrev's proof: from the above relation together with the change of variables between integrals, he concludes that two birationally equivalent smooth Calabi-Yau varieties have the same Weil zeta function.

1.7. Arithmetic vs topological: Milnor fiber and the monodromy conjecture. The poles of the $Z_{\text{Igusa}}(f;s)$ are quite mysterious and it was suspected by Igusa that they are connected with the topology of the complex variety $V_f = \{f=0\} \subset \mathbb{C}^d$, in particular with some aspects of the *Milnor fibration* of f [**Mil68**, **Lê79**]. A beautiful survey about this subject and related problems is [**Sea19**]; see also [**Bud12**].

Consider a non-constant polynomial mapping $f : \mathbb{C}^d \to \mathbb{C}$, and fix a point $x_0 \in V_f$ (which is possibly singular and no necessarily isolated). The *Milnor fibration of f at x_0* is the $\mathcal{C}^\infty$-locally trivial fibration given by the restriction

$$f_| : \mathbb{B}_\varepsilon(x_0) \cap f^{-1}(\mathbb{D}_\eta^*) \longrightarrow \mathbb{D}_\eta^*,$$

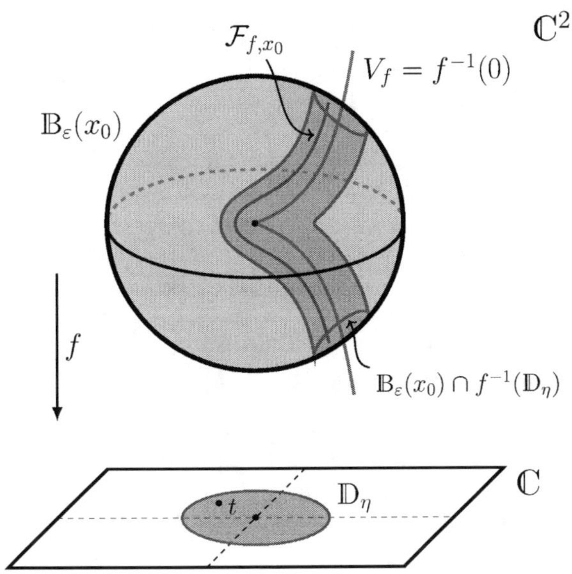

FIGURE 4. MILNOR FIBER AT THE ORIGIN.

where $\mathbb{B}_\varepsilon(x_0)$ is the open ball of radius ε around x_0, and $\mathbb{D}^*_\eta = \{z \in \mathbb{C} \mid |z| < \eta\}$ is the open punctured disk, with $0 < \eta \ll \varepsilon \ll 1$ small enough.

The *Milnor fiber of f at x_0* is any fiber $\mathcal{F}_{f,x_0} := f_{|}^{-1}(t)$ of the previous fibration, for $t \in \mathbb{D}^*_\eta$. Milnor showed that the class of diffeomorphism of $\mathcal{F}_{f,x_0}$ does not depend on t, and each lifting of a small loop around $0 \in \mathbb{D}_\eta$ induces a well defined diffeomorphism (up to isotopy) $\mathcal{F}_{f,x_0} \to \mathcal{F}_{f,x_0}$, called the *geometric monodromy transformation* of the Milnor fiber. The corresponding linear action in the cohomology $T_{x_0} : H^\bullet(\mathcal{F}_{f,x_0}; \mathbb{C}) \to H^\bullet(\mathcal{F}_{f,x_0}; \mathbb{C})$ is called the *complex algebraic monodromy action* of the Milnor fiber. It is well known that $H^q(\mathcal{F}_{f,x_0}; \mathbb{C}) = 0$ for any $q \geq d$.

CONJECTURE 1.51 (IGUSA p-ADIC MONODROMY CONJECTURE). *Let s_0 be a pole of $Z_{\mathrm{Igusa}}(f;s)$ for almost all primes p. Then* $\exp(2\pi i \operatorname{Re}(s_0))$ *is an eigenvalue of the monodromy action T_{x_0} on some level $H^q(\mathcal{F}_{f,x_0}; \mathbb{C})$ at some point $x_0 \in V_f$.*

REMARK 1.52 (A'CAMPO FORMULA). The set of eigenvalues of the monodromy action T_{x_0} is composed with roots of unity and is closed by conjugation. One can study it using the *monodromy zeta function of f at x_0*, which is the alternating product of the characteristic polynomials of T_{x_0} at each level $H^q(F_{x_0}; \mathbb{C})$, i.e.

$$\zeta_{f,x_0}(t) := \prod_{q \geq 0} \det\left(\mathrm{Id} - t \cdot T_{x_0} \mid H^q(F_{x_0}; \mathbb{C})\right)^{(-1)^q}.$$

For any embedded resolution of singularities $h : X \to \mathbb{C}^d$ of V_f, A'CAMPO proved [**A'C75**] the following useful formula:

$$\zeta_{f,x_0}(t) = \prod_{i \in I} \left(1 - t^{N_i}\right)^{\chi\left(E_i^\circ \cap h^{-1}(x_0)\right)}, \tag{1.4}$$

where $\operatorname{div}(h^*f) = \sum_{i\in I} N_i E_i$, and $E_i^\circ := E_i \setminus \left(\bigcup_{j\neq i} E_j\right)$. Note that the collection $(E_i^\circ)_{i\in I}$ is defined such that for any $q \in E_i^\circ$, then h^*f is locally of the form x^N (up to unit) with N constant along the stratum.

EXAMPLE 1.53. In the case of the cusp $f(x,y) = y^2 - x^3$, we already know that $s_0 = -1, -5/6$ are the only two real parts of poles of $Z_{\text{Igusa}}(f;s)$. From the resolution in Example 1.45,

$$\zeta_{f,0}(t) = \frac{(1-t^2)(1-t^3)}{1-t^6} = \frac{1-t}{t^2-t+1},$$

since $E_i \simeq \mathbb{P}^1$ for exceptional divisors, so $\chi(E_i\setminus\{k \text{ pts}\}) = 2-k$. Thus, $t_0 = 1, e^{i\pi/3}$ is a zero and a pole of $\zeta_{f,0}(t)$, respectively.

Using a particular completion over the p-adics for almost all p, DENEF and LOESER [**DL92**] define a specialization of the Igusa zeta function for complex polynomials f, "taking the limit $p \to 1$" in $Z_{\text{Igusa}}(f;s)$. In [**Den91**, Sec. 4.3] the author gives a short explanation of the aforementioned limit for the definition of the topological zeta function.

Let $h : X \to \mathbb{C}^d$ be an embedded resolution of singularities of $V_f = \{f = 0\}$, with numerical data $\{(N_i,\nu_i)\}_{i=0}^r$, and consider the stratification $\{E_I^\circ\}_{I\subset\{0,\dots,r\}}$ as in Remark 1.48.

DEFINITION 1.54. The *topological zeta function of f* is defined by

$$Z_{\text{top}}(f;s) := \sum_{I\subset\{0,\dots,r\}} \chi(E_I^\circ) \prod_{i\in I} \frac{1}{N_i s + \nu_i} \in \mathbb{Q}(s).$$

If $f : (\mathbb{C}^n,0) \to (\mathbb{C},0)$ is a germ of a function at the origin, we define $Z_{\text{top},0}(f;s)$ the *local topological zeta function of f at* 0 by taking $\chi\left(E_I^\circ \cap h^{-1}(0)\right)$ instead of $\chi(E_I^\circ)$ in the above expression.

REMARK 1.55.

(1) $Z_{\text{top}}(f;s)$ does not depend on the chosen resolution (see [**DL92**]), but it is worth noticing that *a priori* it is not defined intrinsically.

(2) Despite of the name, $Z_{\text{top}}(f;s)$ is an *analytical* invariant of f, but *not topological*: a counterexample is given in [**ACLM02a**].

(3) When f is *non-degenerate* with respect to its *Newton polyhedron* $\Gamma(f)$, there exist combinatorial formulas in terms of $\Gamma(f)$ for $\zeta_{f,0}(t)$, $Z_{\text{Igusa}}(f;s)$ and $Z_{\text{top}}(f;s)$, as well as their local versions (see [**Var76, Loe90, DH01**]). These formulas are implemented in `Maple` [**HL00**] and `Sagemath` [**VS12**].

Using this zeta function, one can get a local version of the Monodromy conjecture for germs of non-constant holomorphic functions $f : (\mathbb{C}^n,0) \to (\mathbb{C},0)$.

CONJECTURE 1.56 (MONODROMY CONJECTURE). *Let s_0 be a pole of $Z_{\text{top},0}(f;s)$. Then $\exp(2\pi i s_0)$ is an eigenvalue of the monodromy action T_{x_0} on some level $H^q(\mathcal{F}_{f,x_0};\mathbb{C})$ at some point $x_0 \in V_f$.*

REMARK 1.57.

(1) We should consider the monodromy action T_{x_0} in points $x_0 \in V_f$ others than the origin, since $Z_{\text{top},0}$ also encodes information of the nearby strata

of the singular locus around 0. In the isolated case, it is clear that it suffices to look at the eigenvalues of the monodromy action at the origin.

(2) However, DENEF proved in [**Den93**, Lemma 4.6] that if λ is an eigenvalue of T_{x_0} at $x_0 \in V_f$, then there exist $x_1 \in V_f$ (arbitrarily close to x_0) such that λ appears as zero or pole of $\zeta_{f,x_1}(t)$. Thus, from the practical point of view, to determine all the eigenvalues is equivalent to compute all the possible monodromy zeta functions!

The classical strategy to try to prove (or disprove) the Monodromy conjecture for a particular hypersurface $V_f \subset \mathbb{C}^d$ is as follows. First, compute an explicit embedded resolution of singularities of V_f, or instead, study the possible combinatorial properties appearing in a subfamily of those embedded resolutions. Then, the determinantion of the possible poles of $Z_{\text{Igusa}}(f;s)$ or $Z_{\text{top},0}(f;s)$ from the resolution, excluding as much as possible "fake poles" appearing from the numerical data. Finally, compute all possible monodromy zeta functions using A'Campo formula.

Following the previous strategy, the Monodromy conjecture is proved for some families of singularities e.g.:

- $d = 2$ [**Loe88, Rod04b**] (the existence of a *minimal* embedded resolution of singularities in dimension 2 is one of the key points in this setting),
- $d = 3$ and f homogeneous [**RV01, ACLM02b**],
- superisolated surface singularities [**ACLM02b**],
- f is a product of linear forms (hyperplane arrangements) [**BMT11**],
- f is quasi-ordinary [**ACLM05**],
- f with non-degenerate surface singularities with respect to the Newton polyhedron [**BV16, Loe90**].

It is worth noticing that if we want to study the possible poles using different resolutions of V_f or other birational maps, then in principle the behavior of $Z_{\text{Igusa}}(f;s)$ would be better than $Z_{\text{top},0}(f;s)$ since we can study these transformations via the changes of variables formula. On the other hand, the expressions appearing in $Z_{\text{top},0}(f;s)$ are much easier to study, as well as the determination of "fake poles". The Weak Factorization theorem [**AKMW02, Wło03**] becomes the principal tool to study different birational maps in this case. A combination of both points of view was used e.g. by VEYS in [**Vey97**] for determining zeta functions of curves from *partial resolutions*.

As we will see in the following sections, the use of motivic integration will give a general framework for the study of all these zeta functions, their relations via birational maps and a way to understand some of the connections with other topological and analytical aspects of varieties.

REMARK 1.58 (STRONG MONODROMY CONJECTURE). There exists a *stronger* version of the Monodromy conjecture involving the roots of a polynomial associated to the meromorphic continuation of local zeta functions. The *b-function* or *Bernstein-Sato polynomial* associated to $f \in \mathbb{C}[x_1, \ldots, x_d]$ is the monic polynomial $b_f(s) \in \mathbb{C}[s]$ of smaller degree verifying that there exists a differential operator $P \in \mathbb{C}[s, x_1, \ldots, x_d, \partial_{x_1}, \ldots, \partial_{x_d}]$ such that

$$P \cdot f^{s+1} = b_f(s) f^s,$$

for any $s \in \mathbb{Z}$. It is know [**Kas77**] that the roots of $b_f(s)$ are negative rational numbers. By the above functional equation and using integration by parts, one sees that the poles of $Z_{\text{Igusa}}(f;s)$ lie in $\{\lambda - j \mid b_f(\lambda) = 0,\ j \in \mathbb{Z}_{\geq 0}\}$. The *Strong Monodromy conjecture* says that if s_0 is a pole of $Z_{\text{Igusa}}(f;s)$ then $\operatorname{Re}(s_0)$ is a root of $b_f(s)$. The same is stated for $Z_{\text{top}}(f;s)$ and the respective local versions. This statement is in fact *stronger* in the sense that implies the Monodromy conjecture, since it was proved by MALGRANGE and KASHIWARA in [**Mal83, Kas83**] that any root λ of $b_f(s)$ gives an eigenvalue $\exp(2\pi i\lambda)$ of T_{x_0} at some $x_0 \in V_f$. See [**Den91, Meu16**] for further details.

EXERCISE 1.59. Compute $Z_{\text{top},0}(f_i;s)$ and verify that the (local) Monodromy conjecture holds for $f_1(x,y) = y^2 - x^3$, $f_2(x,y) = x^m + y^m$ $(m \geq 2)$, $f_3(x,y,z) = (x^2+y^3)(x^2y^2+x^6+y^6)$ and $f_4(x,y,z) = y^2 - xz$.

EXERCISE 1.60. Study $Z_{\text{top},x_0}(g;s)$ and $\zeta_{g,x_0}(t)$ for $g(x,y,z) = z^2 - x^3y^2$ over the origin and nearby points. Is the Monodromy conjecture verified in this case? (How V_f looks in transverse sections to the singular set around 0?).

2. Motivic integration

Based on p-adic integration, KONTSEVICH constructed an integral using the arc space $\mathcal{L}(X)$ of $\mathbb{C}[\![t]\!]$-points of a variety. However, one cannot construct a real-valued measure on $\mathcal{L}(X)$ similar to the p-adic case, since $\mathbb{C}((t))$ is not locally compact. Kontsevich's key idea was to define a measure using additive invariants of complex varieties, avoiding Weil's conjectures and relaying directly those invariants by a change of variables formula coming from morphisms between varieties.

The following table summarizes the similarities and differences between p-adic integration in last section and the motivic one presented in the following.

	p-adic integral	(geometric) motivic integral
Functions to study	$f \in \mathbb{Z}[x_1, \dots, x_d]$	f regular over a complex variety X
Arithmetic vs. geometric	sols. of $f = 0$ over the ring $\mathbb{Z}/p^{m+1}\mathbb{Z} \simeq \mathbb{Z}_p/p^{m+1}\mathbb{Z}_p$, i.e. an d-tuple with coord. $a_0 + a_1 p + \cdots + a_m p^m$ $(a_i \in \{0, \dots, p-1\})$	sols. of $f = 0$ over the ring $\mathbb{C}[t]/\left(t^{m+1}\right) \simeq \mathbb{C}[\![t]\!]/\left(t^{m+1}\right)$, i.e. an d-tuple with coord. $a_0 + a_1 t + \cdots + a_m t^m$ $(a_i \in \mathbb{C})$
Domains of integration	$\mathbb{Z}_p^d = \lim_{\leftarrow} \left(\mathbb{Z}/p^{m+1}\mathbb{Z}\right)^d$ (liftings of all sols. mod p^{m+1} with coord. $\sum_{k \geq 0}^{\infty} a_k p^k$)	$\mathcal{L}(X) = \lim_{\leftarrow} \mathcal{L}_m(X)$ (*arcs*: liftings of all sols. mod t^{m+1} with coord. $\sum_{k \geq 0}^{\infty} a_k t^k$)
Algebra of measurable sets	Cylinders	Cylinders/stable sets/semi-algebraic sets of $\mathcal{L}(X)$
Value ring of the measure	$\mathbb{Z}[1/p]$	$\widehat{\mathcal{M}}_{\mathbb{C}}$, a localized and completed *universal ring of additive invariants of complex varieties*
Interesting class of integrable functions	Order of cancellation of f over the p-adic integers	Contact order of an arc along a divisor
Operations	Change of variables/Fubini	Change of variables

2.1. The Grothendieck ring of varieties as universal additive invariant. Denote by $\mathrm{Var}_{\mathbb{C}}$ the category of complex algebraic varieties. It is worth noticing that $\mathrm{Var}_{\mathbb{C}}$ is a *small category*, i.e. the class of objects $\mathrm{Obj}(\mathrm{Var}_{\mathbb{C}})$ forms a set (this follows from the fact that we can identify each algebraic variety with its structural sheaf of rings).

DEFINITION 2.1. Let R be a ring. A map $\lambda : \mathrm{Obj}(\mathrm{Var}_{\mathbb{C}}) \to R$ is an *additive* invariant if for any X, Y complex algebraic varieties:

(1) If X and Y are isomorphic, then $\lambda(X) = \lambda(Y)$.

(2) For any (Zariski) closed subset $F \subset X$, we have $\lambda(X) = \lambda(X \setminus F) + \lambda(F)$.

(3) $\lambda(X \times Y) = \lambda(X) \cdot \lambda(Y)$.

EXAMPLE 2.2. The following are examples of additive invariants:

(1) EULER CHARACTERISTIC:

$$\chi(X) = \sum_{i \geq 0} (-1)^i \dim \mathrm{H}^i(X, \mathbb{Q}) = \sum_{i \geq 0} (-1)^i b_i(X).$$

It is worth noticing that $\chi(\cdot)$ is additive when X is a complex algebraic variety. In general, is the *compactly supported Euler characteristic* $\chi_c(X) = \sum_{i \geq 0} (-1)^i \dim \mathrm{H}^i_c(X, \mathbb{Q})$ which verifies additivity relations. Fox example, $\mathbb{S}^1 = \mathbb{R} \sqcup \{\mathrm{pt}\}$ and $\chi(\mathbb{S}^1) = 0$ but $\chi(\mathbb{R}) = \chi(\mathrm{pt}) = 1$. However, $\chi_c(\mathbb{R}) = -1$.

(2) VIRTUAL HODGE-DELIGNE POLYNOMIAL: To a smooth projective variety X, one can associate

$$H_X(u, v) := \sum_{p,q \geq 0} (-1)^{p+q} h^{p,q}(X) u^p v^q \in \mathbb{Z}[u, v],$$

where $h^{p,q}(X) = \dim H^{p,q}(X, \mathbb{Q})$ are the *Hodge numbers*. In the same way, one can associate the VIRTUAL POINCARÉ POLYNOMIAL

$$P_X(t) = \sum_{i \geq 0} (-1)^i b_i(X) t^i \in \mathbb{Z}[t].$$

Note that $H_X(t, t) = P_X(t)$ and $P_X(1) = \chi(X)$.

(3) COUNTING POINTS: Assume X is defined over $\mathbb{Q}$ and fix p prime, then the map $\mathcal{N}^p(X) = |X(\mathbb{F}_p)|$ is an additive invariant over complex varieties defined over $\mathbb{Q}$.

Any of the previous additive invariants can be considered as the *realization* of a *universal* additive invariant of complex algebraic varieties.

DEFINITION 2.3. The *Grothendieck ring of varieties* $(K_0(\mathrm{Var}_{\mathbb{C}}), +, \cdot)$ is generated by the classes $[X]$, where

$$[X] = [Y], \text{ if } X, Y \in \mathrm{Obj}(\mathrm{Var}_{\mathbb{C}}) \text{ are isomorphic},$$

and relations:

- for any (Zariski) closed subset $F \subset X$, we have: $[X] = [X \setminus F] + [F]$,
- $[X \times Y] = [X] \cdot [Y]$.

The unit elements for addition and multiplication are $0 := [\emptyset]$ and $1 := [\mathrm{pt}]$, respectively. Denote by $\mathbb{L} := [\mathbb{A}^1_{\mathbb{C}}]$ the *Lefschetz motive*.

EXAMPLE 2.4.

(1) $[\mathbb{C}^n] = \mathbb{L}^n$ and $[\mathbb{C}^*] = [\mathbb{C} \setminus \{\mathrm{pt}\}] = [\mathbb{C}] - [\mathrm{pt}] = \mathbb{L} - 1$.

(2) $[\mathbb{CP}^1] = [\mathbb{C} \sqcup \{\infty\}] = \mathbb{L} + 1$. In fact, we know that $\mathbb{CP}^n = \mathbb{C}^n \sqcup \mathbb{CP}^{n-1}$ for any $n \geq 1$, thus

$$[\mathbb{CP}^n] = [\mathbb{C}^n] + [\mathbb{C}^{n-1}] + \cdots + [\mathbb{C}] + [\mathrm{pt}] = \mathbb{L}^n + \mathbb{L}^{n-1} + \cdots + \mathbb{L} + 1.$$

(3) Let $\mathcal{P}_m$ be a pencil of m affine lines at the origin $\mathcal{O}$. Then

$$[\mathcal{P}_m] = [\mathcal{P}_m \setminus \mathcal{O}] + [\mathcal{O}] = m[\mathbb{C}^*] + 1 = m(\mathbb{L} - 1) + 1 = m\mathbb{L} - (m - 1).$$

(4) Take the ordinary cusp $\mathcal{C} : y^2 - x^3 = 0$ in $\mathbb{C}^2$. Using the parametrization $\varphi : \mathbb{C} \to \mathcal{C}$, $\varphi(t) = (t^2, t^3)$, which restrings into an isomorphism $\varphi_{|\mathbb{C}^*}$, we have:

$$[\mathcal{C}] = [\mathcal{C} \setminus \mathcal{O}] + [\mathcal{O}] = [\mathbb{C}^*] + 1 = \mathbb{L}.$$

Note that φ is a bijection of points, but not an isomorphism. However, we have obtain that $[\mathcal{C}] = [\mathbb{C}]$.

(5) A subset $C \subset X$ is called *constructible* if it is the finite disjoint union of locally closed subsets of X. In fact, we can write $C = \bigsqcup_k U_k$ and this well-defines:

$$[C] := \sum_k [U_k].$$

EXERCISE: Prove that $[C]$ does not depend on the chosen decomposition of C.

REMARK 2.5. The above examples give the idea that $K_0(\mathrm{Var}_{\mathbb{C}})$ is a *scissors ring*: there are elements $A \not\simeq B$, but verifying $[A] = [B]$ after *cutting-and-pasting* operations. In fact, any locally trivial fibration in this setting is trivial in $K_0(\mathrm{Var}_{\mathbb{C}})$, as it is shown in the next result.

PROPOSITION 2.6. *The product on $K_0(\mathrm{Var}_{\mathbb{C}})$ extends to Zariski locally trivial fibrations: if $F \overset{i}{\hookrightarrow} X \overset{p}{\to} B$ verifies that for any $x \in B$ there is a Zariski open $x \in U \subset B$ such that $p^{-1}(U) \simeq F \times U$, then $[X] = [F] \cdot [B]$.*

PROOF. It follows from induction over $\dim B$ and operations on $K_0(\mathrm{Var}_{\mathbb{C}})$ (EXERCISE). ☐

REMARK 2.7. It is easy to prove, using successive stratifications on singular sets, that $\{[X] \mid X \text{ smooth and projective}\}$ is a set of generators of $K_0(\mathrm{Var}_{\mathbb{C}})$. In fact, BITTNER [**Bit04**] proved that $K_0(\mathrm{Var}_{\mathbb{C}})$ can be described equivalently by the previous set of generators, subject to the following relation: if $Y \subset X$ is smooth and projective, and $\pi : \mathrm{Bl}_Y(X) \to X$ is the blow-up of X along Y, with exceptional divisor $E = \pi^{-1}(Y)$, then

$$[X] - [Y] = [\mathrm{Bl}_Y(X)] - [E].$$

The above result uses the Weak Factorization theorem.

THEOREM 2.8 (UNIVERSAL PROPERTY). *Let $\lambda : \mathrm{Obj}(\mathrm{Var}_{\mathbb{C}}) \to R$ be an additive invariant. There exists a unique $\tilde{\lambda} : K_0(\mathrm{Var}_{\mathbb{C}}) \to R$ such that the following diagram commutes:*

$$\begin{array}{ccc} \mathrm{Obj}(\mathrm{Var}_{\mathbb{C}}) & \xrightarrow{\ \lambda\ } & R \\ {\scriptstyle [\,\cdot\,]}\downarrow & \nearrow_{\tilde{\lambda}} & \\ K_0(\mathrm{Var}_{\mathbb{C}}) & & \end{array}$$

COROLLARY 2.9. *There exist well defined ring morphisms as follows*

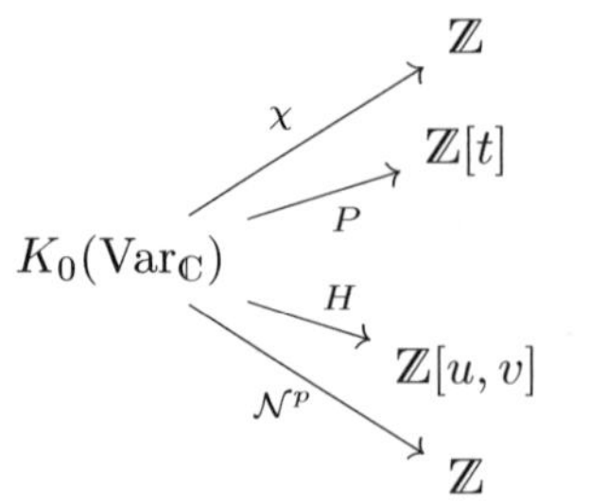

$$\chi([X]) = \chi(X)$$
$$P([X]) = P_X(t)$$
$$H([X]) = H_X(u,v)$$
$$\mathcal{N}^p([X]) = \mathcal{N}^p(X)$$

REMARK 2.10.

(1) The idea is that the class $[X]$ is the "most general" way to "count points" or "measure the size of a variety". Note the analogy

$$p = \left|\mathbb{A}^1_{\mathbb{F}_p}\right| \longleftrightarrow \mathbb{L} = [\mathbb{A}^1_{\mathbb{C}}].$$

(2) In general, a class $[X] \in K_0(\mathrm{Var}_{\mathbb{C}})$ cannot be expressed as a polynomial in $\mathbb{Z}[\mathbb{L}]$. Let $\mathcal{C}_g$ be a smooth projective curve of genus $g > 0$. Since $\mathcal{C}_g$ is a compact Riemann surface, it is known that the Hodge-Deligne polynomial of $\mathcal{C}_g$ is $H_{\mathcal{C}_g}(u,v) = 1 - gu - gv + uv$. However $H(\mathbb{L}^i) = (uv)^i$ for every $i \in \mathbb{N}$.

(3) The ring $K_0(\mathrm{Var}_{\mathbb{C}})$ is quite mysterious and complicated to deal with in general. In the last years, it has been an object of research, in fact:
- (POONER'02): $K_0(\mathrm{Var}_{\mathbb{C}})$ is not a domain, i.e. it contains zero divisors [**Poo02**].
- (BORISOV'14): $\mathbb{L}$ is a zero divisor [**Bor18**] (but not a nilpotent element since $\chi(\mathbb{L}^n) = 1$, for any $n \geq 0$).
- LARSEN-LUNTS CONJECTURE'03 [**LL14**]: If $[X] = [Y]$, then X and Y admit a decomposition into isomorphic locally closed subvarieties. This is TRUE for $\dim X \leq 1$ (LIU-SEBAG'10 [**LS10**]) but FALSE in general (BORISOV'14 [**Bor18**]) .

2.2. Basics on jet spaces and arc spaces. Let X be a complex algebraic variety.

DEFINITION 2.11. Assume that X is affine, i.e. one has $X \subset \mathbb{C}^d$ as the zero locus $X = \{f_1 = \cdots = f_k = 0\}$ for some $f_1, \ldots, f_k \in \mathbb{C}[x_1, \ldots, x_d]$.

- The *m-jet space of* X, denoted by $\mathcal{L}_m(X)$, is the algebraic variety over $\mathbb{C}[t]/(t^{m+1})$ defined by

$$\mathcal{L}_m(X) = \left\{\gamma_m \in \left(\mathbb{C}[t]/(t^{m+1})\right)^d \;\middle|\; f_1(\gamma_m) = \cdots = f_k(\gamma_m) = 0\right\}.$$

  Note that the previous equalities are established modulo t^{m+1}.

- The *(formal) arc space of* X, denoted by $\mathcal{L}(X)$, is the algebraic variety over $\mathbb{C}[\![t]\!]$ defined by

$$\mathcal{L}(X) = \left\{\gamma \in \mathbb{C}[\![t]\!]^d \mid f_1(\gamma) = \cdots = f_k(\gamma) = 0\right\}.$$

For any m-jet γ_m (resp. arc γ), we say that $\gamma_m(0)$ (resp. $\gamma(0)$) is its *origin* on X.

REMARK 2.12.

(1) $\mathcal{L}_m(X)$ and $\mathcal{L}(X)$ are complex algebraic varieties via the identifications:

$$\begin{array}{rcl}\{\mathbb{C}\text{-points of } \mathcal{L}_m(X)\} & = & \{(\mathbb{C}[t]/(t^{m+1}))\text{-points of } X\} \\ \{\mathbb{C}\text{-points of } \mathcal{L}(X)\} & = & \{\mathbb{C}[\![t]\!]\text{-points of } X\}\end{array}$$

Note that $\mathcal{L}(X)$ is infinitely dimensional in general.

(2) The following natural projections modulo t^{n+1}

$$\mathbb{C}[t]/(t^{m+1}) \longrightarrow \mathbb{C}[t]/(t^{n+1}) \quad \text{and} \quad \mathbb{C}[\![t]\!] \longrightarrow \mathbb{C}[\![t]\!]/(t^{n+1}) \simeq \mathbb{C}[t]/(t^{n+1}),$$

for every $n \leq m$, induce natural *truncation maps* between arc and jet spaces

$$\pi_n^m : \mathcal{L}_m(X) \to \mathcal{L}_n(X) \quad \text{and} \quad \pi_n : \mathcal{L}(X) \to \mathcal{L}_n(X).$$

Note that $\pi_n^m = \pi_n^k \circ \pi_k^m$ and $\pi_n = \pi_n^k \circ \pi_k$, for every $n \leq k \leq m$.

EXAMPLE 2.13. Let $X = \mathbb{C}^d$:

$$\begin{aligned}\mathcal{L}_m(\mathbb{C}^d) &= \left\{\left(a_0^{(1)} + \cdots + a_m^{(1)}t^m, \ldots, a_0^{(d)} + \cdots + a_m^{(d)}t^m\right) \;\middle|\; a_j^{(i)} \in \mathbb{C}\right\} \\ &\simeq \mathbb{C}^{d(m+1)}.\end{aligned}$$

REMARK 2.14. For $X \subset \mathbb{C}^d$, looking at the coefficients in $\mathcal{L}_m(X)$, we can identify the variety with a subvariety of $\mathbb{C}^{d(m+1)} \simeq \mathbb{C}^d \times \overset{m+1)}{\cdots} \times \mathbb{C}^d$ such that the truncation maps π_n^m, $n \leq m$, are induced by projections $\pi_n^m : \mathbb{C}^{d(m+1)} \to \mathbb{C}^{d(n+1)}$ on the first $d(n+1)$ components.

EXAMPLE 2.15. Let $X = \{y^2 - x^3 = 0\}$ be the ordinary cusp:

- $\mathcal{L}_0(X) = \{(a_0, b_0) \in \mathbb{C}^2 \mid b_0^2 - a_0^3 = 0\} = X.$
- $\mathcal{L}_1(X) = \{(a_0 + a_1 t, b_0 + b_1 t) \mid (b_0 + b_1 t)^2 - (a_0 + a_1 t)^3 = 0 \mod t^2\}.$

  $= \{(a_0 + a_1 t, b_0 + b_1 t) \mid b_0^2 - a_0^3 = 0 \text{ and } 2b_0 b_1 - 3a_0^2 a_1 = 0\}$

  Taking coefficients, we can see the map $\pi_0^1 : \mathcal{L}_1(X) \to \mathcal{L}_0(X) = X$ induced by the projection $\mathbb{C}^4 \to \mathbb{C}^2 : (a_0, b_0, a_1, b_1) \mapsto (a_0, b_0)$. Note that:
  - The fiber at $(0,0)$ is the whole (a_1, b_1)-plane $W = \{(0,0,a_1,b_1)\} \simeq \mathbb{C}^2$, which is the tangent space of X at the origin $T_{(0,0)}X \simeq \mathbb{C}^2$.
  - For $(a_0, b_0) \neq (0,0)$, the fiber correspond to a line L_{a_0,b_0} passing thought $P_{a_0,b_0} = (a_0, b_0, 0, 0)$ with equation $(2b_0)b_1 - (3a_0^2)a_1 = 0$. Note that $L_{a_0,b_0} \subset P_{a_0,b_0} + W$, corresponds to the tangent line of X at (a_0, b_0) and $T_{(a_0,b_0)}X \simeq \mathbb{C}$.

  Resuming, $\mathcal{L}_1(X)$ is the tangent bundle TX and $\pi_0^1 : TX \to X$ is the natural projection.
- In the same way, $\mathcal{L}_2(X)$ can be seen as a variety in $\mathbb{C}^6$ given by the equations

$$\begin{cases} b_0^2 - a_0^3 = 0 \\ 2b_0 b_1 - 3a_0^2 a_1 = 0 \\ b_1^2 + 2b_0 b_2 - (3a_0 a_1^2 + 3a_0^2 a_2) = 0 \end{cases}$$

  Note that the fiber of π_0^2 at the origin is the plane $\{(0,0,a_1,0,a_2,b_2)\} \simeq \mathbb{C}^3$, but its image by π_1^2 is the line $\{a_0 = b_0 = b_1 = 0\} \subset \mathbb{C}^4$. We deduce that $\pi_1^2 : \mathcal{L}_2(X) \to \mathcal{L}_1(X)$ is not surjective.

However, we can prove that the maps $\pi_m^{m+1} : \mathcal{L}_{m+1}(X) \to \mathcal{L}_m(X)$ are surjective above the non-singular part of $X = \mathcal{L}_0(X)$, moreover, they are fibrations of fiber $\mathbb{C}$.

REMARK 2.16. The previous spaces can be defined for any variety X:

$$\mathcal{L}_m(X) = \operatorname{Hom}\left(\operatorname{Spec}\mathbb{C}[t]/(t^{m+1}), X\right).$$

Then, we have the affine truncation morphisms $\pi_m^{m+1} : \mathcal{L}_{m+1}(X) \to \mathcal{L}_m(X)$ induced by the natural truncation $\mathbb{C}[t]/(t^{m+2}) \twoheadrightarrow \mathbb{C}[t]/(t^{m+1})$, and we define the arc space as an inverse limit

$$\mathcal{L}(X) = \varprojlim \mathcal{L}_m(X) = \operatorname{Hom}\left(\operatorname{Spec}\mathbb{C}[\![t]\!], X\right).$$

Any morphism between varieties $\varphi : Y \to X$ induces well defined morphisms:

$$\varphi_m : \mathcal{L}_m(Y) \to \mathcal{L}_m(X) \quad \text{and} \quad \varphi_\infty : \mathcal{L}(Y) \to \mathcal{L}(X).$$

As $\operatorname{Spec}\mathbb{C}[\![t]\!] = \{(0), (t)\}$, the map $\operatorname{Spec}\mathbb{C}[\![t]\!] \to X$ defines two points: the image of the closed one $\varphi\big((t)\big)$ (the *origin*) and the image of the generic one $\varphi\big((0)\big)$.

Determining the geometry of the varieties $\mathcal{L}_m(X)$ is in general a hard problem. Nevertheless, there exists several practical results about how "complicated" are the fibers of the maps $\pi_n^m : \mathcal{L}_m \to \mathcal{L}_n$.

PROPOSITION 2.17. *Let X be a d-dimensional complex variety. Then:*

(1) $\mathcal{L}_0(X) = X$ *and* $\mathcal{L}_1(X) = TX$.

(2) *If X is smooth, then π_n^m is a Zariski locally trivial fibration with fiber $\mathbb{C}^{d(m-n)}$, for any $n \leq m$. In particular, π_n^m and π_m are surjections and $\mathcal{L}_m(X)$ is smooth of dimension $d(m+1)$.*

(3) *For any $\gamma_m \in \mathcal{L}_m(X)$, the fiber $\left(\pi_m^{m+1}\right)^{-1}(\gamma_m)$ is either empty or isomorphic to $T_{x_0}X$, where $\gamma_m(0) = x_0$.*

(4) *Assume that X is irreducible and consider $X_{\text{reg}} = X \setminus X_{\text{sing}}$. Then the closure of $(\pi_0^m)^{-1}(X_{\text{reg}})$ is an irreducible component of $\mathcal{L}_m(X)$ of dimension $d(m+1)$.*

REMARK 2.18.

(1) If X is singular, then TX is not locally trivial. Also, as we have shown in Example 2.15 that π_n^m is not necessarily surjective.

(2) In [**Mus01**], MUSTAŢĂ obtained several results relating the geometry of the spaces $\mathcal{L}_m(X)$ with the one of X for the locally complete intersection case. In particular, he obtains that $\mathcal{L}_m(X)$ is irreducible for any $m > 0$ if and only if X has rational singularities. It turns out that the he original proof of these results is based on the use of motivic integration! Other results in this direction, as well as formulas for the log-terminal threshold involving the dimensions of jet spaces, were obtained e.g. in [**EM04, EMY03, ELM04**].

EXAMPLE 2.19 ($\mathcal{L}_m(X)$ **vs** $\pi_m(\mathcal{L}(X))$). We know that $\pi_m(\mathcal{L}(X)) \subset \mathcal{L}_m(X)$ but in the singular case not any m-jet can be lifted to an arc of X. Take $X = \{xy = 0\}$ the ordinary node, we have

$$\mathcal{L}(X) = \left\{(x(t), y(t)) \in \mathbb{C}[\![t]\!]^2 \mid x(t)y(t) = 0\right\}.$$

If we take generic arcs $x(t) = \sum_{i\geq 0} a_i t^i$ and $y(t) = \sum_{j\geq 0} b_j t^j$, we have that $(x(t), y(t)) \in \mathcal{L}(X)$ if and only if for any $k \geq 0$:

$$(J_k) \qquad a_0 b_k + a_1 b_{k-1} + \cdots + a_{k-1} b_1 + a_k b_0 = 0$$

Note that, for any $m \geq 0$:

$$\begin{aligned}\mathcal{L}_m(X) &= \left\{(x(t), y(t)) \in \left(\mathbb{C}[t]/(t^{m+1})\right)^2 \mid x(t)y(t) \equiv 0 \mod t^{m+1}\right\} \\ &= \left\{(a_0, b_0, \ldots, a_m, b_m) \in \mathbb{C}^{2(m+1)} \;\middle|\; (J_k) \text{ is verified for any } k = 0, \ldots, m\right\},\end{aligned}$$

since $x(t) \cdot y(t)$ is a generic polynomial of degree t^{2m}. In particular, if we look for $\mathcal{L}_1(X)$, we show that

$$\begin{aligned}\mathcal{L}_1(X) &= \{(a_0 + a_1 t, b_0 + b_1 t) \mid a_0 b_0 = 0, a_0 b_1 + a_1 b_0 = 0\} \\ &= \underbrace{\{a_0 = a_1 = 0\}}_{V_{2,0}} \cup \underbrace{\{a_0 = b_0 = 0\}}_{V_{1,1}} \cup \underbrace{\{b_0 = b_1 = 0\}}_{V_{0,2}}.\end{aligned}$$

Thus $\mathcal{L}_1(X)$ has three irreducible components isomorphic to $\mathbb{C}^2$. What happens with $\pi_1(\mathcal{L}(X))$? If we study the spaces $W_{l,h} = V_{l,h} \cap \pi_1(\mathcal{L}(X))$:

- A 1-jet is in $W_{2,0}$ if it is of the form $\varphi_1 = (0, b_0 + b_1 t)$, verifying that there exists and arc $\tilde{\varphi} = (a_2 t^2 + a_3 t^3 + \ldots, b_2 t^2 + b_3 t^3 + \ldots) \in \mathbb{C}[\![t]\!]^2$ such that $\varphi_1 + \tilde{\varphi}$ verifies (J_k) for any $k \geq 0$. Note that this is automatic for $k = 0, 1$. Now, (J_2) and (J_3) are equivalent to
$$a_2 b_0 = 0 \text{ and } a_2 b_1 + a_3 b_0 = 0,$$
respectively. Taking $\tilde{\varphi} = 0$ for any $b_0, b_1 \in \mathbb{C}$, it is easy to see that (J_k) is verified for any k. Then $W_{2,0} = V_{2,0}$ and $W_{0,2} = V_{0,2}$, by symmetry.
- For $W_{1,1}$, we study the lifts of $(a_1 t, b_1 t)$ in $\mathcal{L}(X)$. In this case, (J_2) is equivalent to $a_1 b_1 = 0$, thus $W_{1,1} \subset W_{2,0} \cap W_{0,2}$. In fact, taking again $\tilde{\varphi} = 0$ as lifted part, we see that $W_{1,1} = V_{1,1} \cap \{a_1 b_1 = 0\}$.

We deduce that $\pi_1(\mathcal{L}(X))$ is a projection over only two of the irreducible components of $\mathcal{L}_1(X)$.

In general, we can prove (Exercise) that $\mathcal{L}_m(X) = \bigcup_{l+h=m+1} V_{l,h}$ where

$$V_{l,h} = \{a_0 = \cdots = a_{l-1} = 0, b_0 = \cdots = b_{h-1} = 0\} \simeq \mathbb{C}^{m+1},$$

and those are exactly the irreducible components of $\mathcal{L}_m(X)$. Moreover, it can be shown that $\pi_m(\mathcal{L}(X)) = V_{m+1,0} \cup V_{0,m+1}$. We deduce that

$$[\mathcal{L}_m(X)] = (m+2)\mathbb{L}^{m+1} - (m+1)\mathbb{L}^m \quad \text{and} \quad [\pi_m(\mathcal{L}(X))] = 2\mathbb{L}^{m+1} - 1.$$

The study of $\pi_m(\mathcal{L}(X))$ was already considered by Nash [**Nas95**] and it is in general a very difficult problem. For a more detailed good survey in arc and n-jet spaces, see [**dF18**]. It should be noticed that $\pi_m(\mathcal{L}(X))$ is a constructible set in $\mathcal{L}_m(X)$, since Greenberg [**Gre66**] proved that there exists a constant $c > 0$ such that the image of π_m is equal to the one of π_m^{cm}, for any $m \geq 0$.

2.3. Motivic measure. Looking at the p-adic case, we look for defining a measure μ normalized over $\mathcal{L}(\mathbb{C})$, i.e. $\mu(\mathcal{L}(\mathbb{C})) = 1$, and with a formula of the type

$$\text{“ } \mu(C) = \frac{|\pi_m C|}{|(\mathbb{A}^d_{\mathbb{C}})^{m+1}|} \text{ ”},$$

when $m \gg 0$ and for any "cylinder" $C \subset \mathcal{L}(X)$, in the spirit of Proposition 1.21. In this setting, the "number of points" does not make sense anymore and we are going to reflect this invariant by the class in the Grothendieck ring.

Thus, in order to give a well defined framework for expressions of the type $[X]/\mathbb{L}^n$, we denote by $\mathcal{M}_{\mathbb{C}}$ the localized ring $K_0(\mathrm{Var}_{\mathbb{C}})[\mathbb{L}^{-1}]$. Remember that the map $K_0(\mathrm{Var}_{\mathbb{C}}) \to \mathcal{M}_{\mathbb{C}}$ is not injective, since $\mathbb{L}$ is known to be a zero divisor, but any additive invariant $\lambda : K_0(\mathrm{Var}_{\mathbb{C}}) \to R$ such that $\lambda(\mathbb{L}) \neq 0$ extends to a map $\mathcal{M}_{\mathbb{C}} \to R[\lambda(\mathbb{L})^{-1}]$. In particular, there exist well defined ring morphisms

$$\mathcal{M}_{\mathbb{C}} \begin{array}{l} \xrightarrow{\chi} \mathbb{Z} \\ \xrightarrow{P} \mathbb{Z}[t,t^{-1}] \\ \xrightarrow{H} \mathbb{Z}[u,v,(uv)^{-1}] \\ \xrightarrow{\mathcal{N}^p} \mathbb{Z}[1/p] \end{array} \quad \text{since} \quad \begin{array}{l} \chi(\mathbb{L}) = 1 \\ P(\mathbb{L}) = t^2 \\ H(\mathbb{L}) = uv \\ \mathcal{N}^p(\mathbb{L}) = p \end{array}$$

2.3.1. *Construction of the measure.* We start generalizing naturally the notion of cylinder.

DEFINITION 2.20. A subset $A \subset \mathcal{L}(X)$ is called a *cylinder* (or *constructible*) if $A = \pi_m^{-1}(C_m)$ for some $m \in \mathbb{Z}_{\geq 0}$ and some constructible set $C_m \subset \mathcal{L}_m(X)$.

REMARK 2.21.

(1) Morally, a cylinder is a set of arcs "$A \simeq C_m \times \mathbb{C}^\infty$" or "$A \simeq \pi_m(A) \times \mathbb{C}^\infty$". The collection of cylinders in $\mathcal{L}(X)$ forms a boolean algebra of sets, i.e. finite unions and complements of cylinders are cylinders, as well as the empty set and $\mathcal{L}(X) = \pi_0^{-1}(X)$.

(2) If we have a partition by constructible sets $X = \bigsqcup_{i=0}^k W_i$, then $\mathcal{L}(X) = \bigsqcup_{i=0}^k \pi_0^{-1}(W_i)$.

PROPOSITION 2.22. *Assume X is smooth and let $A = \pi_{m_0}^{-1}(C_{m_0})$ be a cylinder in $\mathcal{L}(X)$. Then*

$$\frac{[\pi_m(A)]}{\mathbb{L}^{d(m+1)}} \in \mathcal{M}_{\mathbb{C}}$$

is constant for any $m \geq m_0$.

PROOF. Since X is smooth, the maps $\pi_{m_0}^m : \mathcal{L}_m(X) = \pi_m(\mathcal{L}(X)) \longrightarrow \mathcal{L}_{m_0}(X) = \pi_{m_0}(\mathcal{L}(X))$ are locally trivial fibrations with fiber $\mathbb{C}^{d(m-m_0)}$. Thus $[\pi_m(A)] = [C_{m_0}] \cdot [\mathbb{C}^{d(m-m_0)}] = [C_{m_0}] \cdot \mathbb{L}^{d(m-m_0)}$ and the result holds. □

REMARK 2.23. For $A = \mathcal{L}(X)$ with X smooth d-dimensional, note that the previous expression is simply $[X]\mathbb{L}^{-d}$ since $\mathcal{L}(X) = \pi_0^{-1}(X)$.

For arbitrary varieties, the previous stabilizations are not assured.

DEFINITION 2.24. We call $A \subset \mathcal{L}(X)$ *stable* if for some $m_0 \in \mathbb{Z}_{\geq 0}$, we have:

(1) $\pi_{m_0}(A)$ is constructible and $A = \pi_{m_0}^{-1}(\pi_{m_0}(A))$,

(2) for any $m \geq m_0$, the projection $\pi_{m+1}(A) \to \pi_m(A)$ is a piecewise trivial fibration with constant fiber $\mathbb{C}^d$, i.e. there exists a finite partition of $\pi_m(A)$

into locally closed sets S such that any of them admits a open covering $S = \bigcup_k U_k$ verifying $(\pi_m^{m+1})^{-1}(U_k) \simeq U_k \times \mathbb{C}^d$, for any k.

LEMMA 2.25 (DENEF-LOESER, [**DL99**]). *If $A \subset \mathcal{L}(X)$ is a cylinder and $A \cap \mathcal{L}(X_{\text{sing}}) = \emptyset$, then A is stable.*

The result above implies that for any stable A, the limit $\lim_{m\to\infty} \frac{[\pi_m(A)]}{\mathbb{L}^{d(m+1)}}$ exists in $\mathcal{M}_{\mathbb{C}}$. This defines an additive invariant with respect to finite unions and intersections

$$\tilde{\mu}_{\mathcal{L}(X)} : \{\text{Stable subsets of } \mathcal{L}(X)\} \longrightarrow \mathcal{M}_{\mathbb{C}},$$

which is called the *naive motivic measure*. However, in general the stable subsets do not form an algebra of sets, as $\mathcal{L}(X)$ is not stable for general X. Take $X = \{xy = 0\}$, from Example 2.19, we see that the sequence

$$\frac{[\pi_m(\mathcal{L}(X))]}{\mathbb{L}^{d(m+1)}} = \frac{2\mathbb{L}^{m+1} - 1}{\mathbb{L}^{m+1}} = 2 - \frac{1}{\mathbb{L}^{m+1}}$$

does not stabilize. We are going to define a measure $\mu_{\mathcal{L}(X)}$, extending $\tilde{\mu}_{\mathcal{L}(X)}$

2.3.2. *Completion of $\mathcal{M}_{\mathbb{C}}$.* As in the p-adic case, we will consider a completion by a *norm* for which the values $\mathbb{L}^{-m}$ are small. For this, LOOIJENGA [**Loo02**] introduces the notion of *virtual dimension*.

DEFINITION 2.26. An element $\tau \in K_0(\text{Var}_{\mathbb{C}})$ is called *d-dimensional* if there is a finite expression

$$\tau = \sum_i a_i[X_i],$$

with $a_i \in \mathbb{Z}$ and $X_i \in \text{Obj}(\text{Var}_{\mathbb{C}})$ such that $d = \max_i\{\dim X_i\}$, and if there is no such expression with all $\dim X_i \leq d - 1$. We set that the dimension of $[\emptyset]$ is $-\infty$.

The above definition can be extended to elements in $\mathcal{M}_{\mathbb{C}}$ setting $\dim\left(\mathbb{L}^{-1}\right) = -1$.

PROPOSITION 2.27. *The* virtual dimension *map* $\dim : \mathcal{M}_{\mathbb{C}} \to \mathbb{Z} \cup \{-\infty\}$ *is well defined and satisfies, for any $\tau, \tau' \in K_0(\text{Var}_{\mathbb{C}})$:*

(1) $\dim(\tau \cdot \tau') \leq \dim \tau + \dim \tau'$.

(2) $\dim(\tau + \tau') \leq \max\{\dim \tau, \dim \tau'\}$, *with equality if* $\dim \tau \neq \dim \tau'$.

EXERCISE 2.28. Let $A, B \subset \mathcal{L}(X)$ be stable subsets. Show that, if $A \subset B$ then $\dim \tilde{\mu}_{\mathcal{L}(X)}(A) \leq \dim \tilde{\mu}_{\mathcal{L}(X)}(B)$.

We construct the completion with respect the ascending filtration defined by the virtual dimension over $\mathcal{M}_{\mathbb{C}}$:

$$\cdots \subset \mathcal{F}^{m-1}\mathcal{M}_{\mathbb{C}} \subset \mathcal{F}^m\mathcal{M}_{\mathbb{C}} \subset \mathcal{F}^{m+1}\mathcal{M}_{\mathbb{C}} \subset \cdots$$

given by the subgroups

$$\mathcal{F}^m\mathcal{M}_{\mathbb{C}} = \{\tau \in \mathcal{M}_{\mathbb{C}} \mid \dim \tau \leq -m\} = \left\langle \frac{[X]}{\mathbb{L}^i} \;\middle|\; X \in \text{Obj}(\text{Var}_{\mathbb{C}}),\ \dim X - i \leq -m \right\rangle.$$

Note that $\mathcal{F}^m\mathcal{M}_{\mathbb{C}} \cdot \mathcal{F}^n\mathcal{M}_{\mathbb{C}} \subset \mathcal{F}^{m+n}\mathcal{M}_{\mathbb{C}}$.

DEFINITION 2.29. We define the ring

$$\widehat{\mathcal{M}}_{\mathbb{C}} = \varprojlim_m \mathcal{M}_{\mathbb{C}}/\mathcal{F}^m\mathcal{M}_{\mathbb{C}},$$

i.e. the completion of $\mathcal{M}_{\mathbb{C}}$ with respect to the filtration $\mathcal{F}^\bullet\mathcal{M}_{\mathbb{C}}$.

REMARK 2.30.

(1) By definition, a sequence $\left(\frac{[X_k]}{\mathbb{L}^{i_k}}\right)_{k\in\mathbb{N}}$ converges to zero in $\widehat{\mathcal{M}_{\mathbb{C}}}$ if and only if

$$\dim X_k - i_k \underset{k\to\infty}{\longrightarrow} -\infty.$$

(2) As described by BATYREV [**Bat98**], the ring $\widehat{\mathcal{M}_{\mathbb{C}}}$ is the completion with respect the norm

$$\begin{array}{rccc} \delta: & \mathcal{M}_{\mathbb{C}} & \longrightarrow & \mathbb{R}_{\geq 0} \\ & \tau & \longmapsto & \delta(\tau) = e^{\dim \tau} \end{array},$$

setting $\delta(\emptyset) = 0$. This norm is non-archimidean, i.e. for any $\tau, \tau' \in \mathcal{M}_{\mathbb{C}}$:
  (a) $\delta(\tau) = 0$ if and only if $\tau = 0 = [\emptyset]$ in $\mathcal{M}_{\mathbb{C}}$.
  (b) $\delta(\tau \cdot \tau') \leq \delta(\tau) \cdot \delta(\tau')$.
  (c) $\delta(\tau + \tau') \leq \max\{\delta(\tau), \delta(\tau')\}$.

It is worth noticing that this norm is not known to be *multiplicative*, i.e. if the condition (2) is in fact an equality, since $\mathcal{M}_{\mathbb{C}}$ could not be a domain.

EXERCISE 2.31.

(1) Show that a sum $\sum_{i=0}^{\infty} \tau_i$, with $\tau_i \in \mathcal{M}_{\mathbb{C}}$, converges in $\widehat{\mathcal{M}_{\mathbb{C}}}$ if and only if $\tau_i \to 0$.

(2) Fix $N \in \mathbb{Z}$, show that

$$\sum_{i=0}^{\infty} \mathbb{L}^{-Ni} = \frac{1}{1 - \mathbb{L}^{-N}},$$

in $\widehat{\mathcal{M}_{\mathbb{C}}}$.

EXAMPLE 2.32. Revisiting the sequence associated with $X = \{xy = 0\}$, we see that the limit exists in $\widehat{\mathcal{M}_{\mathbb{C}}}$ and

$$\lim_{m\to\infty} \frac{[\pi_m(\mathcal{L}(X))]}{\mathbb{L}^{d(m+1)}} = \lim_{m\to\infty} \left(2 - \frac{1}{\mathbb{L}^{m+1}}\right) = 2.$$

THEOREM 2.33 (DENEF-LOESER, [**DL99**]). *For any cylinder $A \subset \mathcal{L}(X)$, the limit*

$$\mu_{\mathcal{L}(X)}(A) = \lim_{m\to\infty} \frac{[\pi_m(A)]}{\mathbb{L}^{d(m+1)}}$$

exists in $\widehat{\mathcal{M}_{\mathbb{C}}}$. Moreover, the map

$$\begin{array}{rccc} \mu_{\mathcal{L}(X)}: & \{\textit{Cylinders of } \mathcal{L}(X)\} & \longrightarrow & \widehat{\mathcal{M}_{\mathbb{C}}} \\ & A & \longmapsto & \mu_{\mathcal{L}(X)}(A) \end{array},$$

is a σ-additive measure, i.e. for any family $\{A_k\}_{k\in\mathbb{N}}$ of pairwise disjoint cylinders, we have

$$\mu_{\mathcal{L}(X)}\left(\bigsqcup_{k\geq 0} A_k\right) = \sum_{k\geq 0} \mu_{\mathcal{L}(X)}(A_k).$$

The measure $\mu_{\mathcal{L}(X)}$ is called the *motivic measure on* $\mathcal{L}(X)$. When there is not confusion, we denote the measure above simply by $\mu = \mu_{\mathcal{L}(X)}$.

As a consequence of the above theory on measurable sets, we obtain that the arc spaces of subvarieties of X are "negligible" in $\mathcal{L}(X)$.

PROPOSITION 2.34 (DENEF-LOESER, [**DL99**]). *Let $Z \subset X$ be a proper closed subvariety of X. Then $\mu_{\mathcal{L}(X)}(\mathcal{L}(Z)) = 0$.*

REMARK 2.35.

(1) Other authors (for example [**Bat98, Vey06**]) consider another normalization of the motivic measure with an extra factor $\mathbb{L}^d$, such that $\mu(\mathcal{L}(X)) = [X]$. In particular, $\mu(\mathbb{L}^d) = \mathbb{L}^d$.

(2) It is not known whether the natural map $\mathcal{M}_{\mathbb{C}} \to \widehat{\mathcal{M}}_{\mathbb{C}}$ is injective or not. It is easy to see that

$$\ker\left(\mathcal{M}_{\mathbb{C}} \longrightarrow \widehat{\mathcal{M}}_{\mathbb{C}}\right) = \bigcap_{m\in\mathbb{Z}} \mathcal{F}^m\mathcal{M}_{\mathbb{C}}.$$

However, the Hodge polynomial $H : \mathcal{M}_{\mathbb{C}} \to \mathbb{Z}[u, v, (uv)^{-1}]$ factors on the image $\overline{\mathcal{M}_{\mathbb{C}}} \subset \widehat{\mathcal{M}}_{\mathbb{C}}$, i.e. $H(\tau) = 0$ for any $\tau \in \bigcap_{m\geq 0} \mathcal{F}^m\mathcal{M}_{\mathbb{C}}$, since

$$\deg H\left([X_m]\mathbb{L}^{-i_m}\right) = 2\dim X_m - 2i_m \leq -2m \longrightarrow -\infty,$$

where $[X_m]\mathbb{L}^{-i_m} \in \mathcal{F}^m\mathcal{M}_{\mathbb{C}}$ for any $m \in \mathbb{Z}$. As a consequence, the specializations P and χ factor too.

(3) The main point to deal with the singular case is to express $\mathcal{L}(X)\setminus\mathcal{L}(X_{\text{sing}})$ as a union of "complements in $\mathcal{L}(X)$ of smaller and smaller tubular neighborhoods around the singular locus", constructed in terms of the "contact levels" of arcs with respect to X_{sing}. More precisely,

$$\mathcal{L}(X) \setminus \mathcal{L}(X_{\text{sing}}) = \bigcup_{e\geq 0} \mathcal{L}^{(e)}(X),$$

where $\mathcal{L}^{(e)}(X) = \mathcal{L}(X) \setminus \pi_e^{-1}(\mathcal{L}_e(X_{\text{sing}}))$. As it turns out, $A \cap \mathcal{L}^{(e)}(X)$ is stable at any level $m_0 \geq e$ for any cyinder $A \subset \mathcal{L}(X)$, and one can associate a measure to A taking the limit in $\widehat{\mathcal{M}}_{\mathbb{C}}$ when $e \to \infty$. Compare this with OESTERLÉ's result in the case of p-adics (Remark 1.50).

(4) DENEF-LOESER and BATYREV extend the motivic measure for a more general class of *measurable* sets. In particular: ($\mathbb{C}[t]$-)*semi-algebraic sets* of $\mathcal{L}(X)$, defined by finite boolean combinations of conditions involving polynomials, lower coefficients of arcs and (in)equalities using the order ord_t of arcs (see [**DL99**] for more details).

2.4. Motivic integral and change of variables formula. We can now define a *motivic integral* naturally generalizing the previous p-adic integral, and based on the same principle: an integral of a function with countable image values can be expressed as a sum of level values.

DEFINITION 2.36. Let $A \subset \mathcal{L}(X)$ be a cylinder and let $\alpha : A \to \mathbb{Z} \cup \{\infty\}$ be a map such that the fibers $\alpha^{-1}(m) = \{x \in A \mid \alpha(x) = m\}$ are cylinders. We define the *motivic integral of* $\mathbb{L}^{-\alpha}$ *over* A as the expression

$$\int_A \mathbb{L}^{-\alpha} d\mu_{\mathcal{L}(X)} := \sum_{m\in\mathbb{Z}} \mu_{\mathcal{L}(X)}\left(\alpha^{-1}(m)\right)\mathbb{L}^{-m}$$

in $\widehat{\mathcal{M}}_{\mathbb{C}}$, whenever the right-hand side converges. In this case, $\mathbb{L}^{-\alpha}$ is said to be *integrable*.

Note that any map α bounded from below gives an integrable function. Also, one can easily produce examples of integrable functions in this context using characteristic functions over cylinders, e.g. let $\mathcal{C}$ be a finite collection of pairwise disjoint cylinders and $\{a_C\}_{C\in\mathcal{C}} \subset \mathbb{Z}$, then

$$\alpha = \sum_{C\in\mathcal{C}} a_C \mathbf{1}_C$$

induces an integrable function.

EXAMPLE 2.37. For X smooth of dimension d and $\alpha \equiv 0$,

$$\int_{\mathcal{L}(X)} \mathbb{L}^{-0} \mathrm{d}\mu = \mu(\mathcal{L}(X)) = [X]\mathbb{L}^{-d}.$$

We are going to integrate with respect to cylinders expressed using the *order of an arc*

$$\begin{array}{rccl} \operatorname{ord}_t : & \mathbb{C}[\![t]\!] & \longrightarrow & \mathbb{Z}_{\geq 0} \cup \{\infty\} \\ & \gamma & \longmapsto & \operatorname{ord}_t(\gamma) \end{array}$$

defined as

$$\operatorname{ord}_t(\gamma) = \sup\{e \in \mathbb{Z}_{\geq 0} \mid \gamma(t) \in (t^e)\},$$

or equivalently, if there exists a unit $\varphi(t) \in \mathbb{C}[\![t]\!]^\times$ such that $\gamma(t) = t^{\operatorname{ord}_t(\gamma)}\varphi(t)$. Note that

$$\begin{array}{rcl} \operatorname{ord}_t(\gamma) = \infty & \iff & \gamma = 0 \\ \operatorname{ord}_t(\gamma) = 0 & \iff & \gamma(0) \neq 0, \text{i.e. } \gamma \text{ is a unit.} \end{array}$$

EXAMPLE 2.38. Let $X = \mathbb{C}$ and $m \in \mathbb{Z}_{\geq 0}$. Consider the set $A_m = \{\gamma \in \mathcal{L}(\mathbb{C}) \mid \operatorname{ord}_t(\gamma) = m\}$. Then A_m is a cylinder because it can be written as $A_m = \pi_m^{-1}(C_m)$ where $C_m = \{\gamma \in \mathcal{L}_m(\mathbb{C}) \mid \gamma(t) = a_m t^m, \text{ for some } a_m \in \mathbb{C}^*\}$. Therefore

$$\mu(A_m) = [C_m]\,\mathbb{L}^{-(m+1)} = (\mathbb{L}-1)\,\mathbb{L}^{-(m+1)}, \quad \forall m \geq 0.$$

REMARK 2.39. For any cylinders $C_1 \subset \mathbb{C}^{d_1}$ and $C_2 \subset \mathbb{C}^{d_2}$, we have

$$\mu_{\mathcal{L}(\mathbb{C}^{d_1+d_2})}(C_1 \times C_2) = \mu_{\mathcal{L}(\mathbb{C}^{d_1})}(C_1) \cdot \mu_{\mathcal{L}(\mathbb{C}^{d_2})}(C_2).$$

This property is again a consequence of the fact that the truncation map $\pi_{m,n} : \mathcal{L}_m(\mathbb{C}^d) \to \mathcal{L}_n(\mathbb{C}^d)$, for $m \geq n$, is a locally trivial fibration with fiber $\mathbb{C}^{d(m-n)}$. The combination of this property together with Example 2.38 will help us simplify some calculations in explicit examples.

2.4.1. *The contact order of an arc along a divisor.* Let D be an effective Cartier divisor on a smooth algebraic variety X, i.e. a subvariety of X which is locally given by the set of zeros of one equation. One defines the function

$$\begin{array}{rccl} \operatorname{ord}_D : & \mathcal{L}(X) & \longrightarrow & \mathbb{Z}_{\geq 0} \cup \{\infty\} \\ & \gamma & \longmapsto & \operatorname{ord}_t f_D(\gamma) \end{array}$$

where f_D is a local equation of D around the origin $\pi_0(\gamma) = \gamma(0) \in X$, and

$$\operatorname{ord}_t f_D(\gamma) = \operatorname{ord}_t(f_D \circ \gamma)(t).$$

REMARK 2.40.

(1) Note that

$$\begin{aligned} \operatorname{ord}_D(\gamma) = \infty &\iff \gamma \in \mathcal{L}(D_{\text{red}}) \\ \operatorname{ord}_D(\gamma) = 0 &\iff \pi_0(\gamma) \notin D_{\text{red}} \end{aligned}$$

In fact, we can prove that $\operatorname{ord}_D^{-1}(\infty) = \mathcal{L}(D_{\text{red}})$ is not a cylinder, but it is a $\mu_{\mathcal{L}(X)}$-measurable set of measure 0.

(2) If we write $D = \sum_{i=1}^{s} N_i D_i$ as a linear combination of prime divisors, then locally f_D decomposes as a product $f_D = \prod_{i=1}^{s} f_{D_i}^{N_i}$ of defining equations for D_i, hence

$$\operatorname{ord}_D = \sum_{i=1}^{s} N_i \operatorname{ord}_{D_i}.$$

(3) The previous construction can be generalized for a sheaf of ideals $\mathcal{I}$ on X. We define $\operatorname{ord}_{\mathcal{I}} : \mathcal{L}(X) \to \mathbb{Z}_{\geq 0} \cup \{\infty\}$ by:

$$\operatorname{ord}_{\mathcal{I}}(\gamma) = \min \{\operatorname{ord}_t g(\gamma) \mid g \in \mathcal{I}(U),\ U \text{ open affine cover},\ \gamma(0) \in U\}.$$

What about the *m-th contact locus of* D, $\operatorname{ord}_D^{-1}(m)$, for $m \in \mathbb{Z}_{\geq 0}$? Assume that $X = \mathbb{C}^d$, it is easy to see as in the p-adic case that, one has for $m \geq 1$ that:

$$\begin{aligned} \{\gamma \in \mathcal{L}(X) \mid \operatorname{ord}_D(\gamma) \geq m\} &= \{\gamma \in \mathcal{L}(X) \mid \operatorname{ord}_t(f_D \circ \gamma) \geq m\} \\ &= \{\gamma \in \mathcal{L}(X) \mid (f_D \circ \gamma)(t) \equiv 0 \mod t^m\} \\ &= (\pi_{m-1})^{-1}(\mathcal{L}_{m-1}(D)). \end{aligned}$$

The above expression remains true for a general X using coordinate open covers. Thus,

$$\{\gamma \in \mathcal{L}(X) \mid \operatorname{ord}_D(\gamma) = m\} = (\pi_{m-1})^{-1}(\mathcal{L}_{m-1}(D)) \setminus (\pi_m)^{-1}(\mathcal{L}_m(D)),$$

which is clearly a cylinder. Then, if X is smooth:

$$\mu\left(\operatorname{ord}_D^{-1}(m)\right) = \frac{[\mathcal{L}_{m-1}(D)]}{\mathbb{L}^{dm}} - \frac{[\mathcal{L}_m(D)]}{\mathbb{L}^{d(m+1)}}.$$

Note that $\operatorname{ord}_D^{-1}(0) = \pi_0^{-1}(X \setminus D)$. Also, $\mu\left(\operatorname{ord}_D^{-1}(0)\right) = \mathbb{L}^{-d}[X \setminus D]$ whenever X is smooth.

EXAMPLE 2.41. Let $X = \mathbb{C}$ and $D = \operatorname{div}(f) \subset \mathbb{C}$ be the divisor associated with the function $f(x) = x$. We have that $\mathcal{L}_m(D) = \{0\} \subset \mathcal{L}_m(X)$ for any $m \geq 0$ and that

$$\mu(\operatorname{ord}_D^{-1}(m)) = \frac{1}{\mathbb{L}^m} - \frac{1}{\mathbb{L}^{m+1}}$$

In this way, we obtain:

$$\begin{aligned} \int_{\mathcal{L}(\mathbb{C})} \mathbb{L}^{-\operatorname{ord}_D} \mathrm{d}\mu &= \sum_{m \geq 0} \left(\frac{1}{\mathbb{L}^m} - \frac{1}{\mathbb{L}^{m+1}}\right) \cdot \frac{1}{\mathbb{L}^m} = \left(1 - \frac{1}{\mathbb{L}}\right) \sum_{m \geq 0} \frac{1}{\mathbb{L}^{2m}} \\ &= \frac{\mathbb{L} - 1}{\mathbb{L}} \cdot \frac{1}{1 - \mathbb{L}^{-2}} = \frac{\mathbb{L} - 1}{\mathbb{L} - \mathbb{L}^{-1}} = \frac{\mathbb{L}}{\mathbb{L} + 1}. \end{aligned}$$

EXAMPLE 2.42. Let $X = \mathbb{C}$ and D be the divisor associated with the function $f(x) = x^N$, for $N \geq 0$, representing "the origin with multiplicity N". Note that, for any $\gamma \in \mathbb{C}[\![t]\!]$ of order $k \geq 0$, i.e. $\gamma(t) = t^k \varphi(t)$ with $\varphi \in \mathbb{C}[\![t]\!]^\times$,

$$(f \circ \gamma)(t) = t^{kN} \varphi(t)^N.$$

Thus, $N | \operatorname{ord}_D(\gamma)$ for any $\gamma \in \mathcal{L}(\mathbb{C})$, hence $\operatorname{ord}_D^{-1}(m) = \emptyset$ for any $m \neq kN$. One can compute, using Example 2.38:

$$\mu\{\gamma \in \mathcal{L}(\mathbb{C}) \mid \operatorname{ord}_D(\gamma) = kN\} = \mu\{\gamma \in \mathcal{L}(\mathbb{C}) \mid \operatorname{ord}_t(\gamma) = k\} = \frac{\mathbb{L}-1}{\mathbb{L}^{k+1}},$$

for any $k \geq 0$. Then

$$\begin{aligned}\int_{\mathcal{L}(\mathbb{C})} \mathbb{L}^{-\operatorname{ord}_D} \mathrm{d}\mu &= \sum_{k\geq 0} \mu\{\gamma \in \mathcal{L}(\mathbb{C}) \mid \operatorname{ord}_D(\gamma) = kN\} \cdot \frac{1}{\mathbb{L}^{kN}} \\ &= \mathbb{L}^{-1}(\mathbb{L}-1) \sum_{k\geq 0} \frac{1}{\mathbb{L}^{k(N+1)}} = \frac{\mathbb{L}-1}{\mathbb{L}-\mathbb{L}^{-N}}.\end{aligned}$$

(Compare this result with the p-adic one obtained for $\int_{\mathbb{Z}_p} |x|_p^N \, \mathrm{d}\mu$.)

EXERCISE 2.43. Compute the motivic integral associated with $D : x^{N_1} y^{N_2} = 0$ over $\mathbb{C}^2$, where $N_1, N_2 \geq 1$.

EXERCISE 2.44. Let X be a d-dimensional smooth variety and D_0 a smooth divisor. Show that, if $D = N D_0$ for $N \geq 1$, then

$$\int_{\mathcal{L}(X)} \mathbb{L}^{-\operatorname{ord}_D} \mathrm{d}\mu = \mathbb{L}^{-d} \left([X \setminus D_0] + [D_0] \cdot \frac{\mathbb{L}-1}{\mathbb{L}^{N+1}-1} \right).$$

(HINT: Since D is smooth, $\mathcal{L}_m(D)$ is a locally trivial fibration over D with fiber $\mathbb{C}^{(d-1)m}$)

The motivic integral can be defined with respect to a constructible set $W \subset X$ in a natural way:

$$\int_{\mathcal{L}(X)_W} \mathbb{L}^{-\operatorname{ord}_D} \mathrm{d}\mu_{\mathcal{L}(X)},$$

setting $\mathcal{L}(X)_W = \pi_0^{-1}(W)$. In particular, there is a *local* version of the motivic integrals for germs of functions $f : (X, x_0) \to (\mathbb{C}, 0)$, given by

$$\int_{\mathcal{L}(X)_{x_0}} \mathbb{L}^{-\operatorname{ord}_D} \mathrm{d}\mu_{\mathcal{L}(X)},$$

where D is locally defined by f.

2.4.2. *The change of variables formula.* We will continue with more calculations later. First, let us introduce the main tool on motivic integration theory: the *change of variables*. This tool involves the ordinary and relative canonical divisors of varieties and, more generally, the concept of *Jacobian ideal sheaf.*

Let $h : Y \to X$ be a proper birational map, with Y smooth and $d = \dim Y = \dim X$. Recall that the *exceptional locus* of h is the maximal subvariety of Y where h is not an isomorphism.

First, the properness condition in h implies that there is an induced bijection of arcs from $\mathcal{L}(Y)$ to $\mathcal{L}(X)$ away from those contained in the exceptional locus. More precisely, if $E \subset Y$ is the exceptional locus of h and $F = h(E)$, we have a bijection

$$\mathcal{L}(Y) \setminus \mathcal{L}(E) \xleftrightarrow{h_\infty} \mathcal{L}(X) \setminus \mathcal{L}(F).$$

This is due to the *valuative criterion of properness* [**Har77**, Thm. II.4.3].

Now, consider the *sheaf of Kähler differential d-forms* $\Omega_X^d = \bigwedge^d \Omega_X^1$, constructed as follows. The sheaf Ω_X^1 is the unique one such that for any $U \subset X$ with sections $\mathcal{O}_X(U) = \mathbb{C}[x_1, \dots, x_d]/(f_1, \dots, f_r)$, we have

$$\Omega_X^1(U) = \mathcal{O}_X(U) \cdot \frac{\langle \mathrm{d}x_1, \dots, \mathrm{d}x_d \rangle}{(\mathrm{d}f_1, \dots, \mathrm{d}f_r)}.$$

In this case, note that $\Omega_X^2(U)$ is a $\mathcal{O}_X(U)$-module generated by $\{\mathrm{d}x_i \wedge \mathrm{d}x_j\}_{1\leq i<j\leq d}$, with relations given by the ideal generated by $\{\mathrm{d}f_1 \wedge \mathrm{d}x_k, \dots, \mathrm{d}f_r \wedge \mathrm{d}x_k\}_{1\leq k\leq d}$. Local presentations of the sheaf Ω_X^d can be computed inductively. Since Y is smooth, then Ω_Y^d is the usual sheaf of regular differential d-forms.

We are going to define the ideal sheaf $\mathrm{Jac}(h)$ associated with h as follows:

- If X is smooth, then Ω_X^d is locally generated by one element. In this way, we can define the *canonical divisor* K_X as the divisor $\mathrm{div}(\omega)$ for a non-zero top rational differential form $\omega \in \Gamma(\Omega_X^d, X)$. In this case, $\mathrm{Jac}(h)$ is the *discrepancy divisor* (or *relative canonical divisor*) between Y and X, denoted by K_h (see Section 3.1.1).
  Recall that K_h is defined as $K_Y - h^*K_X$, where the representatives K_Y and K_X are chosen such that $h_*K_Y = K_X$, i.e. K_h is locally generated by the ordinary Jacobian determinant with respect to local coordinates.
- For a singular X, the sheaf Ω_X^d is not necessarily locally generated by one element, but we can still compare $h^*(\Omega_X^d)$ with Ω_Y^d. Taking locally a generator ω_Y of Ω_Y^d, then any $h^*(\omega) \in h^*(\Omega_X^d)$ can be written as $h^*(\omega) = g_\omega \cdot \omega_Y$, for some regular function g_ω. Then $\mathrm{Jac}(h)$ is defined as the ideal sheaf which is (locally) generated by those g_ω.

Both K_h and $\mathrm{Jac}(h)$ are supported over the exceptional locus of h.

EXAMPLE 2.45. Let $\mathcal{C} : y^2 - x^3 = 0$ be the ordinary cusp in $\mathbb{C}^2$, and consider the proper birational map $h : \mathbb{C} \to \mathcal{C}$ given by the parametrization $h(u) = (u^2, u^3)$. Taking the coordinate ring $\mathcal{O}_\mathcal{C} = \mathbb{C}[x, y]/(y^2 - x^3)$, it is easy to see that

$$\Omega_\mathcal{C}^1 = \mathcal{O}_\mathcal{C} \cdot \frac{\langle \mathrm{d}x, \mathrm{d}y \rangle}{(2y\mathrm{d}y - 3x^2\mathrm{d}x)}.$$

Thus, the module $h^*\Omega_\mathcal{C}^1$ is generated by $u\mathrm{d}u$, so it is principal in $\Omega_\mathbb{C}^1$ and $\mathrm{Jac}(h) = \langle u \rangle$.

EXAMPLE 2.46. Let X be smooth d-dimensional. In a blowing-up $\pi : Y = \mathrm{Bl}_Z(X) \to X$ over a smooth center $Z \subset X$ of codimension c, the relative canonical divisor is given by $K_\pi = (c-1)E$, where $E = \pi^*Y$ is the exceptional divisor. We compute

$$\begin{aligned}
\int_{\mathcal{L}(Y)} \mathbb{L}^{-\operatorname{ord}_{K_\pi}} \mathrm{d}\mu &= \int_{\mathcal{L}(Y)} \mathbb{L}^{-\operatorname{ord}_{(c-1)E}} \mathrm{d}\mu \\
&= \mathbb{L}^{-d}\left([Y \setminus E] + [E] \cdot \frac{\mathbb{L}-1}{\mathbb{L}^c - 1}\right) \\
&= \mathbb{L}^{-d}\left([Y \setminus E] + \frac{[E]}{[\mathbb{P}^{c-1}]}\right) \\
&= \mathbb{L}^{-d}\left([X \setminus Z] + [Z]\right) \\
&= \mathbb{L}^{-d}[X],
\end{aligned}$$

since $Y \setminus E \simeq X \setminus Z$ by definition and also E is a $\mathbb{P}^{c-1}$-bundle over Z.

The above example gives the idea of how the change or variables should work.

THEOREM 2.47 (DENEF-LOESER, [**DL99**]). *Let $h : Y \to X$ be a proper birational morphism between algebraic varieties X and Y, where Y is not singular. Let $A \subset \mathcal{L}(X)$ be a cylinder and $\alpha : A \to \mathbb{Z} \cup \{\infty\}$ such that $\mathbb{L}^{-\alpha}$ is integrable on A. Then,*

$$\int_A \mathbb{L}^{-\alpha} \mathrm{d}\mu_{\mathcal{L}(X)} = \int_{h^{-1}(A)} \mathbb{L}^{-(\alpha\circ h)-\mathrm{ord}_{\mathrm{Jac}(h)}} \mathrm{d}\mu_{\mathcal{L}(Y)}.$$

In particular, when both X and Y are smooth varieties and $\alpha = \mathrm{ord}_D$ for an effective divisor D, the formula takes the form

$$\int_A \mathbb{L}^{-\operatorname{ord}_D} \mathrm{d}\mu_{\mathcal{L}(X)} = \int_{h^{-1}(A)} \mathbb{L}^{-\operatorname{ord}_{h^*D+K_h}} \mathrm{d}\mu_{\mathcal{L}(Y)},$$

where h^*D is the pull-back of D and $K_h = K_Y - h^*K_X$ is the relative canonical divisor.

REMARK 2.48. Accessible proofs of the change of variables formula for the smooth case can be found in [**Cra04**, **Pop**]. The above result is generalized in [**DL02b**] for both possibly singular X and Y.

2.5. Kontsevich Theorem. As a consequence of the change of variables formula, we are going to prove in an almost straightforward way some strong results on additive invariants, starting with the original *leitmotiv* of motivic integration: Kontsevich's Theorem on Hodge numbers of Calabi-Yau varieties.

Recall that a *Calabi-Yau variety* X is a smooth, projective algebraic variety, admitting a nowhere vanishing regular differential form of maximal degree. An alternative way to formulate this last condition is to say that the canonical divisor K_X is trivial.

THEOREM 2.49 (KONTSEVICH, [**Kon95**]). *Let X and Y be two birationally equivalent Calabi-Yau manifolds. Then $[X] = [Y]$ in $\widehat{\mathcal{M}}_{\mathbb{C}}$, in particular they have the same Hodge numbers.*

PROOF. By Hironaka's desingularization Theorem (more especifically, by the "Elimination of points of indeterminacy" Theorem [**Hir64**, Main Thm. II]), since X and Y are birationally equivalent, there exists a compact smooth variety Z and birational morphisms $h_X : Z \to X$ and $h_Y : Z \to Y$. Note that $K_{h_X} = K_Z - h_X^*K_X = K_Z = K_Z - h_Y^*K_Y = K_{h_Y}$ because of the Calabi-Yau hypothesis. Also, $\mu(\mathcal{L}(X)) = [X]\mathbb{L}^{-d}$ and $\mu(\mathcal{L}(Y)) = [Y]\mathbb{L}^{-d}$ since both X and Y are smooth. Using the change of variables formula

$$\begin{aligned}\mathbb{L}^{-d}[X] = \mu(\mathcal{L}(X)) &= \int_{\mathcal{L}(X)} 1\mathrm{d}\mu_{\mathcal{L}(X)} = \int_{\mathcal{L}(Z)} \mathbb{L}^{-\operatorname{ord}_{K_{h_X}}} \mathrm{d}\mu_{\mathcal{L}(Z)} \\ &= \int_{\mathcal{L}(Z)} \mathbb{L}^{-\operatorname{ord}_{K_{h_Y}}} \mathrm{d}\mu_{\mathcal{L}(Z)} = \int_{\mathcal{L}(Y)} 1\mathrm{d}\mu_{\mathcal{L}(Y)} = \mu(\mathcal{L}(Y)) = \mathbb{L}^{-d}[Y].\end{aligned}$$

Thus, $[X] = [Y]$. We showed that the Hodge polynomial H factorizes over the image of $\mathcal{M}_{\mathbb{C}}$ on $\widehat{\mathcal{M}}_{\mathbb{C}}$, hence $H_X(u, v) = H_Y(u, v)$, which implies that both X and Y have the same Hodge numbers. □

Following the proof above, Kontsevich's Theorem can be generalized in a straightforward manner for K-equivalent varieties.

DEFINITION 2.50. Two compact algebraic varieties are *K-equivalent* if there exists a compact smooth variety Z and birational morphisms $h_X : Z \to X$ and $h_Y : Z \to Y$ such that $h_X^* K_X = h_Y^* K_Y$.

THEOREM 2.51. *Let X and Y be two K-equivalent varieties. Then $[X] = [Y]$ in $\widehat{\mathcal{M}}_{\mathbb{C}}$.*

PROOF. We use exactly the same arguments as in Kontsevich's Theorem above, since the K-equivalence condition implies that $K_{h_X} = K_Z - h_X^* K_X = K_Z - h_Y^* K_Y = K_{h_Y}$. □

The theory of motivic integration allows one to study invariants of algebraic varieties and their relations by the morphisms between them, in particular, resolutions of singularities.

2.6. Useful computations using (embedded) resolutions of singularities. After the change of variables formula, we can easily compute some examples which seem to be very complicated.

EXAMPLE 2.52. Revisiting the ordinary cusp $\mathcal{C} : y^2 - x^3 = 0$ in Example 2.45, we have shown that $\operatorname{Jac}(h) = \langle u \rangle$ where $h : \mathbb{C} \to \mathcal{C}$ is the proper birational map given by the parametrization $h(u) = (u^2, u^3)$. Then, $\operatorname{ord}_{\operatorname{Jac}(h)}$ is simply the order of the divisor associated with the function $f(u) = u$. This case is already studied in Example 2.41. Applying the change of variables formula, we get:

$$\mu(\mathcal{L}(\mathcal{C})) = \int_{\mathcal{L}(\mathcal{C})} \mathbb{L}^0 \mathrm{d}\mu = \int_{\mathcal{L}(\mathbb{C})} \mathbb{L}^{-\operatorname{ord}_t f} \mathrm{d}\mu = \frac{\mathbb{L}}{\mathbb{L}+1}.$$

Compare the previous calculation with the fact that $[\mathcal{C}] = \mathbb{L}$, as we have shown in Example 2.4.

EXERCISE 2.53. Let $\mathcal{P}_k = \{x^k + y^k = 0\}$ be a pencil of $k \geq 1$ lines in $\mathbb{C}^2$. Consider $h : \widetilde{\mathcal{P}_k} \to \mathcal{P}_k$ the resolution of singularities of $\mathcal{P}_k$ induced by the blowing-up of the origin in $\mathbb{C}^2$. Show that $\operatorname{ord}_{\operatorname{Jac}(h)} = 0$ and deduce that $\mu(\mathcal{L}(\mathcal{P}_k)) = k$.

EXERCISE 2.54. For $X = \{z^2 = xy\} \subset \mathbb{C}^3$, show that $\mu(\mathcal{L}(X)) = 1$ and $[X] = \mathbb{L}^2$.

2.6.1. *A combinatorial formula for SNC divisors.* The change of variables is a very useful tool to relay integrals between varieties, but especially to compute motivic integrals from "simpler models" of our varieties. This is the case when we use (embedded) resolution of singularities, as in Igusa's proof in Section 1. In this situation, a simple normal crossing divisor D is involved and we are interested in computing the motivic zeta function of $\mathbb{L}^{-\operatorname{ord}_D}$ for such a D. We obtain the following very useful combinatorial formula.

THEOREM 2.55 (BATYREV [**Bat99b**], CRAW [**Cra04**]). *Let X be a smooth variety and $D = \sum_{i=0}^r N_i D_i$ an effective SNC divisor on X. Consider the natural stratification given by*

$$D_I^\circ = \Big(\bigcap_{i \in I} D_i\Big) \setminus \Big(\bigcup_{j \notin I} D_j\Big),$$

for any $I \subset \{0, \ldots, r\}$, where $D_\emptyset^\circ = X \setminus D$. Then

$$\int_{\mathcal{L}(X)} \mathbb{L}^{-\operatorname{ord}_D} d\mu = \mathbb{L}^{-d} \sum_{I \subset \{0,\ldots,r\}} [D_I^\circ] \prod_{i \in I} \frac{\mathbb{L} - 1}{\mathbb{L}^{N_i+1} - 1} = \mathbb{L}^{-d} \sum_{I \subset \{0,\ldots,r\}} \frac{[D_I^\circ]}{\prod_{i \in I} [\mathbb{P}^{N_i}]}.$$

REMARK 2.56.

(1) The collection $\{D_I^\circ\}_{I \subset \{0,\ldots,r\}}$ is a stratification of X by constructible sets in terms of the regular parts of the irreducible components of D and the successive regular substrata of D_{sing} (see FIGURE 5). In addition, the local equation of D is constant along each D_I°, up to local change of coordinates. This also induces a partition by cylinders of $\mathcal{L}(X)$ given by

$$\mathcal{L}(X) = \bigsqcup_{I \subset \{0,\ldots,r\}} \pi_0^{-1}(D_I^\circ).$$

(2) Roughly speaking, the result above says that once we are in the simple normal crossing situation, the level value

$$\mu\{\gamma \in \mathcal{L}(X) \mid \operatorname{ord}_D(\gamma) = m\} \cdot \mathbb{L}^{-m}$$

becomes constant among the arcs with origin in the different strata D_I°, and only depends on the multiplicities of the divisors meeting at D_I°. Thus, the value of the integral is just the sum of the different constant values weighted by $[D_I^\circ]$, i.e. the "measure of all possible origins" along each stratum.

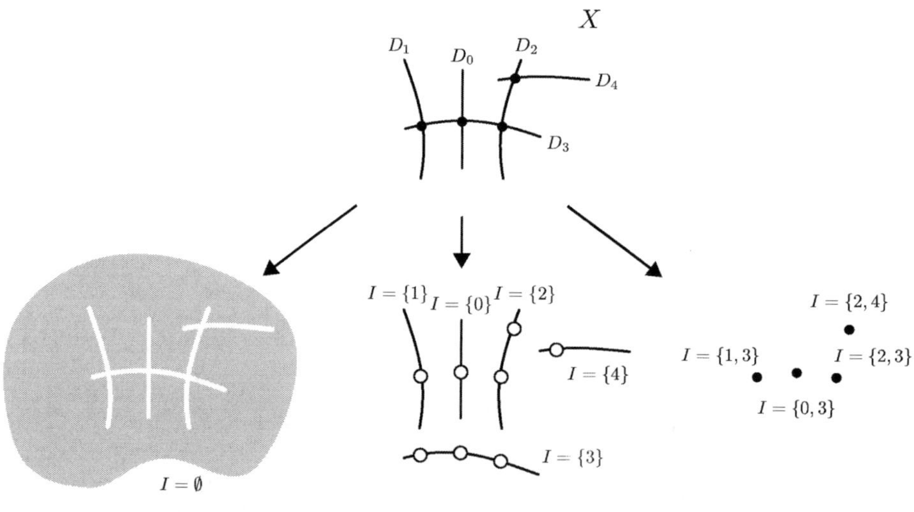

FIGURE 5. Stratification $\{D_I^\circ\}_I$ associated with a normal crossing divisor $D = N_0D_0 + \cdots + N_4D_4$ in X.

COROLLARY 2.57. *Under the same hypothesis of Theorem 2.55, and for any constructible subset $W \subset X$, we have*

$$\int_{\mathcal{L}(X)_W} \mathbb{L}^{-\operatorname{ord}_D} d\mu = \mathbb{L}^{-d} \sum_{I \subset \{0,\ldots,r\}} [D_I^\circ \cap W] \prod_{i \in I} \frac{\mathbb{L} - 1}{\mathbb{L}^{N_i+1} - 1}.$$

In order to prove Theorem 2.55, we first need to compute the measure of $\operatorname{ord}_D^{-1}(m)$, for any $m \in \mathbb{Z}_{\geq 0}$. Note that $\operatorname{ord}_D = \sum_{i=0}^r N_i \operatorname{ord}_{D_i}$, thus for any $\gamma \in \mathcal{L}(X)$ we are going to study the tuples

$$(\operatorname{ord}_{D_0}(\gamma), \ldots, \operatorname{ord}_{D_r}(\gamma)).$$

For any $m \in \mathbb{Z}_{\geq 0}$ and $I \subset \{0, \ldots, r\}$, define the set

$$M_{I,m} := \left\{ (m_0, \ldots, m_r) \in \mathbb{Z}_{\geq 0}^{r+1} \;\middle|\; \sum_j N_j m_j = m \text{ and } m_j = 0 \text{ if and only if } j \notin I \right\}$$

By definition of ord_D, it follows that

$$\gamma \in \pi_0^{-1}(D_I^\circ) \cap \operatorname{ord}_D^{-1}(m) \iff (\operatorname{ord}_{D_0}(\gamma), \ldots, \operatorname{ord}_{D_r}(\gamma)) \in M_{I,m}.$$

Thus, any level set $\operatorname{ord}_D^{-1}(m)$ is partitioned by the sets above:

$$\operatorname{ord}_D^{-1}(m) = \bigsqcup_{I \subset \{0,\ldots,r\}} \; \bigsqcup_{(m_0,\ldots,m_r) \in M_{I,m}} \left(\bigcap_{j=0}^{r} \operatorname{ord}_{D_j}^{-1}(m_j) \right). \tag{2.1}$$

LEMMA 2.58. *Consider $I \subset \{0, \ldots, r\}$. For any $(m_0, \ldots, m_r) \in M_{I,m}$, we have:*

$$\mu \left(\bigcap_{j=0}^{r} \operatorname{ord}_{D_j}^{-1}(m_j) \right) = \mathbb{L}^{-d} [D_I^\circ] (\mathbb{L} - 1)^{|I|} \mathbb{L}^{-\sum_{i \in I} m_i}.$$

PROOF. Recall that $D = \sum_{i=0}^r N_i D_i$ is simple normal crossing if at each point $x \in X$, there is a neighborhood U of x with local coordinates $(x_1, \ldots, x_d)$ such that D is locally defined as

$$f_D = x_1^{N_{i_1}} \cdots x_k^{N_{i_k}} \cdot u(x), \quad u(0) \neq 0, \tag{2.2}$$

for some $1 \leq k \leq d$. Let $X = \bigcup U$ be a covering by finitely many open charts such that D is locally defined as above. This lifts into a finite covering by cylinders $\mathcal{L}(X) = \bigcup \pi_0^{-1}(U)$. It suffices to compute

$$U_{m_0,\ldots,m_r} := \left(\bigcap_{j=0}^{r} \operatorname{ord}_{D_j}^{-1}(m_j) \right) \cap \pi_0^{-1}(U)$$

for a fixed U, where (2.2) holds. Note that, if $I \not\subset \{i_1, \ldots, i_k\}$, then $U \cap D_I^\circ = \emptyset$ and this implies that $U_{m_0,\ldots,m_r} \subset \pi_0^{-1}(U \cap D_I^\circ)$ is also empty.

We assume in the following that $I \subset \{i_1, \ldots, i_k\}$. This implies $|I| \leq k \leq d$. Any arc $\gamma_x \in \pi_0^{-1}(U)$ with origin in x can be represented as a tuple $(\gamma_1(t), \ldots, \gamma_d(t)) \in \mathbb{C}[\![t]\!]^d$ such that $\gamma_i(0) = 0$, for any $i = 1, \ldots, d$. Thus, if we consider $i \in I$:

$$\operatorname{ord}_{D_i}(\gamma) = m_i \iff \pi_{m_i}(\gamma_i(t)) = a_{m_i} t^{m_i} \text{ with } a_{m_i} \in \mathbb{C}^*.$$

Take $n := \max\{m_i \mid i \in I\}$, the truncation of the components of γ_x by π_n produces:

- If $i \in I$: $\pi_n(\gamma_i(t)) = a_{m_i} t^{m_i} + a_{m_i+1} t^{m_i+1} + \cdots + a_n t^n$, with $a_{m_i} \in \mathbb{C}^*$ and $a_j \in \mathbb{C}$ for any $j = m_i + 1, \ldots, n$.
- If $i \notin I$: $\pi_n(\gamma_i(t))$ is a generic polynomial of degree n with zero constant term.

As $|I| \le k$ in (2.2), the projection by π_n of γ_x gives us $|I|$ polynomials of the first type and $d - |I|$ of the second one. Hence, the space of those $\pi_n(\gamma_x)$ is isomorphic to

$$(\mathbb{C}^*)^{|I|} \times \mathbb{C}^{n|I| - \sum_{i\in I} m_i} \times \mathbb{C}^{n(d-|I|)} = (\mathbb{C}^*)^{|I|} \times \mathbb{C}^{nd - \sum_{i\in I} m_i}.$$

We recover the whole $U_{m_0,\dots,m_r}$ by lifting this space at every point of $U \cap D_I^\circ$, obtaining the cylinder

$$U_{m_0,\dots,m_r} = \pi_n^{-1}\left((U \cap D_I^\circ) \times (\mathbb{C}^*)^{|I|} \times \mathbb{C}^{nd - \sum_{i\in I} m_i}\right).$$

Finally,

$$\bigcap_{j=0}^{r} \operatorname{ord}_{D_j}^{-1}(m_j) = \bigcup U_{m_0,\dots,m_r} = \pi_n^{-1}(C_n),$$

where

$$[C_n] = \left[D_I^\circ \times (\mathbb{C}^*)^{|I|} \times \mathbb{C}^{nd - \sum_{i\in I} m_i}\right] = [D_I^\circ](\mathbb{L}-1)^{|I|}\mathbb{L}^{nd - \sum_{i\in I} m_i}.$$

And the result holds:

$$\mu\left(\bigcap_{j=0}^{r} \operatorname{ord}_{D_j}^{-1}(m_j)\right) = \mu(\pi_n^{-1}(C_n)) = [C_n]\mathbb{L}^{d(n+1)} = \mathbb{L}^{-d}[D_I^\circ](\mathbb{L}-1)^{|I|}\mathbb{L}^{-\sum_{i\in I} m_i}.$$

□

REMARK 2.59. In the previous result, even if $\bigcap_{j=0}^{r} \operatorname{ord}_{D_j}^{-1}(m_j)$ does not seem to depend on I, this is in fact the case since we assume that the tuple $(m_0, \dots, m_r)$ belongs to $M_{I,m}$. This implies in particular that $m_j = 0$ for any $j \notin I$.

PROOF OF THEOREM 2.55. Applying the definition, it suffices to use the partition (2.1) of the level sets and Lemma 2.58 to computing our motivic integral:

$$\begin{aligned}
\int_{\mathcal{L}(X)} \mathbb{L}^{-\operatorname{ord}_D} d\mu &= \sum_{m\in\mathbb{Z}_{\ge 0}} \sum_{I\subset\{0,\dots,r\}} \sum_{(m_0,\dots,m_r)\in M_{I,m}} \mu\left(\bigcap_{j=0}^{r} \operatorname{ord}_{D_j}^{-1}(m_j)\right) \mathbb{L}^{-\sum_{i\in I} N_i m_i} \\
&= \mathbb{L}^{-d} \sum_{m\in\mathbb{Z}_{\ge 0}} \sum_{I\subset\{0,\dots,r\}} \sum_{(m_0,\dots,m_r)\in M_{I,m}} [D_I^\circ](\mathbb{L}-1)^{|I|}\mathbb{L}^{-\sum_{i\in I}(N_i+1)m_i} \\
&= \mathbb{L}^{-d} \sum_{I\subset\{0,\dots,r\}} [D_I^\circ](\mathbb{L}-1)^{|I|} \sum_{m\in\mathbb{Z}_{\ge 0}} \sum_{(m_0,\dots,m_r)\in M_{I,m}} \prod_{i\in I} \mathbb{L}^{-(N_i+1)m_i}
\end{aligned}$$

It is easy to see that, for any $(m_0,\dots,m_r)\in\mathbb{Z}_{\geq 0}^{r+1}$, there exist unique $I\subset\{0,\dots,r\}$ and $m\in\mathbb{Z}_{\geq 0}$ such that $(m_0,\dots,m_r)\in M_{I,m}$. Thus, the previous equation becomes

$$\begin{aligned}\int_{\mathcal{L}(X)}\mathbb{L}^{-\operatorname{ord}_D}d\mu &= \mathbb{L}^{-d}\sum_{I\subset\{0,\dots,r\}}[D_I^\circ](\mathbb{L}-1)^{|I|}\sum_{(m_0,\dots,m_r)\in\mathbb{Z}_{\geq 0}^{r+1}}\prod_{i\in I}\mathbb{L}^{-(N_i+1)m_i}\\ &= \mathbb{L}^{-d}\sum_{I\subset\{0,\dots,r\}}[D_I^\circ](\mathbb{L}-1)^{|I|}\cdot\prod_{i\in I}\left(\sum_{m_i>0}\mathbb{L}^{-(N_i+1)m_i}\right)\\ &= \mathbb{L}^{-d}\sum_{I\subset\{0,\dots,r\}}[D_I^\circ](\mathbb{L}-1)^{|I|}\cdot\prod_{i\in I}\left(\frac{1}{1-\mathbb{L}^{-(N_i+1)}}-1\right)\\ &= \mathbb{L}^{-d}\sum_{I\subset\{0,\dots,r\}}[D_I^\circ]\cdot\prod_{i\in I}\frac{\mathbb{L}-1}{\mathbb{L}^{N_i+1}-1}.\end{aligned}$$

□

EXERCISE 2.60. Use the formula above to compute the motivic integral of the divisor associated with $f(x_1,\dots,x_d)=x_1^{N_1}\cdots x_d^{N_d}$ in $\mathbb{C}^d$. (Compare the result with the p-adic case)

2.6.2. *Motivic integral from resolution of singularities.* The latter is very useful for computations using resolutions of singularities. In the smooth case, we get similar formulas for the computation of $[X]$ and the motivic integral associated with a divisor by considering the different multiplicities of the exceptional locus.

PROPOSITION 2.61. *Let $h: Y\to X$ be a proper birational morphism between smooth varieties. Assume that the exceptional locus of h is a SNC divisor with irreducible components $E_0,\dots,E_r$, and that $K_h=\sum_{i=0}^r(\nu_i-1)E_i$, for some $\nu_i\in\mathbb{Z}_{>0}$. Then*

$$[X]=\sum_{I\subset\{0,\dots,r\}}[E_I^\circ]\prod_{i\in I}\frac{\mathbb{L}-1}{\mathbb{L}^{\nu_i}-1}.$$

One can also compute motivic integrals based in the contact order over a divisor from embedded resolutions, following similar arguments.

PROPOSITION 2.62. *Let X be a smooth algebraic variety and let D be an effective Cartier divisor in X. If $h: Y\to X$ is an embedded resolution of singularities of D with numerical data $\{(N_i,\nu_i)\}_{i=0}^r$, i.e. $h^*D=\sum_{i=0}^r N_iE_i$ and $K_h=\sum_{i=0}^r(\nu_i-1)E_i$ being SNC divisors, then*

$$\int_{\mathcal{L}(X)_W}\mathbb{L}^{-\operatorname{ord}_D}d\mu=\mathbb{L}^{-d}\sum_{I\subset\{0,\dots,r\}}[E_I^\circ\cap h^{-1}(W)]\prod_{i\in I}\frac{\mathbb{L}-1}{\mathbb{L}^{N_i+\nu_i}-1},$$

for any constructible set $W\subset X$.

EXERCISE 2.63. From the embedded resolution of the cusp $\mathcal{C}$ in Example 1.45, compute the motivic integral $\int_{\mathcal{L}(\mathbb{C}^2)}\mathbb{L}^{-\operatorname{ord}_{\mathcal{C}}}$.

Note that we automatically obtain formulas for the additive invariants of smooth X just by specializing from the formula above for $[X]$. In particular, by specializing using the Euler characteristic, one has the surprising formula:

$$\chi(X)=\sum_{I\subset\{0,\dots,r\}}\chi(E_I^\circ)\prod_{i\in I}\frac{1}{\nu_i},\tag{2.3}$$

first obtained by Denef and Loeser [**DL92**], using p-adic integration and the Grothendieck-Lefschetz trace formula.

The previous result can be generalized for a singular X, in terms of a *log-resolution of* X, i.e. a proper birational morphism $h : Y \to X$ with Y smooth and such that the exceptional locus of h is a SNC divisor.

Theorem 2.64 (Denef-Loeser, [**DL99**]). *Let $h : Y \to X$ be a log resolution of X and consider $E_0, \ldots, E_r$ the irreducible components of the exceptional locus. Assume that $h^*\Omega^d_X$ is locally principal in Ω^d_Y and that* $\operatorname{div}(\operatorname{Jac}(h)) = \sum_{i=0}^r (b_i - 1)E_i$, *for some $b_i - 1 \in \mathbb{Z}_{\geq 0}$. Then*

$$\mu(\mathcal{L}(X)_W) = \mathbb{L}^{-d} \sum_{I \subset \{0,\ldots,r\}} [E_I^\circ \cap h^{-1}(W)] \prod_{i \in I} \frac{\mathbb{L} - 1}{\mathbb{L}^{b_i} - 1},$$

for any constructible set $W \subset X$.

Note that Example 2.52 is a particular case of the result above. It is worth noticing that the computation of $\mu(\mathcal{L}(X))$ in Theorem 2.64 is constrained to the fact that $h^*\Omega^d_X$ is locally principal in Ω^d_Y. In this case, $\operatorname{div}(\operatorname{Jac}(h))$ is either an effective divisor or the empty set. However, such a resolution always exists by Hironaka's Theorem and we deduce where the element given by the measure of the total space $\mathcal{L}(X)$ lies in $\widehat{\mathcal{M}}_{\mathbb{C}}$.

Corollary 2.65. *The motivic measure $\mu(\mathcal{L}(X))$ is an element of the image of the ring*

$$\mathcal{M}_{\mathbb{C}}\left[\left\{\frac{1}{[\mathbb{P}^a]}\right\}\right]_{a \in \mathbb{Z}_{\geq 0}}$$

in $\widehat{\mathcal{M}}_{\mathbb{C}}$.

This allows us to extend the additive invariants for $\mu(\mathcal{L}(X))$ whenever the image of $\mathbb{P}^a$ by the invariant (for the possibles a) is not zero, as the *Hodge-Deligne polynomial* and the *Euler characteristic* [**DL99**].

We call the *arc-Euler characteristic* of X the value $\chi(\mu(\mathcal{L}(X))) \in \mathbb{Q}$. We can use the last theorem to obtain a similar expression as in the smooth case in (2.3).

Exercise 2.66. Compute the arc-Euler characteristic of the cusp and the ordinary 2-dimensional node.

3. Applications to singularities and the motivic zeta function

We have established a way to measure additive invariants of general varieties, constructing a measure/integration theory (together with a change of variables formula) based on a *virtual universal counting* compatible with (virtual) fibrations by affine spaces, and capturing information from the infinitesimal structure of varieties and divisors via the arc spaces.

Following this spirit, Batyrev introduces a new class of singularity invariants, and Denef and Loeser construct a "universal" zeta function, which gives a new way to study the Monodromy conjecture from an intrinsic point of view of varieties.

Finally, we are going to give new formulas for motivic integrals using partial resolutions involving orbifolds.

3.1. Batyrev's new stringy invariants of mild singularities. The formula for $\chi(X)$ obtained in (2.3) for smooth varieties motivates the development of new (numerical) invariants for "mild" singularities: Batyrev's *stringy invariants* [**Bat98, Bat99a**]. These invariants extend the corresponding classical topological invariants of smooth varieties and they can be defined in terms of data of log resolutions.

The adjective "stringy" comes from the fact that these invariants are placed in the context of String Theory. In fact, BATYREV [**Bat98**] used them to formulate a *topological mirror symmetry test* for singular Calabi-Yau varieties, to give a conjectural definition of stringy Hodge numbers, and prove a version of the McKay correspondence [**Bat99b**]: for any finite $G \subset \mathrm{SL}_d(\mathbb{C})$ we have

$$\chi(\mathbb{C}^d/G) = \#\{\text{conjugacy classes of } G\}.$$

Most of the examples in this section are taken from [**Vey06**].

3.1.1. *"Mild" singularities.* Our aim is to define invariants for which the intrinsic motivic definition could work and be independent of the chosen log resolution. This is the case of *log terminal* singularities. REID's guide [**Rei87**] is a very good tutorial introduction to this study of singularities. See [**KM98**] for a more deep understanding of this subject within the Minimal Model Program.

Let X be a *normal* d-dimensional algebraic variety. In particular:

- X is irreducible.
- X_{sing} has codimension at least 2 in X.
- X has a well defined (Weil) *canonical* divisor K_X (up to linear equivalence): a representative is the divisor $\mathrm{div}(\omega)$ of zeros and poles of a *rational* differential d-form $\omega \in \Omega^d_{\mathbb{C}(X)} = \Omega^d_X \otimes \mathbb{C}(X)$ on X. It is also the Zariski closure of the usual $K_{X_{\text{reg}}}$.

When X is smooth, K_X is a Cartier divisor, but this is not true in general.

Associated to the canonical divisor, we have the *canonical sheaf* ω_X. This sheaf has better birational geometric properties than the Kalhër differentials Ω^d_X (and defines a different concept of "jacobian" between birational maps, as we will see later). It is defined as $\omega_X := j_*(\Omega^d_{X_{\text{reg}}})$ where $j : X_{\text{reg}} \hookrightarrow X$ is the inclusion of the smooth part of X, i.e. the sheaf of rational forms $\omega \in \Omega^d_{\mathbb{C}(X)}$ which are regular over X_{reg}. It holds that $\omega_X = \mathcal{O}_X(K_X)$.

DEFINITION 3.1. A normal variety X is *Gorenstein* if K_X is a Cartier divisor. Equivalently, if the canonical sheaf ω_X is locally generated by one element.

REMARK 3.2. Let $V_f \subset \mathbb{C}^{d+1}$ be a normal hypersurface defined by a non-constant $f \in \mathbb{C}[x_0, \dots, x_d]$. Consider the rational form given by

$$\omega = \frac{\mathrm{d}x_1 \wedge \cdots \wedge \mathrm{d}x_d}{(\partial f/\partial x_0)} \in \Omega^d_{\mathbb{C}(V_f)}.$$

This is regular over any point $p \in X$ such that $(\partial f/\partial x_0)(p) \neq 0$. In fact, since $\Omega^1_{V_f}$ is generated by $\mathrm{d}x_0, \dots, \mathrm{d}x_d$ constrained to $\sum_{i=0}^d (\partial f/\partial x_i)\mathrm{d}x_i = 0$, it is clear that ω admits as a representative

$$\omega = \frac{\mathrm{d}x_0 \wedge \cdots \wedge \widehat{\mathrm{d}x_i} \wedge \cdots \wedge \mathrm{d}x_d}{(\partial f/\partial x_i)} \in \Omega^d_{\mathbb{C}(V_f)},$$

(up to ± 1) in $\Omega^d_{\mathbb{C}(V_f)}$, for any $i=0,\ldots,d$. We obtain that ω is a basis of Ω^d_X at any point in X_{reg}. By normality, ω is also a basis of ω_X. We can conclude that any normal hypersurface is Gorenstein. Normal complete intersections are also Gorenstein.

EXAMPLE 3.3. For $X=\left\{z^2=xy\right\}\subset\mathbb{C}^3$, the differential 2-forms are generated by representative

$$\frac{\mathrm{d}x\wedge\mathrm{d}y}{2z}=\frac{\mathrm{d}x\wedge\mathrm{d}z}{x}=-\frac{\mathrm{d}y\wedge\mathrm{d}z}{y},$$

which is regular in $X_{\text{reg}}=X\setminus 0$.

EXAMPLE 3.4. Let $X=\mathbb{C}^2/\mu_3$ be the quotient by the action $(x,y)\mapsto(\xi x,\xi y)$, for any $\xi\in\mu_3=\left\{z\in\mathbb{C}\mid z^3=1\right\}$. This is a normal surface which can be embedded in $\mathbb{C}^4$ with equations

$$u_1u_3-u_2^2=u_2u_4-u_3^2=u_1u_4-u_2u_3=0.$$

In particular, note that X is not complete intersection. We can see that K_X is represented by $\{u_1=u_2=u_3=0\}$, so K_X is not Cartier.

The Gorenstein condition can be generalized using *mth-tensor products*, in order to obtain a Cartier divisor. The *m-pluricanonical sheaf* is defined by $\omega_X^{[m]}:=j_*((\Omega^d_{X_{\text{reg}}})^{\otimes m})$, i.e. the mth-tensor products of rational forms $\omega_1\otimes\cdots\otimes\omega_m\in(\Omega^d_{\mathbb{C}(X)})^{\otimes m}$ which are regular on X_{reg}. We have that $\omega_X^{[m]}=\mathcal{O}_X(mK_X)$. Using this, most of the theory is extended by considering rational coefficients of divisors.

DEFINITION 3.5. A normal variety X is called $\mathbb{Q}$-*Gorenstein* if mK_X is Cartier for some $m\in\mathbb{Z}_{>0}$. In this case, $\omega_X^{[m]}$ is an invertible sheaf, i.e. a locally free $\mathcal{O}_X$-module of rank 1.

REMARK 3.6. For X Gorenstein, let $h:Y\to X$ be a log resolution of singularities. Since K_X is a Cartier divisor, the pull-back h^*K_X makes sense. Thus, $K_h=K_Y-h^*K_X=\sum_{i=0}^r(\nu_i-1)E_i$, where $\nu_i\in\mathbb{Z}$ is called the *log discrepancy* of the irreducible component E_i. The log resolution h is called *crepant* if $K_h=0$. In the $\mathbb{Q}$-Gorenstein case, the log discrepancies of a log resolution $h:Y\to X$ as above are given by

$$mK_h=mK_Y-h^*(mK_X)=\sum_{i=0}^r m(\nu_i-1)E_i.$$

Thus, $\nu_i\in\frac{1}{m}\mathbb{Z}$.

EXAMPLE 3.7. In the last example, K_X was not Cartier. However, $3K_X$ is Cartier, represented by $\{u_1=0\}$. Taking the standard resolution $h:Y\to X$ with $E\simeq\mathbb{P}^1$, we obtain $K_h=\frac{2}{3}E$.

Now, we can classify numerically the singularities of a $\mathbb{Q}$-Gorenstein variety in terms of the log discrepancies.

DEFINITION 3.8. Let X be a $\mathbb{Q}$-Gorenstein variety, and take $h:Y\to X$ a log resolution of X with log discrepancies $\nu_0,\ldots,\nu_r\in\mathbb{Q}_{\geq 0}$. Then X is called:

(1) *terminal* if $\nu_i>1$ for any $i=0,\ldots,r$.

(2) *canonical* if $\nu_i\geq 1$ for any $i=0,\ldots,r$.

(3) *log terminal* if $\nu_i > 0$ for any $i = 0, \ldots, r$.

(4) *log canonical* if $\nu_i \geq 0$ for any $i = 0, \ldots, r$.

(5) *strictly log canonical* if it is log canonical but not log terminal.

The log terminal singularities are considered "mild", and the singularities which are not log canonical are considered "general". In fact, once we have a log resolution of a variety X with some $\nu_i < 0$, one can produce other log resolutions of X with arbitrary negative ν_i.

REMARK 3.9. In the case of the blowing-up of affine hypersurfaces $V_f \subset \mathbb{C}^{d+1}$, one can use the same techniques as in Remark 3.2 to explicitly computing basis of $\Omega^d_{\mathbb{C}(\widehat{V_f})}$ for the strict transform $\widehat{V_f}$ in the local charts, and then deduce the log discrepancies.

EXAMPLE 3.10.

(1) Assume that X is a surface, i.e. $d = 2$. Then, X is terminal if and only if is smooth, and

$$\begin{array}{rcl} \text{canonical singularity} & \longleftrightarrow & \text{ADE singularity} \\ \text{log terminal singularity} & \longleftrightarrow & \text{Hirzebruch-Jung/quotient singularity} \end{array}$$

(2) Let $X_k = \{x_0^k + \cdots + x_d^k = 0\} \subset \mathbb{C}^{d+1}$ be an affine variety, for $k \geq 0$. We have that the singular locus of X_k is the origin, and we can produce a resolution $h : Y_k = \mathrm{Bl}_0(X_k) \to X_k$ by one blowing-up. In this case, $E = h^{-1}(0)$ is smooth irreducible and isomorphic to the projectivization $\{x_0^k + \cdots + x_d^k = 0\} \subset \mathbb{P}^d$. We can verify (EXERCISE) that $K_h = (d-k)E$, and X_k is

$$\begin{array}{rcl} \text{log terminal} & \Longleftrightarrow & k < d+1 \\ \text{strictly log canonical} & \Longleftrightarrow & k = d+1 \\ \text{not log canonical} & \Longleftrightarrow & k > d+1 \end{array}$$

EXERCISE 3.11. For $h : Y_k = \mathrm{Bl}_0(X_k) \to X_k$ in the previous example, prove that $\mathrm{div}(\mathrm{Jac}(h)) = (d-1)E$, for any $k \geq 0$.

3.1.2. *Stringy invariants.* Let us introduce the invariants associated to mild singularities. It is worth noticing that BATYREV usually takes a different normalization of the motivic measure μ: for any smooth variety X, it is verified that $\mu(\mathcal{L}(X)) = [X]$.

DEFINITION 3.12. Let X be a $\mathbb{Q}$-Gorenstein variety with at most log terminal singularities, and let $h : Y \to X$ be a log resolution, where $E_0, \ldots, E_r$ are the irreducible components of the exceptional locus of h with log discrepancies $\nu_0, \ldots, \nu_r \in \mathbb{Q}_{>0}$. We define:

(1) The *stringy Euler number* of X:

$$\chi_{\mathrm{st}}(X) := \sum_{I \subset \{0,\ldots,r\}} \chi(E_I^\circ) \prod_{i \in I} \frac{1}{\nu_i}.$$

(2) The *stringy Hodge-Deligne polynomial* (or *stringy E-polynomial*) of X:

$$E_{\mathrm{st}}(X) := \sum_{I \subset \{0,\ldots,r\}} H(E_I^\circ) \prod_{i \in I} \frac{uv - 1}{(uv)^{\nu_i} - 1}.$$

(3) The *stringy $\mathcal{E}$-invariant* of X:

$$\mathcal{E}_{\mathrm{st}}(X) := \sum_{I\subset\{0,\dots,r\}} [E_I^\circ] \prod_{i\in I} \frac{\mathbb{L}-1}{\mathbb{L}^{\nu_i}-1}.$$

REMARK 3.13.

(1) It is clear that $\chi_{\mathrm{st}}(X) \in \mathbb{Q}$, also

$$E_{\mathrm{st}}(X) \in \mathbb{Q}(u,v)\left[\left\{\frac{1}{(uv)^a-1}\right\}_{a\in\mathbb{Q}_{>0}}\right],$$

and that $\mathcal{E}_{\mathrm{st}}(X)$ lives in a finite extension of $\widehat{\mathcal{M}}_{\mathbb{C}}$. We have specialization maps:

$$\mathcal{E}_{\mathrm{st}}(X) \underset{H}{\longrightarrow} E_{\mathrm{st}}(X) \underset{u,v\to 1}{\longrightarrow} \chi_{\mathrm{st}}(X).$$

(2) The stringy invariants are generalizations of the classical invariants, since if X is smooth: $\mathcal{E}_{\mathrm{st}}(X) = [X]$ after Theorem 2.55 and hence $\chi_{\mathrm{st}}(X) = \chi(X)$ and $E_{\mathrm{st}}(X) = H(X)$.

(3) Analogously, we can define the *stringy* Poincaré polynomial P_{st}. When E_{st} (resp. P_{st}) is a polynomial, we can define the *stringy Hodge numbers* $h_{\mathrm{st}}^{p,q}$ (resp. the *stringy Betti numbers* b_{st}^k) of X.

EXAMPLE 3.14 (**$\mathcal{E}_{\mathrm{st}}(\cdot)$ vs $\mu(\mathcal{L}(\cdot))$ vs $[\,\cdot\,]$**). Consider $X_k = \{x_0^k + \cdots + x_d^k = 0\} \subset \mathbb{C}^{d+1}$ and its log resolution given in Example 3.10-(2). Note that

$$Y_k \setminus E \simeq X_k \setminus \{0\} \simeq \mathbb{C}^* \times E. \tag{3.1}$$

It follows from the definition that:

$$\mathcal{E}_{\mathrm{st}}(X_k) = [Y_k \setminus E] + [E]\frac{\mathbb{L}-1}{\mathbb{L}^{d+1-k}-1} = [E]\mathbb{L}^{d+1-k}\frac{\mathbb{L}-1}{\mathbb{L}^{d+1-k}-1}.$$

On the other hand, we can compute $\mu(\mathcal{L}(X_k))$ applying Theorem 2.64. Recall that $\operatorname{div}(\operatorname{Jac}(h)) = (d-1)E$, then

$$\mu(\mathcal{L}(X_k)) = \mathbb{L}^{-d}[E]\left(\mathbb{L}-1+\frac{\mathbb{L}-1}{\mathbb{L}^d-1}\right) = \frac{[E]}{[\mathbb{P}^{d-1}]}.$$

Finally, the isomorphism (3.1) implies that $[X_k] = [X_k \setminus \{0\}] + 1 = (\mathbb{L}-1)[E] + 1$.

REMARK 3.15. If X has at most Gorenstein canonical singularities (equivalently $\nu_0,\dots,\nu_r \in \mathbb{Z}_{>0}$), then K_h is an effective normal crossing divisor. In this case, it is easy to see, from Theorem 2.64, that

$$\mathbb{L}^{-d}\mathcal{E}_{\mathrm{st}}(X) = \int_{\mathcal{L}(Y)} \mathbb{L}^{-\operatorname{ord} K_h}\, d\mu.$$

In general, for a $\mathbb{Q}$-Gorenstein log terminal X, the stringy $\mathcal{E}$-invariant $\mathcal{E}_{\mathrm{st}}(X)$ can also be defined intrinsically [**Yas04, DL02b**] by

$$\mathbb{L}^{-d}\mathcal{E}_{\mathrm{st}}(X) = \int_{\mathcal{L}(X)} \mathbb{L}^{-\operatorname{ord}\mathcal{I}_X}\, d\mu,$$

where $\mathcal{I}_X$ is the ideal sheaf on X defined as follows: let Ω_X^d and $\Omega_{X_{\mathrm{reg}}}^d$ be the sheaves of regular differential d-forms on X and X_{reg}, respectively. We have a natural map $\Omega_X^d \to \Omega_{X_{\mathrm{reg}}}^d$, whose image is $\mathcal{I}_X\Omega_{X_{\mathrm{reg}}}^d$. See [**Yas04**, Lemma 1.16]. An alternative way, presented in [**LMVV20**, Secs. 1-2], is to use another measure with respect to the order of the m-pluricanonical sheaf $\omega_X^{[m]}$.

Hence, this intrinsic way of defining $\mathcal{E}_{\rm st}(X)$ in terms of a motivic integral implies that any of the stringy invariants defined above do not depend on the chosen resolution.

This independence is used for proving the following result due to Kontsevich.

THEOREM 3.16 ([**Cra04**, Thm. 3.6]). *Let X be a complex projective variety with at worst Gorenstein canonical singularities. If X admits a crepant resolution $h : Y \to X$, the Hodge numbers of Y are independent of the choice of crepant resolution.*

PROOF. It follows directly from the fact that $K_h = 0$ by hypothesis. Then $\mathcal{E}_{\rm st}(X) = [Y]$ and $E_{\rm st}(X) = H(Y)$, which is independent of the chosen resolution. □

3.2. The motivic zeta function. Let X be a smooth d-dimensional algebraic variety and let $f : X \to \mathbb{C}$ be a non-constant morphism. Consider the Cartier divisor D defined by f in X.

DEFINITION 3.17. The *motivic zeta function of f* is defined as

$$Z_{\rm mot}(f;s) := \int_{\mathcal{L}(X)} \mathbb{L}^{-\operatorname{ord}_D \cdot s} = \sum_{m\geq 0} \mu_{\mathcal{L}(X)}\left(\operatorname{ord}_D^{-1}(m)\right)\cdot \mathbb{L}^{-ms} \in \mathcal{M}_{\mathbb{C}}[\![\mathbb{L}^{-s}]\!].$$

It is straightforward to see that this generalizes the motivic integral of the pair (X, D), which turns out to be $Z_{\rm mot}(f;1)$.

Note that the change of variables also works for functions $\alpha s+\beta : \mathcal{L}(X) \to \mathbb{Z}[s]$ with $\mathbb{L}^{-\alpha}$ and $\mathbb{L}^{-\beta}$ being integrable functions. Hence, Proposition 2.62 can be easily generalized to this case.

THEOREM 3.18 (DENEF-LOESER, [**DL92**]). *Let $h : Y \to X$ be an embedded resolution of singularities of D with numerical data $\{(N_i,\nu_i)\}_{i=0}^r$, i.e. $h^*D = \sum_{i=0}^r N_iE_i$ and $K_h = \sum_{i=0}^r(\nu_i - 1)E_i$ being SNC divisors, then*

$$Z_{\rm mot}(f;s) = \mathbb{L}^{-d} \sum_{I\subset\{0,\ldots,r\}} [E_I^\circ]\prod_{i\in I}\frac{(\mathbb{L}-1)\mathbb{L}^{-(N_is+\nu_i)}}{1-\mathbb{L}^{-(N_is+\nu_i)}}.$$

COROLLARY 3.19. *The motivic zeta function $Z_{\rm mot}(f;s)$ is a rational function in $T = \mathbb{L}^{-s}$ over $\mathcal{M}_{\mathbb{C}}$: there exists a finite subset $S \subset \mathbb{Z}^2_{\geq 1}$ such that*

$$Z_{\rm mot}(f;s) \in \mathcal{M}_{\mathbb{C}}\left[\left\{\frac{\mathbb{L}^{-(as+b)}}{1-\mathbb{L}^{-(as+b)}}\right\}_{(a,b)\in S}\right] \subset \mathcal{M}_{\mathbb{C}}[\![\mathbb{L}^{-s}]\!].$$

EXAMPLE 3.20. Using the embedded resolution of the cusp $\mathcal{C} : f(x,y) = 0$ with $f(x,y) = y^2 - x^3$, it is easy to compute

$$Z_{\rm mot}(f;s) = \mathbb{L}^{-1}(\mathbb{L}-1)\frac{1-\mathbb{L}^{-(s+2)}+\mathbb{L}^{-(2s+2)}-\mathbb{L}^{-(5s+5)}}{\left(1-\mathbb{L}^{-(s+1)}\right)\left(1-\mathbb{L}^{-(6s+5)}\right)}.$$

REMARK 3.21.

(1) As in the p-adic case, $Z_{\mathrm{mot}}(f;s)$ is related with a (motivic) Poincaré series "counting solutions modulo t^{m+1}".

$$Q_{\mathrm{mot}}(f;T) := \sum_{m\geq 0}[\mathcal{L}_m(D)]T^m \in K_0(\mathrm{Var}_{\mathbb{C}})[\![T]\!].$$

In the same way as in Section 1.5, after taking $T=\mathbb{L}^{-s}$ in $Z_{\mathrm{mot}}(f;s)$ one can establish (EXERCISE) the relation

$$Q_{\mathrm{mot}}(f;\mathbb{L}^{-d}T) = \frac{\mathbb{L}^d Z_{\mathrm{mot}}(f;s)-[X]}{T-1}.$$

Then, it follows from last theorem that $Q_{\mathrm{mot}}(f;T)$ is rational on T over $\mathcal{M}_{\mathbb{C}}$.

(2) The previous series are also called $J_f(T)$ by DENEF and LOESER. In [**DL98**], they also introduced the power series $P_f(T)=\sum_{m\geq 0}[\pi_m(\mathcal{L}(D))]T^m$. Note that these series coincide when f defines a smooth hypersurface in X, but they are not equal in general, as we have discussed in Example 2.19.

3.2.1. *Specializations.* Comparing the expression of $Z_{\mathrm{mot}}(f;s)$ obtained in Theorem 3.18 with those for $Z_{\mathrm{top}}(f;s)$ and $Z_{\mathrm{Igusa}}(f;s)$ obtained in Section 1.7, we realize that $Z_{\mathrm{mot}}(f;s)$ is a kind of *universal zeta function*, with the other two as its specializations by applying the morphism $\chi(\cdot)$ and $\mathcal{N}^p(\cdot)$, for $p\gg 0$.

- In order to obtain $Z_{\mathrm{top}}(f;s)$, consider $Z_{\mathrm{mot}}(f;s)$ with $s\in\mathbb{N}$. From Theorem 3.18, we have sums of well defined elements of the form

$$[E_I^\circ]\prod_{i\in I}\frac{(\mathbb{L}-1)\mathbb{L}^{-(N_is+\nu_i)}}{1-\mathbb{L}^{-(N_is+\nu_i)}} = [E_I^\circ]\prod_{i\in I}\frac{\mathbb{L}-1}{\mathbb{L}^{N_is+\nu_i}-1} = [E_I^\circ]\prod_{i\in I}\frac{1}{[\mathbb{P}^{N_is+\nu_i-1}]},$$

in the image of $\mathcal{M}_{\mathbb{C}}\left[\{[\mathbb{P}^a]^{-1}\}_{a\in\mathbb{Z}_{\geq 0}}\right]$ in $\widehat{\mathcal{M}_{\mathbb{C}}}$. Applying the Euler characteristic extended to this ring, we obtain the sum of rational numbers

$$\sum_{I\subset\{0,\dots,r\}}\chi(E_I^\circ)\prod_{i\in I}\frac{1}{N_is+\nu_i}$$

for any $s\in\mathbb{N}$. Thus, $Z_{\mathrm{top}}(f;s)$ is the only function in $\mathbb{Q}(s)$ admitting the previous values for any $s\in\mathbb{N}$, as it is explained in [**Vey06**, Sec. 6.6]. For a more general explanation in terms of morphism of rings on how to specialize to $Z_{\mathrm{top}}(f;s)$, see [**DL98**, Sec. 2.3].

- Roughly speaking, the specialization to $Z_{\mathrm{Igusa}}(f;s)$ comes from extending the $\mathbb{F}_p$-counting measure to

$$\mathcal{N}^p:\mathcal{M}_{\mathbb{C}}\left[\left\{\frac{\mathbb{L}^{-(as+b)}}{1-\mathbb{L}^{-(as+b)}}\right\}_{a,b\in\mathbb{Z}_{\geq 1}}\right]\longrightarrow\mathbb{Q}\left[\left\{\frac{p^{-(as+b)}}{1-p^{-(as+b)}}\right\}_{a,b\in\mathbb{Z}_{\geq 1}}\right],$$

for $p\gg 0$. See [**DL98**, Sec. 2.4] and [**Nic10**, Sec. 5.3] for more details.

- We can also obtain a "finer" zeta function than $Z_{\mathrm{top}}(f;s)$ at the level of Hodge polynomials, the *Hodge zeta function*, specializing by $H(\cdot)$. See for example [**Rod04a**]

REMARK 3.22 (PRACTICAL COMPUTATION OF $Z_{\text{top}}(f;s)$). If we have $Z_{\text{mot}}(f;s)$ described in a form

$$Z_{\text{mot}}(f;s) = \frac{P(\mathbb{L}^{-s})}{\left(1-\mathbb{L}^{-(a_1s+b_1)}\right)\cdots\left(1-\mathbb{L}^{-(a_\ell s+b_\ell)}\right)} \tag{3.2}$$

for some $P \in \mathbb{Z}[\mathbb{L}^{-1}][T]$, then for any $s \in \mathbb{N}$ the above expression can be viewed as a rational fraction $R/S \in \mathbb{Q}(\mathbb{L}^{-1})$, over the "variable $\mathbb{L}^{-1}$". In this way, specializing to $Z_{\text{top}}(f;s)$ is "taking the limit $\mathbb{L}^{-1} \to 1$". We can use l'HÔPITAL RULE on this fraction, and we know by the above (3.2) that the $(\ell+1)$-derivate of S evaluated in $\mathbb{L}^{-1}=1$ is no zero. Moreover,

$$S^{(\ell+1)}(1) = (a_1s+b_1)\cdots(a_\ell s+b_\ell).$$

Then, $Z_{\text{top}}(f;s)$ is the only function in $\mathbb{Q}(s)$ admitting the values $\left(R^{(\ell+1)}/S^{(\ell+1)}\right)(1)$ for any $s \in \mathbb{N}$.

EXAMPLE 3.23. From Example 3.20, it is easy to get for the cusp the following:

$$Z_{\text{top}}(f;s) = \frac{4s+5}{(s+1)(6s+5)}.$$

REMARK 3.24. The above gives us a way to deal with $Z_{\text{top}}(f;s)$ by studying $Z_{\text{mot}}(f;s)$. As it was explained in Section 1.7, $Z_{\text{top}}(f;s)$ is originally defined in terms of a resolution. Also, when we study the Monodromy conjecture for $Z_{\text{top}}(f;s)$, in general we can only compare different resolutions using the Weak Factorization theorem. Now, working with $Z_{\text{mot}}(f;s)$ gives an intrinsic definition and a change of variables formula, which automatically implies that $Z_{\text{top}}(f;s)$ is independent of the chosen resolution, together with a tool to explicitly compare this zeta function between different proper birational morphisms.

3.2.2. *The (motivic) Monodromy conjecture.* In this context, we can translate the Monodromy conjecture (i.e. Conjecture 1.56) in terms of motivic zeta functions. Note that in this case we should be careful when we are speaking about "poles" as $K_0(\text{Var}_{\mathbb{C}})$ is not a domain and we do not know anything about $\mathcal{M}_{\mathbb{C}}$.

CONJECTURE 3.25 (MOTIVIC MONODROMY CONJECTURE). *There exists a finite subset $S \subset \mathbb{Z}^2_{\geq 1}$ verifying*

$$Z_{\text{mot}}(f;s) \in \mathcal{M}_{\mathbb{C}}\left[\mathbb{L}^{-s}, \left\{\frac{1}{1-\mathbb{L}^{-(as+b)}}\right\}_{(a,b)\in S}\right],$$

and such that for each $s_0 = -b/a$ with $(a,b) \in S$, the value $\exp(2\pi i s_0)$ *is an eigenvalue of the monodromy action of f at some point of V_f.*

It is clear that any pole of $Z_{\text{top}}(f;s)$ is also a "pole" of $Z_{\text{mot}}(f;s)$, but the contrary is not true in general: ISHII and KOLLÁR found a counter-example in [**IK03**].

REMARK 3.26.

(1) There also exists a *local* version of the motivic zeta function, consider $f : (X,x_0) \to (\mathbb{C},0)$ a homolorphic map germ on X. We define the *local motivic zeta function* at x_0 by

$$Z_{\text{mot},x_0}(f;s) := \int_{\mathcal{L}(X)_{x_0}} \mathbb{L}^{-\operatorname{ord}_D \cdot s} d\mu,$$

By Proposition 2.62, we have the analogous decomposition

$$Z_{\mathrm{mot},x_0}(f;s) = \mathbb{L}^{-d} \sum_{I\subset\{0,\dots,r\}} [E_I^\circ \cap h^{-1}(x_0)] \prod_{i\in I} \frac{\mathbb{L}-1}{\mathbb{L}^{N_i s+\nu_i}-1}.$$

The above specializes to the *local* topological zeta function $Z_{\mathrm{top},x_0}(f;s)$ by applying $\chi(\cdot)$.

(2) The study of both zeta functions can be extended to the data given by pair (f,ω), with a differential form $\omega \in \Omega_X^d$:

$$\int_{\mathcal{L}(X)} \mathbb{L}^{-\operatorname{ord}_{D_1}\cdot s-\operatorname{ord}_{D_2}} \in \mathcal{M}_{\mathbb{C}}[\![\mathbb{L}^{-s}]\!],$$

where D_1 (resp. D_2) is the divisor associated with f (resp. the zeros and poles of ω), whenever the integral converges. Some authors use this to study the role of the log discrepancies on the Monodromy conjecture [**NV10, NV12, CV17**].

The motivic zeta function has the advantage of capturing the geometrical structure of the varieties and the divisors (in characteristic zero), at the same time that we deal with the uniform behavior of additive invariants. The arc spaces are closely related with the singularities of f and its associated divisor, as well as its infinitesimal structure in general.

In fact, Denef and Loeser constructed in [**DL98, DL02a**] (see also [**DL01**, Sec. 3.5]) a *virtual motivic incarnation* of the Minor fiber of the morphism $f : X \to \mathbb{C}$ at a point $x_0 \in V_f$, as a limit of (an equivariant version of) zeta functions.

A more "topological" construction of the motivic Milnor fiber was construced recently in [**GLM16, GLM18**] via *motivic infinite cyclic covers.* This is made associating an element of the equivariant Grothendieck ring of varieties to a infinite cyclic covers of finite type of a punctured neighborhood of a SNC divisor, and it does not make use of arc spaces.

Other explicit connection of the topology of the Milnor fiber was obtained in [**DL02a**] as follows. Recall that X is a smooth algebraic variety, $f : X \to \mathbb{C}$ is a non-constant morphism and x_0 a point in V_f.

Definition 3.27. Let $T_{x_0} : H^\bullet(\mathcal{F}_{f,x_0};\mathbb{C}) \to H^\bullet(\mathcal{F}_{f,x_0};\mathbb{C})$ be the monodromy action of the Milnor fiber of f at $x_0 \in V_f$. For $m \in \mathbb{Z}_{\geq 0}$, we define the *$m$-Lefschetz number of T_{x_0}* as

$$\Lambda(T_{x_0}^m) := \sum_{q\geq 0}(-1)^q \operatorname{Tr}\left(T_{x_0}^m \mid H^q(\mathcal{F}_{f,x_0};\mathbb{C})\right),$$

for the m-th iterate $T_{x_0}^m$ of T_{x_0} .

Consider the following locally closed subvarieties associated with the local motivic zeta function:

$$\mathcal{X}_{x_0}^m = \{\gamma \in \mathcal{L}_m(X) \mid \pi_0(\gamma) = x_0,\ \operatorname{ord}_t(f\circ\gamma) = m\}.$$

Let $\overline{f}_m : \mathcal{X}_{x_0}^m \to \mathbb{C}^*$ be the morphism sending a m-jet $\gamma \in \mathcal{X}_{x_0}^m$ to the coefficient of t^m in $f(\gamma)$. Considering the natural action of $\mathbb{C}^*$ on $\mathcal{X}_{x_0}^m$ given by $a\cdot\gamma(t) = \gamma(at)$, one has that $\overline{f}_m(a\cdot\gamma) = a^m\overline{f}_m(\gamma)$ and it follows that $\overline{f}_m$ is a locally trivial fibration with geometric monodromy of order m.

Denote by $\mathcal{X}_{x_0}^{m,1}$ the fiber $(\overline{f}_m)^{-1}(1)$, which admits an action of the m-roots of unity μ_m as a restriction of the above $\mathbb{C}^*$-action.

THEOREM 3.28 (DENEF-LOESER, [**DL02a**]). *For every* $m \geq 1$, $\Lambda(T_{x_0}^m) = \chi\left(\mathcal{X}_{x_0}^{m,1}\right)$, *i.e. the m-Lefschetz number of* T_{x_0} *is equal to the Euler characteristic of* $\mathcal{X}_{x_0}^{m,1}$.

The proof in [**DL02a**] is obtained by expressing the class $\left[\mathcal{X}_{x_0}^{m,1}\right] \in \mathcal{M}_{\mathbb{C}}$ in terms of the data of a log-resolution using motivic integration and then comparing formulas with the A'Campo expression of the Lefschetz numbers [**A'C73, A'C75**]. Later in [**HL15**], the authors give a proof without the use of a log-resolution, using ℓ-adic cohomology of non-Archimedean spaces, motivic integration and the Lefschetz fixed point formula for automorphisms of finite order.

3.3. Quotient abelian singularities and embedded $\mathbb{Q}$-resolutions. Even if we have obtained formulas for the motivic integral, zeta functions or stringy invariants via resolution of singularities of pairs (X, D), these models are complicated to obtain in practice. Hence, we are interested in obtaining alternative birational models of (X, D) to derive new formulae and properties.

In [**DL02b**], DENEF and LOESER gave a change of variables formula for proper birational maps $Y \to X$ with Y and X both possibly singular. However, the "jacobian contribution" appearing in this formula is very complicated to compute in general.

Our idea is to take advantage of the analytic properties of *quotient singular spaces* $\mathbb{C}^d/G$ for G a finite abelian group, in order to generalize the study of motivic integrals and zeta functions in the singular case, providing a formula for *partial resolutions* involving such quotients.

3.3.1. *Quotient singularities.* Let $G \subset \mathrm{GL}_d(\mathbb{C})$ be a finite linear group. This group acts naturally in $\mathbb{C}^d$ and defines a normal irreducible affine variety $\mathbb{C}^d/G$ whose points correspond one-to-one with G-orbits in $\mathbb{C}^d$. Moreover, the coordinate ring of $\mathbb{C}^d/G$ is exactly the *ring of G-invariant polynomials* $\mathbb{C}[x_1, \ldots, x_d]^G = \mathbb{C}[w_1, \ldots, w_N]$, which is finitely generated by elements $w_i \in \mathbb{C}[x_1, \ldots, x_d]$ such that $w_i\left(g^{-1} \cdot (x_1, \ldots, x_d)\right) = w_i(x_1, \ldots, x_d)$ for any $g \in G$. This construction comes with natural compatible morphisms:

$$\pi : \mathbb{C}^d \twoheadrightarrow \mathbb{C}^d/G \qquad \text{and} \qquad \pi^* : \mathbb{C}[x_1, \ldots, x_d]^G \hookrightarrow \mathbb{C}[x_1, \ldots, x_d].$$

One can embed $\mathbb{C}^d/G$ in a larger $\mathbb{C}^N$ with coordinates $(u_1, \ldots, u_N)$ via the identification morphism $\phi : \mathbb{C}[u_1, \ldots, u_N] \to \mathbb{C}[w_1, \ldots, w_N]$, with $\phi(u_i) = w_i$. The coordinate ring in this new variables is expressed by

$$\mathbb{C}[x_1, \ldots, x_d]^G \simeq \mathbb{C}[u_1, \ldots, u_N]/\ker(\phi).$$

EXAMPLE 3.29. In Example 3.4 where $X = \mathbb{C}^2/\mu_3$ is the quotient by the action $(x, y) \mapsto (\xi x, \xi y)$ for $\xi \in \mu_3$, it is clear that $\mathbb{C}[x, y]^{\mu_3} = \mathbb{C}[x^3, x^2y, xy^2, y^3]$. By identifying $u_1 = x^3$, $u_2 = x^2y$, $u_3 = xy^2$ and $u_4 = y^3$, we get the embedding $X \hookrightarrow \mathbb{C}^4$ described before. In practice, one defines the ideal $I = \langle u_1 - x^3, \ldots, u_4 - y^3\rangle$ in $\mathbb{C}[u_1, u_2, u_3, u_4] \otimes \mathbb{C}[x, y]^{\mu_3}$ and eliminates the variables x, y using Gröbner basis.

For every finite group $G \subset \mathrm{GL}_d(\mathbb{C})$, denote by G_{big} the normal subgroup generated by all rotations around hyperplanes. Then, $\mathbb{C}^d/G_{\text{big}} \simeq \mathbb{C}^d$ since the G_{big}-invariant polynomials form a polynomial algebra. The choice of a basis in this algebra determines an isomorphism between the group G/G_{big} and another group in $\mathrm{GL}_d(\mathbb{C})$. This motivates the following definition.

DEFINITION 3.30. A finite group $G \subset \mathrm{GL}_d(\mathbb{C})$ is called *small* if no element of G has 1 as an eigenvalue of multiplicity exactly $d-1$.

Thus, the classification of quotient singularities is reduced to the study of small actions. In fact, PRILL [**Pri67**] proved the following well known result: the conjugacy class of a small group G in $\mathrm{GL}_d(\mathbb{C})$ determines $\mathbb{C}^d/G$ up to isomorphism, and vice versa.

It is known that quotient singularities are log-terminal. There also exists a numerical criterium due to REID and TAI to determine whenever $\mathbb{C}^d/G$ is terminal or canonical. See [**Kol13**, Sec. 3.2] for further details about quotient singularities and [**CLS11**, Sec. 11.4] for a toric point of view.

One of the advantages of this kind of singularities is the fact that we can use the "fake coordinates" $[(x_1,\dots,x_d)] \in \mathbb{C}^d/G$ and the ring of G-invariant polynomials $\mathbb{C}[x_1,\dots,x_d]^G$ in order to make calculations, instead of choosing first an embedding $\mathbb{C}^d/G \hookrightarrow \mathbb{C}^N$. Note that $\mathbb{C}^d/G$ has arbitrarily large embedding dimension, for fixed $d>1$.

In particular, we can express the different notions of differentials explained in Sections 2.4.2 and 3.1.1 as follows.

PROPOSITION 3.31. *Let $X = \mathbb{C}^d/G$ be a quotient by a small group $G \subset \mathrm{GL}_d(\mathbb{C})$, with coordinate ring $(\mathcal{O}_{\mathbb{C}^d})^G = \mathbb{C}[w_1,\dots,w_N]$. Then, for any $k \geq 0$:*

(1) $\Omega_X^k \simeq \mathbb{C}[w_1,\dots,w_N] \cdot \langle \mathrm{d}w_{i_1} \wedge \cdots \wedge \mathrm{d}w_{i_k} \rangle_{1 \leq i_1 < \cdots < i_k \leq N}$.

(2) $\omega_X \simeq (\pi_* \Omega_{\mathbb{C}^d}^d)^G$.

A more general version of the second statement can be found in [**Ste77**]. This last result is extremely useful for our computations.

EXAMPLE 3.32. Let $X = \mathbb{C}^2/\mu_2$ be the quotient by the action $(x,y) \mapsto (-x,-y)$ by $\mu_2 = \{\mathrm{id}, -\mathrm{id}\} \subset \mathrm{GL}_2(\mathbb{C})$. The coordinate ring is $\mathbb{C}[x,y]^{\mu_2} = \mathbb{C}[x^2,xy,y^2]$, and the 1-forms Ω_X^1 are generated by $x\mathrm{d}x$, $y\mathrm{d}x + x\mathrm{d}y$, $y\mathrm{d}y$. Note that Ω_X^2 is generated by the forms

$$x^2 \mathrm{d}x \wedge \mathrm{d}y, \quad xy\, \mathrm{d}x \wedge \mathrm{d}y, \quad y^2 \mathrm{d}x \wedge \mathrm{d}y,$$

in particular $\omega = \mathrm{d}x \wedge \mathrm{d}y \notin \Omega_X^2$. However, $\mathrm{d}(-x) \wedge \mathrm{d}(-y) = \mathrm{d}x \wedge \mathrm{d}y$ and ω is in fact a basis of ω_X. We can define an embedding $X \hookrightarrow \mathbb{C}^3$ in coordinates (u,v,w), where X is the hypersurface defined by the equation $v^2 = uw$. Note that ω can be written is this new variables as

$$\mathrm{d}x \wedge \mathrm{d}y = \frac{\mathrm{d}(x^2) \wedge \mathrm{d}(y^2)}{2xy} \longleftarrow \frac{\mathrm{d}u \wedge \mathrm{d}w}{2v},$$

which correspond to a representative of the basis, following Remark 3.2.

REMARK 3.33. If $G \subset \mathrm{SL}_d(\mathbb{C})$, then the canonical volume form $\mathrm{d}x_1 \wedge \cdots \wedge \mathrm{d}x_d$ is G-invariant and then $K_{\mathbb{C}^d/G}$ is trivial, so in particular the quotient $\mathbb{C}^d/G$ is Gorenstein.

In [**DL02b**], DENEF and LOESER present a study of motivic integration on quotient singularities, in order to study the McKay correspondence from motivic integration, following BATYREV's ideas, see also [**Loo02**].

In order to study the space of arcs of $\mathbb{C}^d/G$ using the coordinates in $\mathbb{C}^d$, they use the *space of ramified arcs* $\mathcal{L}^{1/\ell}(\mathbb{C}^d)$, where $\ell = |G|$. These are, roughly speaking, $\mathbb{C}[\![t^{1/\ell}]\!]$-points in $\mathbb{C}^d$ representing arcs in $\mathbb{C}^d/G$ and compatible with the action of the group via primitive ℓth-roots of unity ζ_ℓ. In Example 3.32, if we consider $X = \mathbb{C}^2/\mu_2$ embedded in $\mathbb{C}^3$, one can see that the arc $\gamma(t) = (t,t,t) \in \mathcal{L}(X)$ is in correspondence with the arcs $\varphi_\pm(t) = (\pm t^{1/2}, \pm t^{1/2}) \in \mathcal{L}^{1/2}(\mathbb{C}^2)$ forming an orbit.

They also construct an *orbifold motivic measure* μ^{orb}, "twisting" the usual measure by the jacobian $\mathrm{Jac}(\pi)$ of the projection map $\pi : \mathbb{C}^d \to \mathbb{C}^d/G$, and verifying:

$$\mu^{\mathrm{orb}}(\mathcal{L}(\mathbb{C}^d/G)_0) = \sum_{[g]\in \mathrm{Conj}(G)} \mathbb{L}^{d-\varpi(g)}, \tag{3.3}$$

where the map $\varpi : G \to \mathbb{Q}$ is defined as follows. For any $g \in G$, there exists a basis $\{b_i^g\}_{1\le i\le d}$ of $\mathbb{C}^d$ such that g is represented by the matrix $\mathrm{Diag}\left(\zeta_\ell^{\varepsilon_{g,1}}, \dots, \zeta_\ell^{\varepsilon_{g,d}}\right)$, with $0 \le \varepsilon_{g,i} \le \ell - 1$, $i = 1\dots,d$. Thus, the value $\varpi(g) := \sum_i \varepsilon_{g,i}$ (usually called the *age* of g) does not depend on the chosen basis.

It is worth noticing that the expression (3.3) is true in a quotient $\mathcal{M}_{\mathbb{C}/}$ of the ring $\mathcal{M}_{\mathbb{C}}$ obtained by adding the following relation: *for every linear space V and every finite group $G_V \subset \mathrm{GL}(V)$, we have that $[V] = [V/G_V]$* . This relation is presented as technical and it is known that the specializations $H(\cdot)$ and $\chi(\cdot)$ factorize through it. Nevertheless, LOOIJENGA showed that this relation is true in $\mathcal{M}_{\mathbb{C}}$ if G_V is abelian [**Loo02**, Lemma 5.1]. Note that the above formula implies the McKay correspondence, taking the specialization by the Euler characteristic $\chi(\cdot)$.

3.3.2. *V-manifolds and (embedded) $\mathbb{Q}$-resolutions.* The main idea of the following is to take advantage of the previous theory in quotient singularities to study motivic integrals and the motivic zeta function using *orbifolds* and *partial resolutions.* This generalizes not only the techniques in the previous sections, but also reduces the computations in general.

Let us introduce some notation and notions in the context of $\mathbb{Q}$-resolutions of singularities. We refer to [**MM11**] as main reference text on this subject, see also [**Ste77, AMO14b**].

DEFINITION 3.34. A complex analytic variety X is called a *V-manifold* if $X = \bigcup_k U_k$ such that each open $U_k \simeq \mathbb{C}^d/G_k$, for some finite group $G_k \subset \mathrm{GL}_d(\mathbb{C})$.

We are interested in V-manifolds where the local isotropy groups are abelian. For $\boldsymbol{\ell} = (\ell_1 \dots \ell_r)^\top$ we denote by $\mu_{\boldsymbol{\ell}} = \mu_{\ell_1} \times \cdots \times \mu_{\ell_r}$ the finite abelian group written as a product of finite cyclic groups, that is, μ_{ℓ_i} is the cyclic group of ℓ_i-th roots of unity in $\mathbb{C}$. Consider a matrix of weight vectors

$$A = (a_{ij})_{i,j} = [\, \mathbf{a}_1 \mid \cdots \mid \mathbf{a}_d \,] \in \mathrm{Mat}_{r\times d}(\mathbb{Z}),$$
$$\mathbf{a}_j = (a_{1j} \dots a_{rj})^\top \in \mathrm{Mat}_{r\times 1}(\mathbb{Z}),$$

and the action is given by

$$(x_1, \dots, x_d) \longmapsto (\zeta_{\ell_1}^{a_{11}} \cdots \zeta_{\ell_r}^{a_{r1}}\, x_1, \, \dots \, , \zeta_{\ell_1}^{a_{1d}} \cdots \zeta_{\ell_r}^{a_{rd}}\, x_d).$$

Note that the i-th row of the matrix A can be considered modulo ℓ_i. The set of all orbits $\mathbb{C}^d/\mu_{\boldsymbol{\ell}}$ is called the (*cyclic*) *quotient space of type* $(\boldsymbol{\ell}; A)$ and it is denoted by

$$X(\boldsymbol{\ell}; A) = X\left(\begin{array}{c|ccc} \ell_1 & a_{11} & \cdots & a_{1d} \\ \vdots & \vdots & \ddots & \vdots \\ \ell_r & a_{r1} & \cdots & a_{rd} \end{array}\right).$$

When $r = 1$, the space $X(\ell; a_1, \ldots, a_d)$ is usually denoted by $\frac{1}{\ell}(a_1, \ldots, a_d)$ in the literature. It is proved that any quotient space $\mathbb{C}^d/G$, with $G \subset \mathrm{GL}_d(\mathbb{C})$ finite abelian, is isomorphic to some space of type $(\boldsymbol{\ell}; A)$ with $r \leq d-1$.

It should be noticed that there are a lot of operations over the pair $(\boldsymbol{\ell}; A)$ which induces isomorphisms on the underlying spaces.

EXERCISE 3.35. Show that:

(1) $\frac{1}{\ell}(0, a_2, \ldots, a_d) = \mathbb{C} \times \frac{1}{\ell}(a_2, \ldots, a_d)$.

(2) $\frac{1}{\ell}(a_1, \ldots, a_d) = \frac{1}{k\ell}(ka_1, \ldots, ka_d)$ for any $k > 0$.

REMARK 3.36 (ROW OPERATIONS). The automorphism in $G = \mu_\ell \times \mu_\ell$ given by $(\zeta, \eta) \mapsto (\zeta\eta^{-1}, \eta)$ shows that the identity map induces the following isomorphism:

$$X\left(\begin{array}{c|ccc} \ell & a_{11} & \cdots & a_{1d} \\ \ell & a_{21} & \cdots & a_{2d} \end{array}\right) = X\left(\begin{array}{c|ccc} \ell & a_{11} & \cdots & a_{1d} \\ \ell & a_{21}-a_{11} & \cdots & a_{2d}-a_{1d} \end{array}\right).$$

Moreover, any automorphism $(\zeta, \eta) \mapsto (\zeta^{\alpha_1}\eta^{\beta_1}, \zeta^{\alpha_2}\eta^{\beta_2})$ with $\gcd(\ell, \alpha_1\beta_2 - \alpha_2\beta_1) = 1$ gives in the same way:

$$X\left(\begin{array}{c|ccc} \ell & a_{11} & \cdots & a_{1d} \\ \ell & a_{21} & \cdots & a_{2d} \end{array}\right) = X\left(\begin{array}{c|ccc} \ell & \alpha_1 a_{11} + \alpha_2 a_{21} & \cdots & \alpha_1 a_{1d} + \alpha_2 a_{2d} \\ \ell & \beta_1 a_{11} + \beta_2 a_{21} & \cdots & \beta_1 a_{1d} + \beta_2 a_{2d} \end{array}\right).$$

Thus, we can perform row operations for general $(\boldsymbol{\ell}; A)$ between two rows sharing cyclic group actions of the same order. If two rows have different ℓ_i and ℓ_j, one can always arrive to this situation by re-scaling the rows, using the property of the above exercise.

PROPOSITION 3.37. *The operations* $(\boldsymbol{\ell}; A) \to (\boldsymbol{\ell}'; A')$ *in* TABLE 1 *induce isomorphisms* $\psi : X(\boldsymbol{\ell}; A) \to X(\boldsymbol{\ell}'; A')$.

EXERCISE 3.38. Prove that for $d = 1$ any $X(\boldsymbol{\ell}; A)$ is isomorphic to $\mathbb{C}$.

The action defining $X(\boldsymbol{\ell}; A)$ is free on $(\mathbb{C}^*)^d$ if and only if the associated representation

$$\begin{array}{ccc} \mu_{\boldsymbol{\ell}} & \longrightarrow & \mathrm{GL}_d(\mathbb{C}) \\ (\zeta_{\ell_1}, \ldots, \zeta_{\ell_r}) & \longmapsto & \mathrm{Diag}(\zeta_{\ell_1}^{a_{11}} \cdots \zeta_{\ell_r}^{a_{r1}}, \ldots, \zeta_{\ell_1}^{a_{1d}} \cdots \zeta_{\ell_r}^{a_{rd}}) \end{array}$$

is injective. We can factor the group by the kernel of the action and the isomorphism type does not change. This reduction, together with the condition of being small, motivates the following definition.

DEFINITION 3.39. The type $(\boldsymbol{\ell}; A)$ is said to be *normalized* if the action is free on $(\mathbb{C}^*)^d$ and the group $\mu_{\boldsymbol{\ell}}$ is small as a subgroup of $\mathrm{GL}_d(\mathbb{C})$.

Using the previous operations, we can always reduce $X(\boldsymbol{\ell}; A)$ into its *normalized* form.

PROPOSITION 3.40. *The space* $X(\boldsymbol{\ell}; A)$ *is normalized if and only if the isotropy group of* P *is trivial for all* $P \in \mathbb{C}^d$ *with exactly* $d-1$ *coordinates different from zero.*

REMARK 3.41. In the cyclic case $X(\ell; a_1, \ldots, a_r)$, the order of the isotropy group of any point $P \in \mathbb{C}^d$ with only the ith coordinate different from zero is $\gcd(\ell, a_1, \ldots, \widehat{a_i}, \ldots, a_r)$.

TABLE 1

	Operation $(\boldsymbol{\ell}; A) \to (\boldsymbol{\ell}'; A')$	*Isom.* $\psi : X(\boldsymbol{\ell}; A) \to X(\boldsymbol{\ell}'; A')$
(1)	*Permutation* $\sigma \in \Sigma_d$ *of columns of* A	$[(x_1, \dots, x_d)] \mapsto [(x_{\sigma(1)}, \dots, x_{\sigma(d)})]$
(2)	*Permutation of rows of* $(\boldsymbol{\ell}; A)$	$\psi = \mathrm{id}$
(3)	*Row* $(\ell_i \vert\ a_{i1} \cdots a_{id}) \to (k\ell_i \vert\ ka_{i1} \ \cdots \ ka_{id})$ *for any* $k > 0$	$\psi = \mathrm{id}$
(4)	*Row* $(\ell_i \vert\ a_{i1} \ \cdots \ a_{id}) \to (\ell_i \vert\ ma_{i1} \ \cdots \ ma_{id})$ *for any* $\gcd(\ell_i, m) = 1$.	$\psi = \mathrm{id}$
(5)	*Replace* $a_{ij} \to a_{ij} + k\ell_i$	$\psi = \mathrm{id}$
(6)	*If* $\ell_r = 1$ *then eliminate the last row*	$\psi = \mathrm{id}$
(7)	*Column* $(a_{1d}, \dots, a_{rd})^\top \to (ea_{1d}, \dots, ea_{rd})^\top$, *if* $\gcd(e, a_{id}) = 1$ *for any* $1 \le i \le r$ *and* e *divides* ℓ_1 *and* a_{1j}, $1 \le j \le d-1$.	$[(x_1, \dots, x_d)] \mapsto [(x_1, \dots, x_d^e)]$

EXAMPLE 3.42.

(1) For $d = 2$, one can always reduce any quotient space into the cyclic case by the above operations. The space $X(\ell; a, b)$ is normalized if and only if $\gcd(\ell, a) = \gcd(\ell, b) = 1$. If this is not the case and assuming $\gcd(\ell, a, b) = 1$, one normalizes using the isomorphism

$$\begin{array}{ccc} X(\ell; a, b) & \longrightarrow & X\left(\frac{\ell}{\gcd(\ell,a)\gcd(\ell,b)}; \frac{a}{\gcd(\ell,a)}, \frac{b}{\gcd(\ell,b)}\right) \\ [(x, y)] & \longmapsto & \left[(x^{\gcd(\ell,b)}, y^{\gcd(\ell,a)})\right] \end{array}.$$

(2) For $d = 3$, the quotient space $X(\ell; a, b, c)$ is written in a normalized form if and only if $\gcd(\ell, a, b) = \gcd(\ell, a, c) = \gcd(\ell, b, c) = 1$. As above, one can use isomorphisms $[(x, y, z)] \mapsto [(x, y, z^k)]$ to reduce $X(\ell; a, b, c)$ into its normalized form.

One can find situations illustrating the above operations throughout the remaining examples of this section.

EXAMPLE 3.43. Take $\omega = (q_0, \dots, q_d) \in \mathbb{Z}^d_{>0}$ a weight vector. The *ω-weighted projective space* is the normal irreducible projective V-manifold defined by

$$\mathbb{P}^d_\omega = \frac{\mathbb{C}^{d+1} \setminus 0}{(x_0, \dots, x_d) \underset{\lambda \in \mathbb{C}^*}{\sim} (\lambda^{q_0} x_0, \dots, \lambda^{q_d} x_d)}.$$

It can be proved (EXERCISE) that $\mathbb{P}^d_\omega = \bigcup_{j=0}^d U_j$, where $U_j = \{x_j \neq 0\} \simeq \frac{1}{q_j}(q_0, \dots, \widehat{q_j}, \dots, q_d) = \mathbb{C}^d/\mu_{q_j}$. Also, there is a branched covering

$$\begin{array}{ccc} \mathbb{P}^d & \longrightarrow & \mathbb{P}^d_\omega \\ [x_0, \dots, x_d] & \longmapsto & [x_0^{q_0}, \dots, x_d^{q_d}]_\omega \end{array} .$$

of order $\frac{q_0 \cdots q_d}{\gcd(q_0, \dots, q_d)}$, ramified over the *axis* $\{[x_0, \dots, x_d]_\omega \mid x_0 \cdots x_d = 0\}$. It follows that $\chi(\mathbb{P}^d_\omega) = d + 1$.

Note that the isomorphy class of $\mathbb{P}^d_\omega$ is not completely determined by the weights ω. Taking $r_i := \gcd(q_0, \dots, \widehat{q_i}, \dots, q_d)$, $e_i := r_0 \cdots \widehat{r_i} \cdots r_d$ and $\omega' = (q_0/e_0, \dots, q_d/e_d)$, we have an isomorphism

$$\begin{array}{ccc} \mathbb{P}^d_\omega & \longrightarrow & \mathbb{P}^d_{\omega'} \\ [x_0, \dots, x_d]_\omega & \longmapsto & [x_0^{r_0}, \dots, x_d^{r_d}]_{\omega'} \end{array} .$$

Thus, one can always assume that $\gcd(q_0, \dots, \widehat{q_i}, \dots, q_d) = 1$ for any i. In particular, $\mathbb{P}^1_{(q_0,q_1)} \simeq \mathbb{P}^1$ is smooth.

A complete description of weighted projective spaces was studied by DOLGACHEV in **[Dol82]**.

EXERCISE 3.44. Compute the class $[\mathbb{P}^d_w]$ in $K_0(\mathrm{Var}_\mathbb{C})$.

The construction of these spaces are very useful for several reasons. One of them is that we can generalize the notion of blowing-up by adding *weights*.

DEFINITION 3.45 (WEIGHTED ω-BLOW-UP OF $\mathbb{C}^{d+1}$). Let $\omega = (p_0, \dots, p_k) \in \mathbb{Z}^k_{>0}$ be a weight vector for $1 \leq k \leq d$, and consider the plane $L : x_0 = \dots = x_k = 0$ in $\mathbb{C}^{d+1}$. The *weighted ω-blow-up of $\mathbb{C}^{d+1}$ over the smooth center L* is the V-manifold

$$\widehat{\mathbb{C}}^{d+1}_{L,\omega} := \{((x_0, \dots, x_d), [u_0, \dots, u_k]_\omega) \in \mathbb{C}^{d+1} \times \mathbb{P}^k_\omega \mid (x_0, \dots, x_k) \in \overline{[u_0, \dots, u_k]_\omega}\},$$

together with the birational map $\pi_\omega : \widehat{\mathbb{C}}^{d+1}_{L,\omega} \to \mathbb{C}^{d+1}$ induced by the natural projection $\mathbb{C}^{d+1} \times \mathbb{P}^k_\omega \to \mathbb{C}^{d+1}$. The above condition about the closure means $\exists \lambda \in \mathbb{C}$ such that $x_i = \lambda^{p_i} u_i$, $i = 0, \dots, k$.

REMARK 3.46.

(1) The map π_ω is an isomorphism in $\widehat{\mathbb{C}}^{d+1}_{L,\omega} \setminus E \to \mathbb{C}^{d+1} \setminus L$ with exceptional divisor $E := \pi_\omega^{-1}(L)$ is identified with $\mathbb{P}^k_\omega \times \mathbb{C}^{d-k}$.

(2) When $\omega = (1, \dots, 1)$, we obtain the classic blowing up of $\mathbb{C}^{d+1}$ over L.

(3) As in the classic case, there are $k+1$ charts covering $\widehat{\mathbb{C}}^{d+1}_{L,\omega} = U_0 \cup \dots \cup U_k$ and coming from those of $\mathbb{P}^k_\omega$. Looking at U_0, we define a map $\varphi_0 : \mathbb{C}^{d+1} \to U_0 = \{u_0 \neq 0\}$ as follows

$$\varphi_0(x_0, \dots, x_d) = ((x_0^{p_0}, x_0^{p_1} x_1, \dots, x_0^{p_k} x_k, x_{k+1}, \dots, x_d), [1 : x_1 : \dots : x_d]_\omega),$$

which is surjective, but not injective. For instance, any point of the form $(\zeta^k_{p_0}, 0, \dots, 0)$ have the same image by φ_0. However, one can see (EXERCISE) that it descends into an isomorphism on the quotient

$$\overline{\varphi_0} : \frac{1}{p_0}(-1, p_1, \dots, p_k) \times \mathbb{C}^{d-k} \longrightarrow U_0 = \{u_0 \neq 0\} \subset \widehat{\mathbb{C}}^{d+1}_{L,\omega}.$$

In this chart, E is clearly defined by the equation $x_0 = 0$. Following analogous arguments, one gets explicit descriptions for the rest of the charts.

(4) We will need to do similar weighted ω-blow-ups on quotient spaces $X(\boldsymbol{\ell}, A)$. The way to obtain those kind of transformations is to extend naturally the action of $\mu_{\boldsymbol{\ell}}$ on $\mathbb{C}^{d+1}$ into an action on $\widehat{\mathbb{C}}^{d+1}_{L,\omega}$:

$$\begin{gathered}((x_0,\dots,x_d),[u_0,\dots,u_k]_\omega)\\ \downarrow\\ \left(\left(\prod_{k=1}^{r}\zeta_{\ell_k}^{a_{k0}}\cdot x_0,\dots,\prod_{k=1}^{r}\zeta_{\ell_k}^{a_{kd}}\cdot x_d\right),\left[\prod_{k=1}^{r}\zeta_{\ell_k}^{a_{k0}}\cdot u_0,\dots,\prod_{k=1}^{r}\zeta_{\ell_k}^{a_{kd}}\cdot u_k\right]_\omega\right)\end{gathered}$$

This induces a weighted blow-up $\overline{\pi}_\omega : \widehat{X(\boldsymbol{\ell}, A)}_{L,\omega} \to X(\boldsymbol{\ell}, A)$ given by the commutative diagram

$$\begin{array}{ccc}\widehat{\mathbb{C}}^{d+1}_{L,\omega} & \xrightarrow{\ \pi_\omega\ } & \mathbb{C}^{d+1}\\ \downarrow & & \downarrow\\ \widehat{X(\boldsymbol{\ell}, A)}_{L,\omega}=\widehat{\mathbb{C}}^{d+1}_{L,\omega}/\mu_{\boldsymbol{\ell}} & \xrightarrow[\overline{\pi}_\omega]{} & X(\boldsymbol{\ell}, A)\end{array}$$

Here, the exceptional divisor E of $\overline{\pi}_\omega$ is isomorphic to $E \simeq (\mathbb{P}^k_\omega \times \mathbb{C}^{d-k})/\mu_{\boldsymbol{\ell}}$. The action of $\mu_{\boldsymbol{\ell}}$ respects the charts of $\widehat{\mathbb{C}}^{d+1}_{L,\omega}$, so that $\widehat{X(\boldsymbol{\ell}, A)}_{L,\omega} = \widehat{U}_0 \cup \cdots \widehat{U}_d$, with $\widehat{U}_i = U_i/\mu_{\boldsymbol{\ell}} = \{u_i \neq 0\}$. Explicit expressions of the above transformations and their respective charts are detailed in [**AMO14a**, Sec. 3].

DEFINITION 3.47. Let X be a V-manifold with abelian quotient singularities. A hypersurface D on X is said to have $\mathbb{Q}$-*normal crossings* if it is locally analytically isomorphic to the quotient of a union of coordinate hyperplanes under a diagonal action of a finite abelian group $G \subset \mathrm{GL}_d(\mathbb{C})$.

Let $U = \mathbb{C}^d/G$ be an abelian quotient space. Consider $H \subset U$ an analytic subvariety of codimension one.

DEFINITION 3.48. An *embedded* $\mathbb{Q}$-*resolution* of $(H, 0) \subset (U, 0)$ is a proper analytic map $h : Y \to (U, 0)$ such that

(1) Y is a V-manifold with abelian quotient singularities,

(2) h is an isomorphism over $Y \setminus h^{-1}(H_{\mathrm{sing}})$,

(3) $h^{-1}(H)$ is a hypersurface with $\mathbb{Q}$-normal crossings on Y.

REMARK 3.49. Let $(H, 0)$ be the hypersurface defined by a non-constant analytic germ $f : (U, 0) \to (\mathbb{C}, 0)$ and let $h : Y \to (U, 0)$ be an embedded $\mathbb{Q}$-resolution of $(H, 0)$. Then $h^{-1}(H)$ is locally given by a function of the form

$$x_1^{a_1} \cdots x_d^{a_d} : \mathbb{C}^d/G_0 \longrightarrow \mathbb{C},$$

for some finite abelian group G_0, small and acting diagonally. Moreover, there is a natural finite stratification $Y = \bigsqcup_{k\geq 0} Y_k$ defined as follows: there exist such groups $G_k \subset \mathrm{GL}_n(\mathbb{C})$ as above, $\boldsymbol{N}_k = (N_{1,k}, \dots, N_{d,k}) \in \mathbb{Z}^d_{\geq 0}$, and $\boldsymbol{\nu}_k = (\nu_{1,k}, \dots, \nu_{d,k}) \in$

$\mathbb{Z}^d_{\geq 0}$ such that Y is locally isomorphic around q to $V_k = \mathbb{C}^d/G_k$ and we have in local equations

$$h^{-1}(H) : x_1^{N_{1,k}} \cdots x_d^{N_{d,k}} = 0 \quad \text{and} \quad K_h : x_1^{\nu_{1,k}-1} \cdots x_d^{\nu_{d,k}-1} = 0.$$

Moreover, the data G_k, $\boldsymbol{N}_k$, $\boldsymbol{\nu}_k$ do not depend on the chosen $q \in Y_k$ but only on the stratum Y_k.

EXAMPLE 3.50. For the family of cusps $\mathcal{C}_{p,q} : y^p - x^q = 0$, with $\gcd(p,q) = 1$. Let $\pi_{(p,q)} : \widehat{\mathbb{C}}^2_{(p,q)} \to \mathbb{C}^2$ be the (p,q)-blow-up of $\mathbb{C}^2$ at the origin. We have two charts $\widehat{\mathbb{C}}^2_{(p,q)}$, as it was described above, as follows:

$$\begin{array}{ccc} \widehat{\mathbb{C}}^2_{(p,q)} & \simeq \frac{1}{p}(-1,q) & \cup\ \frac{1}{q}(p,-1) \\ \pi_{(p,q)} \big\downarrow & [(x,y)] & [(x,y)] \\ & \big\downarrow & \big\downarrow \\ \mathbb{C}^2 & (x^p, x^q y) & (xy^p, y^q) \end{array}$$

Thus, $\widehat{\mathbb{C}}^2_{(p,q)}$ contains two origins Q_1 and Q_2, each of them centered in a quotient space. In fact, $\operatorname{Sing} \widehat{\mathbb{C}}^2_{(p,q)} = \{Q_1, Q_2\}$, and the exceptional divisor $E = \pi^{-1}(0)$ is a $\mathbb{P}^1_{(p,q)}$ which is isomorphic to $\mathbb{P}^1$ via the map $[x : y]_{(p,q)} \mapsto [x^q : y^p]$, see FIGURE 6. Note that Q_1 and Q_2 are singularities of $\widehat{\mathbb{C}}^2_{(p,q)}$ but *not* of the exceptional divisor! In fact, E is smooth as an abstract variety.

In the first chart, the total transform is given by

$$x^{pq}(y^p - 1) : \frac{1}{p}(-1,q) \longrightarrow \mathbb{C}.$$

Note that the equation $y^p = 1$ has only one solution in $\frac{1}{p}(-1,q)$. At this point, we have local equation $x^{pq} y = 0$, so the strict transform $\widehat{\mathcal{C}}_{p,q}$ has only one branch (as it should be since $\pi_{(p,q)}$ is an isomorphism outside E) transverse to $E : x^{pq} = 0$. Also in this chart,

$$\pi^*_{(p,q)}(\mathrm{d}x \wedge \mathrm{d}y) = x^{p+q-1} \mathrm{d}x \wedge \mathrm{d}y.$$

The situation in the other chart is analogous to this one.

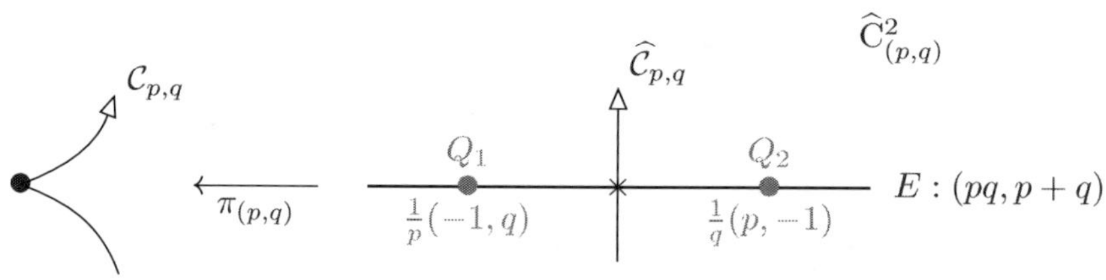

FIGURE 6. Embedded $\mathbb{Q}$-resolution of the cusp $\mathcal{C}_{p,q} : y^p - x^q = 0$.

Since all the charts are written in a normalized form, the proper map $\pi_{(p,q)}$ is an embedded $\mathbb{Q}$-resolution of $\mathcal{C}_{p,q}$.

EXERCISE 3.51.

(1) Compute a usual resolution of $y^{10} - x^3 = 0$ and compare with the above $\mathbb{Q}$-resolution.

(2) Consider now $\mathcal{C}_{p,q} : y^p - x^q = 0$ in $\mathbb{C}^2$ with $\gcd(p,q) = e$ not necessarily one. Prove that the (p', q')-blowing-up of $\mathbb{C}^2$ at the origin, with $p' = p/e$ and $q' = q/e$, is a embedded $\mathbb{Q}$-resolution of $\mathcal{C}_{p,q}$. How many branches $\mathcal{C}_{p,q}$ has? What is the numerical data of the $\mathbb{Q}$-resolution?

REMARK 3.52.

(1) As we will see in other examples, $\mathbb{Q}$-resolutions are very useful in order to compute partial resolution models, which arise with exceptional locus which are "simpler" combinatorially speaking. Moreover, they seem to be unavoidable in order to achieve *faster* resolution processes, extending VILLAMAYOR's work in higher dimensions, see [**McQ20, ATW19**]. Also, they allow us to study some families of singularities containing symmetries in a more effective way, specially for plane curves and the surface case.

(2) A vast geography of examples of embedded $\mathbb{Q}$-resolutions for plane curves and surfaces is detailed through the articles [**MM11, MM13, MM14, AMO14a, AMO14b, LMVV20**].

3.3.3. *A formula for (embedded) $\mathbb{Q}$-resolutions of singularities.* Our aim is to give a formula for the motivic zeta function using $\mathbb{Q}$-resolutions, similar to the one in Theorem 3.18.

This was shown in [**LMVV20**], where the authors study zeta functions and motivic integration in the context of V-manifolds and $\mathbb{Q}$-resolutions. A related study in the context of log geometry was developed in [**BN16**].

Let us first fix some notations. Let $G \subset \mathrm{GL}_d(\mathbb{C})$ be a finite abelian group of order $\ell = |G|$, and let us fix a primitive ℓth-root of unity ζ_ℓ. As in (3.3), any $g \in G$ is represented by $\mathrm{Diag}\left(\zeta_\ell^{\varepsilon_{g,1}}, \dots, \zeta_\ell^{\varepsilon_{g,d}}\right)$, with $0 \le \varepsilon_{g,i} \le \ell - 1$, $i = 1 \dots, d$.

For any tuple $\boldsymbol{k} = (k_1, \dots, k_n) \in \mathbb{Q}^n$, define the map $\varpi_{\boldsymbol{k}} : G \to \mathbb{Q}$ given by

$$\varpi_{\boldsymbol{k}}(g) = \frac{1}{\ell} \sum_{i=1}^{d} k_i \varepsilon_{g,i}. \tag{3.4}$$

Note that this extends the previous notation, since $\varpi(g) = \varpi_{\mathbf{1}}(g)$ with $\mathbf{1} = (1, \dots, 1)$.

Now, let $f : X \to \mathbb{C}$ be a non-constant regular function over a smooth variety X, and consider D the divisor defined by f. Let $h : Y \to X$ be an embedded $\mathbb{Q}$-resolution of D. In this situation, there is a natural finite stratification $Y = \bigsqcup_{k \ge 0} Y_k$ as in Remark 3.49, i.e. by small isotropy groups $G_k \subset \mathrm{GL}_n(\mathbb{C})$ acting diagonally and numerical data $\boldsymbol{N}_k = (N_{1,k}, \dots, N_{d,k}) \in \mathbb{Z}^d_{\ge 0}$, and $\boldsymbol{\nu}_k = (\nu_{1,k}, \dots, \nu_{d,k}) \in \mathbb{Z}^d_{\ge 0}$ corresponding to the multiplicities of $h^*(D)$ and K_h at the points of Y_k.

THEOREM 3.53 ([**LMVV20**, Thm. 4]). *Using the previous notation one has*

$$Z_{\mathrm{mot}}(f; s) = \mathbb{L}^{-d} \sum_{k \ge 0} [Y_k]\, S_{G_k}(\boldsymbol{N}_k, \boldsymbol{\nu}_k; s) \prod_{i=1}^{d} \frac{(\mathbb{L} - 1)\mathbb{L}^{-(N_{i,k}s + \nu_{i,k})}}{1 - \mathbb{L}^{-(N_{i,k}s + \nu_{i,k})}},$$

where

$$S_G(\boldsymbol{N}_k, \boldsymbol{\nu}_k; s) = \sum_{g \in G} \mathbb{L}^{\varpi_{\boldsymbol{N}_k}(g) \cdot s + \varpi_{\boldsymbol{\nu}_k}(g)} \in \mathcal{M}_{\mathbb{C}}[\mathbb{L}^s].$$

EXAMPLE 3.54. If $Y_k = \frac{1}{\ell}(a, b)$ with $\gcd(\ell, a, b) = 1$ and $h^{-1}(V_f)$ and K_h are locally given by $x^{N_1} y^{N_2}$ and $x^{\nu_1 - 1} y^{\nu_2 - 1}$, then the corresponding sum is expressed as

$$S_G(\boldsymbol{N},\boldsymbol{\nu};s)=\sum_{i=0}^{\ell-1}\mathbb{L}^{\frac{[ia]_\ell N_1+[ib]_\ell N_2}{\ell}\cdot s+\frac{[ia]_\ell\nu_1+[ib]_\ell\nu_2}{\ell}}$$

where $[c]_\ell$ denotes the class of c modulo ℓ satisfying $0\leq[c]_\ell\leq\ell-1$. Note that the following relation is satisfied:

$$\frac{c}{\ell}=\left\lfloor\frac{c}{\ell}\right\rfloor+\frac{[c]_\ell}{\ell}.$$

When $\mathbb{L}\to 1$, the sum $S_G(\boldsymbol{N},\boldsymbol{\nu};s)\to\ell=|G|$. Thus, we obtain a formula for the topological zeta function in terms of a $\mathbb{Q}$-resolutions after specializing by the Euler characteristic.

COROLLARY 3.55. *We have*

$$Z_{\text{top}}(f;s)=\sum_{k\geq 0}\chi\left(Y_k\right)\ell_k\prod_{i=1}^{d}\frac{1}{N_{i,k}s+\nu_{i,k}},$$

where ℓ_k denotes the order of the group G_k.

REMARK 3.56.

(1) The above formulas are also valid for the local zeta functions $Z_{\text{mot},x_0}(f;s)$ and $Z_{\text{top},x_0}(f;s)$, adapting the strata in the expressions $[Y_k]$ and $\chi(Y_k)$ by $\left[Y_k\cap h^{-1}(x_0)\right]$ and $\chi\left(Y_k\cap h^{-1}(x_0)\right)$, respectively.

(2) Since $h:Y\to X$ is an isomorphism outside the exceptional locus, the stratification $\{Y_k\}_{k\geq 0}$ is a refinement of the one defined previously by $\{E_I^\circ\}_I$ for usual resolutions, in the sense that we stratify the different intersections between the exceptional components but also by taking into account isotropy groups and multiplicities at each point. However, the interaction between the irreducible components, the isotropy groups and the multiplicities prevent us to give a simpler set-theoretical definition as in $\{E_I^\circ\}_I$.

EXAMPLE 3.57. From the $\mathbb{Q}$-resolution $\pi_{(p,q)}:\widehat{\mathbb{C}}^2_{(p,q)}\to\mathbb{C}^2$ of the family of cusps $\mathcal{C}_{p,q}:y^p-x^q=0$ in Example 3.50, we obtain the following stratification.

Stratum Y_k	Class	$\boldsymbol{N}_k$	$\boldsymbol{\nu}_k$	G_k
$Y_0=\mathbb{C}^2_{(p,q)}\setminus(E\cup\widehat{\mathcal{C}}_{p,q})$	$\mathbb{L}(\mathbb{L}-1)$	$(0,0)$	$(1,1)$	smooth
$Y_1=\widehat{\mathcal{C}}_{p,q}\setminus E$	$\mathbb{L}-1$	$(1,0)$	$(1,1)$	smooth
$Y_2=E\setminus(\widehat{\mathcal{C}}_{p,q}\cup Q_1\cup Q_2)$	$\mathbb{L}-2$	$(pq,0)$	$(p+q,1)$	smooth
$Y_3=E\cap\widehat{\mathcal{C}}_{p,q}$	1	$(pq,1)$	$(p+q,1)$	smooth
$Y_4=Q_1$	1	$(pq,0)$	$(p+q,1)$	$\frac{1}{p}(-1,q)$
$Y_5=Q_2$	1	$(0,pq)$	$(1,p+q)$	$\frac{1}{q}(p,-1)$

Note that $\widehat{\mathcal{C}}_{p,q}$ admit a parametrization by $\mathbb{C}$ similar to the ordinary cusp, since $\gcd(p,q)=1$. Now, it is easy to give an expression for $Z_{\text{top}}(f;s)$ from Corollary 3.55.

$$Z_{\text{top}}(f;s)=\frac{1}{pqs+(p+q)}\left(-1+\frac{1}{s+1}+p+q\right)=\frac{(p+q-1)s+(p+q)}{(pqs+(p+q))(s+1)}.$$

If $p=2,q=3$, we recover exactly the ordinary cusp computed in previous sections. Note also that $Z_{\text{top},0}(f;s)=Z_{\text{top}}(f;s)$, and this due to the fact that f is quasi-homogeneous.

Similarly, using Theorem 3.53 and the stratification described above, one can express $Z_{\text{mot}}(f;s)$ as the product of $\mathbb{L}^{-2}(\mathbb{L}-1)$ and

$$\frac{\mathbb{L}-\mathbb{L}^{-(s+1)}+\mathbb{L}^{-(pqs+(p+q))}\left(S_{G_4}(\boldsymbol{N}_4,\boldsymbol{\nu}_4;s)+S_{G_5}(\boldsymbol{N}_5,\boldsymbol{\nu}_5;s)-2\right)\left(1-\mathbb{L}^{-(s+1)}\right)}{(1-\mathbb{L}^{-(pqs+(p+q))})(1-\mathbb{L}^{-(s+1)})},$$

where

$$S_{G_4}(\boldsymbol{N}_4,\boldsymbol{\nu}_4;s)=1+\sum_{i=1}^{p-1}\mathbb{L}^{\frac{(p-i)pq}{p}s+\frac{(p-i)(p+q)+[iq]_p}{p}}=1+\sum_{i=1}^{p-1}\mathbb{L}^{(p-i)qs+(p+q)+\left\lfloor i\frac{q}{p}\right\rfloor},$$

$$S_{G_5}(\boldsymbol{N}_5,\boldsymbol{\nu}_5;s)=1+\sum_{j=1}^{q-1}\mathbb{L}^{\frac{(q-j)pq}{q}s+\frac{(q-j)(p+q)+[jp]_q}{q}}=1+\sum_{j=1}^{q-1}\mathbb{L}^{(q-j)ps+(p+q)+\left\lfloor j\frac{p}{q}\right\rfloor}.$$

One can recover again $Z_{\text{top}}(f;s)$ by taking $\mathbb{L}\to 1$ above.

EXERCISE 3.58.

(1) Compute $Z_{\text{top}}(f;s)$ for $\mathcal{C}_{p,q}$ without assuming that $\gcd(p,q)=1$.

(2) Give explicit expressions for $Z_{\text{mot}}(f;s)$ for the cusps $\mathcal{C}_{2,3}$, $\mathcal{C}_{10,3}$ and $\mathcal{C}_{6,3}$, using both usual resolutions and $\mathbb{Q}$-resolutions.

Concerning the Monodromy conjecture, another useful aspect about $\mathbb{Q}$-resolutions is that there exist an A'Campo formula in this context, see Theorem 3.59 below. Hence, we can use this theory to study the conjecture for other families of singularities, following the classical strategy described in Section 1.7. The following result was first proved for $d=2$ by VEYS in [**Vey97**].

THEOREM 3.59 ([**MM13**]). *Let $V_f: f=0$ be a hypersurface in $\mathbb{C}^d$ and consider $h: Y\to\mathbb{C}^d$ an embedded $\mathbb{Q}$-resolution of singularities of V_f. Let $Y=\bigcup_{k\geq 0}Y_k$ be the stratification defined in Remark 3.49. Consider the subset of strata $\{Z_m\}_{m\geq 0}\subset\{Y_k\}_{k\geq 0}$ verifying that the associated multiplicities are of the form $\boldsymbol{N}_m=(N_m,0,\ldots,0)$ up to local change of coordinates. Take G_m the small group associated with Z_m, and denote $\ell_m=|G_m|$. Then:*

$$\zeta_{f,x_0}(t)=\prod_{m\geq 0}\left(1-t^{N_m/\ell_m}\right)^{\chi\left(Z_m\cap h^{-1}(x_0)\right)},$$

EXAMPLE 3.60. Applying the previous result to the family of cusps $\mathcal{C}_{p,q}$ with $\gcd(p,q)=1$, it follows that:

$$\zeta_{f,0}(t)=\frac{(1-t^p)(1-t^q)}{1-t^{pq}}.$$

Now, we have shown that the unique non-trivial pole of $Z_{\text{top}}(f;s)$ is $s_0 = -\frac{p+q}{pq}$. Checking $t_0 = e^{2\pi i(p+q)/pq}$ in $\zeta_{f,0}(t)$, one has:

$$\frac{(1-e^{2\pi i(p+q)/q})(1-e^{2\pi i(p+q)/p})}{1-e^{2\pi i(p+q)}}$$

and it is clear that t_0 is a pole of $\zeta_{f,0}(t)$. Otherwise, either $(p+q)/q$ or $(p+q)/p$ is an integer, but this is not possible since p and q are coprimes.

We finish these notes with two examples illustrating the kind of computations we face when we perform $\mathbb{Q}$-resolutions for plane curves and surfaces. For simplicity of exposition and computations, we are going to compute the local version $Z_{\text{top},0}(f;s)$ in both cases.

EXAMPLE 3.61. Let $f(x,y) = (y^p - x^q)(y^m - x^n)$ with $(p,q) = (m,n) = 1$ and $\frac{p}{q} > \frac{m}{n}$. This polynomial define two irreducible curves in $\mathbb{C}^2$

$$\mathcal{C}_1 : y^p - x^q = 0 \quad \text{and} \quad \mathcal{C}_2 : y^m - x^n = 0.$$

We are going to perform a $\mathbb{Q}$-resolution by composing two weighted blow-ups.

The first one is $\pi_{(p,q)} : \mathbb{C}^2_{(p,q)} \to \mathbb{C}^2$ as in Example 3.50. In the first chart, our curve is given by

$$x^{q(p+m)}(y^p - 1)(y^m - x^{pn-qm}) \in \mathcal{O}_{X(p;-1,q)},$$

where the exceptional divisor $E_1 = \pi^{-1}_{(p,q)}(0)$ is of multiplicity $q(p+m)$ given in local equations by $x^{(p+m)q} = 0$, and the strict transform $\widehat{\mathcal{C}_1}$ has only a component in $X(p;-1,q)$ intersecting transversely E_1 in a smooth point. The curve $\widehat{\mathcal{C}_2}$ passes through the singular point of the chart, as is pictured in FIGURE 7. Also, the relative canonical divisor is represented by $x^{p+q-1} = 0$.

We obtain a $\mathbb{Q}$-resolution by performing a ω-blow-up

$$\pi_\omega : \widehat{X(p;-1,q)} \longrightarrow X(p;-1,q),$$

at the origin, with $\omega = (m, pn - qm)$. Following Remark 3.46, one obtains charts $\widehat{X(p;-1,q)} = \widehat{V}_0 \cup \widehat{V}_1$ by studying the action of type μ_p over those of $\mathbb{C}^2_\omega = V_0 \cup V_1$. First, we have the chart $\varphi_0 : X(m;-1,pn-qm) \to V_0$, given by

$$(x,y) \mapsto \left[(x^m, x^{pn-qm}y), [1:y]_\omega\right].$$

Take $(p;-1,q) = (pm;-m,qm)$ and $\eta \in \mu_{pm}$, we see that the action $(x,y) \mapsto (\eta^{-1}x, \eta^{pn}y)$ makes the following diagram commute:

$$\begin{array}{ccc} (x,y) & \longmapsto & (\eta^{-1}x, \eta^{pn}y) \\ \downarrow & & \downarrow \\ (x^m, x^{pn-qm}y) & \longmapsto & (\eta^{-m}x^m, \eta^{qm}x^{pn-qm}y) \end{array}$$

Thus, using the operations described in Remark 3.36 and TABLE 1, it follows that

$$\begin{aligned} \widehat{V}_0 &\overset{\varphi_0}{\cong} X\left(\begin{array}{c|cc} m & -1 & pn-qm \\ pm & -1 & pn \end{array}\right) \overset{(5)}{=} X\left(\begin{array}{c|cc} m & -1 & pn \\ pm & -1 & pn \end{array}\right) \\ &\overset{(7)}{\cong} X\left(\begin{array}{c|cc} m & -p & pn \\ pm & -p & pn \end{array}\right) \overset{(3)}{=} X\left(\begin{array}{c|cc} m & -p & pn \\ m & -1 & n \end{array}\right) \\ &= X\left(\begin{array}{c|cc} m & 0 & 0 \\ m & -1 & n \end{array}\right) = X(m;-1,n), \end{aligned}$$

where the operation **(7)** above is induced by the isomorphism $[(x,y)] \to [(x^p,y)]$. Now, using similar arguments for the other chart, one shows that

$$\widehat{V}_1 \overset{\varphi_1}{\simeq} X\left(\begin{array}{c|cc} pn-qm & m & -1 \\ p(pn-qm) & -pn & q \end{array}\right) \overset{(7)}{\simeq} X\left(\begin{array}{c|cc} pn-qm & m & -p \\ pn-qm & -n & q \end{array}\right)$$

where the last isomorphism is given by $[(x,y)] \to [(x,y^p)]$. By Bezout's identity, let $\alpha,\beta \in \mathbb{Z}$ such that $\alpha m+\beta n = \gcd(m,n) = 1$. The map $(\zeta,\eta) \mapsto (\zeta^\alpha \eta^n, \zeta^{-\beta}\eta^m))$ is an automorphism of the group acting in our space, and it induces an isomorphism

$$\begin{aligned} X\left(\begin{array}{c|cc} pn-qm & m & -p \\ pn-qm & -n & q \end{array}\right) &= X\left(\begin{array}{c|cc} pn-qm & -1 & \alpha p+\beta q \\ pn-qm & 0 & pn-qm \end{array}\right) \\ &= X(pn-qm;\ -1, \alpha p+\beta q). \end{aligned}$$

Finally, we can express the weighted blow-up in normalized charts as:

$$\begin{array}{ccc} \widehat{X(p;-1,q)} \simeq & X(m;-1,n) \cup & X(pn-qm;\ -1,\alpha p+\beta q) \\ \pi_\omega \downarrow & [(x^p,y)] & [(x,y^p)] \\ & \updownarrow & \updownarrow \\ X(p;-1,q) & [(x^m, x^{pn-qm}y)] & [(xy^m, y^{pn-qm})] \end{array}$$

The local equations of the total transform become

$$x^{m(q+n)}(y^m-1) : X(m;-1,n) \longrightarrow \mathbb{C} \quad \text{and}$$

$$x^{q(p+m)}y^{m(q+n)}(1-x^{pn-qm}) : X(pn-qm;\ -1,\alpha p+\beta q) \longrightarrow \mathbb{C}$$

The exceptional divisor $E_2 = \pi_\omega^{-1}(0)$ is a quotient space , $\mathbb{P}^1_\omega/\mu_p$, which is isomorphic to $\mathbb{P}^1$ via the map

$$\begin{array}{ccc} \mathbb{P}^1_\omega/\mu_p & \longrightarrow & \mathbb{P}^1 \\ [x:y]_\omega & \longmapsto & [x^{pn-qm} : y^m] \end{array}.$$

Now, E_2 has order $m(p+n)$, intersecting the strict transform $\widehat{\mathcal{C}}_2$ in a smooth point, and also E_1 at the origin of the second chart (see FIGURE 7). Pulling-back by $h = \pi_\omega \circ \pi_{(p,q)}$, one sees that the relative canonical divisor is of the form $K_h = (p+q-1)E_1 + (m+n-1)E_2$. Note that all the previous quotient spaces are normalized, and $h : \widehat{X(p;-1,q)} \to \mathbb{C}^2$ is an embedded $\mathbb{Q}$-resolution of $\mathcal{C}_1 + \mathcal{C}_2$.

We are going to compute $Z_{\text{top},0}(f;s)$. Define

$$(N_{E_1}, \nu_{E_1}) = (q(p+m), p+q) \quad \text{and} \quad (N_{E_2}, \nu_{E_2}) = (m(q+n), m+n)\,.$$

Following FIGURE 7, we stratify the exceptional set as:

$$h^{-1}(0) = E_1^* \sqcup E_2^* \sqcup P_1 \sqcup P_2 \sqcup Q_1 \sqcup Q_2 \sqcup Q,$$

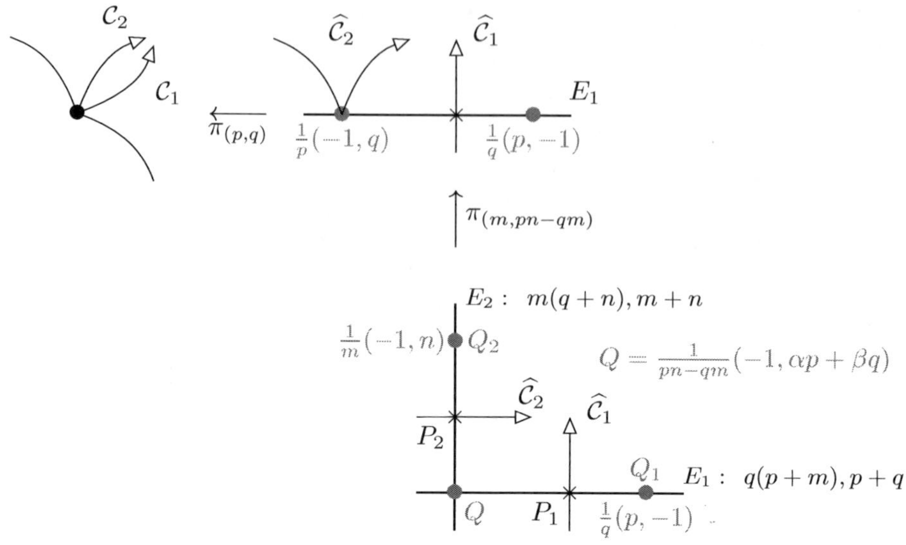

FIGURE 7. Embedded $\mathbb{Q}$-resolution of $\mathcal{C}_1 : y^p - x^q = 0$ and $\mathcal{C}_2 : y^m - x^n = 0$.

where $E_i^* \simeq \mathbb{P}^1 \setminus \{3 \text{ pts}\}$. So $\chi(E_i^*) = -1$, and we can compute the contribution of every stratum to $Z_{\text{top},0}(f;s)$ using the isotropy groups:

$$E_1^* : (-1)\frac{1}{N_{E_1}s + \nu_{E_1}}, \qquad E_2^* : (-1)\frac{1}{N_{E_2}s + \nu_{E_2}},$$

$$P_1 : \frac{1}{(s+1)(N_{E_1}s + \nu_{E_1})}, \qquad P_2 : \frac{1}{(s+1)(N_{E_2}s + \nu_{E_2})},$$

$$Q_1 : \frac{q}{N_{E_1}s + \nu_{E_1}}, \qquad Q_2 : \frac{m}{N_{E_2}s + \nu_{E_2}}, \qquad Q : \frac{pn - qm}{(N_{E_1}s + \nu_{E_1})(N_{E_3}s + \nu_{E_2})}$$

Then, simplifying one has that $Z_{\text{top},0}(f;s)$ is:

$$\frac{\big(qm(\nu_{E_1} + \nu_{E_2}) - N_{E_1} - N_{E_2}\big)s^2 + \big((qm-1)(\nu_{E_1} + \nu_{E_2}) + \nu_{E_1}\nu_{E_2}\big)s + \nu_{E_1}\nu_{E_2}}{(s+1)(N_{E_1}s + \nu_{E_1})(N_{E_2}s + \nu_{E_2})}.$$

EXERCISE 3.62. Verify that the Monodromy conjecture holds for the last example. (HINT: Study first in what cases we can have poles of multiplicity 2 or 3, then use the partial fraction decomposition to determine the poles).

EXAMPLE 3.63. Let $g(x, y, z) = x^p + y^q + z^r$ with $p, q, r \in \mathbb{N}$. Assume that (p, q, r) are pairwise coprimes, and take $\omega = (qr, pr, pq)$. Let D be the divisor in $\mathbb{C}^3$ defined by g, with an isolated singularity at the origin. This is called the 2-dimensional *Brieskorn-Pham singularity*.

Let $\pi_\omega : \widehat{\mathbb{C}}^3_\omega \to \mathbb{C}^3$ be the ω-blow-up of $\mathbb{C}^3$ at the origin, with the following description by charts:

$$
\begin{array}{ccccccc}
\widehat{\mathbb{C}}^3_\omega & \simeq & \frac{1}{qr}(-1,pr,pq) & \cup & \frac{1}{pr}(qr,-1,pq) & \cup & \frac{1}{pq}(qr,pr,-1) \\
 & & [(x,y,z)] & & [(x,y,z)] & & [(x,y,z)] \\
\pi_\omega \downarrow & & \downarrow & & \downarrow & & \downarrow \\
\mathbb{C}^2 & & (x^{qr}, x^{pr}y, x^{pq}z) & & (xy^{qr}, y^{pr}, y^{pq}z) & & (xz^{qr}, yz^{pr}, z^{pq})
\end{array}
$$

Hence, the exceptional divisor $E = \pi_\omega^{-1}(0)$ is isomorphic to $\mathbb{P}^2_\omega$. Note that $\operatorname{Sing}(\widehat{\mathbb{C}}^3_\omega)$ are three lines in generic position located in E, which can be identified with $L_u \cup L_v \cup L_w \subset \mathbb{P}^2_\omega$, the corresponding axis in coordinates $[u:v:w]_\omega$. However, it should be noticed that $\mathbb{P}^2_\omega$ is a normal surface and one can check that $L_u \cup L_v \cup L_w$ are not in the singular set of $\mathbb{P}^2_\omega$ as an abstract variety. In fact, $\operatorname{Sing}(\mathbb{P}^2_\omega)$ is contained in the union of the three origins.

Focusing in the last chart, one can study the isotropy group for $P = (a,b,c) \in \mathbb{C}^3$:

$$(\mu_{pq})_P = \left\{\zeta_{pq}^k \in \mu_{pq} \mid \left(\zeta_{pq}^{kqr}a, \zeta_{pq}^{kpr}b, \zeta_{pq}^{-k}c\right) = (a,b,c)\right\}.$$

It is clear that $(\mu_{pq})_P$ is trivial if and only if either $c = 0$ and $a, b \neq 0$, or $c \neq 0$. Also, $(\mu_{pq})_P$ is the whole μ_{pq} if and only if P is the origin. In any point of the form $P_1 = (a,0,0)$ with $a \neq 0$,

$$(\mu_{pq})_{P_1} = \left\{\zeta_{pq}^k \in \mu_{pq} \mid kqr \equiv 0 \mod pq\right\} = \left\{\zeta_{pq}^k \in \mu_{pq} \mid kr \equiv 0 \mod p\right\}.$$

Since r is invertible modulo p, there are q solutions $k = 0, p, 2p, \dots, (q-1)p$ in μ_{pq}. Hence, $|(\mu_{pq})_{P_1}| = q$. Analogously for $P_2 = (0,b,0)$ with $b \neq 0$, one has $|(\mu_{pq})_{P_2}| = p$.

In the same chart, the total transform has local equations:

$$z^{pqr}\left(x^p + y^q + 1\right) : \frac{1}{pq}(qr, pr, -1) \longrightarrow \mathbb{C}$$

where $E : z = 0$ and $\widehat{D} : x^p + y^q + 1$ is the strict transform. Also,

$$\pi^*(\mathrm{d}x \wedge \mathrm{d}y \wedge \mathrm{d}z) = z^{qr+pr+pq-1}\mathrm{d}x \wedge \mathrm{d}y \wedge \mathrm{d}z.$$

Note that $\widehat{D}$ intersects both the singular lines $x = z = 0$ and $y = z = 0$ at exactly one point. In fact, the intersection $\widehat{D} \cap E$ can be identified with the curve $\mathcal{C} : x^p + y^q + z^r = 0$ in $\mathbb{P}^2_\omega$. Moreover, by Remark 3.43, the map

$$\begin{array}{ccc} \mathbb{P}^2_\omega & \longrightarrow & \mathbb{P}^2 \\ [x:y:z]_\omega & \longmapsto & [x^p : y^q : z^r] \end{array},$$

is an isomorphism and send $\mathcal{C}$ into a projective line. Thus, both $\widehat{D}$ and E are smooth surfaces, intersecting in $\mathbb{Q}$-normal crossings, and thus $\pi_\omega : \widehat{\mathbb{C}}^3_\omega \to \mathbb{C}^3$ is an embedded $\mathbb{Q}$-resolution.

Define $(N_E, \nu_E) = (pqr, qr + pr + pq)$. Following FIGURE 8, we can describe the stratification of the exceptional locus $E = \bigcup_{k\geq 0} E_k$ as in TABLE 2. Then, simplifying:

$$Z_{\mathrm{top},0}(g;s) = \frac{(\nu_E - p - q - r - 1)s + \nu_E}{(s+1)(N_E s + \nu_E)}$$

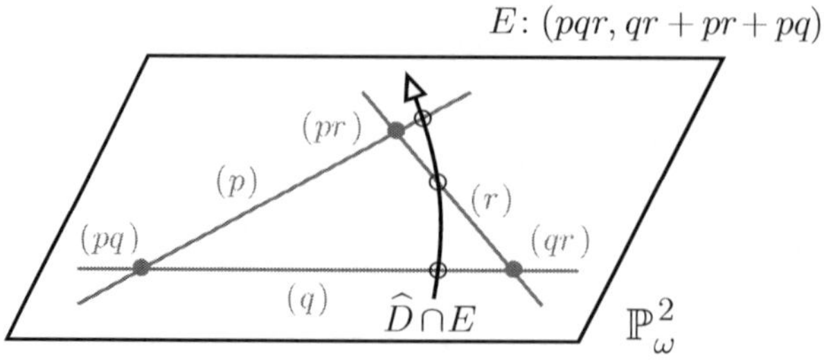

FIGURE 8. Section in the exception divisor E of the embedded $\mathbb{Q}$-resolution of $g(x,y,z) = x^p + y^q + z^r = 0$.

Stratum E_k	$\chi(E_k)$	$\boldsymbol{N}_k$	$\lvert G_k\rvert$
$E_0 = E \setminus \widehat{D}$	$\chi(\mathbb{P}^2_\omega \setminus \mathcal{C}) = 3 - \chi(\mathbb{P}^1) = 1$	$(N_E, 0, 0)$	smooth
$E_1 = \widehat{\mathcal{C}} \setminus (L_u \cup L_v \cup L_w)$	$\chi(\mathbb{P}^1 \setminus \{3 \text{ pts}\}) = -1$	$(N_E, 1, 0)$	smooth
$E_2 = L_u \setminus (\widehat{\mathcal{C}} \cup L_v \cup L_w)$	-1	$(N_E, 0, 0)$	r
$E_3 = L_v \setminus (\widehat{\mathcal{C}} \cup L_u \cup L_w)$	-1	$(N_E, 0, 0)$	q
$E_4 = L_w \setminus (\widehat{\mathcal{C}} \cup L_u \cup L_v)$	-1	$(N_E, 0, 0)$	p
$E_5 = L_u \cap \widehat{\mathcal{C}}$	1	$(N_E, 1, 0)$	r
$E_6 = L_v \cap \widehat{\mathcal{C}}$	1	$(N_E, 1, 0)$	q
$E_7 = L_w \cap \widehat{\mathcal{C}}$	1	$(N_E, 1, 0)$	p
$E_8 = L_u \cap L_v$	1	$(N_E, 0, 0)$	qr
$E_9 = L_u \cap L_w$	1	$(N_E, 0, 0)$	pq
$E_{10} = L_v \cap L_w$	1	$(N_E, 0, 0)$	pr

TABLE 2. Stratification of the exception divisor in the embedded $\mathbb{Q}$-resolution of $g(x,y,z) = x^p + y^q + z^r = 0$, following FIGURE 8. Note that $\boldsymbol{\nu}_k = (\nu_E, 1, 1)$, for any stratum E_k.

EXERCISE 3.64.

(1) Prove that the Monodromy conjecture holds for the Brieskorn-Pham singularity.

(2) Compute $Z_{\mathrm{mot},0}(g;s)$ for $p=2$, $q=3$, $r=5$.

(3) Compute $Z_{\mathrm{top},0}(g;s)$ when p,q,r are not necessarily pairwise coprime. (An embedded $\mathbb{Q}$-resolution can be found in [**MM11**, Ex. IV.2.6]).

EXERCISE 3.65 (ISHII-KOLLÁR [**IK03**]). Let $f(x_1,x_2,x_3,x_4,x_5) = x_1^3 + x_2^3 + x_3^3 + x_4^3 + x_5^6$. Compute $Z_{\mathrm{top}}(f;s)$ and $Z_{\mathrm{mot}}(f;s)$ and prove that $Z_{\mathrm{mot}}(f;s)$ has a pole that is not an actual pole of $Z_{\mathrm{top}}(f;s)$.

EXERCISE 3.66. Revisit Exercises 1.59 and 1.60 using embedded $\mathbb{Q}$-resolutions (note that usual resolutions are also $\mathbb{Q}$-resolutions).

REMARK 3.67. More sophisticated examples can be found in [**LMVV20**]. In that paper, the authors deal with zeta functions in a more general context: they define the motivic zeta function $Z_{\text{mot}}(D_1, D_2; s)$ of two $\mathbb{Q}$-Cartier divisors D_1 and D_2 on a $\mathbb{Q}$-Gorenstein variety. The idea is that D_1 and D_2 correspond to divisors associated with a pair (f, ω) as in Remark 3.26.

The function $Z_{\text{mot}}(D_1, D_2; s)$ is defined using a motivic integral as in the smooth case, but instead of using the usual motivic measure $\mu_{\mathcal{L}(X)}$, a $\mathbb{Q}$-Gorenstein motivic measure $\mu_{\mathcal{L}(X)}^{\text{QGor}}$ is introduced. This mesure is similar to one defined by YASUDA [**Yas04**] in the Gorenstein case, and also to the Denef-Loeser orbifold measure μ^{orb} [**DL02b**], and it describes $Z_{\text{mot}}(D_1, D_2; s)$ as an element in $\widehat{\mathcal{M}}_{\mathbb{C}}[\mathbb{L}^{1/m}][\![\mathbb{L}^{-s/m}]\!]$, for some $m > 0$.

The main point about taking integrals with respect to $\mu_{\mathcal{L}(X)}^{\text{QGor}}$ is that the "jacobian part" of the change of variables formula can be expressed only in terms of the order of the relative canonical divisor, instead of the order of the jacobian ideal [**LMVV20**, Thm. 2]. In this way, the formula in Theorem 3.53 holds also for $\mathbb{Q}$-resolutions $h : Y \to X$ of $\mathbb{Q}$-Cartier divisors D_1 and D_2 on a $\mathbb{Q}$-Gorenstein variety X. This gives a way to study Batyrev's stringy invariants using $\mathbb{Q}$-resolutions, and also the Monodromy conjecture in a more general context.

Acknowledgments

The author would like to thank the anonymous referee, as well as Edwin LEÓN-CARDENAL, Jorge MARTÍN-MORALES and José I. COGOLLUDO-AGUSTÍN, for their valuable comments and suggestions which helped to improve these notes. Also, to Farid TARI for proposing me to give a first mini-course on this subject at ICMC-USP in São Carlos. Finally, to Marcelo SAIA for giving me the opportunity to present a second one at the *Thematic Program on Singularity Theory 2020* held at IMPA.

References

[A'C73] Norbert A'Campo, *Le nombre de Lefschetz d'une monodromie* (French), Nederl. Akad. Wetensch. Proc. Ser. A **76** = Indag. Math. **35** (1973), 113–118. MR0320364

[A'C75] Norbert A'Campo, *La fonction zêta d'une monodromie* (French), Comment. Math. Helv. **50** (1975), 233–248, DOI 10.1007/BF02565748. MR371889

[ACLM02a] E. Artal Bartolo, P. Cassou-Noguès, I. Luengo, and A. Melle Hernández, *The Denef-Loeser zeta function is not a topological invariant*, J. London Math. Soc. (2) **65** (2002), no. 1, 45–54, DOI 10.1112/S0024610701002848. MR1875134

[ACLM02b] E. Artal Bartolo, P. Cassou-Noguès, I. Luengo, and A. Melle Hernández, *Monodromy conjecture for some surface singularities* (English, with English and French summaries), Ann. Sci. École Norm. Sup. (4) **35** (2002), no. 4, 605–640, DOI 10.1016/S0012-9593(02)01100-X. MR1981174

[ACLM05] ———, *Quasi-ordinary power series and their zeta functions*, Mem. Amer. Math. Soc. **178** (2005), no. 841, vi+85. MR2172403

[AKMW02] Dan Abramovich, Kalle Karu, Kenji Matsuki, and Jarosław Włodarczyk, *Torification and factorization of birational maps*, J. Amer. Math. Soc. **15** (2002), no. 3, 531–572, DOI 10.1090/S0894-0347-02-00396-X. MR1896232

[Alu07] Paolo Aluffi, *Celestial integration, stringy invariants, and Chern-Schwartz-MacPherson classes*, Real and complex singularities, Trends Math., Birkhäuser, Basel, 2007, pp. 1–13, DOI 10.1007/978-3-7643-7776-2_1. MR2280127

[AMO14a] Enrique Artal Bartolo, Jorge Martín-Morales, and Jorge Ortigas-Galindo, *Cartier and Weil divisors on varieties with quotient singularities*, Internat. J. Math. **25** (2014), no. 11, 1450100, 20, DOI 10.1142/S0129167X14501006. MR3285300

[AMO14b] Enrique Artal Bartolo, Jorge Martín-Morales, and Jorge Ortigas-Galindo, *Intersection theory on abelian-quotient V-surfaces and $\mathbf{Q}$-resolutions*, J. Singul. **8** (2014), 11–30, DOI 10.5427/jsing.2014.8b. MR3193225

[ATW19] D. Abramovich, M. Temkin, and J. Włodarczyk, *Functorial embedded resolution via weighted blowings up*, 2019, Preprint, avaiable at arXiv:1906.07106 [math.AG].

[Bat98] Victor V. Batyrev, *Stringy Hodge numbers of varieties with Gorenstein canonical singularities*, Integrable systems and algebraic geometry (Kobe/Kyoto, 1997), World Sci. Publ., River Edge, NJ, 1998, pp. 1–32. MR1672108

[Bat99a] Victor V. Batyrev, *Birational Calabi-Yau n-folds have equal Betti numbers*, New trends in algebraic geometry (Warwick, 1996), London Math. Soc. Lecture Note Ser., vol. 264, Cambridge Univ. Press, Cambridge, 1999, pp. 1–11, DOI 10.1017/CBO9780511721540.002. MR1714818

[Bat99b] Victor V. Batyrev, *Non-Archimedean integrals and stringy Euler numbers of log-terminal pairs*, J. Eur. Math. Soc. (JEMS) **1** (1999), no. 1, 5–33, DOI 10.1007/PL00011158. MR1677693

[Bit04] Franziska Bittner, *The universal Euler characteristic for varieties of characteristic zero*, Compos. Math. **140** (2004), no. 4, 1011–1032, DOI 10.1112/S0010437X03000617. MR2059227

[Bli11] Manuel Blickle, *A short course on geometric motivic integration*, Motivic integration and its interactions with model theory and non-Archimedean geometry. Volume I, London Math. Soc. Lecture Note Ser., vol. 383, Cambridge Univ. Press, Cambridge, 2011, pp. 189–243. MR2885337

[BMT11] Nero Budur, Mircea Mustaţă, and Zach Teitler, *The monodromy conjecture for hyperplane arrangements*, Geom. Dedicata **153** (2011), 131–137, DOI 10.1007/s10711-010-9560-1. MR2819667

[BN16] Emmanuel Bultot and Johannes Nicaise, *Computing motivic zeta functions on log smooth models*, Math. Z. **295** (2020), no. 1-2, 427–462, DOI 10.1007/s00209-019-02342-5. MR4100007

[Bor18] Lev A. Borisov, *The class of the affine line is a zero divisor in the Grothendieck ring*, J. Algebraic Geom. **27** (2018), no. 2, 203–209, DOI 10.1090/jag/701. MR3764275

[Bud12] Nero Budur, *Singularity invariants related to Milnor fibers: survey*, Zeta functions in algebra and geometry, Contemp. Math., vol. 566, Amer. Math. Soc., Providence, RI, 2012, pp. 161–187, DOI 10.1090/conm/566/11220. MR2858923

[BV16] Bart Bories and Willem Veys, *Igusa's p-adic local zeta function and the monodromy conjecture for non-degenerate surface singularities*, Mem. Amer. Math. Soc. **242** (2016), no. 1145, vii+131, DOI 10.1090/memo/1145. MR3498149

[Cam16] Jean-Baptiste Campesato, *An inverse mapping theorem for blow-Nash maps on singular spaces*, Nagoya Math. J. **223** (2016), no. 1, 162–194, DOI 10.1017/nmj.2016.29. MR3572744

[Cam17] Jean-Baptiste Campesato, *On a motivic invariant of the arc-analytic equivalence* (English, with English and French summaries), Ann. Inst. Fourier (Grenoble) **67** (2017), no. 1, 143–196. MR3612328

[CFKP19] Jean-Baptiste Campesato, Toshizumi Fukui, Krzysztof Kurdyka, and Adam Parusiński, *Arc spaces, motivic measure and Lipschitz geometry of real algebraic sets*, Math. Ann. **374** (2019), no. 1-2, 211–251, DOI 10.1007/s00208-019-01805-8. MR3961309

[CL08] Raf Cluckers and François Loeser, *Constructible motivic functions and motivic integration*, Invent. Math. **173** (2008), no. 1, 23–121, DOI 10.1007/s00222-008-0114-1. MR2403394

[CLNS18] Antoine Chambert-Loir, Johannes Nicaise, and Julien Sebag, *Motivic integration*, Progress in Mathematics, vol. 325, Birkhäuser/Springer, New York, 2018, DOI 10.1007/978-1-4939-7887-8. MR3838446

[CLS11] David A. Cox, John B. Little, and Henry K. Schenck, *Toric varieties*, Graduate Studies in Mathematics, vol. 124, American Mathematical Society, Providence, RI, 2011, DOI 10.1090/gsm/124. MR2810322

[Cra04] Alastair Craw, *An introduction to motivic integration*, Strings and geometry, Clay Math. Proc., vol. 3, Amer. Math. Soc., Providence, RI, 2004, pp. 203–225. MR2103724

[CV17] Thomas Cauwbergs and Willem Veys, *Monodromy eigenvalues and poles of zeta functions*, Bull. Lond. Math. Soc. **49** (2017), no. 2, 342–350, DOI 10.1112/blms.12003. MR3656302

[Den91] J. Denef, *Local zeta functions and Euler characteristics*, Duke Math. J. **63** (1991), no. 3, 713–721, DOI 10.1215/S0012-7094-91-06330-1. MR1121152

[Den93] J. Denef, *Degree of local zeta functions and monodromy*, Compositio Math. **89** (1993), no. 2, 207–216. MR1255694

[dF18] Tommaso de Fernex, *The space of arcs of an algebraic variety*, Algebraic geometry: Salt Lake City 2015, Proc. Sympos. Pure Math., vol. 97, Amer. Math. Soc., Providence, RI, 2018, pp. 169–197. MR3821149

[dFLNU07] Tommaso de Fernex, Ernesto Lupercio, Thomas Nevins, and Bernardo Uribe, *Stringy Chern classes of singular varieties*, Adv. Math. **208** (2007), no. 2, 597–621, DOI 10.1016/j.aim.2006.03.005. MR2304329

[DH01] Jan Denef and Kathleen Hoornaert, *Newton polyhedra and Igusa's local zeta function*, J. Number Theory **89** (2001), no. 1, 31–64, DOI 10.1006/jnth.2000.2606. MR1838703

[DL92] J. Denef and F. Loeser, *Caractéristiques d'Euler-Poincaré, fonctions zêta locales et modifications analytiques* (French), J. Amer. Math. Soc. **5** (1992), no. 4, 705–720, DOI 10.2307/2152708. MR1151541

[DL98] Jan Denef and François Loeser, *Motivic Igusa zeta functions*, J. Algebraic Geom. **7** (1998), no. 3, 505–537. MR1618144

[DL99] Jan Denef and François Loeser, *Germs of arcs on singular algebraic varieties and motivic integration*, Invent. Math. **135** (1999), no. 1, 201–232, DOI 10.1007/s002220050284. MR1664700

[DL01] Jan Denef and François Loeser, *Geometry on arc spaces of algebraic varieties*, European Congress of Mathematics, Vol. I (Barcelona, 2000), Progr. Math., vol. 201, Birkhäuser, Basel, 2001, pp. 327–348. MR1905328

[DL02a] Jan Denef and François Loeser, *Lefschetz numbers of iterates of the monodromy and truncated arcs*, Topology **41** (2002), no. 5, 1031–1040, DOI 10.1016/S0040-9383(01)00016-7. MR1923998

[DL02b] Jan Denef and François Loeser, *Motivic integration, quotient singularities and the McKay correspondence*, Compositio Math. **131** (2002), no. 3, 267–290, DOI 10.1023/A:1015565912485. MR1905024

[Dol82] Igor Dolgachev, *Weighted projective varieties*, Group actions and vector fields (Vancouver, B.C., 1981), Lecture Notes in Math., vol. 956, Springer, Berlin, 1982, pp. 34–71, DOI 10.1007/BFb0101508. MR704986

[ELM04] Lawrence Ein, Robert Lazarsfeld, and Mircea Mustaţă, *Contact loci in arc spaces*, Compos. Math. **140** (2004), no. 5, 1229–1244, DOI 10.1112/S0010437X04000429. MR2081163

[EM04] Lawrence Ein and Mircea Mustaţă, *Inversion of adjunction for local complete intersection varieties*, Amer. J. Math. **126** (2004), no. 6, 1355–1365. MR2102399

[EMY03] Lawrence Ein, Mircea Mustaţă, and Takehiko Yasuda, *Jet schemes, log discrepancies and inversion of adjunction*, Invent. Math. **153** (2003), no. 3, 519–535, DOI 10.1007/s00222-003-0298-3. MR2000468

[Fic05] Goulwen Fichou, *Motivic invariants of arc-symmetric sets and blow-Nash equivalence*, Compos. Math. **141** (2005), no. 3, 655–688, DOI 10.1112/S0010437X05001168. MR2135282

[Fic17] Goulwen Fichou, *On Grothendieck rings and algebraically constructible functions*, Math. Ann. **369** (2017), no. 1-2, 761–795, DOI 10.1007/s00208-017-1564-9. MR3694660

[GLM16] Manuel González Villa, Anatoly Libgober, and Laurenţiu Maxim, *Motivic infinite cyclic covers*, Adv. Math. **298** (2016), 413–447, DOI 10.1016/j.aim.2016.04.019. MR3505746

[GLM18] Manuel González Villa, Anatoly Libgober, and Laurenţiu Maxim, *Motivic zeta functions and infinite cyclic covers*, Local and global methods in algebraic geometry, Contemp. Math., vol. 712, Amer. Math. Soc., [Providence], RI, [2018] ©2018, pp. 117–141, DOI 10.1090/conm/712/14345. MR3832402

[Gre66] Marvin J. Greenberg, *Rational points in Henselian discrete valuation rings*, Inst. Hautes Études Sci. Publ. Math. **31** (1966), 59–64. MR207700

[Hal50] Paul R. Halmos, *Measure Theory*, D. Van Nostrand Company, Inc., New York, N. Y., 1950. MR0033869

[Har77] Robin Hartshorne, *Algebraic geometry*, Graduate Texts in Mathematics, No. 52, Springer-Verlag, New York-Heidelberg, 1977. MR0463157

[Hir64] Heisuke Hironaka, *Resolution of singularities of an algebraic variety over a field of characteristic zero. I, II*, Ann. of Math. (2) **79** (1964), 109–203; ibid. (2) **79** (1964), 205–326, DOI 10.2307/1970547. MR0199184

[HL00] M. Hoornaert and D. Loots, *Computer program written in Maple for the calculation of Igusa's local zeta function*, 2000, Available at `http://www.wis.kuleuven.ac.be/algebra/kathleen.htm`.

[HL15] Ehud Hrushovski and François Loeser, *Monodromy and the Lefschetz fixed point formula* (English, with English and French summaries), Ann. Sci. Éc. Norm. Supér. (4) **48** (2015), no. 2, 313–349, DOI 10.24033/asens.2246. MR3346173

[Igu74] Jun-ichi Igusa, *Complex powers and asymptotic expansions. I. Functions of certain types*, J. Reine Angew. Math. **268(269)** (1974), 110–130, DOI 10.1515/crll.1974.268-269.110. MR347753

[Igu00] Jun-ichi Igusa, *An introduction to the theory of local zeta functions*, AMS/IP Studies in Advanced Mathematics, vol. 14, American Mathematical Society, Providence, RI; International Press, Cambridge, MA, 2000, DOI 10.1090/amsip/014. MR1743467

[IK03] Shihoko Ishii and János Kollár, *The Nash problem on arc families of singularities*, Duke Math. J. **120** (2003), no. 3, 601–620, DOI 10.1215/S0012-7094-03-12034-7. MR2030097

[Kas83] Masaki Kashiwara, *Holonomic systems of linear differential equations with regular singularities and related topics in topology*, Algebraic varieties and analytic varieties (Tokyo, 1981), Adv. Stud. Pure Math., vol. 1, North-Holland, Amsterdam-New York, 1983, pp. 49–54. MR715645

[Kas77] Masaki Kashiwara, *B-functions and holonomic systems. Rationality of roots of B-functions*, Invent. Math. **38** (1976/77), no. 1, 33–53, DOI 10.1007/BF01390168. MR430304

[KM98] János Kollár and Shigefumi Mori, *Birational geometry of algebraic varieties*, Cambridge Tracts in Mathematics, vol. 134, Cambridge University Press, Cambridge, 1998. With the collaboration of C. H. Clemens and A. Corti; Translated from the 1998 Japanese original, DOI 10.1017/CBO9780511662560. MR1658959

[Kol13] János Kollár, *Singularities of the minimal model program*, Cambridge Tracts in Mathematics, vol. 200, Cambridge University Press, Cambridge, 2013. With a collaboration of Sándor Kovács, DOI 10.1017/CBO9781139547895. MR3057950

[Kon95] M. Kontsevich, *Lecture at Orsay.*

[KP03] S. Koike and A. Parusiński, *Motivic-type invariants of blow-analytic equivalence* (English, with English and French summaries), Ann. Inst. Fourier (Grenoble) **53** (2003), no. 7, 2061–2104. MR2044168

[LC22] E. León-Cardenal, *Archimedean zeta functions and oscillatory integrals*, p-Adic Analysis, Arithmetic and Singularities (Carlos Galindo, Alejandro Melle Hernández, Julio José Moyano-Fernández, and Wilson A. Zúñiga-Galindo, eds.), Contemporary Mathematics, vol. 778, Amer. Math. Soc., Providence, RI, 2022, to appear.

[Lê79] Lê Dũng Tráng, *Sur les cycles évanouissants des espaces analytiques* (French, with English summary), C. R. Acad. Sci. Paris Sér. A-B **288** (1979), no. 4, A283–A285. MR524794

[LL14] M. Larsen and V. A. Lunts, *Rationality of motivic zeta function and cut-and-paste problem*, 2014, Preprint, avaiable at `arXiv:abs/1410.7099 [math.AG]`.

[LMVV20] Edwin León-Cardenal, Jorge Martín-Morales, Willem Veys, and Juan Viu-Sos, *Motivic zeta functions on $\mathbb{Q}$-Gorenstein varieties*, Adv. Math. **370** (2020), 107192, 34, DOI 10.1016/j.aim.2020.107192. MR4100755

[Loe88] F. Loeser, *Fonctions d'Igusa p-adiques et polynômes de Bernstein* (French), Amer. J. Math. **110** (1988), no. 1, 1–21, DOI 10.2307/2374537. MR926736

[Loe90] François Loeser, *Fonctions d'Igusa p-adiques, polynômes de Bernstein, et polyèdres de Newton* (French), J. Reine Angew. Math. **412** (1990), 75–96, DOI 10.1515/crll.1990.412.75. MR1079002

[Loo02] Eduard Looijenga, *Motivic measures*, Astérisque **276** (2002), 267–297. Séminaire Bourbaki, Vol. 1999/2000. MR1886763

[LS10] Qing Liu and Julien Sebag, *The Grothendieck ring of varieties and piecewise isomorphisms*, Math. Z. **265** (2010), no. 2, 321–342, DOI 10.1007/s00209-009-0518-7. MR2609314

[LZ19] Edwin León-Cardenal and W. A. Zúñiga-Galindo, *An introduction to the theory of local zeta functions from scratch* (English, with English and Spanish summaries), Rev. Integr. Temas Mat. **37** (2019), no. 1, 45–76, DOI 10.18273/revint. MR3920742

[Mal83] B. Malgrange, *Polynômes de Bernstein-Sato et cohomologie évanescente* (French), Analysis and topology on singular spaces, II, III (Luminy, 1981), Astérisque, vol. 101, Soc. Math. France, Paris, 1983, pp. 243–267. MR737934

[McQ20] Michael McQuillan, *Very functorial, very fast, and very easy resolution of singularities*, Geom. Funct. Anal. **30** (2020), no. 3, 858–909, DOI 10.1007/s00039-020-00523-7. MR4135673

[Meu16] Diane Meuser, *A survey of Igusa's local zeta function*, Amer. J. Math. **138** (2016), no. 1, 149–179, DOI 10.1353/ajm.2016.0006. MR3462883

[Mil68] John Milnor, *Singular points of complex hypersurfaces*, Annals of Mathematics Studies, No. 61, Princeton University Press, Princeton, N.J.; University of Tokyo Press, Tokyo, 1968. MR0239612

[MM11] J. Martín-Morales, *Embedded* $\mathbf{Q}$*-resolutions and Yomdin-Lê surface singularities*, Ph.D thesis dissertation at Universidad de Zaragoza (2011), Available at `http://zaguan.unizar.es/record/6870`.

[MM13] Jorge Martín-Morales, *Monodromy zeta function formula for embedded* $\mathbf{Q}$*-resolutions*, Rev. Mat. Iberoam. **29** (2013), no. 3, 939–967, DOI 10.4171/RMI/745. MR3090142

[MM14] Jorge Martín-Morales, *Embedded* $\mathbf{Q}$*-resolutions for Yomdin-Lê surface singularities*, Israel J. Math. **204** (2014), no. 1, 97–143, DOI 10.1007/s11856-014-1078-z. MR3273453

[Mus01] Mircea Mustaţă, *Jet schemes of locally complete intersection canonical singularities*, Invent. Math. **145** (2001), no. 3, 397–424, DOI 10.1007/s002220100152. With an appendix by David Eisenbud and Edward Frenkel. MR1856396

[Nas95] John F. Nash Jr., *Arc structure of singularities*, Duke Math. J. **81** (1995), no. 1, 31–38 (1996), DOI 10.1215/S0012-7094-95-08103-4. A celebration of John F. Nash, Jr. MR1381967

[Nic10] Johannes Nicaise, *An introduction to p-adic and motivic zeta functions and the monodromy conjecture*, Algebraic and analytic aspects of zeta functions and L-functions, MSJ Mem., vol. 21, Math. Soc. Japan, Tokyo, 2010, pp. 141–166. MR2647606

[NV10] András Némethi and Willem Veys, *Monodromy eigenvalues are induced by poles of zeta functions: the irreducible curve case*, Bull. Lond. Math. Soc. **42** (2010), no. 2, 312–322, DOI 10.1112/blms/bdp128. MR2601558

[NV12] András Némethi and Willem Veys, *Generalized monodromy conjecture in dimension two*, Geom. Topol. **16** (2012), no. 1, 155–217, DOI 10.2140/gt.2012.16.155. MR2872581

[Oes82] Joseph Oesterlé, *Réduction modulo* p^n *des sous-ensembles analytiques fermés de* $\mathbf{Z}_p^N$ (French), Invent. Math. **66** (1982), no. 2, 325–341, DOI 10.1007/BF01389398. MR656627

[Ost16] Alexander Ostrowski, *Über einige Lösungen der Funktionalgleichung* $\psi(x)\cdot\psi(x) = \psi(xy)$ (German), Acta Math. **41** (1916), no. 1, 271–284, DOI 10.1007/BF02422947. MR1555153

[Poo02] Bjorn Poonen, *The Grothendieck ring of varieties is not a domain*, Math. Res. Lett. **9** (2002), no. 4, 493–497, DOI 10.4310/MRL.2002.v9.n4.a8. MR1928868

[Pop] M. Popa, *Course on modern aspects of the cohomological study of varieties*, Course notes, Available at `http://www.math.northwestern.edu/~mpopa/571/index.html`.

[Pri67] David Prill, *Local classification of quotients of complex manifolds by discontinuous groups*, Duke Math. J. **34** (1967), 375–386. MR210944

[PV22] N. Potemans and W. Veys, *Introduction to p-adic Igusa zeta functions*, p-Adic Analysis, Arithmetic and Singularities (Carlos Galindo, Alejandro Melle Hernández, Julio José Moyano-Fernández, and Wilson A. Zúñiga-Galindo, eds.), Contemporary Mathematics, vol. 778, Amer. Math. Soc., Providence, RI, 2022, to appear

[Rei87] Miles Reid, *Young person's guide to canonical singularities*, Algebraic geometry, Bowdoin, 1985 (Brunswick, Maine, 1985), Proc. Sympos. Pure Math., vol. 46, Amer. Math. Soc., Providence, RI, 1987, pp. 345–414. MR927963

[Rod04a] B. Rodrigues, *Geometric determination of the poles of highest and second highest order of Hodge and motivic zeta functions*, Nagoya Math. J. **176** (2004), 1–18, DOI 10.1017/S0027763000008977. MR2108122

[Rod04b] B. Rodrigues, *On the monodromy conjecture for curves on normal surfaces*, Math. Proc. Cambridge Philos. Soc. **136** (2004), no. 2, 313–324, DOI 10.1017/S0305004103007382. MR2040577

[RV01] B. Rodrigues and W. Veys, *Holomorphy of Igusa's and topological zeta functions for homogeneous polynomials*, Pacific J. Math. **201** (2001), no. 2, 429–440, DOI 10.2140/pjm.2001.201.429. MR1875902

[Sea19] José Seade, *On Milnor's fibration theorem and its offspring after 50 years*, Bull. Amer. Math. Soc. (N.S.) **56** (2019), no. 2, 281–348, DOI 10.1090/bull/1654. MR3923346

[Ste77] J. H. M. Steenbrink, *Mixed Hodge structure on the vanishing cohomology*, Real and complex singularities (Proc. Ninth Nordic Summer School/NAVF Sympos. Math., Oslo, 1976), Sijthoff and Noordhoff, Alphen aan den Rijn, 1977, pp. 525–563. MR0485870

[Var76] A. N. Varchenko, *Zeta-function of monodromy and Newton's diagram*, Invent. Math. **37** (1976), no. 3, 253–262, DOI 10.1007/BF01390323. MR424806

[Vey97] Willem Veys, *Zeta functions for curves and log canonical models*, Proc. London Math. Soc. (3) **74** (1997), no. 2, 360–378, DOI 10.1112/S0024611597000130. MR1425327

[Vey06] Willem Veys, *Arc spaces, motivic integration and stringy invariants*, Singularity theory and its applications, Adv. Stud. Pure Math., vol. 43, Math. Soc. Japan, Tokyo, 2006, pp. 529–572, DOI 10.2969/aspm/04310529. MR2325153

[Vil89] Orlando Villamayor, *Constructiveness of Hironaka's resolution*, Ann. Sci. École Norm. Sup. (4) **22** (1989), no. 1, 1–32, DOI 10.24033/asens.1573. MR985852

[VS12] J. Viu-Sos, *Computer program calculating the (local) Igusa and Topological zeta functions of a non-degenerated polynomial with respect to his Newton polyhedron, written in Sage*, 2012, Available at `https://jviusos.github.io/sage.html`.

[Wło03] Jarosław Włodarczyk, *Toroidal varieties and the weak factorization theorem*, Invent. Math. **154** (2003), no. 2, 223–331, DOI 10.1007/s00222-003-0305-8. MR2013783

[Yas04] Takehiko Yasuda, *Twisted jets, motivic measures and orbifold cohomology*, Compos. Math. **140** (2004), no. 2, 396–422, DOI 10.1112/S0010437X03000368. MR2027195

Dpto. de Matemáticas e Informática, ETSI Caminos Canales y Puertos, Universidad Politécnica de Madrid, C\Prof. Aranguren 3, 28040 Madrid, Spain

Email address: `jviusos@math.cnrs.fr`

URL: `https://jviusos.github.io/`

Contemporary Mathematics
Volume **778**, 2022
https://doi.org/10.1090/conm/778/15657

p-Adic analysis: A quick introduction

W. A. Zúñiga-Galindo

ABSTRACT. These notes aim to provide a fast introduction to p-adic analysis assuming basic knowledge in algebra and analysis. The text corresponds to the lecture notes for a Mini-Course in the L. Santaló Research Summer School 2019. Palacio de la Magdalena, Santander, Spain, June 24-28, 2019.

CONTENTS

1. Introduction

These notes aim to provide a fast introduction to p-adic analysis assuming basic knowledge in algebra and analysis. We cannot provide detailed proofs. For an in-depth discussion, the reader may consult [**1**], [**23**], [**24**], see also [**12**], [**14**], [**21**], [**22**]. We focus on basic aspects of analysis involving complex-valued functions.

In the last thirty years p-adic analysis has received great attention due to its connections with physics, biology, cryptography, and several mathematical theories, see e.g. [**1**], [**2**], [**17**], [**18**], [**15**], [**16**], [**24**], [**25**], and the references therein. As a consequence of all this, nowadays, p-adic analysis is having a tremendous expansion. Let us mention a couple examples. First, the developing of the theory of p-adic pseudodifferential equations, which is a theory connected with several fields, see e.g. [**1**], [**18**], [**15**], [**16**], [**24**], [**25**] and the reference therein. Second, the deep

2020 *Mathematics Subject Classification.* Primary 43A70, 46S10, 11S40, 60G53.
The author was partially supported by UTRGV through the Debnath Endowed Professorship.

connection between local zeta functions and string amplitudes, see e.g. [**5**], [**6**], and the references therein.

The L. Santaló Research School 2019 aims to provide an introduction to the area of local zeta functions. In the Archimedean case, $\mathbb{K} = \mathbb{R}$ or $\mathbb{C}$, the study of local zeta functions was initiated by Gel'fand and Shilov [**11**]. The meromorphic continuation of the local zeta functions was established, independently, by Atiyah [**3**] and Bernstein [**4**], see also [**10**, Theorem 5.5.1 and Corollary 5.5.1]. The main motivation was that the meromorphic continuation of Archimedean local zeta functions implies the existence of fundamental solutions (i.e. Green functions) for differential operators with constant coefficients. It is important to mention here, that in the p-adic framework, the existence of fundamental solutions for pseudodifferential operators is also a consequence of the fact that the Igusa local zeta functions admit a meromorphic continuation, see [**18**, Chapter 10] and [**25**, Chapter 5]. In the 70s, Igusa developed a uniform theory for local zeta functions over local fields of characteristic zero [**10**]. For an introduction to the basic aspects of local zeta functions the reader may consult [**20**].

2. p-Adic numbers: essential facts

2.1. Basic facts. In this section we summarize the basic aspects of the field of p-adic numbers, for an in-depth discussion the reader may consult [**1**, **12**–**14**, **21**–**23**] and [**24**].

DEFINITION 2.1. Let F be a field. A norm (or an absolute value) on F is a real-valued function, $|\cdot|$, satisfying

(i) $|x| = 0 \Leftrightarrow x = 0$;
(ii) $|xy| = |x||y|$;
(iii) $|x + y| \leq |x| + |y|$ (*triangle inequality*), for any $x, y \in F$.

DEFINITION 2.2. A norm $|\cdot|$ is called non-Archimedean (or ultrametric), if it satisfies

$$|x + y| \leq \max\{|x|, |y|\}. \tag{2.1}$$

Notice that (2.1) implies the triangle inequality.

EXAMPLE 2.3. The trivial norm is defined as

$$|x|_{\text{trivial}} = \begin{cases} 1 & \text{if } x \neq 0 \\ 0 & \text{if } x = 0. \end{cases}$$

From now on we will work only with non-trivial norms.

DEFINITION 2.4. Let p be a fixed prime number, and let x be a nonzero rational number. Then, $x = p^k \frac{a}{b}$, with $p \nmid ab$, and $k \in \mathbb{Z}$. The p-adic absolute value (or p-adic norm) of x is defined as

$$|x|_p = \begin{cases} p^{-k} & \text{if } x \neq 0 \\ 0 & \text{if } x = 0. \end{cases}$$

EXERCISE 2.5. The function $|\cdot|_p$ is a non-Archimedean norm on $\mathbb{Q}$. In addition, show that $|x + y|_p = \max\{|x|_p, |y|_p\}$ when $|x|_p \neq |y|_p$.

We set $\mathbb{R}_+ := \{x \in \mathbb{R};\ x \geq 0\}$. We denote by $\mathbb{N}$ the set of non-negative integers.

DEFINITION 2.6. Let X be a non-empty set. A distance, or metric, on X is a function $d : X \times X \to \mathbb{R}_+$ satisfying the following properties:

(i) $d(x,y) = 0$ if and only if $x = y$;
(ii) $d(x,y) = d(y,x)$;
(iii) $d(x,y) \leq d(x,z) + d(z,y)$ for any x, y, z in X.

The pair (X,d) is called a metric space.

EXAMPLE 2.7. Let F be a field endowed with a norm $|\cdot|$. The distance $d(x,y) := |x-y|$, for x, y in F, is called the induced distance by $|\cdot|$. The pair (F,d) is a metric space.

DEFINITION 2.8. Let (X,d) be a metric space. The metric d is called non-Archimedean if

$$d(x,y) \leq \max\{d(x,z), d(z,y)\} \text{ for any } x,\ y,\ z \in X.$$

EXAMPLE 2.9. Take $X = \mathbb{Q}$, and d the distance induced by the p-adic norm $|\cdot|_p$, for a fixed prime p. Then d is non-Archimedean.

DEFINITION 2.10. Let (X,d) be a metric space. A sequence $\{a_i\}_{i\in\mathbb{N}}$ in X is called a Cauchy sequence, if for any $\epsilon > 0$ there exists N such that $d(a_m, a_n) < \epsilon$ whenever both $m > N$, $n > N$.

DEFINITION 2.11. Two metrics d_1 and d_2 on a set X are called equivalent if a sequence is Cauchy with respect to d_1 if and only if it is Cauchy with respect to d_2. We say that two norms are equivalent if they induce equivalent metrics.

EXERCISE 2.12. Let α be a fixed positive real number. For $x \in \mathbb{Q}$, we define $\|x\| = |x|_\infty^\alpha$, where $|\cdot|_\infty$ denotes the standard absolute value. Show that $\|\cdot\|$ is a norm if and only if $\alpha \leq 1$, and that in that case it is equivalent to the norm $|\cdot|_\infty$.

THEOREM 2.13 (Ostrowski, [**1**, Theorem 1.3.2]). *Any non trivial absolute value on $\mathbb{Q}$ is equivalent to $|\cdot|_p$ or to the standard absolute value $|\cdot|_\infty$.*

REMARK 2.14. (i) Let F be a field endowed with a norm $|\cdot|$. We introduce a topology on F by giving a basis of open sets consisting of the *open balls* $B_r(a)$ with center a and radius $r > 0$:

$$B_r(a) = \{x \in F; |x-a| < r\}.$$

(ii) A sequence of points $\{x_i\}_{i\in\mathbb{N}} \subset F$ is called *Cauchy* if

$$|x_m - x_n| \to 0, \qquad m,n \to \infty.$$

(iii) A field F with a non trivial absolute value $|\cdot|$ is said to be *complete* if any Cauchy sequence $\{x_i\}_{i\in\mathbb{N}}$ has a limit point $x^* \in F$, i.e. if $|x_n - x^*| \to 0$, $n \to \infty$. This is equivalent to the fact that (F,d), with $d(x,y) = |x-y|$, is *a complete metric space.*

(iv) Let (X,d), (Y,D) be two metric spaces. A bijection $\rho : X \to Y$ satisfying

$$D(\rho(x), \rho(x')) = d(x,x'),\ x,\ x' \in X,$$

is called an *isometry*.

The following fact is well-known, see e.g. [**19**].

THEOREM 2.15. *Let (M, d) be a metric space. There exists a complete metric space $(\widetilde{M}, \widetilde{d})$, such that M is isometric to a dense subset of $\widetilde{M}$. This space $\widetilde{M}$ is unique up to isometries, that is, if $\widetilde{M_0}$ is a complete metric space having M as a dense subspace, then $\widetilde{M_0}$ is isometric to $\widetilde{M}$.*

EXERCISE 2.16. Let $(F, |\cdot|)$ be a valued field, where $|\cdot|$ is a non-Archimedean absolute value. Assume that F is complete with respect to $|\cdot|$. Then, the series $\sum_{k\geq 0} a_k$, $a_k \in F$ converges if an only if $\lim_{k\to\infty} |a_k| = 0$.

2.2. The field of p-adic numbers. We set

$$\mathbb{Q}_p := \left\{ x = p^{\gamma} \sum_{i=0}^{\infty} x_i p^i;\ \gamma \in \mathbb{Z},\ x_i \in \{0, 1, \dots, p-1\},\ x_0 \neq 0 \right\} \cup \{0\},$$

and $ord(x) := \gamma$, $|x|_p := p^{-\gamma}$ for $x \in \mathbb{Q}_p \smallsetminus \{0\}$. We set $|0|_p := 0$ and $ord(0) := +\infty$.

LEMMA 2.17. *With the above notation, the following assertions hold true:*

(i) *$\mathbb{Q}_p$ is a field of characteristic zero;*
(ii) *$|\cdot|_p : \mathbb{Q}_p \to \mathbb{R}_+$ is a norm;*
(iii) *$(\mathbb{Q}_p, |\cdot|_p)$ is a complete metric space;*
(iv) *$\mathbb{Q}$ is dense in $\mathbb{Q}_p$;*
(v) *the completion of $(\mathbb{Q}, |\cdot|_p)$ is $(\mathbb{Q}_p, |\cdot|_p)$.*

PROOF. (i) We set

$$\mathbb{Z}_p := \left\{ x \in \mathbb{Q}_p;\ x = p^{\gamma} \sum_{i=0}^{\infty} x_i p^i; \gamma \in \mathbb{N},\ x_0 \neq 0 \right\}.$$

We first show that $\mathbb{Z}_p$ is a ring. Take

$$x = p^{\alpha} \sum_{i=0}^{\infty} x_i p^i \text{ with } \alpha \in \mathbb{N},\ x_0 \neq 0, \quad y = p^{\beta} \sum_{i=0}^{\infty} y_i p^i \text{ with } \beta \in \mathbb{N},\ y_0 \neq 0.$$

And set $\gamma = \min\{\alpha, \beta\}$ and

$$z = p^{\gamma} \sum_{i=0}^{\infty} z_i p^i \text{ with } \gamma \in \mathbb{N},\ z_0 \neq 0.$$

We now define the digits z_is by the following recursive formulae:

$$p^{\gamma} \sum_{i=0}^{L-1} z_i p^i \equiv p^{\alpha} \sum_{i=0}^{L-1} x_i p^i + p^{\beta} \sum_{i=0}^{L-1} y_i p^i \bmod p^L \text{ for } L \geq \gamma + 1. \tag{2.2}$$

Here $A \equiv B \bmod p^L$ means p^L divides $A - B$. Now, we define $x + y = z$. Notice that (2.2) determines uniquely all the digits z_i's. For the product xy, we set

$$w = p^{\alpha+\beta} \sum_{i=0}^{\infty} w_i p^i \text{ with } w_0 \neq 0.$$

Then the digits w_is are uniquely determined by the following recursive formulae:

$$p^{\alpha+\beta} \sum_{i=0}^{L-1} w_i p^i \equiv \left(p^{\alpha} \sum_{i=0}^{L-1} x_i p^i \right) \left(p^{\beta} \sum_{i=0}^{L-1} y_i p^i \right) \bmod p^L \text{ for } L \geq \alpha + \beta + 1. \tag{2.3}$$

Now we define $xy = w$. It is not difficult to verify that $(\mathbb{Z}_p, +, \cdot)$ is a commutative ring with unity. Furthermore, by using (2.3) one verifies that $xy = 0$ implies that

$x = 0$ or $y = 0$. This means that $\mathbb{Z}_p$ is a domain. Finally, we notice that the field of fractions of $\mathbb{Z}_p$ is precisely $\mathbb{Q}_p$. In order to verify this assertion it is necessary to use Exercise 2.19.

(ii) By definition $|x|_p = 0$ if and only if $x = 0$. Now given $x = p^{\alpha}\widetilde{x}$, with $|\widetilde{x}|_p = 1$ and $y = p^{\beta}\widetilde{y}$, with $|\widetilde{y}|_p = 1$, see Exercise 2.19-(iii), we have $xy = p^{\alpha+\beta}\widetilde{z}$, with $|\widetilde{z}|_p = 1$, see Exercise 2.19-(ii). Then $|xy|_p = |x|_p\,|y|_p$. Finally, if x,$y \in \mathbb{Z}_p$, then $|x+y|_p \le p^{-\min\{\alpha,\beta\}} = \max\left\{|x|_p\,,\ |y|_p\right\}$, by using the definition of the sum operation. In the general case, with $\alpha \le \beta$,

$$\begin{aligned}|x+y|_p &= \left|p^{\alpha}\left(\widetilde{x}+p^{\beta-\alpha}\widetilde{y}\right)\right|_p = |p^{\alpha}|_p\left|\widetilde{x}+p^{\beta-\alpha}\widetilde{y}\right|_p \le p^{-\alpha}\max\left\{|\widetilde{x}|_p\,, \left|p^{\beta-\alpha}\widetilde{y}\right|_p\right\}\\ &= p^{-\alpha}\max\left\{1, p^{-(\beta-\alpha)}\right\} = p^{-\alpha} = \max\left\{|x|_p\,,\ |y|_p\right\}.\end{aligned}$$

(iii) Let $\left\{x^{(m)}\right\}_{m\in\mathbb{N}}$ be a Cauchy sequence in $\mathbb{Q}_p$, with

$$x^{(m)} = p^{\gamma_m}\sum_{i=0}^{\infty} x_i^{(m)}p^i.$$

Then, given L there exists $M \in \mathbb{N}$ such that

$$\left|x^{(m)} - x^{(n)}\right|_p < p^{-L} \text{ for } m \ge n > M \Leftrightarrow ord(x^{(m)} - x^{(n)}) > L \text{ for } m \ge n > M,$$

which implies the existence of $\gamma \in \mathbb{Z}$, $x_i \in \{0, 1, \dots, p-1\}$ for $i = \gamma, \dots, L$ such that

$$x^{(m)} = \sum_{i=\gamma}^{L} x_i p^i + p^{L+1}e^{(m)} = x_L + p^{L+1}e^{(m)}, \text{ for } m > M.$$

Then there exists $x = p^{\gamma}\sum_{i=0}^{\infty} x_i p^i = \lim_{L\to\infty} x_L \in \mathbb{Q}_p$ such that

$$\left|x - x^{(m)}\right|_p < p^{-L} \text{ for } m > M.$$

Which implies that $x^{(m)} \to x$.

(iv) Any $x \in \mathbb{Q}_p \smallsetminus \{0\}$ can be written as $x = p^{\gamma}\widetilde{x}$, with $\widetilde{x} \in \mathbb{Z}_p$ and $|\widetilde{x}|_p = 1$. Given p^{-L}, with $L \in \mathbb{N}$, there exists $\frac{a}{b} \in \mathbb{Q}$ such that $\left|x - \frac{a}{b}\right|_p < p^{-L}$. We take $b^{-1} = p^{\gamma}$ and $a \in \mathbb{Z}$ satisfying $|a - \widetilde{x}|_p < p^{-L+\gamma}$.

(v) It follows from (i)-(iii) by using Theorem 2.15. □

As a consequence of Lemma 2.17, the field of p-adic numbers $\mathbb{Q}_p$ is the completion of $\mathbb{Q}$ with respect to the distance induced by $|\cdot|_p$. Also, by Lemma 2.17, any p-adic number $x \neq 0$ has a unique representation of the form

$$x = p^{\gamma}\sum_{i=0}^{\infty} x_i p^i,$$

where $\gamma = \gamma(x) \in \mathbb{Z}$, $x_i \in \{0, 1, \dots, p-1\}$, $x_0 \neq 0$. The integer γ is called the *p-adic order of x,* and it will be denoted as $ord(x)$. By definition $ord(0) = +\infty$.

REMARK 2.18. The unit ball

$$\mathbb{Z}_p = \{x \in \mathbb{Q}_p;\ |x|_p \le 1\} = \{x \in \mathbb{Q}_p;\ x = \sum_{i=i_0}^{\infty} x_i p^i, i_0 \ge 0\},$$

is a domain of principal ideals. Any ideal of $\mathbb{Z}_p$ has the form

$$p^m\mathbb{Z}_p = \{x \in \mathbb{Z}_p; x = \sum_{i \geq m} x_i p^i\}, \ m \in \mathbb{N}.$$

Indeed, let $I \subseteq \mathbb{Z}_p$ be an ideal. Set $m_0 = \min_{x \in I} ord(x) \in \mathbb{N}$, and let $x_0 \in I$ such that $ord(x_0) = m_0$. Then $I = x_0\mathbb{Z}_p$. From a geometric point of view, the ideals of the form $p^m\mathbb{Z}_p$, $m \in \mathbb{Z}$, constitute a fundamental system of neighborhoods around the origin in $\mathbb{Q}_p$.

The *residue field* of $\mathbb{Q}_p$ is $\mathbb{Z}_p/p\mathbb{Z}_p \cong \mathbb{F}_p$ (the finite field with p elements). The *group of units* of $\mathbb{Z}_p$ is

$$\mathbb{Z}_p^{\times} = \{x \in \mathbb{Z}_p; |x|_p = 1\}.$$

EXERCISE 2.19. (i) Show that $x = x_0 + x_1 p + \cdots \in \mathbb{Z}_p$ is a unit if and only if $x_0 \neq 0$. (ii) $\mathbb{Z}_p^{\times}$ is an Abelian group under the multiplication operation defined in the proof of Lemma 2.17-(i). (iii) If $x \in \mathbb{Q}_p \setminus \{0\}$, then $x = p^m u$, $m \in \mathbb{Z}$, $u \in \mathbb{Z}_p^{\times}$. **Hint**. Suppose that $x_0 \neq 0$. One has to show the existence of an element $y = y_0 + y_1 p + \cdots \in \mathbb{Z}_p^{\times}$ such that $xy = 1$. The digits y_is are determined recursively by using the formulae $xy \equiv 1 \bmod p^L$, $L \geq 1$. Notice that $x_0 y_0 \equiv 1 \bmod p$ has a solution $y_0 \in \{1, \ldots, p-1\}$.

EXAMPLE 2.20. The formula $\frac{-1}{(p-1)} = \sum_{i=0}^{\infty} p^i$ holds true in $\mathbb{Q}_p$, i.e.

$$-1 = (p-1) + (p-1)p + (p-1)p^2 + \cdots$$

Indeed, set

$$\begin{aligned} z^{(n)} &:= (p-1) + (p-1)p + \cdots + (p-1)p^n \\ &= (p-1)\frac{p^{n+1}-1}{p-1} = p^{n+1} - 1. \end{aligned}$$

Then $\lim_{n \to \infty} z^{(n)} = \lim_{n \to \infty} p^{n+1} - 1 = 0 - 1 = -1$, since $|p^{n+1}|_p = p^{-n-1}$.

2.3. Topology of $\mathbb{Q}_p$. Define

$$B_r(a) = \{x \in \mathbb{Q}_p; \ |x - a|_p \leq p^r\}, \ r \in \mathbb{Z},$$

as the *ball with center a and radius p^r*, and

$$S_r(a) = \{x \in \mathbb{Q}_p; \ |x - a|_p = p^r\}, \ r \in \mathbb{Z}$$

as the *sphere with center a and radius p^r*.

The topology of $\mathbb{Q}_p$ is quite different from the usual topology of $\mathbb{R}$. First of all, since $|\cdot|_p : \mathbb{Q}_p \to \{p^m; \, m \in \mathbb{Z}\} \cup \{0\}$, the radii are always integer powers of p, for the sake of brevity we just use the power in the notation $B_r(a)$ and $S_r(a)$. On the other hand, since the powers of p and zero form a discrete set in $\mathbb{R}$, in the definition of $B_r(a)$ and $S_r(a)$ we can always use '$\leq$'. Indeed,

$$\{x \in \mathbb{Q}_p; \ |x - a|_p < p^r\} = \{x \in \mathbb{Q}_p; \ |x - a|_p \leq p^{r-1}\} = B_{r-1}(a).$$

REMARK 2.21. Notice that $B_r(a) = a + p^{-r}\mathbb{Z}_p$ and $S_r(a) = a + p^{-r}\mathbb{Z}_p^{\times}$.

We declare that the $B_r(a)$, $r \in \mathbb{Z}$, $a \in \mathbb{Q}_p$, are open subsets. These sets form a basis for the topology of $\mathbb{Q}_p$.

PROPOSITION 2.22. *$S_r(a)$, $B_r(a)$ are open and closed sets in the topology of $\mathbb{Q}_p$.*

PROOF. We first show that $S_r(a)$ is open. Notice that

$$S_r(a) = \bigsqcup_{i\in\{1,\dots,p-1\}} a + p^{-r}i + p^{-r+1}\mathbb{Z}_p = \bigsqcup_{i\in\{1,\dots,p-1\}} B_{(r-1)}(a + p^{-r}i),$$

and consequently $S_r(a)$ is an open set.

In order to show that $S_r(a)$ is closed, we take a sequence $\{x_n\}_{n\in\mathbb{N}}$ of points of $S_r(a)$ converging to $\widetilde{x}_0 \in \mathbb{Q}_p$. We must show that $\widetilde{x}_0 \in S_r(a)$. Note that $x_n = a + p^{-r}u_n$, $u_n \in \mathbb{Z}_p^\times$. Since $\{x_n\}_{n\in\mathbb{N}}$ is a Cauchy sequence, we have

$$|x_n - x_m|_p = p^r|u_n - u_m|_p \to 0, \qquad n, m \to \infty,$$

and consequently $\{u_n\}_{n\in\mathbb{N}}$ is also Cauchy. Since $\mathbb{Q}_p$ is complete $u_n \to \widetilde{u}_0$. Then $x_n \to a + p^{-r}\widetilde{u}_0$, so in order to conclude our proof we must verify that $\widetilde{u}_0 \in \mathbb{Z}_p^\times$. Because u_m is arbitrarily close to $\widetilde{u}_0$, their p-adic expansions must agree up to a big power of p, hence $\widetilde{u}_0 \in \mathbb{Z}_p^\times$.

A similar argument shows that $B_r(a)$ is closed. □

LEMMA 2.23. *If $b \in B_r(a)$ then $B_r(b) = B_r(a)$, i.e. any point of the ball $B_r(a)$ is its center.*

PROOF. Let $x \in B_r(b)$, then

$$|x - a|_p = |x - b + b - a|_p \le \max\{|x - b|_p, |b - a|_p\} \le p^r,$$

i.e. $B_r(b) \subseteq B_r(a)$. Since $a \in B_r(b)$ (i.e. $|b - a|_p = |a - b|_p \le p^r$), we can repeat the previous argument to show that $B_r(a) \subseteq B_r(b)$. □

EXERCISE 2.24. Show that any two balls in $\mathbb{Q}_p$ are either disjoint or one is contained in another.

EXERCISE 2.25. Show that the boundary of any ball is the empty set.

EXERCISE 2.26. The set of balls of $\mathbb{Q}_p$ is countable. **Hint**. Given ball $B_r(a) = a + p^{-r}\mathbb{Z}_p$, with $a \in \mathbb{Q}_p$, $r \in \mathbb{Z}$, there is a rational number $\widetilde{a} \in \mathbb{Q}$ such that $B_r(a) = \widetilde{a} + p^{-r}\mathbb{Z}_p$. Then $B_r(a)$ is uniquely determined by the pair $(\widetilde{a}, r)$.

EXERCISE 2.27. Every open set in $\mathbb{Q}_p$ is a union at most of a countable set of disjoint balls.

A topological space X is called *disconnected* if it can be represented as a union of two disjoint nonempty open subsets. Otherwise we say that X is a *connected* topological space. A subset A of X is called *disconnected* if it can be represented as

$$A = (Y_1 \cap A) \bigsqcup (Y_2 \cap A),$$

where Y_1, Y_2 are nonempty open subsets of X, $Y_1 \cap A \ne \emptyset$, $Y_2 \cap A \ne \emptyset$, and $\bigsqcup$ denotes disjoint union. A topological space X is called *totally disconnected* if the only connected sets are the empty set and the points $\{a\}$, $a \in X$.

EXERCISE 2.28. The space $\mathbb{Q}_p$ is totally disconnected. **Hint**. Given a set $A \ne \{a\}$, with $a \in A$, show that there is a ball $B_r(a)$ such that $B_r(a) \cap A \ne A$. Now use that

$$A = (B_r(a) \cap A) \bigsqcup ((\mathbb{Q}_p \setminus B_r(a)) \cap A).$$

A subset $K \subset X$ is called *compact* if each of its open covers contains a finite subcover. A topological space X is called *locally compact* if every point of X has a compact neighborhood.

THEOREM 2.29. [**1**, Theorem 1.8.5] *A set $K \subset \mathbb{Q}_p$ is compact if and only if it is closed and bounded in $\mathbb{Q}_p$.*

REMARK 2.30. $\mathbb{Q}_p$ is a fractal. More precisely, $\mathbb{Q}_p$ is homeomorphic to a Cantor-like set of $\mathbb{R}$, see e.g. [**1**, Section 1.9] and the references therein.

2.4. The $n-$dimensional p-adic space. We extend the p-adic norm to $\mathbb{Q}_p^n$ by taking

$$\|x\|_p := \max_{1\leq i\leq n} |x_i|_p, \text{ for } x = (x_1, \dots, x_n) \in \mathbb{Q}_p^n.$$

We define $ord(x) = \min_{1\leq i\leq n}\{ord(x_i)\}$, then $\|x\|_p = p^{-ord(x)}$. The metric space $(\mathbb{Q}_p^n, \|\cdot\|_p)$ is a separable complete ultrametric space (here, separable means that $\mathbb{Q}_p^n$ contains a countable dense subset, which is $\mathbb{Q}^n$).

For $r \in \mathbb{Z}$, denote by $B_r^n(a) = \{x \in \mathbb{Q}_p^n \ ; \ \|x - a\|_p \leq p^r\}$ *the ball of radius p^r with center at* $a = (a_1, \dots, a_n) \in \mathbb{Q}_p^n$, and take $B_r^n(0) := B_r^n$. Note that $B_r^n(a) = B_r(a_1) \times \cdots \times B_r(a_n)$, where $B_r(a_i) := \{x_i \in \mathbb{Q}_p \ ; \ |x_i - a_i|_p \leq p^r\}$ is the one-dimensional ball of radius p^r with center at $a_i \in \mathbb{Q}_p$. The ball B_0^n equals the product of n copies of $B_0 = \mathbb{Z}_p$. We also denote by $S_r^n(a) = \{x \in \mathbb{Q}_p^n \ ; \ \|x - a\|_p = p^r\}$ *the sphere of radius p^r with center at* $a = (a_1, \dots, a_n) \in \mathbb{Q}_p^n$, and take $S_r^n(0) := S_r^n$. We notice that $S_0^1 = \mathbb{Z}_p^\times$ (the group of units of $\mathbb{Z}_p$), but $\left(\mathbb{Z}_p^\times\right)^n \subsetneq S_0^n$.

As a topological space $(\mathbb{Q}_p^n, \|\cdot\|_p)$ is totally disconnected. Two balls in $\mathbb{Q}_p^n$ are either disjoint or one is contained in the other. As in the one dimensional case, a subset of $\mathbb{Q}_p^n$ is compact if and only if it is closed and bounded in $\mathbb{Q}_p^n$. Since the balls and spheres are both open and closed subsets in $\mathbb{Q}_p^n$, one has that $(\mathbb{Q}_p^n, \|\cdot\|_p)$ is a locally compact topological space. For further details, the reader may consult for instance [**1**], [**24**], [**21**].

3. Integration in $\mathbb{Q}_p^n$

3.1. Measure theory: A basic dictionary. The notion of measure of a set is a mathematical abstraction of the naive notions of the length of a segment, the area of a plane figure, and the volume of a body.

Let X be a non-empty set. We want to introduce a notion of measure for a class of subsets of X. A suitable class is a σ-algebra of subsets of X. Denote by $\mathcal{P}(X)$ the power set of X, then a subset $\Sigma \subset \mathcal{P}(X)$ is called a σ-algebra, if it satisfies the following properties:

(i) $X \in \Sigma$;

(ii) Σ is closed under complementation: if $A \in \Sigma$ then $A^c := X \smallsetminus A \in \Sigma$;

(iii) Σ is closed under countable unions: if $A_i \in \Sigma$ for $i \in \mathbb{N}$, then $\cup_{i\in\mathbb{N}} A_i \in \Sigma$.

Notice that it follows from the above definition that $\varnothing \in \Sigma$, and that Σ is closed under countable intersections. The elements of Σ are called *measurable sets*, which means that we can assign a measure to these sets. The pair (X, Σ) is called a *measurable space*. Assume that (Y, Λ) is another measurable space and that $f : (X, \Sigma) \to (Y, \Lambda)$ is a function between measurable spaces. The function f is called a *measurable function* if the preimage of every measurable set is measurable.

EXAMPLE 3.1. Let X be a non-empty set. the following are some simple examples of σ-algebras.

(i) $\Sigma = \{X, \varnothing\}$, this is the trivial σ-algebra;

(ii) $\Sigma = \mathcal{P}(X)$, this is the discrete σ-algebra;

(iii) $\Sigma = \{X, \varnothing, A, A^c\}$ is the σ-algebra generated by the subset A.

EXAMPLE 3.2. Let F be a family of subsets of X. Then there exists a unique smallest σ-algebra $\sigma(F)$ which contains any set in F. The σ-algebra $\sigma(F)$ is called the σ-algebra generated by F, it agrees with intersection of all the σ-algebras containing F. If (X, d) is a metric space, the σ-algebra generated by the open balls is called the Borel σ-algebra of X.

DEFINITION 3.3. Let (X, Σ) be a *measurable space. A function* $\mu : \Sigma \to [0, +\infty]$ is called a *measure* if it satisfies the following properties:
(i) $\mu(\varnothing) = 0$;
(ii) for any countable collection A_i, $i \in \mathbb{N}$, of pairwise disjoint sets in Σ,

$$\mu\left(\bigsqcup_{i\in\mathbb{N}} A_i\right) = \sum_{i\in\mathbb{N}} \mu(A_i).$$

Let μ be a measure on (X, Σ). The following are some basic properties of μ:
(i) monotonicity: if A_1 and A_2 are measurable sets with $A_1 \subset A_2$, then $\mu(A_1) \leq \mu(A_2)$;
(ii) subadditivity: for any countable collection A_i, $i \in \mathbb{N}$, of measurable sets in Σ,

$$\mu\left(\bigcup_{i\in\mathbb{N}} A_i\right) \leq \sum_{i\in\mathbb{N}} \mu(A_i);$$

(iii) continuity from below: if A_i, $i \in \mathbb{N}$, are measurable sets in Σ such that $A_i \subset A_{i+1}$ for all i, then the union of the sets A_i is measurable, and

$$\mu\left(\bigcup_{i\in\mathbb{N}} A_i\right) = \lim_{i\to\infty} \mu(A_i);$$

(iv) continuity from above: if A_i, $i \in \mathbb{N}$, are measurable sets in Σ such that $A_{i+1} \subset A_i$ for all i, then the intersection of the sets A_i is measurable. In addition, if A_0 has finite measure, then

$$\mu\left(\bigcap_{i\in\mathbb{N}} A_i\right) = \lim_{i\to\infty} \mu(A_i).$$

3.2. The Haar measure.

THEOREM 3.4. [**9**, Theorem B. Section 58] *Let $(G, \cdot)$ be a locally compact topological Abelian group. There exists a regular Borel measure μ_{Haar} (called a Haar measure of G), unique up to multiplication by a positive constant, such that $\mu_{Haar}(U) > 0$ for every non empty open set U, and $\mu_{Haar}(x \cdot E) = \mu_{Haar}(E)$, for every Borel set E.*

NOTATION 1. We will denote the Haar measure by dx, then $\mu_{\text{Haar}}(U) = \int_U dx$.

EXERCISE 3.5. Prove that $(\mathbb{Q}_p, +)$, respectively $(\mathbb{Q}_p^\times, \cdot)$, are locally compact topological groups. Since $\mathbb{Q}_p$, respectively $\mathbb{Q}_p^\times = \mathbb{Q}_p \backslash \{0\}$, are metric spaces, the continuity of the sum, respectively of the product, means that if $x_n \to x$, and $y_n \to y$, then $x_n + y_n \to x + y$, respectively $x_n y_n \to xy$.

Since $(\mathbb{Q}_p, +)$ is a locally compact topological group, by Theorem 3.4 there exists a measure dx, which is invariant under translations, i.e. $\int_B dx = \int_{a+B} dx$. We denote this fact symbolically as $d(x+a) = dx$. If we normalize this measure by the condition $\int_{\mathbb{Z}_p} dx = 1$, then dx is unique.

In the n-dimensional case, $(\mathbb{Q}_p^n, +)$ is also locally compact topological group. We denote by $d^n x$ the Haar measure normalized by the condition $\int_{\mathbb{Z}_p^n} d^n x = 1$. This measure agrees with the product measure $dx_1 \cdots dx_n$, and it also satisfies that $d^n(x+a) = d^n x$, for $a \in \mathbb{Q}_p^n$.

The balls of $\mathbb{Q}_p^n$ generate the Borel σ-algebra of $\mathbb{Q}_p^n$. The measure $d^n x$ assigns to each open compact subset U a nonnegative real number $\int_U d^n x$, which satisfies

$$\int_{\cup_{k=1}^{\infty} U_k} d^n x = \sum_{k=1}^{\infty} \int_{U_k} d^n x, \tag{3.1}$$

for all compact open subsets U_k in $\mathbb{Q}_p^n$, which are pairwise disjoint, and sucht that $\cup_{k=1}^{\infty} U_k$ is still compact.

REMARK 3.6. Let $\mathcal{B}(\mathbb{Q}_p^n)$ be the Borel σ-algebra on $\mathbb{Q}_p^n$. Let $d^n x$ be the normalized Haar measure on $\left(\mathbb{Q}_p^n, \mathcal{B}(\mathbb{Q}_p^n)\right)$. The Haar measure of a Borel set A is denoted as $\mu_{\text{Haar}}^{(n)}(A)$. The fact that $d^n x$ is a regular measure means that for any measurable subset A of $\mathbb{Q}_p^n$ is holds that

$$\begin{aligned}\mu_{\text{Haar}}^{(n)}(A) &= \sup\left\{\mu_{\text{Haar}}^{(n)}(F)\,; F \subset A,\ F \text{ compact and measurable}\right\}\\ &= \inf\left\{\mu_{\text{Haar}}^{(n)}(G); G \supset A,\ G \text{ open and measurable}\right\}.\end{aligned}$$

3.3. Integration of locally constant functions. A function $\varphi : \mathbb{Q}_p^n \to \mathbb{C}$ is said to be *locally constant* if for every $x \in \mathbb{Q}_p^n$ there exists an open compact subset U, containing x, and such that $f(x) = f(u)$ for all $u \in U$.

EXERCISE 3.7. Every locally constant function is continuous.

Any locally constant function $\varphi : \mathbb{Q}_p^n \to \mathbb{C}$ can be expressed as a linear combination of characteristic functions of the form

$$\varphi(x) = \sum_{k=1}^{\infty} c_k 1_{U_k}(x), \tag{3.2}$$

where $c_k \in \mathbb{C}$,

$$1_{U_k}(x) = \begin{cases} 1 & \text{if} \quad x \in U_k \\ 0 & \text{if} \quad x \notin U_k, \end{cases}$$

and $U_k \subseteq \mathbb{Q}_p^n$ is an open compact for every k. Indeed, there exists a covering $\{U_i\}_{i \in \mathcal{N}}$ of $\mathbb{Q}_p^n$ such that each U_i is open compact and $\varphi\,|_{U_i}$ is a constant function. Since $\left(\mathbb{Q}_p^n, \|\cdot\|_p\right)$ is a separable metric space any open cover has a countable subcover, consequently we may take $\mathcal{N} = \mathbb{N}$.

Let $\varphi : \mathbb{Q}_p^n \to \mathbb{C}$ be a locally constant function as in (3.2). Assume that $A = \bigsqcup_{i=1}^{k} U_i$, the symbol $\bigsqcup$ means disjoint union, i.e. the sets U_i are pairwise disjoint, with U_i open compact. Then, we define

$$\int_A \varphi(x)\, d^n x = c_1 \int_{U_1} d^n x + \cdots + c_k \int_{U_k} d^n x. \tag{3.3}$$

We recall that given a function $\varphi : \mathbb{Q}_p^n \to \mathbb{C}$ the *support of* φ is the set

$$Supp(\varphi) = \overline{\{x \in \mathbb{Q}_p^n \ : \ \varphi(x) \neq 0\}}.$$

A locally constant function with compact support is called a *test function* or a *Bruhat-Schwartz function*. These functions form a $\mathbb{C}$-vector space denoted as $\mathcal{D} := \mathcal{D}\left(\mathbb{Q}_p^n\right)$. From (3.3) one has that the mapping

$$\begin{array}{ccc} \mathcal{D} & \longrightarrow & \mathbb{C} \\ \varphi & \longmapsto & \int_{\mathbb{Q}_p^n} \varphi \, d^n x, \end{array} \tag{3.4}$$

is a well-defined linear functional.

3.4. Integration of continuous functions with compact support. We now extend the integration to a larger class of functions. Let U be an open compact subset of $\mathbb{Q}_p^n$. We denote by $C(U, \mathbb{C})$ the space of all the complex-valued continuous functions supported on U, endowed with the supremum norm, i.e. for $\varphi \in C(U, \mathbb{C})$, we set

$$\|\varphi\| := \sup_{x \in U} |\varphi(x)|.$$

We denote by $C_0(\mathbb{Q}_p^n, \mathbb{C})$ the space of all the complex-valued continuous functions vanishing at infinity, endowed also with the supremum norm. The function φ vanishes at infinity, if given $\varepsilon > 0$, there exists a compact subset K such that $|\varphi(x)| < \varepsilon$, if $x \notin K$.

It is known that $\mathcal{D}$ is dense in $C_0(\mathbb{Q}_p^n, \mathbb{C})$, see e.g. [**23**, Proposition 1.3]. We identify $C(U, \mathbb{C})$ with a subspace of $C_0(\mathbb{Q}_p^n, \mathbb{C})$, therefore $\mathcal{D}$ is dense in $C(U, \mathbb{C})$.

Now, the functional (3.4) satisfies

$$\left| \int_U \varphi \, d^n x \right| \leq \sup_{x \in U} |\varphi(x)| \int_U d^n x, \text{ for } \varphi \in \mathcal{D},$$

and then it has a unique extension to $C(U, \mathbb{C})$. Indeed, given $f \in C(U, \mathbb{C})$ by the density of $\mathcal{D}$ there is sequence $\{f_m\}_{m \in \mathbb{N}}$ in $\mathcal{D}$ approaching f in the supremum norm. Now we take

$$\int_{\mathbb{Q}_p^n} f(x) \ d^n x := \lim_{m \to \infty} \int_{\mathbb{Q}_p^n} f_m(x) \ d^n x. \tag{3.5}$$

Notice that the convergence of the sequence in the right-hand side of (3.5) follows from

$$\left| \int_{\mathbb{Q}_p^n} (f(x) - f_m(x)) \ d^n x \right| \leq \int_{\mathbb{Q}_p^n} |f(x) - f_m(x)| \ d^n x \leq C(U) \|f - f_m\| \to 0$$
$$\text{as } m \to \infty.$$

More generally, if $f_m \overrightarrow{\|\cdot\|} f$, with f, $f_m \in C(U, \mathbb{C})$ for $m \in \mathbb{N}$, then (3.5) holds true.

3.4.1. *Some remarks on uniform convergence.* We recall the notion of uniform convergence. Let E be a non-empty set. Let $f_n : E \to \mathbb{C}$, $n \in \mathbb{N}$ be a sequence of complex-valued functions. We say that the sequence $\{f_n\}_{n \in \mathbb{N}}$ is uniformly convergent on E with limit $f : E \to \mathbb{C}$, if for every $\epsilon > 0$, there exists a natural number N such that for all $n > N$ and any $x \in E$, $|f_n(x) - f(x)| < \epsilon$, which is equivalent to say for every $\epsilon > 0$, there exists a natural number N such that for all $n > N$, $\sup_{x \in E} |f_n(x) - f(x)| < \epsilon$.

The Weierstrass M-test is a very useful criterion for determining the uniform convergence of sequences. Let $\{f_n\}_{n\in\mathbb{N}}$ be a sequence of functions $f_n : E \to \mathbb{C}$ and let M_n be a sequence of positive real numbers such that $|f_n(x)| < M_n$ for all $x \in E$ and $n \in \mathbb{N}$. If $\sum_n M_n$ converges, then $\sum_n f_n$ converges uniformly on E.

3.4.2. *Some remarks on convergent power series.* Let $(F, |\cdot|)$ be a complete field. Let us denote by $F[[z_1, \ldots, z_n]]$, the ring of formal power series with coefficients in F. An element of this ring has the form

$$\sum_i c_i z^i = \sum_{(i_1,\ldots,i_n)\in\mathbb{N}^n} c_{i_1,\ldots,i_n} z_1^{i_1} \ldots x z_n^{i_n},$$

where $i = (i_1, \ldots, i_n) \in \mathbb{N}^n$, and the $c_{i_1,\ldots,i_n}$'s are in F.

A formal series $\sum_i c_i z^i$ is said to be convergent if there exists a positive real number r such that $\sum_i c_i a^i$ converges for any $a = (a_1, \ldots, a_n) \in F^n$ satisfying $\max_i |a_i| < r$. The convergent series form a subring of $F[[z_1, \ldots, z_n]]$, which will be denoted as $F\langle\langle z_1, \ldots, z_n\rangle\rangle$.

If for $\sum_i c_i z^i$ there exists $\sum_i c_i^{(0)} z^i \in \mathbb{R}\langle\langle z_1, \ldots, z_n\rangle\rangle$ (a real convergent series) such that $|c_i| \le c_i^{(0)}$ for all $i \in \mathbb{N}^n$, we say that $\sum_i c_i^{(0)} z^i$ is a *dominant series* for $\sum_i c_i z^i$ and write

$$\sum c_i z^i << \sum c_i^{(0)} x^i.$$

EXERCISE 3.8. A formal power series is convergent if and only if it has a dominant series.

EXAMPLE 3.9. We set $k = (k_1, \ldots, k_n) \in \mathbb{N}^n$ and $|k| := k_1 + \ldots + k_n$. Let $f(z) = \sum_k c_k z^k$ be a complex convergent power series on $\max_i |x_i| < r$. This series has a dominant series, and by the Weierstrass M-test, the sequence $\sum_{|k|\le m} c_k z^k$ converges uniformly to $f(z)$ on $\max_i |z_i| < r$.

We construct a 'radial-type function' on $|x_i|_p \le p^L < r$ for $i = 1, \ldots, n$, i.e. on the ball B_L^n, by taking

$$f(|x_1|_p, \ldots, |x_n|_p) = \sum_k c_k |x|_p^k := \sum_{(k_1,\ldots,k_n)} c_{(k_1,\ldots,k_n)} \prod_{i=1}^n |x_i|_p^{k_i},$$

for $x = (x_1, \ldots, x_n) \in B_L^n$. Notice that

$$\sup_{x\in B_L^n} \left| f(|x_1|_p, \ldots, |x_n|_p) - \sum_{|k|\le m} c_{(k_1,\ldots,k_n)} \prod_{i=1}^n |x_i|_p^{k_i} \right|$$

$$\le \sup_{\max_i |z_i| < r} \left| f(z_1, \ldots, z_n) - \sum_{|k|\le m} c_{(k_1,\ldots,k_n)} \prod_{i=1}^n z_i^{k_i} \right|,$$

and consequently $\sum_{|k|\le m} c_{(k_1,\ldots,k_n)} \prod_{i=1}^n |x_i|_p^{k_i}$ converges uniformly to $f(|x_1|_p, \ldots, |x_n|_p)$ on B_L^n. Then

$$\int_{B_L^n} f(|x_1|_p, \ldots, |x_n|_p) d^n x = \lim_{m\to\infty} \sum_{|k|\le m} c_{(k_1,\ldots,k_n)} \int_{B_L^n} \prod_{i=1}^n |x_i|_p^{k_i} \, d^n x$$

$$= \lim_{m\to\infty} \sum_{|k|\le m} c_{(k_1,\ldots,k_n)} \prod_{i=1}^n \int_{B_L} |x_i|_p^{k_i} \, dx_i. \tag{3.6}$$

3.5. The change of variables formula in dimension one. Let us start by establishing the formula:

$$d\left(ax\right)=\left|a\right|_p dx,\ a\in\mathbb{Q}_p^{\times}, \tag{3.7}$$

which means

$$\int_{aU} dx=\left|a\right|_p\int_U dx,$$

for every Borel set $U\subseteq\mathbb{Q}_p$, for instance an open compact subset. Indeed, consider

$$\begin{array}{cccc} T_a: & \mathbb{Q}_p & \longrightarrow & \mathbb{Q}_p \\ & x & \longmapsto & ax, \end{array}$$

with $a\in\mathbb{Q}_p^{\times}$. T_a is a topological and algebraic isomorphism. Then $U\mapsto\int_{aU}dx$ is a Haar measure for $(\mathbb{Q}_p,+)$, and by the uniqueness of such measure, there exists a positive constant $C(a)$ such that $\int_{aU}dx=C(a)\int_U dx$. To compute $C(a)$ we can pick any open compact set, for instance $U=\mathbb{Z}_p$. Now we show that

$$\int_{a\mathbb{Z}_p} dx=C(a)=\left|a\right|_p.$$

Let us consider first the case $a\in\mathbb{Z}_p$, i.e. $a=p^l u$, $l\in\mathbb{N}$, $u\in\mathbb{Z}_p^{\times}$. Fix a system of representatives $\{b\}$ of $\mathbb{Z}_p/p^l\mathbb{Z}_p$ in $\mathbb{Z}_p$, then $\mathbb{Z}_p=\bigsqcup_{b\in\mathbb{Z}_p/p^l\mathbb{Z}_p}\left(b+p^l\mathbb{Z}_p\right)$, and

$$1=\int_{\mathbb{Z}_p} dx=\sum_{b\in\mathbb{Z}_p/p^l\mathbb{Z}_p}\int_{b+p^l\mathbb{Z}_p} dx=\sum_{b\in\mathbb{Z}_p/p^l\mathbb{Z}_p}\int_{p^l\mathbb{Z}_p} dx=\#\left(\mathbb{Z}_p/p^l\mathbb{Z}_p\right)\int_{p^l\mathbb{Z}_p} dx,$$

i.e. $\left|a\right|_p=p^{-l}=\int_{a\mathbb{Z}_p} dx$. The case $a\notin\mathbb{Z}_p$ is treated in a similar way.

Now, for $f:U\to\mathbb{C}$, where U is a Borel set, we have

$$\int_U f\left(x\right)dx=\left|a\right|_p\int_{a^{-1}U-a^{-1}b} f\left(ay+b\right)\,dy,\ \text{for any } a\in\mathbb{Q}_p^{\times}, b\in\mathbb{Q}_p.$$

The formula follows by changing variables as $x=ay+b$, and by using $dx=d\left(ay+b\right)=d(ay)=\left|a\right|_p dy$, since the Haar measure is invariant under translations.

EXAMPLE 3.10. For any $r\in\mathbb{Z}$,

$$\int_{B_r} dx=\int_{p^{-r}\mathbb{Z}_p} dx=p^r\int_{\mathbb{Z}_p} dy=p^r.$$

EXAMPLE 3.11. For any $r\in\mathbb{Z}$,

$$\int_{S_r} dx=\int_{B_r} dx-\int_{B_{r-1}} dx=p^r-p^{r-1}=p^r\left(1-p^{-1}\right).$$

EXAMPLE 3.12. Take $U=\mathbb{Z}_p\smallsetminus\{0\}$. Then

$$\int_U dx=\int_{\mathbb{Z}_p} dx=1.$$

Notice that U is not compact, since the sequence $\{p^n\}_{n\in\mathbb{N}}\subseteq U$ converges to $0\notin U$. Now, by using

$$\mathbb{Z}_p\smallsetminus\{0\}=\bigsqcup_{j=0}^{\infty}\left\{x\in\mathbb{Z}_p;\ \left|x\right|_p=p^{-j}\right\},$$

and by changing variables as $x = p^j y$, $dx = p^{-j} dy$, we have

$$\int_{\mathbb{Z}_p \smallsetminus \{0\}} dx = \sum_{j=0}^{\infty} \int_{p^j \mathbb{Z}_p^\times} dx = \left(\sum_{j=0}^{\infty} p^{-j} \right) \int_{\mathbb{Z}_p^\times} dy = \frac{1-p^{-1}}{1-p^{-1}} = 1.$$

This calculation shows that $\mathbb{Z}_p \smallsetminus \{0\}$ has Haar measure 1. In particular, $\{0\}$ has Haar measure 0.

NOTATION 2. For $s \in \mathbb{C}$, $a > 0$, we set $a^s := e^{s \ln a}$.

EXAMPLE 3.13. Set

$$Z(s) := \int_{\mathbb{Z}_p \smallsetminus \{0\}} |x|_p^s \, dx, \; s \in \mathbb{C} \text{ for } \operatorname{Re}(s) > -1.$$

We prove that $Z(s)$ has a meromorphic continuation to the whole complex plane as a rational function of p^{-s}.

Indeed,

$$\begin{aligned} Z(s) &= \int_{\mathbb{Z}_p \smallsetminus \{0\}} |x|_p^s \, dx = \sum_{j=0}^{\infty} \int_{|x|_p = p^{-j}} |x|_p^s \, dx = \sum_{j=0}^{\infty} p^{-js} \int_{|x|_p = p^{-j}} dx \\ &= \left(1 - p^{-1}\right) \sum_{j=0}^{\infty} p^{-j(s+1)} \text{ (here we need the hypothesis } \operatorname{Re}(s) > -1) \\ &= \frac{\left(1-p^{-1}\right)}{1-p^{-1-s}}, \quad \text{for } \operatorname{Re}(s) > -1. \end{aligned}$$

We now note that the right hand-side is defined for any complex number $\operatorname{Re}(s) \neq -1$, therefore, it gives a meromorphic continuation of $Z(s)$ to the half-plane $\operatorname{Re}(s) < -1$. Thus, we have shown that integral $Z(s)$ has a meromorphic continuation to the whole $\mathbb{C}$ with simple poles satisfying $\operatorname{Re}(s) = -1$.

EXAMPLE 3.14. Take $p \neq 2$. We compute

$$Z(s, x^2 - 1) = \int_{\mathbb{Z}_p} \left|x^2 - 1\right|_p^s \, dx, \quad \text{for} \quad \operatorname{Re}(s) > -1.$$

Let us take $\{0, 1, \ldots, p-1\} \subset \mathbb{Z} \subset \mathbb{Z}_p$ as a system of representatives of $\mathbb{F}_p \simeq \mathbb{Z}_p / p\mathbb{Z}_p$. Then $\mathbb{Z}_p = \bigsqcup_{j=0}^{p-1} (j + p\mathbb{Z}_p)$, and

$$\begin{aligned} Z(s, x^2 - 1) &= \sum_{j=0}^{p-1} \int_{j + p\mathbb{Z}_p} \left|(x-1)(x+1)\right|_p^s dx \\ &= p^{-1} \sum_{j=0}^{p-1} \int_{\mathbb{Z}_p} \left|(j - 1 + py)(j + 1 + py)\right|_p^s dy, \quad (x = j + py). \end{aligned}$$

Let us consider first the integrals in which $j \mp 1 + py \in \mathbb{Z}_p^\times$, i.e. the reduction mod p of $j \mp 1$ is a nonzero element of $\mathbb{F}_p$. In this case,

$$\int_{\mathbb{Z}_p} \left|(j - 1 + py)(j + 1 + py)\right|_p^s dy = 1,$$

and since $p \neq 2$ there are exactly $p-2$ of those js, then

$$\begin{aligned} Z(s,x^2-1) &= (p-2)\,p^{-1} + p^{-1}\int_{\mathbb{Z}_p} |py\,(2+py)|_p^s\,dy + p^{-1}\int_{\mathbb{Z}_p} |(-2+py)\,py|_p^s\,dy \\ &= (p-2)\,p^{-1} + 2p^{-1-s}\int_{\mathbb{Z}_p} |y|_p^s\,dy = (p-2)\,p^{-1} + 2p^{-1-s}\frac{1-p^{-1}}{1-p^{-1-s}}. \end{aligned}$$

EXERCISE 3.15. Take $q\,(x) = \prod_{i=1}^{r}(x-\alpha_i)^{e_i} \in \mathbb{Z}_p\,[x]$, $\alpha_i \in \mathbb{Z}_p$, $e_i \in \mathbb{N} \smallsetminus \{0\}$. Assume that $\alpha_i \not\equiv \alpha_j \mod p$. By using the methods presented in Examples 3.13 and 3.14 compute the integral

$$Z(s,q(x)) = \int_{\mathbb{Z}_p \smallsetminus \{0\}} |q\,(x)|_p^s \;dx.$$

3.6. Improper integrals. Our next task is the integration of functions that do not have compact support. A function $f : \mathbb{Q}_p^n \to \mathbb{C}$ is said to be *locally integrable*, $f \in L^1_{loc}$, if

$$\int_K f\,(x)\,d^n x < \infty \text{ for every compact subset } K \subset \mathbb{Q}_p^n.$$

DEFINITION 3.16 (Improper Integral). A function $f \in L^1_{loc}$ is said to be integrable in $\mathbb{Q}_p^n$, if

$$\lim_{l\to+\infty}\int_{B_l^n} f\,(x)\,d^n x = \lim_{l\to+\infty}\sum_{j=-\infty}^{l}\int_{S_j^n} f\,(x)\,d^n x$$

exists. If the limit exists, it is denoted as $\int_{\mathbb{Q}_p^n} f\,(x)\,d^n x$, and we say that the *improper integral exists.* Note that in this case,

$$\int_{\mathbb{Q}_p^n} f\,(x)\,d^n x = \sum_{j=-\infty}^{+\infty}\int_{S_j^n} f\,(x)\,d^n x.$$

EXAMPLE 3.17. The function $|x|_p$ is locally integrable but not integrable.

EXAMPLE 3.18. Let $f : \mathbb{Q}_p \to \mathbb{C}$ be a *radial function* i.e. $f(x) = f(|x|_p)$. If $\sum_{j=-\infty}^{+\infty} f\,(p^j)\,p^j < \infty$. Then

$$\int_{\mathbb{Q}_p} f\left(|x|_p\right)dx = \sum_{j=-\infty}^{+\infty}\int_{|x|_p=p^j} f\left(|x|_p\right)dx = \left(1-p^{-1}\right)\sum_{j=-\infty}^{+\infty} f\,(p^j)\,p^j. \tag{3.8}$$

EXERCISE 3.19. Generalize formula (3.8) to the n-dimensional case.

EXERCISE 3.20. By using $\sum_{r=0}^{+\infty} rp^{-r} = \frac{p}{(p-1)^2}$, show that

$$\int_{\mathbb{Z}_p} \ln\left(|x|_p\right)\,dx = -\frac{\ln p}{p-1}.$$

3.7. Further remarks on integrals of continuous functions with compact support.

EXAMPLE 3.21 (Continuation of Example 3.9). With the notation given in Example 3.9 and using formula (3.6) and Example 3.13, we have

$$\begin{aligned}\int_{B_L^n} f(|x_1|_p,\ldots,|x_n|_p)d^n x &= \lim_{m\to\infty}\sum_{|k|\le m} c_{(k_1,\ldots,k_n)}\int_{B_L^n}\prod_{i=1}^n |x_i|_p^{k_i}\, d^n x\\ &= \lim_{m\to\infty}\sum_{|k|\le m} c_{(k_1,\ldots,k_n)}\prod_{i=1}^n\int_{B_L}|x_i|_p^{k_i}\, dx_i\\ &= \lim_{m\to\infty}\sum_{|k|\le m} c_{(k_1,\ldots,k_n)}\prod_{i=1}^n p^{L+Lk_i}\int_{\mathbb{Z}_p}|y_i|_p^{k_i}\, dx_i\\ &= \lim_{m\to\infty}\sum_{|k|\le m} c_{(k_1,\ldots,k_n)}\prod_{i=1}^n p^{L+Lk_i}\frac{(1-p^{-1})}{1-p^{-1-k_i}}\\ &= \left(1-p^{-1}\right)^n\sum_{k\in\mathbb{N}^n}\frac{p^{L\left(n+\sum_{i=1}^n k_i\right)}c_{(k_1,\ldots,k_n)}}{\prod_{i=1}^n\left(1-p^{-1-k_i}\right)}.\end{aligned}$$

EXAMPLE 3.22. Let $f(x)=e^{-|x|_p}\Omega\left(|x|_p\right)$, where $\Omega\left(|x|_p\right)$ denotes the characteristic function of $\mathbb{Z}_p$. We compute first $\int f(x)dx$ by using Example 3.18:

$$\begin{aligned}\int_{\mathbb{Q}_p} e^{-|x|_p}\Omega\left(|x|_p\right)dx &= \int_{\mathbb{Z}_p} e^{-|x|_p}dx = \sum_{j=0}^{\infty}\int_{p^j\mathbb{Z}_p^\times} e^{-|x|_p}dx\\ &= \sum_{j=0}^{\infty} e^{-p^{-j}}\int_{p^j\mathbb{Z}_p^\times} dx = \sum_{j=0}^{+\infty} p^{-j}\left(1-p^{-1}\right)e^{-p^{-j}}.\end{aligned}$$

We now compute the integral by using Example 3.21:

$$\int_{\mathbb{Z}_p} e^{-|x|_p}dx = \lim_{M\to\infty}\sum_{k=0}^{M}\frac{(-1)^k}{k!}\int_{\mathbb{Z}_p}|x|_p^k\,dx = \left(1-p^{-1}\right)\sum_{k=0}^{\infty}\frac{(-1)^k}{k!\left(1-p^{-1-k}\right)}.$$

Consequently,

$$\sum_{j=0}^{\infty} p^{-j}\left(1-p^{-1}\right)e^{-p^{-j}} = \left(1-p^{-1}\right)\sum_{k=0}^{\infty}\frac{(-1)^k}{k!\left(1-p^{-1-k}\right)}.$$

We invite the reader to verify this identity directly .

EXERCISE 3.23. Consider the integral

$$I(s)=\int_{\mathbb{Z}_p\smallsetminus\{0\}} e^{-|x|_p^s}dx \text{ for } \mathrm{Re}(s)>0.$$

Show that $I(s)$ has a meromorphic continuation to the whole complex plane with poles belonging to the set $\cup_{j=1}^{\infty}\left\{\mathrm{Re}(s)=\frac{-1}{j}\right\}$. Is the meromorphic continuation a rational function of p^{-s}? **Hint**. Recall that the uniform limit of holomorphic functions is holomorphic.

4. Change of variables formula

Let $U \subseteq \mathbb{Q}_p^n$ be *an open subset.* A function $h : U \to \mathbb{Q}_p$ is said to be *analytic in* U, if for every $b = (b_1, \ldots, b_n) \in U$ there exists an open subset $\widetilde{U} \subset U$, with $b \in \widetilde{U}$, and a convergent power series such that $h(x) = \sum_{i \in \mathbb{N}^n} a_i (x-b)^i$ for $x \in \widetilde{U}$, with $i = (i_1, \ldots, i_n)$ and $(x-b)^i = \prod_{j=1}^n (x_j - b_j)^{i_j}$. In this case, $\frac{\partial}{\partial x_l} h(x) = \sum_{i \in \mathbb{N}^n} a_i \frac{\partial}{\partial x_l} (x-b)^i$ is a convergent power series, see e.g. [**1**, Proposition 2.2.6]. Here $\frac{\partial}{\partial x_l}$ denotes the standard the derivative of a series.

Let U, V open subsets in $\mathbb{Q}_p^n$. A mapping $H : U \to V$, $H = (H_1, \ldots, H_n)$ is called *analytic* if each H_i is analytic. The mapping H is said to be *bi-analytic* if H and H^{-1} are analytic.

THEOREM 4.1. *Let $K_0, K_1 \subset \mathbb{Q}_p^n$ be open compact subsets, and consider a bi-analytic map $H = (H_1, \ldots, H_n) : K_1 \to K_0$ such that*

$$\det \left[\frac{\partial H_i}{\partial y_j}(z) \right] \neq 0, \textit{ for any } z \in K_1.$$

If f is a continuous function on K_0, then

$$\int_{K_0} f(x) \; d^n x = \int_{K_1} f(H(y)) \left| \det \left[\frac{\partial H_i}{\partial y_j}(y) \right] \right|_p d^n y, \qquad (x = H(y)).$$

For the proof of this theorem the reader may consult [**10**, Prop. 7.4.1] or [**8**, Section 10.1.2].

EXAMPLE 4.2. Set $U := \mathbb{Q}_p \smallsetminus \{\frac{-d}{c}\}$ and $V := \mathbb{Q}_p \smallsetminus \{\frac{a}{c}\}$, where a, b, c, $d \in \mathbb{Q}_p$, with $c \neq 0$. Consider the function

$$\begin{array}{ccc} U & \to & V \\ \\ x & \to & y = \frac{ax+b}{cx+d}, \end{array}$$

this is an analytic function in U, with inverse $x = \frac{dy-b}{-cy+a}$, which is analytic in V. Assume that $ad - bc \neq 0$, and take $\varphi : \mathbb{Q}_p \to \mathbb{C}$ a Bruhat-Schwartz function with support contained in V, then

$$\int_V \varphi(y) \, dy = \int_U \varphi \left(\frac{ax+b}{cx+d} \right) \frac{|ad-bc|_p}{|cx+d|_p^2} dx.$$

EXAMPLE 4.3. Compute $\int_{B_r^n} d^n x$, where $B_r^n = \left\{ x \in \mathbb{Q}_p^n ; \|x\|_p \leq p^r \right\}$. We first recall that

$$B_r^n = p^{-r}\mathbb{Z}_p^n = \underbrace{\begin{array}{c} p^{-r}\mathbb{Z}_p \times \ldots \times p^{-r}\mathbb{Z}_p \\ n - copies \end{array}}.$$

By changing variables as $x_i = p^{-r} y_i$ for $i = 1, \ldots, n$, we have $d^n x = p^{nr} d^n y$, and

$$\int_{B_r^n} d^n x = p^{nr} \int_{\mathbb{Z}_p^n} d^n x = p^{nr}.$$

EXERCISE 4.4. Show that

$$\int_{S_r^n} d^n x = p^{nr} \left(1 - p^{-n} \right).$$

EXERCISE 4.5. Set

$$Z(x,t) := (1-p^{-n})\|x\|_p^{-n}\sum_{k=0}^{\infty} q^{-kn}e^{-at(q^{-k}\|x\|_p^{-1})^{\alpha}} - \|x\|_p^{-n}e^{-at(p\|x\|_p^{-1})^{\alpha}},$$

for and $x \in \mathbb{Q}_p^n \smallsetminus \{0\}$, $t>0$, $a>0$, $\alpha>0$. Show the formula

$$Z(x,t) = \sum_{m=1}^{\infty}\frac{(-1)^m}{m!}\frac{1-p^{\alpha m}}{1-p^{-\alpha m-n}}(at)^m\|x\|_p^{-\alpha m-n}.$$

Hint. Use the technique presented in Examples 3.9, 3.21. First show that

$$(1-p^{-n})\sum_{k=0}^{\infty} q^{-kn}e^{-at(q^{-k}\|x\|_p^{-1})^{\alpha}} = \int_{\mathbb{Z}_p^n} e^{-at\|y\|_p^{\alpha}\|x\|_p^{-\alpha}}d^n y.$$

EXERCISE 4.6. Generalize Example 3.13 to $\mathbb{Q}_p^n$, i.e. to show that the integral

$$Z(s) = \int_{\mathbb{Z}_p^n\smallsetminus\{0\}} \|x\|_p^s\, d^n x, \text{ for } \operatorname{Re}(s) > -n,$$

admits an analytic continuation to the whole complex plane as a rational function of p^{-s}.

EXERCISE 4.7. Let α be a real number. Show that

$$I(\alpha) = \int_{\mathbb{Z}_p^n\smallsetminus\{0\}} \frac{1}{\|x\|_p^{\alpha}}d^n x < \infty \text{ if } \alpha < n.$$

EXERCISE 4.8. Let β be a real number. Show that

$$J(\beta) = \int_{\mathbb{Q}_p^n\smallsetminus\mathbb{Z}_p^n} \frac{1}{\|x\|_p^{\beta}}d^n x < \infty \text{ if } \beta > n.$$

EXAMPLE 4.9. Take $N \geq 4$ and complex variables s_{1j} and $s_{(N-1)j}$ for $2 \leq j \leq N-2$ and s_{ij} for $2 \leq i < j \leq N-2$. Put $\boldsymbol{s} := (s_{ij}) \in \mathbb{C}^d$, where $d = \frac{N(N-3)}{2}$ denotes the total number of indices ij. The Koba-Nielsen local zeta functions is defined as follows:

$$Z^{(N)}(\boldsymbol{s}) = \int_{\mathbb{Q}_p^{N-3}} \prod_{i=2}^{N-2} |x_j|_p^{s_{1j}}\,|1-x_j|_p^{s_{(N-1)j}} \prod_{2\leq i<j\leq N-2} |x_i-x_j|_p^{s_{ij}} \prod_{i=2}^{N-2} dx_i,$$

where $\prod_{i=2}^{N-2} dx_i$ is the normalized Haar measure on $\mathbb{Q}_p^{N-3}$. If $N=4$, we have

$$Z^{(4)}(\boldsymbol{s}) = \int_{\mathbb{Q}_p} |x|_p^{s_{12}}\,|1-x|_p^{s_{32}}\, dx. \tag{4.1}$$

Notice that in integral (4.1) does not contain a test function, and consequently its convergence is not direct. In order to regularize it, we proceed as follows:

$$\begin{aligned} Z^{(4)}(s_{12},s_{32}) &= \int_{\mathbb{Z}_p} |x|_p^{s_{12}}\,|1-x|_p^{s_{32}}\, dx + \int_{\mathbb{Q}_p\smallsetminus\mathbb{Z}_p} |x|_p^{s_{12}}\,|1-x|_p^{s_{32}}\, dx \\ &=: Z_0^{(4)}(s_{12},s_{32}) + Z_1^{(4)}(s_{12},s_{32}). \end{aligned}$$

To study integral $Z_0^{(4)}(s_{12},s_{32})$, we use that $\mathbb{Z}_p = \bigsqcup_{j=0}^{p-1} j+p\mathbb{Z}_p$, to get

$$Z_0^{(4)}(s_{12},s_{32}) = \sum_{j=0}^{p-1}\int_{j+p\mathbb{Z}_p} |x|_p^{s_{12}}\,|1-x|_p^{s_{32}}\, dx =: \sum_{j=0}^{p-1} Z_{0,j}^{(4)}(s_{12},s_{32}).$$

Now, for $j \neq 0,1$,

$$Z_{0,j}^{(4)}(s_{12},s_{32}) = p^{-1}.$$

In the case $j=0$,

$$Z_{0,0}^{(4)}(s_{12},s_{32}) = \int_{p\mathbb{Z}_p} |x|_p^{s_{12}}\, dx = p^{-1-s_{12}}\left(\frac{1-p^{-1}}{1-p^{-1-s_{12}}}\right).$$

In the case $j=1$,

$$\begin{aligned} Z_{0,1}^{(4)}(s_{12},s_{32}) &= \int_{1+p\mathbb{Z}_p} |1-x|_p^{s_{32}}\, dx = p^{-1-s_{32}}\int_{\mathbb{Z}_p} |y|_p^{s_{32}}\, dy \\ &= p^{-1-s_{32}}\left(\frac{1-p^{-1}}{1-p^{-1-s_{32}}}\right). \end{aligned}$$

Consequently,

$$Z_0^{(4)}(s_{12},s_{32}) = (p-2)\,p^{-1} + p^{-1-s_{12}}\left(\frac{1-p^{-1}}{1-p^{-1-s_{12}}}\right) + p^{-1-s_{32}}\left(\frac{1-p^{-1}}{1-p^{-1-s_{32}}}\right).$$

Notice that $Z_0^{(4)}(s_{12},s_{32})$ is holomorphic, and consequently the underlying integrals converge, in

$$\text{(4.2)} \qquad \operatorname{Re}(s_{12}) > -1 \text{ and } \operatorname{Re}(s_{32}) > -1.$$

To study $Z_1^{(4)}(s_{12},s_{32})$, we first notice that

$$Z_1^{(4)}(s_{12},s_{32}) = \int_{\mathbb{Q}_p\smallsetminus\mathbb{Z}_p} |x|_p^{s_{12}}\, |1-x|_p^{s_{32}}\, dx = \int_{\mathbb{Q}_p\smallsetminus\mathbb{Z}_p} |x|_p^{s_{12}+s_{32}}\, dx.$$

If $\operatorname{Re}(s_{12}) + \operatorname{Re}(s_{32}) + 1 < 0$, by Exercise 4.8, integral $Z_1^{(4)}(s_{12},s_{32})$ exists and it can be computed as

$$\begin{aligned} Z_1^{(4)}(s_{12},s_{32}) &= \sum_{j=1}^{\infty} \int_{|x|_p=p^j} |x|_p^{s_{12}+s_{32}}\, dx = \left(1-p^{-1}\right)\sum_{j=1}^{\infty} p^{j(s_{12}+s_{32}+1)} \\ &= \left(1-p^{-1}\right)\frac{p^{s_{12}+s_{32}+1}}{1-p^{s_{12}+s_{32}+1}}. \end{aligned}$$

Thus $Z_1^{(4)}(s_{12},s_{32})$ is holomorphic in

$$\text{(4.3)} \qquad \operatorname{Re}(s_{12}) + \operatorname{Re}(s_{32}) + 1 < 0.$$

Finally, it is not difficult to see that conditions (4.2)–(4.3) define an open set in $\mathbb{C}$.

EXERCISE 4.10. Consider the integral

$$J(s) = \int_{\mathbb{Q}_p^n} e^{-\|x\|_p^2}\, \|x\|_p^s\, d^n x \text{ for } \operatorname{Re}(s) > 0.$$

Show that $J(s)$ has a meromorphic continuation to the whole complex plane. Is the meromorphic continuation a rational function of p^{-s}? **Hint**. Express $J(s) = J_1(s) + J_2(s)$, where

$$J_1(s) := \int_{\mathbb{Q}_p^n\setminus\mathbb{Z}_p^n} e^{-\|x\|_p^2}\, \|x\|_p^s\, d^n x, \quad J_2(s) := \int_{\mathbb{Z}_p^n} e^{-\|x\|_p^2}\, \|x\|_p^s\, d^n x.$$

Show that $J_1(s)$ is holomorphic function in $\mathbb{C}$, see Lemma below. For $\mathrm{Re}(s)>0$ justify that

$$J_2(s)=\sum_{j=0}^{\infty}\frac{(-1)^j}{j!}\int_{\mathbb{Z}_p^n}\|x\|_p^{s+2j}\,d^nx.$$

LEMMA 4.11 ([**10**, Lemma 5.3.1]). *Let U be an open subset of $\mathbb{C}$ and let $f:\mathbb{Q}_p^n\times U\to\mathbb{C}$ be a continuous function on $\mathbb{Q}_p^n\times U$ having the following properties: (i) if K is a compact subset of U, there exists an integrable function $\varphi_K\geq 0$ on $\mathbb{Q}_p^n$ satisfying $|f(x,s)|\leq\varphi_K(x)$ for all $(x,s)\in\mathbb{Q}_p^n\times K$; $f(x,\cdot)$ is a holomorphic on U for every $x\in\mathbb{Q}_p^n$. Then $\int_{\mathbb{Q}_p^n}f(x,s)\,d^nx$ defines a holomorphic function on U.*

5. Additive characters

Given a nonzero p-adic number

$x=x_{-m}p^{-m}+x_{-m+1}p^{-m+1}+\ldots+x_{-1}p^{-1}+x_0+xp+\ldots$ with $x_{-m}\neq 0$ and $m>0$,

we define its *fractional part as*

$$\{x\}_p=x_{-m}p^{-m}+x_{-m+1}p^{-m+1}+\ldots+x_{-1}p^{-1}\in\mathbb{Q}.$$

If $x\in\mathbb{Z}_p$, we set $\{x\}_p:=0$. Now the function

$$\chi_p(x):=\exp 2\pi i\{x\}_p$$

is called the *standard additive character* of $\mathbb{Q}_p$ (more precisely of $(\mathbb{Q}_p,+)$). Notice that

$$\chi_p:(\mathbb{Q}_p,+)\to(S,\cdot)$$

is a continuous homomorphism from $(\mathbb{Q}_p,+)$ into the unit complex circle considered as a multiplicative group, i.e. χ_p satisfies the following:

(i) $|\chi_p(x)|=1$ for $x\in\mathbb{Q}_p$;
(ii) $\chi_p(x+y)=\chi_p(x)\chi_p(y)$ for $x,\,y\in\mathbb{Q}_p$;
(iii) $\overline{\chi_p(x)}=\frac{1}{\chi_p(x)}=\chi_p(-x)$ for $x\in\mathbb{Q}_p$, where the bar means complex conjugate;
(iv) $\chi_p(x)\not\equiv 1$ for $x\in\mathbb{Q}_p\smallsetminus\mathbb{Z}_p$.

EXAMPLE 5.1. Let r be an integer. To show that

$$\int_{B_r}\chi_p(\xi x)\,dx=\begin{cases}p^r & \text{if}\quad |\xi|_p\leq p^{-r}\\ \\ 0 & \text{if}\quad |\xi|_p\geq p^{-r+1}.\end{cases}$$

If $|\xi|_p\leq p^{-r}$, then $|\xi x|_p\leq 1$ which means that $\xi x\in\mathbb{Z}_p$ and thus $\chi_p(\xi x)\equiv 1$,

$$\int_{B_r}\chi_p(\xi x)\,dx=\int_{B_r}dx=p^r.$$

If $|\xi|_p\geq p^{-r+1}$, there exists $x_0\in S_r$, i.e. $|x_0|_p=p^r$, such that $|x_0\xi|_p\geq p$, i.e. $x_0\xi\in\mathbb{Q}_p\smallsetminus\mathbb{Z}_p$ and thus we may assume that $\chi_p(x_0\xi)\neq 1$. We now change variables as $x=y+x_0$, notice that y runs through $p^{-r}\mathbb{Z}_p$, to get

$$\int_{|x|_p\leq p^r}\chi_p(\xi x)\,dx=\chi_p(\xi x_0)\int_{|y|_p\leq p^r}\chi_p(\xi y)\,dy,$$

which implies the announced formula.

EXERCISE 5.2. Let r be an integer. For $x = (x_1, \ldots, x_n)$, $\xi = (\xi_1, \ldots, \xi_n) \in \mathbb{Q}_p^n$, we set $x \cdot \xi := \sum_i x_i \xi_i$. To show that

$$\int_{B_r^n} \chi_p (x \cdot \xi)\, d^n x = \begin{cases} p^{rn} & \text{if} \quad \|\xi\|_p \leq p^{-r} \\ 0 & \text{if} \quad \|\xi\|_p \geq p^{-r+1}. \end{cases}$$

Hint: remember that $x = p^{ord(x)} \widetilde{x}$, with $\|\widetilde{x}\|_p = 1$ and $\xi = p^{ord(\xi)} \widetilde{\xi}$, with $\left\|\widetilde{\xi}\right\|_p = 1$. Thus $x \cdot \xi = p^{ord(x)+ord(\xi)} \widetilde{x} \cdot \widetilde{\xi}$.

EXERCISE 5.3. Let r be an integer. To show that

$$\int_{S_r^n} \chi_p (x \cdot \xi)\, dx = \begin{cases} p^{nr}\left(1 - p^{-n}\right) & \text{if} \quad \|\xi\|_p \leq p^{-r} \\ -p^{rn-n} & \text{if} \quad \|\xi\|_p = p^{-r+1} \\ 0 & \text{if} \quad \|\xi\|_p \geq p^{-r+2}. \end{cases} \tag{5.1}$$

Hint: use Example 5.1 and

$$\int_{S_r^n} \chi_p (\xi x)\, d^n x = \int_{B_r^n} \chi_p (\xi x)\, d^n x - \int_{B_{r-1}^n} \chi_p (\xi x)\, d^n x.$$

EXERCISE 5.4. Extend the formula given in Example 3.18 to n-dimensional case, i.e. for radial functions $f : \mathbb{Q}_p^n \to \mathbb{C}$ i.e. $f(x) = f(\|x\|_p)$.

LEMMA 5.5. *(i) Let $f : \mathbb{R} \to \mathbb{C}$ be a function such that $\sum_{r=0}^{\infty} |f(p^{-r})| p^{-nr} < \infty$. Then*

$$\int_{\mathbb{Q}_p^n} f\left(\|x\|_p\right) \chi_p (\xi \cdot x)\, d^n x = \frac{\left(1 - p^{-n}\right)}{\|\xi\|_p^n} \sum_{r=0}^{\infty} f\left(\frac{p^{-r}}{\|\xi\|_p}\right) p^{-nr} - \frac{1}{\|\xi\|_p^n} f\left(\frac{p}{\|\xi\|_p}\right) \tag{5.2}$$

for $\xi \neq 0$, in the sense of improper integrals. (ii) If $f\left(\|x\|_p\right)$ is integrable, then $f\left(\|x\|_p\right) \chi_p (\xi \cdot x)$ is integrable with respecto x for any fixed ξ and the formula (5.2) *holds true.*

PROOF. (ii) Suppose that $\|\xi\|_p = p^N$, by using (5.1),

$$\begin{aligned}
\int_{\mathbb{Q}_p^n} f\left(\|x\|_p\right) \chi_p (\xi \cdot x)\, d^n x &= \sum_{r=-\infty}^{\infty} f(p^r) \int_{S_r^n} \chi_p (\xi \cdot x)\, d^n x \\
&= \sum_{r=-\infty}^{-N} f(p^r) p^{nr}\left(1 - p^{-n}\right) - p^{-nN} f(p^{-N+1}) \\
&= \frac{\left(1 - p^{-n}\right)}{\|\xi\|_p^n} \sum_{r=0}^{\infty} f\left(\frac{p^{-r}}{\|\xi\|_p}\right) p^{-nr} - \frac{1}{\|\xi\|_p^n} f\left(\frac{p}{\|\xi\|_p}\right).
\end{aligned}$$

(ii) Notice that

$$\left\|f\left(\|-\|_p\right) \chi_p (\xi \cdot -)\right\|_1 = \|f\|_1 = \int_{\mathbb{Q}_p^n} \left|f\left(\|x\|_p\right)\right| d^n x < \infty$$

implies that

$$\int_{\mathbb{Q}_p^n \smallsetminus p\mathbb{Z}_p^n} \left| f\left(\|x\|_p \right) \right| d^n x = \sum_{r=0}^{\infty} \int_{S_r^n} \left| f\left(\|x\|_p \right) \right| d^n x = \sum_{r=0}^{\infty} \left| f\left(p^{-r} \right) \right| p^{-nr} \leq \|f\|_1 .$$

□

By using this formula with $f \equiv 1$, we get that

$$\int_{\mathbb{Q}_p^n} \chi_p\left(\xi \cdot x\right) dx = \begin{cases} \infty & \text{if} \quad \xi = 0 \\ \\ 0 & \text{if} \quad \xi \neq 0 \end{cases} =: \delta\left(\xi\right), \text{ the Dirac distribution.}$$

The Dirac distribution (or Dirac delta function) is the Fourier transform of the constant function 1.

EXERCISE 5.6. Set

$$Z(x,t) = \int_{\mathbb{Q}_p^n} \chi_p(x \cdot \xi) e^{-at\|\xi\|_p^{\alpha}} \, d^n\xi, \text{ for } t > 0 \text{ and } x \in \mathbb{Q}_p^n.$$

Show the formula

$$Z(x,t) = (1 - p^{-n}) \|x\|_p^{-n} \sum_{k=0}^{\infty} q^{-kn} e^{-at(q^{-k}\|x\|_p^{-1})^{\alpha}} - \|x\|_p^{-n} e^{-at(p\|x\|_p^{-1})^{\alpha}},$$

for $t > 0$ and $x \in \mathbb{Q}_p^n \smallsetminus \{0\}$.

6. Fourier Analysis on $\mathbb{Q}_p^n$

6.1. Some function spaces. For $1 \leq \rho < \infty$, we denote by $L^{\rho} := L^{\rho}\left(\mathbb{Q}_p^n, d^n x\right)$, the $\mathbb{C}$-vector space of all the complex-valued and Borel measurable functions f satisfying

$$\|f\|_{\rho} := \left\{ \int_{\mathbb{Q}_p^n} |f(x)|^{\rho} \, d^n x \right\}^{\frac{1}{\rho}} < \infty.$$

For $\rho = \infty$, $f \in L^{\infty} := L^{\infty}\left(\mathbb{Q}_p^n, d^n x\right)$, if

$$\|f\|_{\infty} := \text{ess sup}_{x \in \mathbb{Q}_p^n} \; |f(x)| < \infty. \tag{6.1}$$

The condition appearing on the right-hand side in (6.1) means that function f is bounded almost everywhere, i.e. this condition may be false in a set of measure zero. L^{ρ} is a Banach space if we identify functions f and g satisfying $f(x) = g(x)$ almost everywhere. This identification implies that $\|\cdot\|_{\rho}$ is norm, and that L^{ρ} is a complete metric space for the distance induced by $\|\cdot\|_{\rho}$.

We denote by $C_0 := C_0(\mathbb{Q}_p^n, \mathbb{C})$ the $\mathbb{C}$-vector space of continuous functions on $\mathbb{Q}_p^n$ that vanish at infinity endowed with the L^{∞}-norm. The condition 'vanishing at infinity' means that for any $\epsilon > 0$ there exists a compact subset $K \subset \mathbb{Q}_p^n$ such that

$$|f(x)| < \epsilon \text{ for } x \in \mathbb{Q}_p^n \smallsetminus K.$$

REMARK 6.1. Lebesgue's dominated convergence theorem. Let $f_m : \mathbb{Q}_p^n \to \mathbb{C}$, $m \in \mathbb{N}$, be a sequence of complex-valued Borel measurable functions. Suppose that the sequence converges pointwise to a function f and that there exists an integrable function g such that $|f_m(x)| \leq g(x)$ for any $x \in \mathbb{Q}_p^n$ and all m. Then f is integrable and

$$\lim_{m\to\infty} \int_{\mathbb{Q}_p^n} f_m(x)\, d^n x = \int_{\mathbb{Q}_p^n} \lim_{m\to\infty} f_m(x)\, d^n x = \int_{\mathbb{Q}_p^n} f(x) d^n x.$$

EXERCISE 6.2. Let $I = (a, b) \subset \mathbb{R}$ and let $f(x,t)$ be a continuous function on $\mathbb{Q}_p^n \times I$ satisfying the following conditions: (i) $f(\cdot, t) \in L^1$ for any $t \in I$; (ii) $\frac{\partial f(x,t)}{\partial t}$ is continuous in t for any $x \in \mathbb{Q}_p^n$; (iii) $\left|\frac{\partial f(x,t)}{\partial t}\right| \leq h(x) \in L^1$ for every $t \in I$. Then

$$\frac{\partial}{\partial t}\int_{\mathbb{Q}_p^n} f(x,t) d^n x = \int_{\mathbb{Q}_p^n} \frac{\partial}{\partial t} f(x,t) d^n x.$$

Hint. For every t, $t' \in I$, with $t' > t$, $f(x,t') = f(x,t) + (t'-t)\frac{\partial}{\partial t} f(x,t)$.

6.2. The Fourier transform. Given both $x = (x_1, \ldots, x_n) \in \mathbb{Q}_p^n$ and $\xi = (\xi_1, \ldots, \xi_n) \in \mathbb{Q}_p^n$, we define

$$x \cdot \xi = x_1\xi_1 + \ldots + x_n\xi_n.$$

If $f \in L^1$ its Fourier transform is the function $\widehat{f}$ defined by

$$\widehat{f}(\xi) = \int_{\mathbb{Q}_p^n} f(x)\, \chi_p(x\cdot\xi)\, d^n x.$$

We also use the notation $\mathcal{F}_{x\to\xi}(f)$, $\mathcal{F}(f)$ to denote the Fourier transform of f.

LEMMA 6.3. *(i) The mapping $f \to \widehat{f}$ is a bounded linear mapping from L^1 to L^∞ satisfying $\left\|\widehat{f}\right\|_\infty \leq \|f\|_1$.*

(ii) If $f \in L^1$, then $\widehat{f}$ is uniformly continuous.

PROOF. (i)

$$\left|\widehat{f}(\xi)\right| = \left|\int_{\mathbb{Q}_p^n} f(x)\, \chi_p(x\cdot\xi)\, d^n x\right| \leq \int_{\mathbb{Q}_p^n} |f(x)|\, d^n x = \|f\|_1.$$

(ii) Notice that

$$\widehat{f}(\xi+h) - \widehat{f}(\xi) = \int_{\mathbb{Q}_p^n} f(x)\, \chi_p(x\cdot\xi)\, \{\chi_p(x\cdot h) - 1\}\, d^n x,$$

and since $|f(x)\, \chi_p(x\cdot\xi)\, \{\chi_p(x\cdot h) - 1\}| \leq 2\,|f(x)| \in L^1$, by using the dominated convergence theorem,

$$\lim_{h\to 0}\left|\widehat{f}(\xi+h) - \widehat{f}(\xi)\right| \leq \int_{\mathbb{Q}_p^n} |f(x)| \lim_{h\to 0} |\chi_p(x\cdot h) - 1|\, d^n x = 0,$$

i.e. for any $\epsilon > 0$, there is $\delta > 0$, such that for any ξ', $\xi \in \mathbb{Q}_p^n$, with $\|\xi' - \xi\|_p = \|h\|_p < \delta$, it holds that $\left|\widehat{f}(\xi') - \widehat{f}(\xi)\right| < \epsilon$. □

EXERCISE 6.4. Let $f : \mathbb{R}_+ \to \mathbb{R}$ be a function such that $f(\|\xi\|_p)$ satisfies the following properties: (i) $f(\|\xi\|_p) = \lambda$ for $\|\xi\|_p \leq p^r$, where λ, $r \in \mathbb{Z}$ are fixed real numbers; (ii) $f(\|\xi\|_p)$ is a non-decreasing function in $\|\xi\|_p$; (iii) there exist positive constants C_0, C_1, γ_0, γ_1 and $l \in \mathbb{N}$ such that

$$C_0\|\xi\|_p^{\gamma_0} \leq f(\|\xi\|_p) \leq C_1\|\xi\|_p^{\gamma_1} \text{ for } \|\xi\|_p \geq p^l > p^r.$$

We now set

$$Z(x,t;f,\lambda,r) := Z(x,t) = \int_{\mathbb{Q}_p^n} \chi_p(-x\cdot\xi)\, e^{-t\left(f(\|\xi\|_p)-\lambda\right)} d^n\xi$$

$$= \int_{p^{-r}\mathbb{Z}_p^n} \chi_p(-x\cdot\xi)\, d^n\xi + e^{t\lambda}\int_{\mathbb{Q}_p^n\smallsetminus p^{-r}\mathbb{Z}_p^n} \chi_p(-x\cdot\xi)\, e^{-tf(\|\xi\|_p)} d^n\xi,$$

for $t>0$, and $x\in\mathbb{Q}_p^n$. Show the following assertions: first, for each $t>0$ fixed, $Z(x,t)\in L^1(\mathbb{Q}_p^n)\cap C(\mathbb{Q}_p^n)$; $Z(x,t)\geq 0$ for any x, $t\geq 0$. **Hint.** Use that $\mathbb{Q}_p^n = \bigsqcup_{k=0}^{\infty} S_k^n$. Second, show the formulae:

$$Z(x,t) = \Omega\left(p^r\|x\|_p\right)\left\{p^{rn} - \frac{e^{-tf(p\|x\|_p^{-1})}}{\|x\|_p^n} + \sum_{j=-r+1}^{ord(x)} p^{jn} e^{-tf(p^j)}\right\},$$

for $x\neq 0$ and $t>0$, and

$$Z(0,t) = p^{rn} + \sum_{j=-r+1}^{\infty} p^{jn} e^{-tf(p^j)},\ \text{for } t>0.$$

REMARK 6.5. The translation operator $\mathcal{T}_h$, $h\in\mathbb{Q}_p^n$, is defined by $(\mathcal{T}_h f)(x) = f(x-h)$. If $f\in L^1$, then $\widehat{(\mathcal{T}_h f)}(\xi) = \chi_p(\xi\cdot h)\,\widehat{f}(\xi)$, and $\mathcal{F}_{x\to\xi}(\chi_p(x\cdot h)\, f(x)) = \left(\mathcal{T}_{-h}\widehat{f}\right)(\xi) = \widehat{f}(\xi+h)$.

We denote by $\Delta_k(x) := \Omega\left(p^{-k}\|x\|_p\right)$ the characteristic function of the ball $B_k^n = p^{-k}\mathbb{Z}_p^n$ for $k\in\mathbb{Z}$. A locally constant function with compact support (a test function) $\varphi:\mathbb{Q}_p^n\to\mathbb{C}$ is a linear combination of characteristic functions of balls, then

$$\varphi(x) = \sum_{i=1}^{m} c_i \mathcal{T}_{h_i}\Delta_{k_i}(x).$$

Notice that $\mathcal{T}_{h_i}\Delta_{k_i}(x)$ is the characteristic function of the ball $h_i + p^{-k_i}\mathbb{Z}_p^n$. We denote by $\mathcal{D}(\mathbb{Q}_p^n)$ the $\mathbb{C}$–vector space of test functions (the Bruhat-Schwartz).

We set $\delta_k(x) = p^{kn}\Omega\left(p^k\|x\|_p\right)$ for $k\in\mathbb{Z}$. Notice that δ_k satisfies

$$\int_{\mathbb{Q}_p^n} \delta_k(x)\, d^n x = 1 \text{ for any } k\in\mathbb{Z}.$$

EXERCISE 6.6. To show that $\widehat{\Delta}_k(\xi) = \delta_k(\xi)$. Hence if $\varphi(x) = \sum_{i=1}^m c_i\mathcal{T}_{h_i}\Delta_{k_i}(x)$ is a test function, then $\widehat{\varphi}(\xi) = \sum_{i=1}^m c_i\chi_p(\xi\cdot h_i)\,\delta_k(\xi)$. Consequently, $\widehat{\varphi}(\xi)$ is also a test function.

REMARK 6.7. Notice that $\lim_{k\to\infty}\Delta_k(x) = 1$, and that

$$\lim_{k\to\infty}\delta_k(x) = \begin{cases} \infty & \text{if} \quad x=0 \\ 0 & \text{if} \quad x\neq 0. \end{cases}$$

EXERCISE 6.8. Set $h_t(x) := \int_{\mathbb{Q}_p}\chi_p(-\xi x)\, e^{-t|\xi|_p^{\alpha}} d\xi$, for $t>0$, $\alpha>0$. Show that $h_t(\xi)$ is a continuous function in ξ for $t>0$ fixed.

PROPOSITION 6.9 ([**23**, Chap. I, Proposition 1.3]). *$\mathcal{D}(\mathbb{Q}_p^n)$ is dense in C_0 as well as in L^ρ, $1\leq\rho<\infty$.*

PROPOSITION 6.10 (Riemann-Lebesgue Theorem). *If $f \in L^1$, then $\widehat{f}(\xi) \to 0$ as $\|\xi\|_p \to \infty$.*

PROOF. Given $\epsilon > 0$ and $f \in L^1(\mathbb{Q}_p^n)$, there exists $g_\epsilon \in \mathcal{D}(\mathbb{Q}_p^n)$ such that $\|f - g_\epsilon\|_1 < \epsilon$, cf. Proposition 6.9 . We now use that $\widehat{g_\epsilon}$ has compact support. For $\xi \notin \text{supp}\, \widehat{g_\epsilon}$ we have

$$\left|\widehat{f}(\xi)\right| = \left|\widehat{f}(\xi) - \widehat{g_\epsilon}(\xi)\right| \leq \left\|\widehat{(f - g_\epsilon)}\right\|_\infty \leq \|f - g_\epsilon\|_1 < \epsilon.$$

□

DEFINITION 6.11. Given $f,\ g : \mathbb{Q}_p^n \to \mathbb{C}$ their convolution is the function

$$\begin{aligned} h(x) = f(x) * g(x) &= \int_{\mathbb{Q}_p^n} f(x - z) g(z) d^n z \\ &= \int_{\mathbb{Q}_p^n} f(z) g(x - z) d^n z, \end{aligned}$$

when the defining integral exists.

REMARK 6.12. Young's inequality. Assume that $f \in L^\rho$, $g \in L^\sigma$ and $\frac{1}{\rho} + \frac{1}{\sigma} = \frac{1}{\gamma} + 1$ with $1 \leq \rho, \sigma, \gamma \leq \infty$. Then $\|f * g\|_\gamma \leq \|f\|_\rho \|g\|_\sigma$.

The following proposition is left as an exercise to the reader.

PROPOSITION 6.13. *If $f \in L^\rho$, $1 \leq \rho \leq \infty$, and $g \in L^1$, then $f * g \in L^\rho$ and $\|f * g\|_\rho \leq \|f\|_\rho \|g\|_1$.*

REMARK 6.14. Fubini's theorem. Let $f : \mathbb{Q}_p^{n+m} \to \mathbb{C}$ be a function such that the repeated integral

$$\int_{\mathbb{Q}_p^n} \left(\int_{\mathbb{Q}_p^m} f(x, y) d^m y \right) d^n x$$

exists, then $f \in L^1\left(\mathbb{Q}_p^{n+m}, d^{n+m} x\right)$, and the following formulae hold:

$$\begin{aligned} \int_{\mathbb{Q}_p^n} \left(\int_{\mathbb{Q}_p^m} f(x, y) d^m y \right) d^n x &= \int_{\mathbb{Q}_p^{n+m}} f(x, y) d^{n+m}(x, y) \\ &= \int_{\mathbb{Q}_p^m} \left(\int_{\mathbb{Q}_p^n} f(x, y) d^n x \right) d^m y. \end{aligned}$$

LEMMA 6.15. *If $f,\ g \in L^1$, then $\widehat{f * g} = \widehat{f}\, \widehat{g}$.*

PROOF. By Proposition 6.13, $f * g \in L^1$. The formula for the Fourier transform follows from Fubini's theorem. We invite the reader to verify this calculation. □

LEMMA 6.16. *If $f,\ g \in L^1$, then*

$$\int_{\mathbb{Q}_p^n} \widehat{f}(y) g(y)\, d^n y = \int_{\mathbb{Q}_p^n} f(y) \widehat{g}(y)\, d^n y.$$

PROOF. By using that $\left\|\widehat{f}\right\|_{\infty} \leq \|f\|_1 < \infty$, Fubini's theorem and the definition of Fourier transform, we have

$$\int_{\mathbb{Q}_p^n} \widehat{f}(y) g(y)\, d^n y = \int_{\mathbb{Q}_p^n} \left\{ \int_{\mathbb{Q}_p^n} \chi_p(x \cdot y) f(x)\, d^n x \right\} g(y)\, d^n y$$
$$= \int_{\mathbb{Q}_p^n} \left\{ \int_{\mathbb{Q}_p^n} \chi_p(x \cdot y) g(y)\, d^n y \right\} f(x)\, d^n x = \int_{\mathbb{Q}_p^n} f(x) \widehat{g}(x)\, d^n x.$$

□

6.3. The Fourier transform on the space of test functions. Let $\varphi : \mathbb{Q}_p^n \to \mathbb{C}$ be a locally constant function, this means that for each $x \in \mathbb{Q}_p^n$, there exists an integer $l = l(x)$ such that

$$\varphi(x + x') = \varphi(x) \text{ for any } x' \in B_l^n. \tag{6.2}$$

Since $B_l^n = \left\{ x \in \mathbb{Q}_p^n; \|x\|_p \leq p^l \right\} = p^{-l}\mathbb{Z}_p^n$, condition (6.2) is equivalent to

$$\varphi\,|_{x+p^{-l}\mathbb{Z}_p^n} \equiv \varphi(x).$$

If φ is a test function, then supp φ is open compact, and consequently there exist a finite number of integers l_i and a finite number of points z_i in $\mathbb{Q}_p^n$ such that

$$\text{supp } \varphi = \bigsqcup_{i=1}^{r} \left(z_i + p^{-l_i}\mathbb{Z}_p^n \right).$$

We set

$$k := \max_{1 \leq i \leq r} -l_i.$$

Then $z_i + p^{-l_i}\mathbb{Z}_p^n \supset z_i + p^k\mathbb{Z}_p^n$ (i.e. $B_{-k}^n(z_i) \subset B_{l_i}^n(z_i)$) and

$$\varphi\,|_{x+p^k\mathbb{Z}_p^n} \equiv \varphi(x) \quad \text{for any } x \in \text{supp } \varphi.$$

This means that φ is constant on the cosets of $p^k\mathbb{Z}_p^n$ (i.e. on the cosets of $\mathbb{Q}_p^n/p^k\mathbb{Z}_p^n$). The integer k is called the *index of local constancy* of φ. We now use the fact that supp φ is compact, which means that it is closed and bounded, then there exists an integer m such that

$$\text{supp } \varphi \subset p^m\mathbb{Z}_p^n.$$

Naturally, for any $x \in$supp φ, $x + p^k\mathbb{Z}_p^n \subset p^m\mathbb{Z}_p^n$, which implies that $k \geq m$.

In conclusion, $\varphi \in \mathcal{D}(\mathbb{Q}_p^n)$ if and only if there exist integers k, m, with $k \geq m$, such that φ is constant on the cosets of $p^k\mathbb{Z}_p^n$ (i.e. on the cosets of $p^m\mathbb{Z}_p^n/p^k\mathbb{Z}_p^n$) and is supported on $p^m\mathbb{Z}_p^n$. These functions form $\mathbb{C}$-vector space denoted as $\mathcal{D}_k^m(\mathbb{Q}_p^n) := \mathcal{D}_k^m$. We fix a set of representatives Is of $p^m\mathbb{Z}_p^n/p^k\mathbb{Z}_p^n =: G_{m,k}$, then the characteristic functions of the balls $I + p^k\mathbb{Z}_p^n$, $I \in G_{m,k}$ span $\mathcal{D}_k^m$, i.e.

$$\left\{ \Omega\left(p^k \|x - I\|_p \right) \right\}_{I \in G_{m,k}}$$

is a basis for $\mathcal{D}_k^m$. Notice that the dimension of $\mathcal{D}_k^m$ is $\#G_{m,k} = p^{(m-k)n}$.

LEMMA 6.17. $\mathcal{F}(\mathcal{D}_k^m) \subset \mathcal{D}_{-m}^{-k}$.

PROOF. Take $\varphi \in \mathcal{D}_k^m$, since $\varphi(x) = \sum_{I \in G_{m,k}} c_I \Omega\left(p^k \|x - I\|_p\right)$, and $\mathcal{F}$ is a linear operator, we may assume that $\varphi(x) = \Omega\left(p^k \|x - I\|_p\right)$. Then

$$\begin{aligned}\mathcal{F}_{x \to \xi}\left(\Omega\left(p^k \|x - I\|_p\right)\right) &= \int_{I + p^k \mathbb{Z}_p^n} \chi_p(\xi \cdot x)\, d^n x = p^{-nk} \chi_p(\xi \cdot I) \int_{\mathbb{Z}_p^n} \chi_p\left(p^k \xi \cdot y\right) d^n y \\ &= p^{-nk} \chi_p(\xi \cdot I)\, \Omega\left(p^{-k} \|\xi\|_p\right).\end{aligned}$$

Which means that $\mathcal{F}_{x \to \xi}\left(\Omega\left(p^k \|x - I\|_p\right)\right)$ is supported in $p^{-k}\mathbb{Z}_p^n$. Finally, we verify that $-m$ is the index of local constancy of $\mathcal{F}_{x \to \xi}\left(\Omega\left(p^k \|x - I\|_p\right)\right)$. By using that $p^{-k} < \|I\|_p \le p^{-m}$ we have

$$\chi_p(\xi \cdot I)\,|_{\xi_0 + p^{-m}\mathbb{Z}_p^n} = \chi_p(\xi_0 \cdot I),$$

for any $\xi_0 \in p^{-k}\mathbb{Z}_p^n$, and consequently $p^{-nk} \chi_p(\xi \cdot I)\, \Omega\left(\left\|p^k \xi\right\|_p\right) \in \mathcal{D}_{-m}^{-k}$. □

THEOREM 6.18 (see e.g. [**1**, Theorem 4.8.2]). *The map*

$$\begin{array}{ccc}\mathcal{D}(\mathbb{Q}_p^n) & \to & \mathcal{D}(\mathbb{Q}_p^n) \\ \varphi & \to & \widehat{\varphi},\end{array} \tag{6.3}$$

where $\widehat{\varphi}(\xi) = \int_{\mathbb{Q}_p^n} \chi_p(\xi \cdot x) \varphi(x)\, d^n x$, *is a well-defined linear operator, with inverse given by*

$$\varphi(x) = \int_{\mathbb{Q}_p^n} \chi_p(-\xi \cdot x)\, \widehat{\varphi}(\xi)\, d^n \xi.$$

In other words, the mapping (6.3) *is a isomorphism of* $\mathbb{C}$*-vector spaces on* $\mathcal{D}(\mathbb{Q}_p^n)$.

6.4. The inverse Fourier transform. One expects that the inverse Fourier transform be given by

$$f(x) = \int_{\mathbb{Q}_p^n} \chi_p(-\xi \cdot x)\, \widehat{f}(\xi)\, d^n \xi = \int_{\mathbb{Q}_p^n} \overline{\chi_p(\xi \cdot x)}\, \widehat{f}(\xi)\, d^n \xi.$$

This formula does not always make sense since $\widehat{f}$ is not necessarily in L^1 when $f \in L^1$.

EXERCISE 6.19. Show that the Fourier transform of $f(x) = \Omega\left(|x|_p\right) \ln\left(\frac{1}{|x|_p}\right)$ is

$$\widehat{f}(\xi) = \begin{cases} \frac{p^{-1} \ln p}{1 - p^{-1}} & \text{if} \quad |\xi|_p \le 1 \\ \\ \left(\frac{\ln p}{1 - p^{-1}}\right) |\xi|_p^{-1} & \text{if} \quad |\xi|_p > 1. \end{cases}$$

DEFINITION 6.20. If g is locally integrable and $k \in \mathbb{Z}$, we define

$$A_k g = \int_{\mathbb{Q}_p^n} g(x)\, \Delta_k(x)\, d^n x = \int_{\|x\|_p \le p^k} g(x)\, d^n x.$$

Notice that if $g \in L^1$, then $A_k g \to \int_{\mathbb{Q}_p^n} g(x)\, d^n x$ as $k \to \infty$ (why?). Now, the limit $\lim_{k \to \infty} A_k g$ may exist even though $\int_{\mathbb{Q}_p^n} g(x)\, d^n x$ does not exist.

EXERCISE 6.21. To show the following fact: if $f \in L^1$, $k \in \mathbb{Z}$, then

$$A_k\left(\widehat{f}\ \overline{\chi_p(x\cdot)}\right) = \int_{\mathbb{Q}_p^n} \widehat{f}(\xi)\ \overline{\chi_p(x\cdot\xi)}\ \Delta_k(\xi)\, d^n\xi$$
$$= \int_{\mathbb{Q}_p^n} f(y)\,\delta_k(y-x)\, d^n y = p^{nk}\int_{\|x-y\|_p \le p^{-k}} f(y)\, d^n y.$$

DEFINITION 6.22. Let $f : \mathbb{Q}_p^n \to \mathbb{C}$ be locally integrable. A point $x \in \mathbb{Q}_p^n$ is called a regular point of f if

$$f_k(x) := p^{nk}\int_{\|x-y\|_p \le p^{-k}} f(y)\, d^n y \to f(x) \text{ as } k \to \infty.$$

THEOREM 6.23 ([**23**, Theorem 1.14]). *Let f be a locally integrable function. There exists a zero measure subset $L = L(f)$ such that any $x \in \mathbb{Q}_p^n \smallsetminus L$ is a regular point of f.*

EXERCISE 6.24. Assume that f is locally integrable and continuous at x. Show that

$$p^{nk}\int_{\|x-y\|_p \le p^{-k}} f(y)\, d^n y \to f(x) \text{ as } k \to \infty.$$

COROLLARY 6.25. *If $f \in L^1$, then*

$$A_k\left(\widehat{f}\ \overline{\chi_p(x\cdot)}\right) = \int_{\mathbb{Q}_p^n} \widehat{f}(\xi)\,\chi_p(-x\cdot\xi)\,\Delta_k(\xi)\, d^n\xi \to f(x)$$

almost everywhere. In particular, it converges at each point of continuity of f.

PROOF. The corollary follows from Exercise 6.21, Theorem 6.23 and Exercise 6.24. □

THEOREM 6.26. *If f and $\widehat{f}$ are both integrable then f is equal a.e. to a continuous function. With f modified (on a set of measure zero) to be continuous, we have*

$$f(x) = \int_{\mathbb{Q}_p^n} \chi_p(-\xi\cdot x)\,\widehat{f}(\xi)\, d^n\xi \quad \textit{for all } x \in \mathbb{Q}_p^n. \tag{6.4}$$

PROOF. If $\widehat{f}$ is integrable, then $A_k\left(\widehat{f}\ \chi_p(-\xi\cdot)\right)$ converges to a continuous function:

$$\int_{\mathbb{Q}_p^n} \chi_p(-\xi\cdot x)\,\widehat{f}(\xi)\, d^n\xi. \tag{6.5}$$

By Corollary 6.25, $f(x)$ agrees with (6.5) a.e. and consequently f is continuous almost everywhere. By modifying f on a set of measure zero we obtain formula (6.4). □

EXERCISE 6.27. Define

$$\mathcal{L}(\mathbb{Q}_p^n) := \left\{ f : \mathbb{Q}_p^n \to \mathbb{C} :\ f,\, \widehat{f} \in L^1 \text{ and } f,\, \widehat{f} \text{ are continuous.} \right\}$$

Then $\mathcal{L}(\mathbb{Q}_p^n) \underset{\to}{\mathcal{F}} \mathcal{L}(\mathbb{Q}_p^n)$ is an isomorphism of $\mathbb{C}$-vector spaces. In particular, the formula

$$f(0) = \int_{\mathbb{Q}_p^n} \widehat{f}(\xi)\, d^n\xi$$

holds true.

COROLLARY 6.28. *If $f,\, g \in L^1$ and $\widehat{f} = \widehat{g}$ a.e., then $f(x) = g(x)$ a.e.*

PROOF. Since $\widehat{(f-g)} = 0$, by Theorem 6.26, $(f-g)(x) = 0$ a.e. $\square$

REMARK 6.29. Monotone convergence lemma (Levi's monotone convergence theorem). Let $h_k : \mathbb{Q}_p^n \to [0,\infty]$, $k \in \mathbb{N}$, be a sequence of non-negative Borel measurable functions satisfying

$$0 \leq h_k(x) \leq h_{k+1}(x) \leq \infty \quad \text{for any } x \in \mathbb{Q}_p^n.$$

Assume that $h : \mathbb{Q}_p^n \to [0,\infty]$ is the pointwise limit of the sequence $\{h_k\}_{k\in\mathbb{N}}$. Then h is Borel measurable and

$$\lim_{k\to\infty} \int_{\mathbb{Q}_p^n} h_k(x)\, d^n x = \int_{\mathbb{Q}_p^n} h(x) d^n x.$$

The integral $\int_{\mathbb{Q}_p^n} h(x) d^n x$ could be equal to $+\infty$.

COROLLARY 6.30. *If $f \in L^1$, $\widehat{f} \geq 0$ and f is continuous at zero, then $\widehat{f} \in L^1$ and $f(x) = \int_{\mathbb{Q}_p^n} \chi_p(-\xi\cdot x)\, \widehat{f}(\xi)\, d^n\xi$ at each regular point of f. In particular, $f(0) = \int_{\mathbb{Q}_p^n} \widehat{f}(\xi)\, d^n\xi$.*

PROOF. We need only to show that $\widehat{f} \in L^1$. Since f and $\Delta_k \in L^1$, by Lemma 6.16 and Exercise 6.6, we have

$$\int_{\mathbb{Q}_p^n} \widehat{f}(\xi)\, \Delta_k(\xi)\, d^n\xi = \int_{\mathbb{Q}_p^n} f(\xi)\, \delta_k(\xi)\, d^n\xi.$$

By Theorem 6.23,

$$f(0) = \lim_{k\to\infty} \int_{\mathbb{Q}_p^n} f(\xi)\, \delta_k(\xi)\, d^n\xi = \lim_{k\to\infty} \int_{\mathbb{Q}_p^n} \widehat{f}(\xi)\, \Delta_k(\xi)\, d^n\xi.$$

Finally, by using the fact that $\widehat{f} \geq 0$, and monotone convergence lemma, we have $\widehat{f} \in L^1$. $\square$

7. The L^2-theory

THEOREM 7.1. *If $f \in L^1 \cap L^2$, then $\left\|\widehat{f}\right\|_2 = \|f\|_2$.*

PROOF. We set $g(x) := \overline{f}(-x)$, then $\widehat{g} = \overline{\widehat{f}}$. Since $f,\, g \in L^1$, $f * g \in L^1$ and

$$\widehat{f * g} = \widehat{f}\widehat{g} = \left|\widehat{f}\right|^2 \geq 0,$$

cf. Proposition 6.13 and Lemma 6.15.

Now, since $f,\, g \in L^2$, then $f * g$ is continuous, indeed, by using the Cauchy-Schwarz inequality,

$$|(f*g)(x+y) - (f*g)(x)| = \left| \int_{\mathbb{Q}_p^n} \{f(x+y-z) - f(x-z)\}\, g(z) d^n z \right|$$

$$\leq \sqrt{\int_{\mathbb{Q}_p^n} |g(z)|^2\, d^n z}\, \sqrt{\int_{\mathbb{Q}_p^n} |f(x+y-z) - f(x-z)|^2\, d^n z}$$

$$= \|g\|_2 \sqrt{\int_{\mathbb{Q}_p^n} |f(u+y) - f(u)|^2\, d^n z} = \|g\|_2\, \|f(\cdot + y) - f(\cdot)\|_2 \to 0 \text{ as } y \to 0,$$

by the dominated convergence theorem and the fact that

$$|f(u+y)-f(u)|^2 \leq 4\max\left\{|f(u+y)|^2, |f(u)|^2\right\} \leq 4\left\{|f(u+y)|^2 + |f(u)|^2\right\}.$$

We now apply Corollary 6.30 to $f*g$, with $\widehat{f*g} \geq 0$, to get that $\widehat{f*g} = \left|\widehat{f}\right|^2 \in L^1$ and

$$(f*g)(0) = \int_{\mathbb{Q}_p^n} \left|\widehat{f}(\xi)\right|^2 d^n\xi = \int_{\mathbb{Q}_p^n} g(-z)\, f(z) d^n z = \int_{\mathbb{Q}_p^n} |f(z)|^2 d^n z.$$

□

REMARK 7.2. Let $(Y, \|\cdot\|_Y)$ be a Banach space, this means that $(Y, \|\cdot\|_Y)$ is a normed complex space such that Y is a complete metric space for the distance induced by $\|\cdot\|_Y$. Let $(X, \|\cdot\|_X)$ be a complex normed space, and let $D(T)$ be a subspace of $(X, \|\cdot\|_X)$. Let $T : D(T) \to Y$ be a linear bounded operator, i.e. T satisfies $T(\alpha x + \beta y) = \alpha T(x) + \beta T(y)$ for any α, $\beta \in \mathbb{C}$, and any x, $y \in D(T)$, and

$$\|T\| := \sup_{x\in D(T)} \frac{\|T(x)\|_Y}{\|x\|_X} < \infty.$$

Then T has an extension $\widetilde{T} : \overline{D(T)} \to Y$, where $\overline{D(T)}$ denotes the closure of $D(T)$ in $(X, \|\cdot\|_X)$, with the same norm $\left\|\widetilde{T}\right\| = \|T\|$. If $\overline{D(T)} = X$, i.e. if $D(T)$ is dense in X, then $\widetilde{T}$ is unique.

REMARK 7.3. Let $T : X \to Y$ be a linear operator. Then T is continuous if and only if T is bounded (i.e. $\|T\| := \sup_{x\in X} \frac{\|T(x)\|_Y}{\|x\|_X} < \infty$).

From this theorem, it follows that the mapping

$$\begin{array}{ccc} L^1 \cap L^2 & \to & L^2 \\ f & \to & \widehat{f} \end{array}$$

is an L^2-isometry on $L^1 \cap L^2$, which is a dense subspace of L^2 (Why?). Thus, this mapping has an extension to an L^2-isometry from L^2 into L^2. We now extend the Fourier transform to L^2.

DEFINITION 7.4. For $f \in L^2$, let

$$f_k := f\Delta_k, \text{ for } k \in \mathbb{N}, \tag{7.1}$$

and

$$\widehat{f}(\xi) := \lim_{k\to\infty} \widehat{f_k}(\xi) = \lim_{k\to\infty} \int_{\|x\|_p \leq p^k} \chi_p(\xi\cdot x)\, f(x) d^n x,$$

where the limit is taken in L^2.

LEMMA 7.5. *If f, $g \in L^2$, then*

$$\int_{\mathbb{Q}_p^n} \widehat{f}\; g d^n y = \int_{\mathbb{Q}_p^n} f\; \widehat{g} d^n y.$$

PROOF. We first notice that $f_k \overset{L^2}{\to} f$ and $g_k \overset{L^2}{\to} g$, and that f_k, $g_k \in L^1 \cap L^2$ for every k. Hence, by Lemma 6.16,

$$\int_{\mathbb{Q}_p^n} f_k\; \widehat{g_k} d^n x = \int_{\mathbb{Q}_p^n} \widehat{f_k}\; g_k d^n x.$$

Now, by using Theorem 7.1 and the Cauchy-Schwarz inequality,

$$\left|\int_{\mathbb{Q}_p^n} f_k\ \widehat{g_k} d^n x\right| \leq \|f_k\|_2 \|\widehat{g_k}\|_2 = \|f_k\|_2 \|g_k\|_2 ,$$

which means that the bilinear form

$$(f_k, g_k) \to \int_{\mathbb{Q}_p^n} f_k\ \widehat{g_k} d^n x$$

is bounded (and consequently continuous) in each variable in L^2, then

$$\int_{\mathbb{Q}_p^n} f_k\ \widehat{g_k} d^n x \overset{L^2}{\rightarrow} \int_{\mathbb{Q}_p^n} f\ \widehat{g} d^n x.$$

□

THEOREM 7.6. *The Fourier transform is unitary in L^2.*

PROOF. We have to show that the Fourier transform is a bijective linear mapping that preserves the L^2-norm. We already know that $f \underset{\rightarrow}{\mathcal{F}} \widehat{f}$ is a linear L^2-isometry. It remains to show that it is onto. By contradiction, we assume that $\mathcal{F}$ is not onto. Notice that $\mathcal{F}(L^2)$ is closed in L^2 because $\left\|\widehat{f}\right\|_2 = \|f\|_2$, if $\mathcal{F}(L^2) \neq L^2$, then by general theory of Hilbert spaces, $\mathcal{F}(L^2)$ has an orthogonal complement $\mathcal{F}(L^2)^{\perp}$ such that $L^2 = \mathcal{F}(L^2) \oplus \mathcal{F}(L^2)^{\perp}$, where for any $g \in \mathcal{F}(L^2)^{\perp}$ and any $\widehat{f} \in \mathcal{F}(L^2)$, $\left\langle \widehat{f}, g \right\rangle = 0$. Then, there exists $\overline{h} \in L^2$, $\left\|\overline{h}\right\|_2 \neq 0$ such that

$$\left\langle \widehat{f}, \overline{h} \right\rangle = \int_{\mathbb{Q}_p^n} \widehat{f} h d^n x = 0 \text{ for any } f \in L^2.$$

By using Lemma 7.5, $\widehat{h} = 0$, but $\left\|\overline{h}\right\|_2 = \|h\|_2 = \left\|\widehat{h}\right\|_2 = 0$, cf. Theorem 7.1, which contradicts $\left\|\overline{h}\right\|_2 \neq 0$. □

EXERCISE 7.7. Show that $\left\|\widehat{f}\right\|_2 = \|f\|_2$ for $f \in L^2$ is equivalent to for any f, $g \in L^2$,

$$\langle f, g\rangle = \left\langle \widehat{f}, \widehat{g} \right\rangle \text{ i.e. } \int_{\mathbb{Q}_p^n} f\ \overline{g} d^n x = \int_{\mathbb{Q}_p^n} \widehat{f}\ \overline{\widehat{g}} d^n x.$$

EXERCISE 7.8. For $f \in L^2$, we set $\widetilde{f}$ for the reflection of f which is defined as $\widetilde{f}(x) = f(-x)$. Show that if $f \in L^2$, then

$$\mathcal{F}^{-1}(f) = \mathcal{F}\left(\widetilde{f}\right).$$

THEOREM 7.9. *If $f \in L^2$, then*

$$\lim_{k\to\infty} \int_{\|x\|_p \leq p^k} \chi_p(\xi \cdot x) f(x) d^n x = \widehat{f}(\xi) \quad \textit{almost everywhere.}$$

PROOF. We use that Δ_k, f, $\widehat{f} \in L^2$, jointly with Lemma 7.5 and Exercise 7.7 to get

$$\begin{aligned}\lim_{k\to\infty}\int_{\|x\|_p\le p^k}\chi_p(\xi\cdot x)\,f(x)d^nx &= \lim_{k\to\infty}\int_{\mathbb{Q}_p^n} f(x)\overline{\chi_p(-\xi\cdot x)\,\Delta_k(x)}d^nx \\ = \lim_{k\to\infty}\int_{\mathbb{Q}_p^n}\widehat{f}(z)\overline{\delta_k(z-\xi)}d^nz &= \lim_{k\to\infty} p^{nk}\int_{\|z-\xi\|_p\le p^{-k}}\widehat{f}(z)d^nz.\end{aligned}$$

Now, since $\widehat{f} \in L^2$, then $\widehat{f} \in L^1_{loc}$, the result follows from Theorem 6.23. □

REMARK 7.10. (i) The Fourier transform can be extended to $L^1 + L^2$, which means that if $f = f_1+f_2$, with $f_1 \in L^1$, $f_2 \in L^2$, then $\widehat{f} = \widehat{f_1}+\widehat{f_2}$, where the Fourier transforms are defined in L^1 and L^2 respectively. The function $\widehat{f}$ is well-defined in L^1_{loc}.

(ii) If $f \in L^1$, $g \in L^\rho$, $\rho \in [1,2]$, then $\widehat{f*g} = \widehat{f}\,\widehat{g}$ a.e., cf. [**23**, Theorem 2.7].

8. $\mathcal{D}$ as a topological vector space

We define a topology on $\mathcal{D}$ as follows. We say that a sequence $\{\varphi_j\}_{j\in\mathbb{N}}$ of functions in $\mathcal{D}$ converges to zero, if the two following conditions hold:

(C1) there are two fixed integers k and m such that each φ_j is constant on the cosets of $p^k\mathbb{Z}_p^n$ and is supported on $p^m\mathbb{Z}_p^n$, i.e. $\varphi_j \in \mathcal{D}_k^m$;

(C2) $\varphi_j \to 0$ uniformly.

$\mathcal{D}$ endowed with the above topology becomes a topological vector space.

We recall that $\mathcal{D}_k^m$ is $\mathbb{C}$-vector space of dimension $N_{m.k} := \#\left(p^m\mathbb{Z}_p^n/p^k\mathbb{Z}_p^n\right)$. Given $c = (c_1,\ldots,c_{N_{m.k}}) \in \mathbb{C}^{N_{m.k}}$, we set $\|c\|_{\mathbb{C}} = \max_i |c_i|$. Then $\left(\mathbb{C}^{N_{m.k}}, \|\cdot\|_{\mathbb{C}}\right)$ is a Banach space and

$$\left(\mathbb{C}^{N_{m.k}}, \|\cdot\|_{\mathbb{C}}\right) \simeq \mathcal{D}_k^m \text{ as topological spaces, (why?).}$$

A key fact is that uniform convergence in $\mathcal{D}_k^m$ agrees with the convergence in the supremum norm (L^∞-norm), which in turn agrees with the convergence in the $\|\cdot\|_{\mathbb{C}}$-norm.

EXERCISE 8.1. Show that $\mathcal{D}$ is a complete and separable topological vector space.

THEOREM 8.2. *The map*

$$\begin{array}{ccc}\mathcal{D}(\mathbb{Q}_p^n) & \to & \mathcal{D}(\mathbb{Q}_p^n) \\ \varphi & \to & \widehat{\varphi},\end{array} \tag{8.1}$$

where $\widehat{\varphi}(\xi) = \int_{\mathbb{Q}_p^n}\chi_p(\xi\cdot x)\,\varphi(x)\,d^nx$, *is a homeomorphism of topological vector spaces , with inverse given by*

$$\varphi(x) = \int_{\mathbb{Q}_p^n}\chi_p(-\xi\cdot x)\,\widehat{\varphi}(\xi)\,d^n\xi.$$

PROOF. We already know that (8.1) is an isomorphism of vector spaces. It remains to show that the continuity of $\mathcal{F}$ and $\mathcal{F}^{-1}$. Let $\varphi_j \xrightarrow{\mathcal{D}} \varphi$, i.e. $\varphi_j, \varphi \in \mathcal{D}_k^m$ for some integers m, k, and $\varphi_j \xrightarrow{\text{unif.}} \varphi$. Since $\widehat{\varphi_j}, \widehat{\varphi} \in \mathcal{D}_{-m}^{-k}$, in order to show that

$\widehat{\varphi_j} \underset{\rightarrow}{\mathcal{D}} \widehat{\varphi}$, it is sufficient to establish that $\widehat{\varphi_j} \xrightarrow{\text{unif.}} \widehat{\varphi}$, i.e. $\|\widehat{\varphi_j} - \widehat{\varphi}\|_\infty \to 0$ as $j \to \infty$. By using that

$$\begin{aligned} \|\widehat{\varphi_j} - \widehat{\varphi}\|_\infty &\leq \int_{\mathbb{Q}_p^n} |\varphi_j - \varphi| \, d^n x \leq \|\varphi_j - \varphi\|_\infty \int_{p^m \mathbb{Z}_p^n} d^n x \\ &= p^{-mn} \|\varphi_j - \varphi\|_\infty \to 0 \text{ as } j \to \infty. \end{aligned}$$

The continuity of the inverse Fourier transform is established by using the same argument. □

9. The space of distributions on $\mathbb{Q}_p^n$

The $\mathbb{C}$-vector space $\mathcal{D}'\left(\mathbb{Q}_p^n\right) := \mathcal{D}'$ of all continuous linear functionals on $\mathcal{D}$ is called the *Bruhat-Schwartz space of distributions* . We endow $\mathcal{D}'$ with the weak topology, i.e. a sequence $\{T_j\}_{j\in\mathbb{N}}$ in $\mathcal{D}'$ converges to T if

$$\lim_{j\to\infty} T_j(\varphi) = T(\varphi) \text{ for any } \varphi \in \mathcal{D}.$$

EXERCISE 9.1. Define the map

$$\begin{array}{ccc} \mathcal{D}' \times \mathcal{D} & \to & \mathbb{C} \\ (T, \varphi) & \to & T(\varphi). \end{array}$$

Then (T, φ) is a bilinear form which is continuous in T and φ separately. We call this map the pairing between $\mathcal{D}'$ and $\mathcal{D}$. From now on we will use (T, φ) instead of $T(\varphi)$.

EXERCISE 9.2. If $f \in L^\rho$, $1 \leq \rho \leq \infty$, then f induces a distribution. More precisely,

$$(f, \varphi) = \int_{\mathbb{Q}_p^n} f\varphi d^n x$$

REMARK 9.3. If T is a distribution and g is a locally integrable function such that

$$(T, \varphi) = \int_{\mathbb{Q}_p^n} g\varphi d^n x \text{ for all } \varphi \in \mathcal{D},$$

we identify T with function g. In this case, some authors say that T is a regular distribution.

EXAMPLE 9.4. (i) The distribution $(\delta, \varphi) = \varphi(0)$ is called the Dirac distribution.

(ii) $(1, \varphi) = \int_{\mathbb{Q}_p^n} \varphi d^n x$.

LEMMA 9.5. *Every linear functional on $\mathcal{D}$ is continuous, i.e. $\mathcal{D}'$ agrees with the algebraic dual of $\mathcal{D}$.*

PROOF. Let $T : \mathcal{D} \to \mathbb{C}$ be a linear functional, and let $\{\varphi_j\}_{j\in\mathbb{N}}$ be a sequence of test functions converging to 0. Then $\varphi_j \in \mathcal{D}_k^m$, for all j, consequently

$$\varphi_j(x) = \sum_{I \in G_{m,k}} \varphi_j(I) \, \Omega\left(p^k \|x - I\|_p\right),$$

and $\max_{I\in G_{m,k}} |\varphi_j(I)| \to 0$ as $j \to \infty$. Then

$$T\varphi_j(x) = \sum_{I \in G_{m,k}} \varphi_j(I) \, T\left(\Omega\left(p^k \|x - I\|_p\right)\right) = \sum_{I \in G_{m,k}} c_{k,I} \, \varphi_j(I),$$

and

$$|T\varphi_j(x)| \leq \sum_{I\in G_{m,k}} |c_{k,I}| \, |\varphi_j(I)| \leq \left\{ \sum_{I\in G_{m,k}} |c_{k,I}| \right\} \max_{I\in G_{m,k}} |\varphi_j(I)| \to 0,$$

as $j \to \infty$. □

We set

$$\Gamma_p^{(n)}(\alpha) := \frac{1-p^{\alpha-n}}{1-p^{-\alpha}}, \text{ for } \alpha \in \mathbb{C} \smallsetminus \{0\}.$$

The function

$$k_\alpha(x) = \frac{\|x\|_p^{\alpha-n}}{\Gamma_p^{(n)}(\alpha)}, \quad \text{for } x \in \mathbb{Q}_p^n, \ \operatorname{Re}(\alpha) > 0, \, \alpha \neq n,$$

is called *the multi-dimensional Riesz Kernel*. It determines a distribution from $\mathcal{D}'$, which admits a meromorphic continuation to all $\alpha \in \mathbb{C} \smallsetminus \left\{ n + \frac{2\pi\sqrt{-1}}{\ln p}\mathbb{Z} \right\}$ given by

$$(k_\alpha(x), \varphi(x)) = \frac{1-p^{-n}}{1-p^{\alpha-n}}\varphi(0) + \frac{1-p^{-\alpha}}{1-p^{\alpha-n}} \int_{\|x\|_p>1} \|x\|_p^{\alpha-n} \varphi(x)\, d^n x$$

$$+\frac{1-p^{-\alpha}}{1-p^{\alpha-n}} \int_{\|x\|_p\leq 1} \|x\|_p^{\alpha-n} (\varphi(x)-\varphi(0))\, d^n x. \tag{9.1}$$

In the case $\alpha = 0$, by passing to the limit in (9.1), we obtain

$$(k_0(x), \varphi(x)) := \lim_{\alpha\to 0} (k_\alpha(x), \varphi(x)) = \varphi(0),$$

i.e., $k_0(x) = \delta(x)$, the Dirac delta function.

EXERCISE 9.6. With the above notation, for $\operatorname{Re}(\alpha) > 0$, the following formulas hold true:

$$(k_{-\alpha}(x), \varphi(x)) = \frac{1-p^{\alpha}}{1-p^{-\alpha-n}} \int_{\mathbb{Q}_p^n} (\varphi(x)-\varphi(0)) \frac{d^n x}{\|x\|_p^{\alpha+n}}, \tag{9.2}$$

and

(9.3)

$$(k_\alpha(x), \varphi(x)) = \frac{1-p^{-\alpha}}{1-p^{\alpha-n}} \int_{\mathbb{Q}_p^n} \|x\|_p^{\alpha-n} \varphi(x)\, d^n x, \ \text{ for } \alpha \in \mathbb{C} \smallsetminus \left\{ n + \frac{2\pi\sqrt{-1}}{\ln p}\mathbb{Z} \right\}.$$

10. The Fourier transform on $\mathcal{D}'$

DEFINITION 10.1. For $T \in \mathcal{D}'$ its Fourier transform, denoted as $\mathcal{F}(T)$ or $\widehat{T}$, is the distribution defined as

$$(\mathcal{F}(T), \varphi) = (T, \mathcal{F}(\varphi)) \text{ for all } \varphi \in \mathcal{D}.$$

Notice that since $\varphi \to \mathcal{F}(\varphi)$ is a homeomorphism of $\mathcal{D}$, $\mathcal{F}(T)$ is well-defined.

EXAMPLE 10.2. $\widehat{\delta} = 1$. Indeed,

$$\left(\widehat{\delta}, \varphi\right) = (\delta, \widehat{\varphi}) = \widehat{\varphi}(0) = \int_{\mathbb{Q}_p^n} \varphi d^n x = (1, \varphi) \text{ for all } \varphi \in \mathcal{D}.$$

DEFINITION 10.3. The inverse Fourier transform of $T \in \mathcal{D}'$, denoted as $\mathcal{F}^{-1}(T)$ or $\check{T}$, is defined as

$$\left(\mathcal{F}^{-1}(T), \varphi\right) = \left(T, \mathcal{F}^{-1}(\varphi)\right) \text{ for all } \varphi \in \mathcal{D}.$$

REMARK 10.4. Notice that $\mathcal{F}^{-1}(T) \in \mathcal{D}'$ and that $\mathcal{F}^{-1}(\mathcal{F}(T)) = T$ for any $T \in \mathcal{D}'$.

LEMMA 10.5. *The map $T \to \mathcal{F}(T)$ is a homeomorphism of $\mathcal{D}'$ onto $\mathcal{D}'$.*

PROOF. The onto part follows from Remark 10.4. For the continuity, we proceed as follows. Let $T_j \underset{\to}{\mathcal{D}'} T$, i.e. $(T_j, \varphi) \to (T, \varphi)$ for all $\varphi \in \mathcal{D}$. Then

$$\left(\widehat{T_j}, \varphi\right) = (T_j, \widehat{\varphi}) \to (T, \widehat{\varphi}) \text{ for all } \varphi \in \mathcal{D},$$

i.e.

$$\left(\widehat{T_j}, \varphi\right) \to \left(\widehat{T}, \varphi\right) \text{ for all } \varphi \in \mathcal{D}.$$

□

DEFINITION 10.6. Linear change of variables for distributions. For $A \in GL_n(\mathbb{Q}_p^n)$ and $b \in \mathbb{Q}_p^n$, we define

$$(T(Ax+b), \varphi) = \frac{1}{|\det A|_p} \left(T, \varphi\left(A^{-1}(y-b)\right)\right).$$

EXAMPLE 10.7. Recall that $\mathcal{T}_b(\varphi(x)) = \varphi(x-b)$ for $b \in \mathbb{Q}_p^n$ and $\varphi \in \mathcal{D}$. Then, for $G \in \mathcal{D}'$,

$$(\mathcal{T}_b G, \varphi) = (G, \mathcal{T}_{-b}\varphi).$$

EXAMPLE 10.8. The reflection operator, denoted as $\widetilde{\bullet}$, acting on $\varphi \in \mathcal{D}$ is defined as $\widetilde{\varphi}(x) = \varphi(-x)$. Then for $G \in \mathcal{D}'$, $\left(\widetilde{G}, \varphi\right) = (G, \widetilde{\varphi})$.

EXERCISE 10.9. Show that

$$G \to \mathcal{F}(G),\ G \to \mathcal{F}^{-1}(G),\ G \to \mathcal{T}_b G,\ G \to \widetilde{G}$$

are homeomorphisms of $\mathcal{D}'$ onto $\mathcal{D}'$.

DEFINITION 10.10. For $G \in \mathcal{D}'$ and $\theta \in \mathcal{D}$, we define the distribution θG by

$$(\theta G, \varphi) = (G, \theta\varphi) \text{ for all } \varphi \in \mathcal{D}.$$

EXERCISE 10.11. For ψ, $\varphi \in \mathcal{D}$, $G \in \mathcal{D}'$, the maps

$$G \to \varphi G, \quad \psi \to \psi\varphi$$

are continuous maps from $\mathcal{D}'$ into $\mathcal{D}'$ and from $\mathcal{D}$ into $\mathcal{D}$, respectively.

EXERCISE 10.12. Show that $\mathcal{F}_{x\to\xi}(\chi_p(x\cdot\xi_0)) = \delta(\xi - \xi_0)$ in $\mathcal{D}'$.

EXAMPLE 10.13. We want to compute $\mathcal{F}^{-1}_{\xi\to x}\left(\frac{1}{\|\xi\|_p^\alpha + m^2}\right)$, where $\alpha > 0$ and $m > 0$. Notice that if $\alpha > n$, then $\frac{1}{\|\xi\|_p^\alpha + m^2} \in L^1$ (Why?). In the general case, $\frac{1}{\|\xi\|_p^\alpha + m^2} \notin L^1$. If we use the notion of improper integral, we have

$$\int_{\mathbb{Q}_p^n} \frac{\chi(-\xi\cdot x)}{\|\xi\|_p^\alpha + m^2} d^n\xi = \sum_{j=-\infty}^{\infty} \int_{\|\xi\|_p = p^j} \frac{\chi(-\xi\cdot x)}{\|\xi\|_p^\alpha + m^2} d^n\xi.$$

The exact mathematical meaning of this formula is

$$\sum_{j=-\infty}^{N} \int_{\|\xi\|_p = p^j} \frac{\chi(-\xi\cdot x)}{\|\xi\|_p^\alpha + m^2} d^n\xi \underset{\to}{\mathcal{D}'} \mathcal{F}^{-1}_{\xi\to x}\left(\frac{1}{\|\xi\|_p^\alpha + m^2}\right) \text{ as } N \to \infty.$$

Indeed, for any test function φ, we have

$$\sum_{j=-\infty}^{N}\int_{\mathbb{Q}_p^n}\int_{\|\xi\|_p=p^j}\frac{\chi(-\xi\cdot x)}{\|\xi\|_p^{\alpha}+m^2}\varphi(x)\,d^n\xi d^n x=\int_{\mathbb{Q}_p^n}\int_{\|\xi\|_p\leq p^N}\frac{\chi(-\xi\cdot x)}{\|\xi\|_p^{\alpha}+m^2}\varphi(x)\,d^n\xi d^n x$$
$$=\int_{\|\xi\|_p\leq p^N}\frac{1}{\|\xi\|_p^{\alpha}+m^2}\left\{\int_{\mathbb{Q}_p^n}\chi(-\xi\cdot x)\varphi(x)\,d^n x\right\}d^n\xi=\int_{\|\xi\|_p\leq p^N}\frac{\mathcal{F}^{-1}(\varphi)(\xi)}{\|\xi\|_p^{\alpha}+m^2}d^n\xi$$
$$=\int_{\mathbb{Q}_p^n}\frac{\mathcal{F}^{-1}(\varphi)}{\|\xi\|_p^{\alpha}+m^2}d^n\xi=\left(\frac{1}{\|\xi\|_p^{\alpha}+m^2},\mathcal{F}^{-1}(\varphi)\right)\text{ for } N \text{ sufficiently large.}$$

PROPOSITION 10.14. *For* $\alpha\in\mathbb{C}\smallsetminus\left\{n+\frac{2\pi\sqrt{-1}}{\ln p}\mathbb{Z}\right\}$, $\mathcal{F}_{x\to\xi}\left(k_\alpha(x)\right)=\|\xi\|_p^{-\alpha}$ *in* $\mathcal{D}'$.

PROOF. In $\mathcal{D}'$, by using Exercise 5.3, we have

$$\begin{aligned}\mathcal{F}_{x\to\xi}\left(k_\alpha(x)\right)&=\lim_{N\to\infty}\sum_{j=-\infty}^{N}\int_{\|x\|_p=p^j}k_\alpha(x)\chi_p(x\cdot\xi)\,d^n x\\&=\lim_{N\to\infty}\sum_{j=-\infty}^{N}\frac{p^{j(\alpha-n)}}{\Gamma_p^{(n)}(\alpha)}\int_{\|x\|_p=p^j}\chi_p(x\cdot\xi)\,d^n x\\&=\lim_{N\to\infty}\sum_{j=-\infty}^{N}\frac{p^{j(\alpha-n)}}{\Gamma_p^{(n)}(\alpha)}\begin{cases}p^{jn}\left(1-p^{-n}\right)&\text{if }\ ord(\xi)\geq j\\-p^{jn-n}&\text{if }\ ord(\xi)=j-1\\0&\text{if }\ ord(\xi)\leq j-2.\end{cases}\end{aligned}$$

For $\xi\neq 0$ there is $N>ord(\xi)$, then

$$\mathcal{F}_{x\to\xi}\left(k_\alpha(x)\right)=\sum_{j=-\infty}^{ord(\xi)}\frac{\left(1-p^{-n}\right)p^{j\alpha}}{\Gamma_p^{(n)}(\alpha)}-\frac{p^{-n}p^{(ord(\xi)+1)\alpha}}{\Gamma_p^{(n)}(\alpha)}=\|\xi\|_p^{-\alpha}.$$

□

LEMMA 10.15. *If* $\alpha>0$, *then*

$$\|x\|_p^{\alpha}=\frac{1-p^{\alpha}}{1-p^{\alpha-n}}\int_{\mathbb{Q}_p^n}\|y\|_p^{\alpha-n}\left\{\chi_p(x\cdot y)-1\right\}d^n y\ \text{ in } \mathcal{D}'.$$

PROOF. Take ψ, $\varphi\in\mathcal{D}$ with $\psi=\mathcal{F}^{-1}(\varphi)$. Then by Proposition 10.14 and formula (9.2), we have

$$\begin{aligned}\left(\|x\|_p^{\alpha},\varphi\right)&=\Gamma_p^{(n)}(\alpha+n)\left(k_{\alpha+n},\varphi\right)=\Gamma_p^{(n)}(\alpha+n)\left(\mathcal{F}_{x\to\xi}(k_{\alpha+n}),\psi(\xi)\right)\\&=\Gamma_p^{(n)}(\alpha+n)\left(\|\xi\|_p^{-\alpha-n},\psi(\xi)\right)=\left(k_{-\alpha},\psi\right)\\&=\frac{1-p^{\alpha}}{1-p^{-\alpha-n}}\int_{\mathbb{Q}_p^n}\left(\psi(x)-\psi(0)\right)\frac{d^n x}{\|x\|_p^{\alpha+n}}\\&=\frac{1-p^{\alpha}}{1-p^{-\alpha-n}}\int_{\mathbb{Q}_p^n}\frac{1}{\|x\|_p^{\alpha+n}}\left\{\int_{\mathbb{Q}_p^n}\left(\chi_p(-\xi\cdot x)-1\right)\varphi(\xi)\,d^n\xi\right\}d^n x.\end{aligned}$$

Now if $\varphi(\xi) = 0$ for $\|\xi\|_p > p^N$, then

$$\int_{\mathbb{Q}_p^n} (\chi_p(-\xi \cdot x) - 1)\, \varphi(\xi)\, d^n\xi = 0 \text{ for } \|x\|_p \leq p^N,$$

and consequently we may use Fubini's theorem to get

$$(\|x\|_p^\alpha, \varphi) = \frac{1 - p^\alpha}{1 - p^{-\alpha - n}} \int_{\mathbb{Q}_p^n} \varphi(\xi) \left\{ \int_{\mathbb{Q}_p^n} \frac{\chi_p(\xi \cdot x) - 1}{\|x\|_p^{\alpha+n}}\, d^n x \right\} d^n\xi.$$

□

DEFINITION 10.16. We say that a locally constant function f belongs to $\mathcal{U}_{loc}(\mathbb{Q}_p^n) := \mathcal{U}_{loc}$ if and only if f is constant on the cosets of $p^k\mathbb{Z}_p^n$ for some $k \in \mathbb{Z}$. Equivalently, there is an integer k such that for any $x \in \mathbb{Q}_p^n$,

$$f(x + x') = f(x) \text{ for any } x' \in B_{-k}^n. \tag{10.1}$$

The largest $-k$ such that (10.1) is verified is called the exponent (or parameter) of local constancy of f.

EXAMPLE 10.17. (i) $\mathcal{D} \subset \mathcal{U}_{loc}$. (ii) Consider $f \in \mathcal{U}_{loc}$ such that $|f(y)|\, e^{-t\|y\|_p^\alpha} \in L^1$ for $t > 0$. Then

$$h(x) = \int_{\mathbb{Q}_p^n} \{f(x - y) - f(x)\}\, e^{-t\|y\|_p^\alpha}\, d^n y \in \mathcal{U}_{loc}.$$

PROPOSITION 10.18. *$f \in \mathcal{U}_{loc}$ if and only if $f \in \mathcal{D}'$ and there is $k \in \mathbb{Z}$ such that $\mathcal{T}_x f = f$ for any $x \in p^k\mathbb{Z}_p^n$.*

PROOF. Since $\mathcal{U}_{loc} \subset L^1_{loc}$, any $f \in \mathcal{U}_{loc}$ gives rise to an element from $\mathcal{D}'$ satisfying $(\mathcal{T}_x f, \varphi) = (f, \varphi)$ for any $x \in p^k\mathbb{Z}_p^n$.

Conversely, assume that $f \in \mathcal{D}'$, and $\mathcal{T}_x f = f$, for all $x \in p^k\mathbb{Z}_p^n$. We set

$$h(y) = p^{kn}(f, \mathcal{T}_y \Delta_{-k}).$$

Then $h(y) \in \mathcal{U}_{loc}$. Indeed, since for all $x \in p^k\mathbb{Z}_p^n$,

$$(\mathcal{T}_x h)(y) = p^{kn}(f, \mathcal{T}_{y-x}\Delta_{-k}) = p^{kn}(\mathcal{T}_x f, \mathcal{T}_y \Delta_{-k}) = p^{kn}(f, \mathcal{T}_y \Delta_{-k}) = h(y).$$

Now we show that $f = h$ in $\mathcal{D}'$. It is sufficient to verify

$$(h, \varphi) = (f, \varphi) \text{ i.e. } p^{kn} \int_{\mathbb{Q}_p^n} (\mathcal{T}_{-y} f, \Delta_{-k})\, \varphi(y)\, d^n y = (f, \varphi),$$

for φ equals to the characteristic function of a sufficiently small ball $J + p^l\mathbb{Z}_p^n$. Thus, we may assume that $J + p^l\mathbb{Z}_p^n \subset J + p^k\mathbb{Z}_p^n$, with $l \geq k$. Then

$$\left(h, 1_{J+p^l\mathbb{Z}_p^n}\right) = h(J)p^{-nl}$$

and consequently,

$$\begin{aligned}
\left(h, 1_{J+p^l\mathbb{Z}_p^n}\right) &= p^{kn} \int_{J+p^l\mathbb{Z}_p^n} (\mathcal{T}_{-y} f, \Delta_k)\, d^n y = p^{kn-nl} \int_{\mathbb{Z}_p^n} (\mathcal{T}_{-J-p^l u} f, \Delta_k)\, d^n u \\
&= p^{kn-nl} \int_{\mathbb{Z}_p^n} (\mathcal{T}_{-p^l u} f, \mathcal{T}_J \Delta_k)\, d^n u = p^{kn-nl} \int_{\mathbb{Z}_p^n} (f, \mathcal{T}_J \Delta_k)\, d^n u \\
&= p^{kn-nl}(f, \mathcal{T}_J \Delta_k) \\
&= h(J)p^{-nl}.
\end{aligned}$$

□

Definition 10.19. For $T \in \mathcal{D}'$ and $\varphi \in \mathcal{D}$, $T * \varphi$ is defined by $\widehat{T * \varphi} = \widehat{\varphi}\, \widehat{T}$.

Exercise 10.20. Show that for $T \in \mathcal{D}'$ and $\varphi \in \mathcal{D}$, the map $T \to T * \varphi$ is a continuous map from $\mathcal{D}'$ into $\mathcal{D}'$. See Exercise 10.11.

Exercise 10.21. To show the following formulas for φ, $\psi \in \mathcal{D}$, $G \in \mathcal{D}'$ and $x \in \mathbb{Q}_p^n$:

(i) $\widetilde{(\mathcal{T}_{-x}\varphi)} = \mathcal{T}_x \widetilde{\varphi}$;
(ii) $\widetilde{(\mathcal{T}_{-x}G)} = \mathcal{T}_x \widetilde{G}$;
(iii) $\widetilde{(\varphi * \psi)} = \widetilde{\varphi} * \widetilde{\psi}$;
(iv) $\widetilde{(\widetilde{\varphi} * \psi)} = \varphi * \widetilde{\psi}$.

Proposition 10.22 ([**23**, Theorem 3.15]). *The following three characterizations of $\varphi * G \in \mathcal{D}'$ for $\varphi \in \mathcal{D}$, $G \in \mathcal{D}'$ are equivalent:*
(i) $\widehat{\varphi * G} = \widehat{\varphi}\, \widehat{G}$;
(ii) $(\varphi * G, \psi) = (G, \widetilde{\varphi} * \psi)$ *for* $\psi \in \mathcal{D}$;
(iii) $\varphi * G$ *belongs to* $\mathcal{U}_{loc}$ *and agrees with function*

$$g(x) = (G, T_x \widetilde{\varphi}) = (G(y), \varphi(x - y)).$$

Exercise 10.23. Assume that $\varphi \in \mathcal{U}_{loc}$ such that

$$\int_{\|y\|_p > 1} \|y\|_p^{\alpha - n} \, |\varphi(y)| \, d^n y < \infty.$$

Then $k_\alpha * \varphi$ is a function given by

$$(k_\alpha * \varphi)(x) = \frac{1 - p^{-n}}{1 - p^{\alpha - n}} \varphi(x) + \frac{1 - p^{-\alpha}}{1 - p^{\alpha - n}} \int_{\|y\|_p > 1} \|y\|_p^{\alpha - n} \varphi(x - y) \, d^n y$$
$$+ \frac{1 - p^{-\alpha}}{1 - p^{\alpha - n}} \int_{\|y\|_p \leq 1} \|y\|_p^{\alpha - n} (\varphi(x - y) - \varphi(x)) \, d^n y.$$

Hint. Recall that φ is necessarily continuous.

Exercise 10.24. Assume that $\mathrm{Re}(\alpha) > 0$, then for $\varphi \in \mathcal{D}$, following formulas hold true:

$$(k_{-\alpha} * \varphi)(x) = \frac{1 - p^{\alpha}}{1 - p^{-\alpha - n}} \int_{\mathbb{Q}_p^n} (\varphi(x - y) - \varphi(x)) \frac{d^n y}{\|y\|_p^{\alpha + n}}$$
$$= \frac{1 - p^{\alpha}}{1 - p^{-\alpha - n}} \int_{\mathbb{Q}_p^n} (\varphi(y) - \varphi(x)) \frac{d^n x}{\|y - x\|_p^{\alpha + n}},$$

and

$$(k_\alpha * \varphi)(x) = \frac{1 - p^{-\alpha}}{1 - p^{\alpha - n}} \int_{\mathbb{Q}_p^n} \|y\|_p^{\alpha - n} \varphi(x - y) \, d^n y, \text{ for } \alpha \in \mathbb{C} \smallsetminus \left\{ n + \frac{2\pi\sqrt{-1}}{\ln p} \mathbb{Z} \right\}.$$

Exercise 10.25. For $f \in \mathcal{U}_{loc}$, $G \in \mathcal{D}'$, we define $fG \in \mathcal{D}'$ as $(fG, \varphi) = (G, f\varphi)$ for $\varphi \in \mathcal{D}$. Show that $G \to fG$ is a continuous map of $\mathcal{D}'$ into $\mathcal{D}'$.

Definition 10.26. A distribution $G \in \mathcal{D}'$ has compact support if there is a $k \in \mathbb{Z}$ such that $\Delta_k G = G$.

We denote by $\mathcal{D}'_{comp}$ the $\mathbb{C}$-vector space of distributions with compact support.

THEOREM 10.27. *$f \in \mathcal{U}_{loc}$ if and only if $\widehat{f}$ has compact support. In particular, if $f \in \mathcal{D}'$ then $\mathcal{T}_x f = f$ for all $x \in p^k\mathbb{Z}_p^n$ if and only if $\Delta_k \widehat{f} = \widehat{f}$.*

PROOF. Assume that $f \in \mathcal{U}_{loc}$ and that $\mathcal{T}_x f = f$ (i.e. $f\mid_{x+p^k\mathbb{Z}_p^n} \equiv f(x)$). By using Proposition 10.22-(iii), we have

$$\mathcal{F}^{-1}\left(\Delta_k \widehat{f}\right) = \mathcal{F}^{-1}(\Delta_k) * f = \mathcal{F}(\Delta_k) * f = \delta_k * f$$

$$= p^{nk}\int_{\|x-y\|_p \le p^{-k}} f(y)d^n y = p^{nk}\int_{x+p^k\mathbb{Z}_p^n} f(y)d^n y = p^{nk} f(x)\int_{x+p^k\mathbb{Z}_p^n} d^n y = f(x),$$

i.e. $\Delta_k \widehat{f} = \widehat{f}$.

Now suppose that $\widehat{f} \in \mathcal{D}'$ satisfying $\Delta_k \widehat{f} = \widehat{f}$. Then for any $\varphi \in \mathcal{D}$ and for all $x \in p^k\mathbb{Z}_p^n$,

$$(\mathcal{T}_x f, \varphi) = (f, \mathcal{T}_{-x}\varphi) = \left(\mathcal{F}(f), \mathcal{F}^{-1}(\mathcal{T}_{-x}\varphi)\right)$$

$$= \left(\mathcal{F}(f)(\xi), \chi_p(x\cdot\xi)\mathcal{F}^{-1}(\varphi)(\xi)\right) = \left(\Delta_k(\xi)\widehat{f}(\xi), \chi_p(x\cdot\xi)\mathcal{F}^{-1}(\varphi)(\xi)\right)$$

$$= \left(\widehat{f}(\xi), \Delta_k(\xi)\chi_p(x\cdot\xi)\mathcal{F}^{-1}(\varphi)(\xi)\right) = \left(\widehat{f}, \mathcal{F}^{-1}(\varphi)\right) = (f, \varphi).$$

□

DEFINITION 10.28. Given $F \in \mathcal{D}'(\mathbb{Q}_p^n)$ and $G \in \mathcal{D}'(\mathbb{Q}_p^m)$, their *direct product* $F \times G$ is defined by the formula

$$(F(x) \times G(y), \varphi(x,y)) = (F(x), (G(y), \varphi(x,y))) \text{ for } \varphi(x,y) \in \mathcal{D}\left(\mathbb{Q}_p^{n+m}\right).$$

The direct product of distributions is well defined due to the fact that any $\varphi(x,y) \in \mathcal{D}\left(\mathbb{Q}_p^{n+m}\right)$ can be expressed as

$$\varphi(x,y) = \sum_{i=1}^{L} \theta_i(x)\psi_i(y) \text{ where } \theta_i(x) \in \mathcal{D}(\mathbb{Q}_p^n), \ \psi_i(y) \in \mathcal{D}(\mathbb{Q}_p^m).$$

Which in turn is a consequence of the fact that

$$B_r^l(a) = B_r(a_1) \times \cdots \times B_r(a_l)$$

for any $l \in \mathbb{N} \smallsetminus \{0\}$, $a \in \mathbb{Q}_p^l$ and $r \in \mathbb{Z}$. Furthermore, the direct product is commutative: $F \times G = G \times F$. Notice that in the case $G = 1$, we have

$$\left(F(x), \int_{\mathbb{Q}_p^m} \varphi(x,y)\, d^m y\right) = \int_{\mathbb{Q}_p^m} (F(x), \varphi(x,y))\, d^m y.$$

EXAMPLE 10.29. $\delta(x_1) \times \delta(x_2) \times \cdots \times \delta(x_n) = \delta(x)$, $x = (x_1, \ldots, x_n)$.

EXERCISE 10.30. Show that the direct product of distributions is continuous with respect to the joint factors.

PROPOSITION 10.31. *Let $f \in \mathcal{D}'$ with $\Delta_k f = f$. Then $\widehat{f}$ is a locally constant function given by*

$$\widehat{f}(\xi) = (f(\xi), \Delta_k(\xi)\chi_p(x\cdot\xi)).$$

PROOF. By using the commutativity of the direct product of distributions with $G = 1$ and Theorem 10.27,

$$\begin{aligned}\left(\widehat{f},\varphi\right) &= (f,\widehat{\varphi}) = (\Delta_k f,\widehat{\varphi}) = (f,\Delta_k\widehat{\varphi}) \\ &= \left(f(\xi), \int_{\mathbb{Q}_p^n} \Delta_k(\xi)\,\chi_p(x\cdot\xi)\,\varphi(x)\,d^n x\right) \\ &= \int_{\mathbb{Q}_p^n} \left(f(\xi), \Delta_k(\xi)\,\chi_p(x\cdot\xi)\right)\varphi(x)\,d^n x,\end{aligned}$$

□

DEFINITION 10.32. If T, $W \in \mathcal{D}'\left(\mathbb{Q}_p^n\right)$ and W has compact support, then $T* W \in \mathcal{D}'\left(\mathbb{Q}_p^n\right)$ is defined by

$$\widehat{T*W} = \widehat{T}\,\widehat{W}.$$

EXERCISE 10.33. If $W \in \mathcal{D}'\left(\mathbb{Q}_p^n\right)$ has compact support, then the map $T \to T*W$ is a continuous map from $\mathcal{D}'\left(\mathbb{Q}_p^n\right)$ into $\mathcal{D}'\left(\mathbb{Q}_p^n\right)$.

Hint. Notice that the following mappings are continuous:

$$\begin{array}{ccc}\mathcal{D}'\left(\mathbb{Q}_p^n\right) & \to & \mathcal{D}'\left(\mathbb{Q}_p^n\right) \\ \widehat{T} & \to & \widehat{T}\,\widehat{W},\end{array}$$

and

$$\begin{array}{ccc}\mathcal{D}'\left(\mathbb{Q}_p^n\right) & \xrightarrow{\text{mult. }\widehat{W}} & \mathcal{D}'\left(\mathbb{Q}_p^n\right) \\ \widehat{T} & \to & \widehat{T}\,\widehat{W} \\ \updownarrow \mathcal{F}^{-1} & & \updownarrow \mathcal{F}^{-1} \\ \mathcal{D}'\left(\mathbb{Q}_p^n\right) & \xrightarrow{*W} & \mathcal{D}'\left(\mathbb{Q}_p^n\right). \\ T & \to & T*W\end{array}$$

THEOREM 10.34. *(i) Let $T_1,\ldots,T_s \in \mathcal{D}'\left(\mathbb{Q}_p^n\right)$ all but (at most) one being in $\mathcal{U}_{loc}$. Then $T_1\cdots T_s \in \mathcal{D}'\left(\mathbb{Q}_p^n\right)$ is well-defined as a commutative and associative product.*

(ii) Let $W_1,\ldots,W_s \in \mathcal{D}'\left(\mathbb{Q}_p^n\right)$ all but (at most) one being in $\mathcal{D}'_{comp}$. Then $W_1*\cdots*W_s \in \mathcal{D}'\left(\mathbb{Q}_p^n\right)$ is well-defined as a commutative and associative convolution product.

EXAMPLE 10.35. (i) Take $\phi \in \mathcal{D}$, then the associated distribution has compact support.

(ii) $\delta \in \mathcal{D}'$ has compact support (Why?).

(iii) Let $T \in \mathcal{D}'$, then $\delta * T = T$. Indeed,

$$\widehat{\delta * T} = \widehat{\delta}\,\widehat{T} = 1\widehat{T} = \widehat{T}.$$

(iv) Let $T \in \mathcal{D}'$, then $\delta_k * T \in \mathcal{D}'$, and

$$\lim_{k\to\infty} \delta_k * T = T \text{ in } \mathcal{D}'.$$

Indeed,

$$\lim_{k\to\infty} \widehat{\delta_k * T} = \lim_{k\to\infty} \widehat{\delta_k}\widehat{T} = \lim_{k\to\infty} \Delta_k\widehat{T} = \widehat{T} \text{ in } \mathcal{D}'.$$

THEOREM 10.36. *$\mathcal{D}$ is dense in $\mathcal{D}'$. More precisely, every distribution is the weak limit of a sequence of test functions.*

PROOF. Take $T \in \mathcal{D}'$, then $\delta_k * T \in \mathcal{U}_{loc}$, $\Delta_l(\delta_k * T) \in \mathcal{D}$ for any l, k. We now show that

$$\lim_{l\to\infty}\left(\lim_{k\to\infty}(\Delta_l\widehat{(\delta_k * T)})\right) = \widehat{T} \text{ in } \mathcal{D}'.$$

Indeed, $\Delta_l\widehat{(\delta_k * T)} = \delta_l * \widehat{(\delta_k * T)} = \delta_l * \Delta_k\widehat{T}$. For a fixed Δ_k, the map

$$\begin{array}{ccc} \mathcal{D}' & \to & \mathcal{D}' \\ G & \to & \delta_l * G \end{array}$$

is continuous. Then $\lim_{l\to\infty}\delta_l * \Delta_k\widehat{T} = \delta * \Delta_k\widehat{T} = \Delta_k\widehat{T}$. Now the map

$$\begin{array}{ccc} \mathcal{D}' & \to & \mathcal{D}' \\ G & \to & \Delta_k G \end{array}$$

is continuous. Consequently $\lim_{k\to\infty}\Delta_k\widehat{T} = \widehat{T}$, and

$$\lim_{l\to\infty}\left(\lim_{k\to\infty}(\Delta_l\widehat{(\delta_k * T)})\right) = \lim_{l\to\infty}\delta_l * \widehat{T} = \widehat{T} \text{ in } \mathcal{D}'.$$

Now by using the fact that the Fourier transform is continuous in $\mathcal{D}'$, we have

$$\lim_{l\to\infty}\left(\widehat{\lim_{k\to\infty}(\Delta_l(\delta_k * T))}\right) = \widehat{T}.$$

Finally, by using that the Fourier transform is an isomorphism in $\mathcal{D}'$,

$$\lim_{l\to\infty}\left(\lim_{k\to\infty}(\Delta_l(\delta_k * T))\right) = T.$$

□

10.1. Operations on distributions. The multiplication and the convolution between arbitrary distributions is a subtle concept. Here review these operations following [**24**, Sections 6, 7].

DEFINITION 10.37. Given $F, G \in \mathcal{D}'(\mathbb{Q}_p^n)$, their convolution $F * G$ is defined by

$$(F * G, \varphi) = \lim_{k\to\infty}(F(x) \times G(y), \Delta_k(x)\varphi(x+y)) \tag{10.2}$$

if the limit exists for all $\varphi \in \mathcal{D}(\mathbb{Q}_p^n)$.

THEOREM 10.38. *(i) If $F * G$ exists, then $G * F$ exists and $F * G = G * F$. (ii) If $F, G \in \mathcal{D}'(\mathbb{Q}_p^n)$ and supp$F \subset B_N^n$ (i.e. if $\Delta_N F = F$), then the convolution $F * G$ exists, and it is given by the formula*

$$(F * G, \varphi) = (F(x) \times G(y), \Delta_N(x)\varphi(x+y)) \text{ for } \varphi \in \mathcal{D}(\mathbb{Q}_p^n).$$

(iii) In the case in which $F = \psi \in \mathcal{D}(\mathbb{Q}_p^n)$, $\psi * G$ is a locally constant function given by

$$(\psi * G)(z) = (G(y), \psi(z-y)).$$

PROOF. (i) We notice that without loss of generality in (10.2), we can assume that $\varphi(x) = \Omega\left(p^{-l}\|x-j\|_p\right)$ for some integer l and some $j \in \mathbb{Q}_p^n$.

CLAIM. For k sufficiently large, we have $\Omega\left(p^{-k}\|x-j\|_p\right)=\Omega\left(p^{-k}\|x\|_p\right)$ (i.e. $\Delta_k(x-j)=\Delta_k(x)$) and

(10.3)

$$\Omega\left(p^{-k}\|x-j\|_p\right)\Omega\left(p^{-l}\|x+y-j\|_p\right)=\Omega\left(p^{-k}\|y\|_p\right)\Omega\left(p^{-l}\|x+y-j\|_p\right).$$

Indeed,

$$\begin{aligned}(F*G,\varphi)&=\lim_{k\to\infty}\left(F(x)\times G(y),\Omega\left(p^{-k}\|x\|_p\right)\Omega\left(p^{-l}\|x+y-j\|_p\right)\right)\\&=\lim_{k\to\infty}\left(F(x)\times G(y),\Omega\left(p^{-k}\|x-j\|_p\right)\Omega\left(p^{-l}\|x+y-j\|_p\right)\right)\\&=\lim_{k\to\infty}\left(G(y)\times F(x),\Omega\left(p^{-k}\|y\|_p\right)\Omega\left(p^{-l}\|x+y-j\|_p\right)\right)\\&=\lim_{k\to\infty}\left(G(y)\times F(x),\Delta_k(y)\Omega\left(p^{-l}\|x+y-j\|_p\right)\right)=(G*F,\varphi).\end{aligned}$$

In order to establish (10.3), we may assume without loss of generality that $k>l$. Now $\Omega\left(p^{-k}\|x-j\|_p\right)\Omega\left(p^{-l}\|x+y-j\|_p\right)=1$ if and only if

$$\|x-j\|_p\leq p^k \quad\text{and}\quad \|x+y-j\|\leq p^l. \tag{10.4}$$

Now, if $\|y\|_p>p^k$, then by the ultrametric property of $\|\cdot\|_p$, $\|x+y-j\|=\|y\|_p\leq p^l$. Which is impossible. Consequently (10.4) implies that

$$\|y\|_p\leq p^k \quad\text{and}\quad \|x+y-j\|\leq p^l. \tag{10.5}$$

Now $\Omega\left(p^{-k}\|y\|_p\right)\Omega\left(p^{-l}\|x+y-j\|_p\right)$ if and only if (10.5) holds true. If $\|x-j\|_p>p^k$, then since $\|y\|_p\leq p^k$, $\|x+y-j\|=\|x-j\|_p\leq p^l$. Which is impossible. Then (10.4) holds true.

(ii) It follows from

$$\begin{aligned}(F*G,\varphi)&=\lim_{k\to\infty}\left(F(x)\times G(y),\Delta_k(x)\varphi(x+y)\right)\\&=\lim_{k\to\infty}\left(\Delta_N(x)F(x)\times G(y),\Delta_k(x)\varphi(x+y)\right)\\&=\lim_{k\to\infty}\left(F(x)\times G(y),\Delta_k(x)\Delta_N(x)\varphi(x+y)\right)\\&=\left(F(x)\times G(y),\Delta_N(x)\varphi(x+y)\right).\end{aligned}$$

(iii) By using (ii), with $\Delta_N\psi=\psi$, we have

$$\begin{aligned}(\psi*G,\varphi)&=\left(\psi(x)\times G(y),\Delta_N(x)\varphi(x+y)\right)\\&=\int_{\mathbb{Q}_p^n}\left(G(y),\Delta_N(x)\varphi(x+y)\right)\psi(x)\,d^nx\\&=\left(G(y),\int_{\mathbb{Q}_p^n}\psi(x)\varphi(x+y)\,d^nx\right)=\left(G(y),\int_{\mathbb{Q}_p^n}\psi(z-y)\varphi(z)\,d^nz\right)\\&=\int_{\mathbb{Q}_p^n}\left(G(y),\psi(z-y)\right)\varphi(z)\,d^nz.\end{aligned}$$

The fact that $(G(y),\psi(y-z))\in\mathcal{U}_{loc}$ was established in Proposition 10.22. □

DEFINITION 10.39. Given $T,G\in\mathcal{D}'\left(\mathbb{Q}_p^n\right)$, their product $T\cdot G$ is defined by

$$(FT\cdot G,\varphi)=\lim_{k\to\infty}\left(G,(T*\delta_k)\varphi\right)$$

if the limit exists for all $\varphi \in \mathcal{D}\left(\mathbb{Q}_p^n\right)$. Alternatively,

$$T \cdot G = \lim_{k \to \infty} \left(T * \delta_k\right) G \text{ in } \mathcal{D}'\left(\mathbb{Q}_p^n\right).$$

THEOREM 10.40. *The existence of the product $T \cdot G$ is equivalent to the existence of $\mathcal{F}[T] * \mathcal{F}[G]$. In addition, $\mathcal{F}[T \cdot G] = \mathcal{F}[T] * \mathcal{F}[G]$ and $\mathcal{F}[T * G] = \mathcal{F}[T] \cdot \mathcal{F}[G]$. If the product $T \cdot G$ exists then the product $G \cdot T$ exists and they are equal.*

PROOF. The result follows from the following facts: $(T * \delta_k) G \in \mathcal{D}'$ (because $T * \delta_k \in \mathcal{U}_{loc}$), $\Delta_k \widehat{T} * \widehat{G}$ (because $\Delta_k \widehat{T}$ has compact support), and that

$$\mathcal{F}\left(\left(T * \delta_k\right) G\right) = \Delta_k \widehat{T} * \widehat{G}.$$

By taking the limit $k \to \infty$, and using the continuity of the Fourier transform, we have

$$\lim_{k \to \infty} \mathcal{F}\left(\left(T * \delta_k\right) G\right) = \mathcal{F}\left(\lim_{k \to \infty} \left(T * \delta_k\right) G\right) = \lim_{k \to \infty} \Delta_k \widehat{T} * \widehat{G}.$$

Which implies that the existence of the product $T \cdot G$ is equivalent to the existence of $\mathcal{F}[T] * \mathcal{F}[G]$. Finally, if $T \cdot G$ exists, by uisng the fcat that the convolution of dsitributions is commutative, we have

$$\widehat{T \cdot G} = \widehat{T} * \widehat{G} = \widehat{G} * \widehat{T} = \widehat{G \cdot T}.$$

□

EXERCISE 10.41. Given $f(x)$ a non-constant polynomial in $\mathbb{Q}_p[x_1, \ldots, x_n]$, the complex power attached to it (also called the Igusa local zeta function of f) is the distribution $|f|_p^s$, $s \in \mathbb{C}$, defined as

$$\left(|f|_p^s, \varphi\right) = \int_{\mathbb{Q}_p \smallsetminus f^{-1}(0)} |f(x)|_p^s \varphi(x)\, d^n x, \text{ for } \operatorname{Re}(s) > 0,$$

where $a^s := e^{s \ln a}$ for $a > 0$. Show that for $\operatorname{Re}(\alpha)$, $\operatorname{Re}(\beta) > 0$,

$$|f|_p^\alpha |f|_p^\beta = |f|_p^{\alpha+\beta} \text{ in } \mathcal{D}'.$$

Hint. If $f(x) \neq 0$, then for a k sufficiently large,

$$|f(x+y)|_p = |f(x)|_p \text{ for any } y \in p^k \mathbb{Z}_p^n.$$

This fact follows from the Taylor formula for a polynomial.

References

[1] S. Albeverio, A. Yu. Khrennikov, and V. M. Shelkovich, *Theory of p-adic distributions: linear and nonlinear models*, London Mathematical Society Lecture Note Series, vol. 370, Cambridge University Press, Cambridge, 2010, DOI 10.1017/CBO9781139107167. MR2641698

[2] Vladimir Anashin and Andrei Khrennikov, *Applied algebraic dynamics*, De Gruyter Expositions in Mathematics, vol. 49, Walter de Gruyter & Co., Berlin, 2009, DOI 10.1515/9783110203011. MR2533085

[3] M. F. Atiyah, *Resolution of singularities and division of distributions*, Comm. Pure Appl. Math. **23** (1970), 145–150, DOI 10.1002/cpa.3160230202. MR256156

[4] I. N. Bernšteĭn, *Modules over a ring of differential operators. An investigation of the fundamental solutions of equations with constant coefficients* (Russian), Funkcional. Anal. i Priložen. **5** (1971), no. 2, 1–16. MR0290097

[5] Miriam Bocardo-Gaspar, H. García-Compeán, and W. A. Zúñiga-Galindo, *Regularization of p-adic string amplitudes, and multivariate local zeta functions*, Lett. Math. Phys. **109** (2019), no. 5, 1167–1204, DOI 10.1007/s11005-018-1137-1. MR3946489

[6] M. Bocardo-Gaspar, H. García-Compeán, and W. A. Zúñiga-Galindo, *On p-adic string amplitudes in the limit p approaches to one*, J. High Energy Phys. **8** (2018), 043, front matter+22, DOI 10.1007/jhep08(2018)043. MR3866346

[7] M. Bocardo-Gaspar, Willem Veys, and W. A. Zúñiga-Galindo, *Meromorphic continuation of Koba-Nielsen string amplitudes*, J. High Energy Phys. **9** (2020), 138, 43, DOI 10.1007/jhep09(2020)138. MR4190286

[8] N. Bourbaki, *Éléments de mathématique. Fasc. XXXVI. Variétés différentielles et analytiques. Fascicule de résultats (Paragraphes 8 à 15)* (French), Actualités Scientifiques et Industrielles [Current Scientific and Industrial Topics], No. 1347, Hermann, Paris, 1971. MR0281115

[9] Paul R. Halmos, *Measure Theory*, D. Van Nostrand Company, Inc., New York, N. Y., 1950. MR0033869

[10] Jun-ichi Igusa, *An introduction to the theory of local zeta functions*, AMS/IP Studies in Advanced Mathematics, vol. 14, American Mathematical Society, Providence, RI; International Press, Cambridge, MA, 2000, DOI 10.1090/amsip/014. MR1743467

[11] I. M. Gel'fand and G. E. Shilov, *Generalized functions. Vol. I: Properties and operations*, Academic Press, New York-London, 1964. Translated by Eugene Saletan. MR0166596

[12] Fernando Q. Gouvêa, *p-adic numbers*, 2nd ed., Universitext, Springer-Verlag, Berlin, 1997. An introduction, DOI 10.1007/978-3-642-59058-0. MR1488696

[13] Svetlana Katok, *p-adic analysis compared with real*, Student Mathematical Library, vol. 37, American Mathematical Society, Providence, RI; Mathematics Advanced Study Semesters, University Park, PA, 2007, DOI 10.1090/stml/037. MR2298943

[14] Neal Koblitz, *p-adic numbers, p-adic analysis, and zeta-functions*, 2nd ed., Graduate Texts in Mathematics, vol. 58, Springer-Verlag, New York, 1984, DOI 10.1007/978-1-4612-1112-9. MR754003

[15] Anatoly N. Kochubei, *Pseudo-differential equations and stochastics over non-Archimedean fields*, Monographs and Textbooks in Pure and Applied Mathematics, vol. 244, Marcel Dekker, Inc., New York, 2001, DOI 10.1201/9780203908167. MR1848777

[16] Kozyrev S. V., Methods and Applications of Ultrametric and p-Adic Analysis: From Wavelet Theory to Biophysics, *Proc. Steklov Inst. Math.* 274 (2011), Suppl. 1, 1-84.

[17] Andrei Khrennikov, *Non-Archimedean analysis: quantum paradoxes, dynamical systems and biological models*, Mathematics and its Applications, vol. 427, Kluwer Academic Publishers, Dordrecht, 1997, DOI 10.1007/978-94-009-1483-4. MR1746953

[18] Andrei Yu. Khrennikov, Sergei V. Kozyrev, and W. A. Zúñiga-Galindo, *Ultrametric pseudodifferential equations and applications*, Encyclopedia of Mathematics and its Applications, vol. 168, Cambridge University Press, Cambridge, 2018, DOI 10.1017/9781316986707. MR3793137

[19] Serge Lang, *Real and functional analysis*, 3rd ed., Graduate Texts in Mathematics, vol. 142, Springer-Verlag, New York, 1993, DOI 10.1007/978-1-4612-0897-6. MR1216137

[20] Edwin León-Cardenal and W. A. Zúñiga-Galindo, *An introduction to the theory of local zeta functions from scratch* (English, with English and Spanish summaries), Rev. Integr. Temas Mat. **37** (2019), no. 1, 45–76, DOI 10.18273/revint. MR3920742

[21] Alain M. Robert, *A course in p-adic analysis*, Graduate Texts in Mathematics, vol. 198, Springer-Verlag, New York, 2000, DOI 10.1007/978-1-4757-3254-2. MR1760253

[22] W. H. Schikhof, *Ultrametric calculus: An introduction to p-adic analysis*, Cambridge Studies in Advanced Mathematics, vol. 4, Cambridge University Press, Cambridge, 2006. Reprint of the 1984 original [MR0791759]. MR2444734

[23] M. H. Taibleson, *Fourier analysis on local fields*, Princeton University Press, Princeton, N.J.; University of Tokyo Press, Tokyo, 1975. MR0487295

[24] V. S. Vladimirov, I. V. Volovich, and E. I. Zelenov, *p-adic analysis and mathematical physics*, Series on Soviet and East European Mathematics, vol. 1, World Scientific Publishing Co., Inc., River Edge, NJ, 1994, DOI 10.1142/1581. MR1288093

[25] W. A. Zúñiga-Galindo, *Pseudodifferential equations over non-Archimedean spaces*, Lecture Notes in Mathematics, vol. 2174, Springer, Cham, 2016, DOI 10.1007/978-3-319-46738-2. MR3586737

University of Texas Rio Grande Valley, School of Mathematical & Statistical Sciences, One West University Blvd, Brownsville, Texas 78520

Email address: `wilson.zunigagalindo@utrgv.edu`

Part II: Articles

Contemporary Mathematics
Volume **778**, 2022
https://doi.org/10.1090/conm/778/15659

On maximal order poles of generalized topological zeta functions

Enrique Artal Bartolo and Manuel González Villa

Abstract. We show some examples of topological zeta functions associated to an isolated plane curve singular point and an allowed, in the sense due to Némethi and Veys in 2012, differential form that have several poles of order two. This is in contrast to the case of the standard differential form where, as showed by Veys for plane curves in 1995, and by Nicaise and Xu in general twenty one years later, there is always at most one pole of maximal order.

1. Introduction

W. Veys determined all poles of the local topological zeta function of an isolated singular point of a plane curve [**Vey95**]. In particular, he showed [**Vey95**, Theorem 4.2] that there is at most one pole of order two, and, if there is such a pole of order two, it is the largest pole, and its value is the opposite of the log canonical threshold of the singular point. Later, Veys conjetured that analogous statements hold for arbitrary dimension [**LV99**, Conjecture 0.2], and proved, together with A. Laeremans, the result for polynomials that are non-degenerate with respect to their Newton polyhedron at the origin. They also noticed that a double pole must be of the form $-\frac{1}{n}$ for some $n \in \mathbb{N}\setminus\{0\}$ [**LV99**, Corollary 3.4]. Finally, the conjecture has recently been established by J. Nicaise and C. Xu [**NX16**, Theorem 3.5].

F. Loeser already proved [**Loe88**] that if s_0 is a pole of order two of topological zeta functions of a singular point of a reduced plane curve, then the monodromy operator of the Milnor fibration of the singular point has a Jordan block of size two for the eigenvalue $\exp(2\pi i s_0)$. This fact has been generalized, answering a question of C.T.C. Wall [**Wal04**], for non reduced plane curves by A. Melle-Hernández, T. Torelli and W. Veys [**MTV10**].

Motivated by the remark that the poles of the topological zeta function determine only a few eigenvalues of the monodromy of the singular point, A. Némethi and Veys [**NV12**] proposed to study generalized topological zeta functions. Generalized

2020 *Mathematics Subject Classification.* Primary 32S40, 14B05.

Key words and phrases. Topological zeta functions, maximal order poles, plane curves.

The first named author was partially supported by Grant PID2020-114750GB-C31 funded by MCIN/AEI/10.13039/501100011033 and Departamento de Ciencia, Universidad y Sociedad del Conocimiento of the Gobierno de Aragón (E22 20R: “Álgebra y Geometría”); the second named author was partially supported by Grant PID2020-114750GB-C33 funded by MCIN/AEI/10.13039/501100011033.

topological zeta functions are associated to a function and an *allowed* differential form. The collection of allowed differential forms for a given funcion f must verify the following three conditions:

(1) The standard differential form is allowed.
(2) If s_0 is a pole of the generalized zeta function associated to the given function and an allowed differential form, then $\exp(2\pi i s_0)$ is an eigenvalue of the monodromy.
(3) Any eigenvalue is of the form $\exp(2\pi i s_0)$ for a pole s_0 of the generalized zeta function associated to some allowed form.

In this note, we show some examples of generalized zeta functions associated to an isolated plane curve singular point and an allowed form that have several poles of order two. We also discuss on some examples which combinations of order two poles can appear and find examples where the double pole is not of the form $-\frac{1}{n}$.

2. Generalized local topological zeta functions

Let us define the local topological zeta function associated to an isolated singular point of a plane curve defined by $f : \mathbb{C}^2 \to \mathbb{C}$ and a differential form ω.

Let $\pi : X \to \mathbb{C}^2$ be an embedded resolution of $f^{-1}(0) \cup \operatorname{div}(\omega)$. We will consider only holomorphic forms ω such that an embedded resolution of $f^{-1}(0)$ is also a resolution of $f^{-1}(0) \cup \operatorname{div}(\omega)$ and the branching components of both resolutions coincide. We denote by E_i, $i \in S$, the irreducible components (exceptional divisors and strict transforms) of the inverse image $\pi^{-1}(f^{-1}(0) \cup \operatorname{div}(\omega))$. We denote by N_i and $\nu_i - 1$ the multiplicities of E_i in the divisor of $\pi^* f$ and $\pi^*\omega$, respectively. The family $\{(N_i, \nu_i)\}_{i \in S}$ is called the numerical data of the resolution (X, π, f, ω). We consider the stratification of X in locally closed subsets given by the subsets

$$\mathring{E}_I := \left(\bigcap_{i\in I} E_i\right) \setminus \left(\bigcup_{j \notin I} E_j\right) \quad \text{for } I \subset S.$$

DEFINITION 2.1. The (local) topological zeta function of (f, ω) at $0 \in \mathbb{C}$ is

$$Z_{\text{top}}(f, \omega; s) := \sum_{I \subset S} \chi(\mathring{E}_I \cap \pi^{-1}(0)) \prod_{i \in I} \frac{1}{\nu_i + sN_i} \in \mathbb{Q}(s).$$

EXAMPLE 2.2. Let us consider the following two families of cuspidal singular points of complex plane curve.

- $g_a = (x - ay^2)^2 + y^5$, defines a $(2, 5)$ cusp tangent to $\{x = 0\}$ for any $a \in \mathbb{C}$,
- $h_b = (y - bx^2)^3 + x^7$, defines a $(3, 7)$ cusp tangent to $\{y = 0\}$ for any $b \in \mathbb{C}$.

For any choice $a, b \in \mathbb{C}$ with $a, b \neq 1$, and $a \neq b$, the product

$$f_{a,b} := g_1 \cdot g_a \cdot g_b \cdot h_1 \cdot h_{-1}$$

defines an isolated singular point of a plane curve with 5 branches. The dual resolution graph and the numerical data of $f_{a,b}$ and the differential form $\omega = dxdy$ are shown in Figure 1. Notice that the default numerical data of arrows are $(1, 1)$.

The Milnor number of $f_{ab}^{-1}(0)$ at the origin is 188 and the monodromy operator associated to its Milnor fibration has 9 Jordan blocks of size 2. The eigenvalues

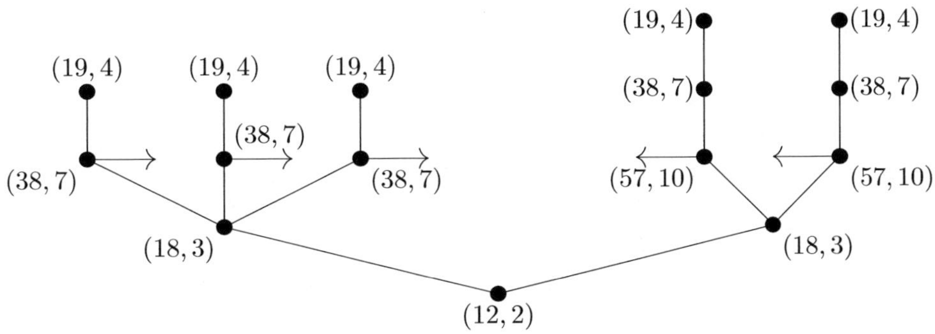

FIGURE 1. Dual graph of an embedded resolution of $f_{a,b}$.

(resp. the eigenvalues corresponding to the Jordan blocks of size 2) are given by the roots of

$$\frac{(t-1)(t^{57}-1)^2(t^{38}-1)^3(t^{18}-1)^3}{(t^{19}-1)^5} \qquad \text{(resp. } \frac{(t^3-1)(t^6-1)(t^2-1)^2}{(t-1)^4}\text{)}.$$

The local topological zeta function has a pole $(-\frac{1}{6})$ of order two, due to the subgraph formed by the three vertices with decorations $(12,2)$ and $(18,3)$ and the edges between them, and it is given by

$$Z_{\text{top}}(f_{ab},\omega;s) = \frac{70+1051s+5138s^2+7864s^3-1368s^4}{(57\,s+10)(38\,s+7)(6\,s+1)^2(s+1)}.$$

3. Examples with several poles of order two

The following examples show how to produce several poles of order two considering allowed differential forms.

EXAMPLE 3.1. Let us consider f_{ab} from Example 2.2 together with the differential form $\omega_1 = x^3dxdy$. Assuming that $a \neq 0$, the dual resolution graph is in Figure 2, showing also the numerical data of f_{ab} and the differential form ω_1. Again, the default numerical data of (solid) arrows are $(1,1)$; dashed rows correspond to the strict transform of (ω_1).

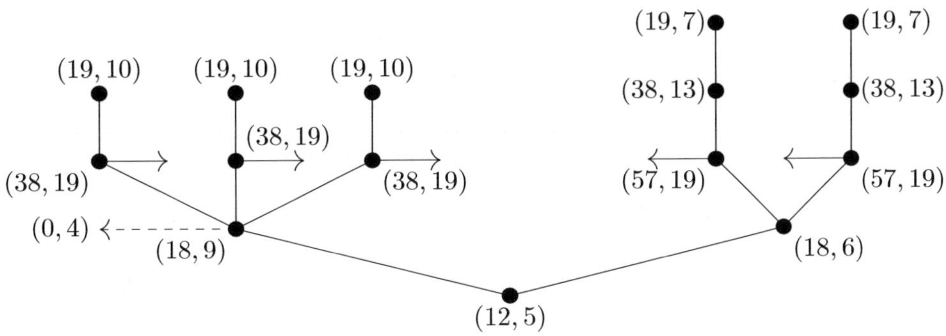

FIGURE 2. Dual graph of an embedded resolution of f_{ab} and ω_1.

Now, the local topological zeta function has two poles ($-\frac{1}{2}$ and $-\frac{1}{3}$) of order two, and it is given by

$$Z_{\text{top}}(f_{ab}, \omega_1; s) = \frac{57 + 357\,s + 625\,s^2 + 31\,s^3 - 486\,s^4}{228\,(3\,s+1)^2(2\,s+1)^2(s+1)}.$$

The pole $-\frac{1}{2}$ is due to the the vertices with decorations $(18, 9)$ and $(38, 19)$ and the edges between them. The pole $-\frac{1}{3}$ is due to the the vertices with decorations $(18, 6)$ and $(57, 19)$ and the edges between them.

EXAMPLE 3.2. Let us consider f_{ab} from Example 2.2 together with the differential form

$$\omega_2 = x(x-y^2)^2(x-ay^2)^2(x-by^2)^2y^2(y-x^2)^4(y+x^2)^4dxdy.$$

Assuming that $ab \neq 0$, the dual resolution graph and the numerical data of f_{ab} and the differential form ω_2 are in Figure 3. Again, the default numerical data of (solid) arrows are $(1, 1)$ and dashed arrows correspond to the strict transform of (ω_2).

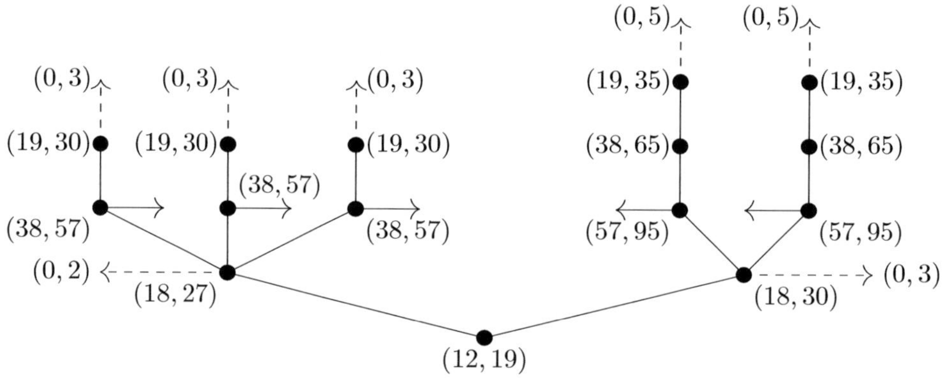

FIGURE 3. Dual graph of an embedded resolution of f_{ab} and ω_2.

In this example, the local topological zeta function has two poles ($-\frac{3}{2}$ and $-\frac{5}{3}$) of order two, and it is given by

$$Z_{\text{top}}(f_{ab}, \omega_2; s) = -\frac{16734\,s^4 + 88541\,s^3 + 168881\,s^2 + 134709\,s + 36195}{1710\,(3\,s+5)^2(2\,s+3)^2(s+1)}.$$

The pole $-\frac{3}{2}$ is due to the the vertices with decorations $(18, 27)$ and $(38, 57)$ and the edges between them. The pole $-\frac{5}{3}$ is due to the the vertices with decorations $(18, 30)$ and $(57, 95)$ and the edges between them.

REMARK 3.3. It is not hard to prove that it is impossible to find any pair of double poles. First, if the pole attached to the first exceptional component is double, no other double pole exists. This is the case of the poles of the form $-\frac{1}{6} - k$, and $-\frac{5}{6} - k$, for some $k \in \mathbb{Z}_{>0}$. Notice that $-\frac{1}{6}$ (resp. $-\frac{5}{6}$) is attained with the standard differential form (resp. with the form $\omega_3 = (xy)^4dxdy$), see Examples 2.2, and 4.1.

Moreover, it is not possible to simultaneously achieve double poles of the forms $-\frac{1}{3} - k$ and $-\frac{2}{3} - k$, because 3 does not divide $\gcd(18, 38)$. The other combinations are attained in Examples 3.1, and 3.2. However, not all choices for k are possible. For instance, one can check that double poles $-\frac{2}{3}$ and $-\frac{3}{2}$ cannot happen simultaneously.

4. Poles distinct from $-\frac{1}{n}$

The following examples show how to produce poles of order two distinct from $-\frac{1}{n}$ considering allowed differential forms.

EXAMPLE 4.1. Let us consider consider f_{ab} from Example 2.2 together with the differential form $\omega_3 = (xy)^4 dxdy$. The dual resolution graph and the numerical data of f_{ab} and the differential form ω_2 are in Figure 4. Again, the default numerical data of (solid) arrows are $(1,1)$ and dashed arrows correspond to the strict transform of (ω_3).

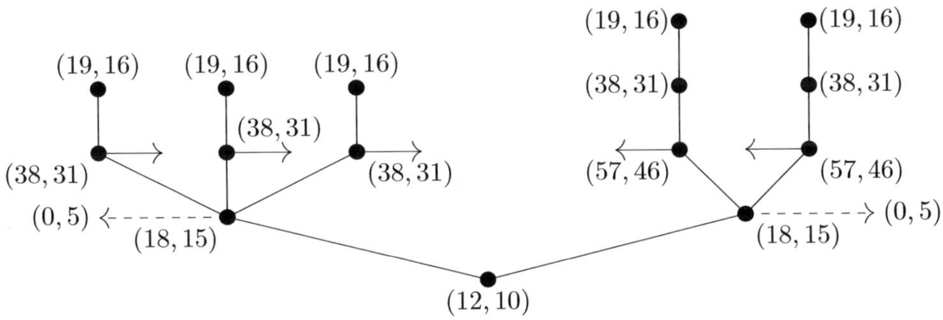

FIGURE 4. Dual graph of an embedded resolution of f_{ab} and ω_3.

The local topological zeta function has a pole $(-5/6)$ of order two, and it is given by

$$Z_{\text{top}}(f_{ab}, \omega_3; s) = \frac{-41496\, s^4 - 91616\, s^3 - 55926\, s^2 + 1559\, s + 7130}{5\,(6\, s+5)^2 (38\, s+31)(57\, s+46)(s+1)}.$$

EXAMPLE 4.2. Let us consider $g_{p,q} = (y^p + x^q)(y^q + x^p)$, $1 < p < q$, $\gcd(p,q) = 1$, $\omega = (xy)^{a-1} dxdy$, $a \geq 1$. An embedded resolution of $(g_{p,q}, \omega)$ has too many irreducible components, but in [**AMO14**, Example 4.6] we have an embedded **Q**-resolution with few strata, see Figure 5, where $[\bullet]$ stands for the order of the group associated to the 0-dimensional stratum, see the explanation below.

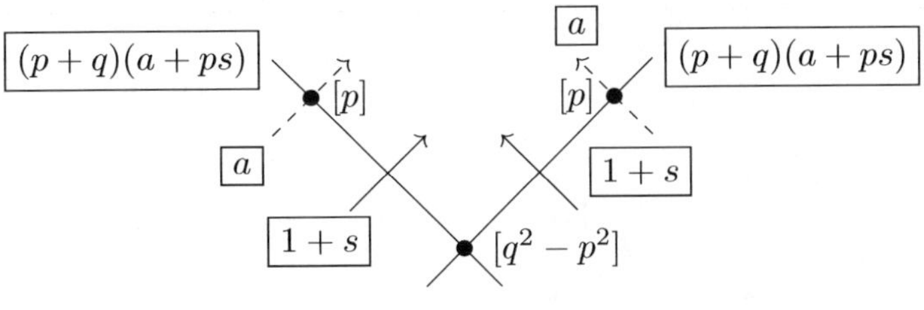

FIGURE 5. An embedded **Q**-resolution of $g_{p,q}$.

The concept of embedded **Q**-resolution $\pi : X \to \mathbb{C}^2$ was introduced in the work [**Mar14**]. For dimension 2, we mean that a finite set of points in the exceptional divisor may be quotient singularities for some cyclic group; the preimage of the total divisor satisfies a natural condition of **Q**-normal crossings. In this case there are also some strata $p_1^i, \dots, p_i^{n_i}$ such that $(X, p_j^i) \cong (\mathbb{C}^2, 0)/G_j^i$ where G_j^i is a group of order $m_i^j > 1$ with a small action on $\mathbb{C}^2$. The one-dimensional strata will be of the form $\mathring{E}_i = E_i \setminus \left(\bigcup_{j\neq i} E_j \cup \{p_1^1, \dots, p_{n_i}^i\}\right)$. The intersection of two components $E_i \cap E_j$ has also associated an order $m_{i,j} \geq 1$. Following [**Vey97**] we have

$$Z_{\text{top}}(f, \omega; s) = \sum_{i\in I} \frac{\chi(\mathring{E}_i) + \sum_{j=1}^{n_i} m_i^j}{\nu_i + sN_i} + \sum_{E_i\cap E_j\neq\emptyset} \frac{m_{i,j}}{(\nu_i + sN_i)(\nu_j + sN_j)} \in \mathbb{Q}(s).$$

Applying the formula we obtain

$$Z_{\text{top}}(g_{p,q}, \omega; s) = \frac{q-p}{q+p}\frac{1}{(a+ps)^2} + \frac{2}{(a-p)(q+p)}\left(\frac{1}{1+s} + \frac{(ap-p^2-1)a}{a+ps}\right).$$

In particular, for $a \not\equiv 0 \mod p$, we obtain that $-\frac{a}{p}$ is a double pole.

References

[AMO14] E. Artal Bartolo, J. Martín-Morales, and J. Ortigas-Galindo, *Intersection theory on abelian-quotient V-surfaces and **Q**-resolutions*, J. Singul. **8** (2014), 11–30, DOI 10.5427/jsing.2014.8b. MR3193225

[Loe88] F. Loeser, *Fonctions d'Igusa p-adiques et polynômes de Bernstein* (French), Amer. J. Math. **110** (1988), no. 1, 1–21, DOI 10.2307/2374537. MR926736

[LV99] A. Laeremans and W. Veys, *On the poles of maximal order of the topological zeta function*, Bull. London Math. Soc. **31** (1999), no. 4, 441–449, DOI 10.1112/S0024609398005566. MR1687560

[Mar14] J. Martín-Morales, *Embedded **Q**-resolutions for Yomdin-Lê surface singularities*, Israel J. Math. **204** (2014), no. 1, 97–143, DOI 10.1007/s11856-014-1078-z. MR3273453

[MTV10] A. Melle-Hernández, T. Torrelli, and W. Veys, *Monodromy Jordan blocks, b-functions and poles of zeta functions for germs of plane curves*, J. Algebra **324** (2010), no. 6, 1364–1382, DOI 10.1016/j.jalgebra.2010.07.022. MR2671810

[NV12] A. Némethi and W. Veys, *Generalized monodromy conjecture in dimension two*, Geom. Topol. **16** (2012), no. 1, 155–217, DOI 10.2140/gt.2012.16.155. MR2872581

[NX16] J. Nicaise and C. Xu, *Poles of maximal order of motivic zeta functions*, Duke Math. J. **165** (2016), no. 2, 217–243, DOI 10.1215/00127094-3165648. MR3457672

[Vey95] W. Veys, *Determination of the poles of the topological zeta function for curves*, Manuscripta Math. **87** (1995), no. 4, 435–448, DOI 10.1007/BF02570485. MR1344599

[Vey97] W. Veys, *Zeta functions for curves and log canonical models*, Proc. London Math. Soc. (3) **74** (1997), no. 2, 360–378, DOI 10.1112/S0024611597000130. MR1425327

[Wal04] C. T. C. Wall, *Singular points of plane curves*, London Mathematical Society Student Texts, vol. 63, Cambridge University Press, Cambridge, 2004, DOI 10.1017/CBO9780511617560. MR2107253

Departamento de Matemáticas-IUMA, Universidad de Zaragoza, c/ Pedro Cerbuna 12, 50009 Zaragoza, Spain

Email address: `artal@unizar.es`

Centro de Investigaciones Matemáticas, A.C., Callejón Jalisco s/n, Col. Valenciana, C.P. 36023 Guanajuato, Gto., Mexico

Email address: `manuel.gonzalez@cimat.mx`

Contemporary Mathematics
Volume **778**, 2022
https://doi.org/10.1090/conm/778/15660

Local invariants of minimal generic curves on rational surfaces

José Ignacio Cogolludo-Agustín, Tamás László, Jorge Martín-Morales, and András Némethi

ABSTRACT. Let $(C,0)$ be a reduced curve germ in a normal surface singularity $(X,0)$. The main goal is to recover the delta invariant $\delta(C)$ of the abstract curve $(C,0)$ from the topology of the embedding $(C,0) \subset (X,0)$. We give explicit formulae whenever $(C,0)$ is *minimal generic* and $(X,0)$ is rational (as a continuation of some previous papers by the authors).

Additionally, in this case, we prove that if $(X,0)$ is a quotient singularity, then $\delta(C)$ only admits the values $r-1$ or r, where r is the number or irreducible components of $(C,0)$. ($\delta(C) = r-1$ realizes the extremal lower bound, valid only for 'ordinary r–tuples'.)

1. Introduction

The present note is a natural continuation of the manuscripts [**8**, **9**]. The main setup is the following. We fix a complex normal surface singularity $(X,0)$, and an embedded reduced curve germ $(C,0)$ on it. We wish to connect the analytic invariants of the abstract curve $(C,0)$, eg. its delta invariant $\delta(C)$, with the embedded geometry of the pair $(C,0) \subset (X,0)$. The goal is to recover $\delta(C)$ from the embedded topology of the pair under certain restrictions (eg. when $(X,0)$ is rational).

One can attack this problem in two different ways, either analytically as in [**8**], or via topological tools, as in [**9**]. The first one is based on vanishing theorems (valid from a certain bound of Chern classes), the second one is based on the machinery of multivariable zeta (Poincaré) functions/series, their counting functions and surgery formulae. These formulae are valid again only for certain bound of the

2020 *Mathematics Subject Classification.* Primary 14B05, 32Sxx; Secondary 14E15.

Key words and phrases. Normal surface singularities, delta invariant of curves, Poincaré series, periodic constant, twisted duality, rational surface singularities, Weil divisors, Riemann-Roch.

The first and third authors are partially supported by PID2020-114750GB-C31 and Departamento de Ciencia, Universidad y Sociedad del Conocimiento del Gobierno de Aragón (Grupo de referencia E22_20R "Álgebra y Geometría"). The third author is also partially supported by FQM-333 "Junta de Andalucía". The second and fourth authors are supported by NKFIH Grant "Élvonal (Frontier)" KKP 126683. The second author was also supported by ERCEA Consolidator Grant 615655 - NMST, and partially by the Basque Government through the BERC 2018-2021 program and Gobierno Vasco Grant IT1094-16, by the Spanish Ministry of Science, Innovation and Universities: BCAM Severo Ochoa accreditation SEV-2017-0718.

corresponding Chern classes (in this case one relies on asymptotically true formulae, and the bounds from which their validity is true has geometrical consequences).

In both theories, these bounds impose some restrictions on the topology of the curve $(C, 0)$, in such cases the discussion is optimal. In this article we call such curves *minimal generic curves* of $(X, 0)$.

In the body of the paper we determine $\delta(C)$ for any 'minimal generic curve germ' $(C, 0)$ embedded into a rational surface singularity $(X, 0)$ from the topology of the embedding. The proof is based on a surgery formula of the multivariable topological Poincaré series and several recent developments in the area (see eg. [**8, 9, 17–20**]).

Surprisingly, it turns out that the minimality condition imposed by being 'minimal generic' is reflected as an extremal behavior in the value of the delta invariant as well.

Recall that if $(C, 0)$ is an abstract reduced curve germ with r irreducible components, then $\delta(C) \geq r - 1$, and $\delta(C) = r - 1$ is realized for the very special family of 'ordinary r–tuples'. They are analytically isomorphic with the union of the r axes of $(\mathbb{C}^r, 0)$, or with the union of r smooth generic lines in $(\mathbb{C}^r, 0)$.

It turns out that 'minimal generic' curve germs on a cyclic quotient singularity are all ordinary r–tuples, a fact firstly noticed in [**11**] (a work, which partly motivated our work). In the present note we indicate several proofs of this fact (of rather different nature).

But what is even more surprising is that this behavior almost survives whenever $(X, 0)$ is an arbitrary quotient singularity.

Our main result states that in such a case (i.e. if $(X, 0)$ is quotient singularity and $(C, 0)$ is a minimal generic curve germ on it) $\delta(C) \in \{r - 1, r\}$. ($\delta(C) = r$ for $r \geq 3$ means that $(C, 0)$ is isomorphic with r lines in $(\mathbb{C}^{r-1}, 0)$ in generic position.)

The paper is organized as follows. In the preliminary section 2 we recall all the needed material (with additional comments and remarks, in order to help the understanding of the main connections and ideas) regarding normal surface singularities. We discuss the definition of *minimal generic curves*, their connection with special minimal Chern classes (rational divisors), and the related Laufer algorithm. Then we continue with a material related with multivariable series, the topological and analytical Poincaré series, counting functions, and finally the surgery formula. We refer to [**8, 9, 17–20**] for more details. Here we also discuss in short the connection with the kappa invariant, which was historically one of the major motivations, cf. [**7, 11**].

In section 3 we discuss the case when $(X, 0)$ is a cyclic quotient singularity. A formula for the delta invariant of minimal generic curves is already known (see [**11**]), but we present several alternative proofs using different techniques, which shed some light into the general problem.

Section 4 is devoted to developing the theory of minimal generic curves on rational surface singularities with star-shaped minimal resolution graphs (here in some parts the rationality is not even needed). In Theorem 4.5 we provide a formula for the delta invariant of minimal generic curves on rational surface singularities with a star-shaped graph.

In section 5 the main result is applied to quotient surface singularities. As a result, in Theorem 5.1 we describe the formula for the delta invariant of minimal generic curves on quotient singularities based only on the information provided by the dual graph.

Section 6 is devoted to proving Theorem 5.1 and describing the exceptional cases where the general formula does not apply. The different nature of these cases should definitely be better understood.

2. Preliminaries

2.1. Topological invariants of surface singularities.

2.1.1. *Normal singularities, good resolutions, and dual graphs.* Let us consider a complex normal surface singularity $(X, 0)$, that is, a complex surface germ such that its local ring of functions $\mathcal{O}_{X,0}$ is integrally closed in its field of fractions. Equivalently, this means that any bounded holomorphic function on $X \setminus \{0\}$ can be holomorphically extended to X.

Let $\pi : \tilde{X} \to X$ be a *resolution*, that is, a proper holomorphic map from a smooth surface $\tilde{X}$ to a given representative of $(X, 0)$ such that π is biholomorphic outside $\pi^{-1}(\{0\})$. Moreover, we will require π to be a *good resolution*, which means that the exceptional set $\pi^{-1}(\{0\})$ is a union of smooth curves that intersect each other transversally.

A good resolution for normal surfaces always exists, but is not unique. Given a good resolution, we define its *dual graph*, as a decorated graph, say Γ, whose set of vertices, V, is in bijection with $\{E_v\}_{v\in V}$ the irreducible components of the exceptional set of π. Two vertices u, v are connected in Γ if and only if $E_u \cap E_v \neq \emptyset$. Moreover, each vertex is decorated with two numbers, namely $E_v^2, g(E_v)$, where E_v^2 is the self-intersection of the compact Riemann surface E_v in $\tilde{X}$ and $g(E_v)$ is its genus.

Associated to the dual graph Γ there is an *intersection matrix* $M = (m_{u,v})_{u,v\in V}$ given as the intersection matrix of the curves $\{E_v\}_{v\in V}$ in $\tilde{X}$, that is, $m_{u,v} := (E_u, E_v)$. This matrix is given as the numbers E_v^2 in the diagonal entries $m_{v,v}$ and the incidence matrix coefficients of Γ outside the diagonal.

2.1.2. *The link of a normal surface singularity.* If one assumes $(X, 0)$ is embedded in an affine germ $(\mathbb{C}^N, 0)$, then the link Σ of $(X, 0)$ is defined as the intersection of X with a small enough $(2N-1)$-dimensional sphere centered in 0. The topology of Σ does not depend on the embedding or the small-enough sphere. Moreover, $(X, 0)$ is a cone over Σ and hence $\Sigma \sim X \setminus \{0\} \cong \tilde{X} \setminus \pi^{-1}(0)$, $\Sigma \cong \partial\tilde{X}$, where $\sim$ represents the same homotopy type.

In this paper we assume that *the link Σ is a rational homology sphere* ($\mathbb{Q}\mathrm{HS}^3$ for short). This is equivalent to asking the dual graph Γ of a good resolution to be a connected tree, where all E_v's are rational, that is $g(E_v) = 0$ for all $v \in V$.

2.1.3. *Divisor lattice structure.* Define the lattice L as $H_2(\tilde{X}, \mathbb{Z})$. It is generated by the exceptional divisors E_v, $v \in V$, that is, $L = \oplus_{v\in V}\mathbb{Z}\langle E_v\rangle$. In the homology exact sequence of the pair $(\tilde{X}, \Sigma)$ one has $H_2(\Sigma, \mathbb{Z}) = 0$, $H_1(\tilde{X}, \mathbb{Z}) = 0$, hence the exact sequence has the form:

$$0 \to L \to H_2(\tilde{X}, \Sigma, \mathbb{Z}) \to H_1(\Sigma, \mathbb{Z}) \to 0. \tag{1}$$

Set $L' := \mathrm{Hom}(H_2(\tilde{X}, \mathbb{Z}), \mathbb{Z})$. Lefschetz-Poincaré duality $H_2(\tilde{X}, \Sigma, \mathbb{Z}) \cong H^2(\tilde{X}, \mathbb{Z})$ defines a perfect pairing $L \otimes H_2(\tilde{X}, \Sigma, \mathbb{Z}) \to \mathbb{Z}$. Hence L' can be identified with $H_2(\tilde{X}, \Sigma, \mathbb{Z})$. By (1) $L'/L \cong H_1(\Sigma, \mathbb{Z})$, which will be denoted by H. Note that since Σ is a $\mathbb{Q}\mathrm{HS}^3$, H is a finite abelian group. In fact, even if Σ is not $\mathbb{Q}\mathrm{HS}^3$, since the intersection form on L is non–degenerate, $H_2(\tilde{X},\mathbb{Z}) \to H_2(\tilde{X},\Sigma,\mathbb{Z})$ is injective, and $L'/L = \mathrm{Tors}(H_1(\Sigma,\mathbb{Z}))$.

As mentioned above, since the intersection form is non–degenerate, L' embeds into $L_{\mathbb{Q}} := L \otimes \mathbb{Q}$, and it can be identified with the rational cycles $\{\ell' \in L_{\mathbb{Q}} \mid (\ell', L)_{\mathbb{Q}} \in \mathbb{Z}\}$, where $(\,,\,)$ denotes the intersection form on L and $(\,,\,)_{\mathbb{Q}}$ its extension to $L_{\mathbb{Q}}$. Hence, in the sequel we regard L' as $\oplus_{v\in V}\mathbb{Z}\langle E_v^*\rangle$, the lattice generated by the rational cycles $E_v^* \in L_{\mathbb{Q}}$, $v \in V$, where $(E_u^*, E_v)_{\mathbb{Q}} = -\delta_{u,v}$ (Kronecker delta) for any $u, v \in V$. The inclusion $L \subset L'$ in the bases $\{E_v\}_v$ and $\{E_v^*\}_v$ is given by the intersection matrix of L as follows:

$$E_v = -\sum_{u\in V}(E_v, E_u)E_u^*.$$

Given an element $\ell' \in L'$ we denote by $[\ell'] \in H$ its class in H. In fact, $L' \subset L_{\mathbb{Q}}$ sits naturally in the superlattice $\frac{1}{d}L$, where $d := |H|$ is the order of H and L is identified with its integral points. The inclusion $L' \subset L_{\mathbb{Q}}$ is given by $-M^{-1}$, where M is the intersection matrix of Γ. The matrix $-M^{-1}$ has positive entries, since M is negative definite. Note that the coefficients of the cycles E_v^*'s, are the columns of $-M^{-1}$.

The elements E_v^* have the following geometrical interpretation as well: consider $\gamma_v \subset \tilde{X}$ a curvette associated with E_v, that is, a smooth irreducible curve in $\tilde{X}$ intersecting E_v transversally. Then $\pi^*\pi_*(\gamma_v) = \gamma_v + E_v^*$ (see 2.2.4 for the notation π^*).

2.1.4. *The canonical divisor and canonical cycle.* Let $K_{\tilde{X}}$ be the canonical divisor in the smooth surface $\tilde{X}$. The canonical divisor in X is defined as $K_X := \pi_*(K_{\tilde{X}})$. Note that $K_\pi := K_{\tilde{X}} - \pi^*(K_X)$ has support on the exceptional set $\pi^{-1}(0)$. The divisor K_π is called the relative canonical divisor of π, and it is determined topologically by the adjunction formula, which imposes the following linear system

$$(K_\pi + E_v, E_v) = -2, \text{ for all } v \in V \tag{2}$$

In particular $K_\pi \in L'$. In some cases, it is more convenient to use the (anti)canonical cycle $Z_K := -K_\pi$. Using (2), Z_K can be written as

$$Z_K = E - \sum_{v\in V}(2 - \mathrm{val}_v)E_v^*, \tag{3}$$

where $E = \sum_{v\in V} E_v$ and val_v is the valency of v.

Recall that by the Riemann-Roch theorem $\chi(\mathcal{O}_\ell) = -\frac{1}{2}(\ell, \ell - Z_K)$. This motivates the definition $\chi(\ell') := -\frac{1}{2}(\ell', \ell' - Z_K) \in \mathbb{Q}$ for any $\ell' \in L'$.

2.1.5. *The H–partition of L'.* The lattice L' admits a partition parametrized by the group H, where for any $h \in H$ one sets

$$L'_h = \{\ell' \in L' \mid [\ell'] = h\} \subset L'. \tag{4}$$

Note that $L'_0 = L$. Given $h \in H$ one can define $r_h := \sum_v r_v E_v \in L'_h$ as the unique element of L'_h such that $0 \le r_v < 1$ for all v. Equivalently, $r_h = \sum_v \{l'_v\}E_v$ for any $\ell' = \sum_v l'_v E_v \in L'_h$, where $0 \le \{\cdot\} < 1$ represents the usual fractional part.

2.1.6. *Examples. Quotient singularities.* An infinite family of examples of normal surface singularities with a $\mathbb{Q}\mathrm{HS}^3$-link is provided by the quotient $X = \mathbb{C}^2/G$ of $\mathbb{C}^2$ by the action of a small finite subgroup G of $\mathrm{GL}(2,\mathbb{C})$. These are called *quotient surface* singularities. Note that $H = G/[G,G]$ is finite, hence any such $(X, 0)$ has a $\mathbb{Q}\mathrm{HS}^3$-link.

If $G_1, G_2 \subset \mathrm{GL}(2,\mathbb{C})$, then $X_1 = \mathbb{C}^2/G_1$ and $X_2 = \mathbb{C}^2/G_2$ are isomorphic if and only if G_1 and G_2 are conjugated. In particular, it can happen that G_1 and G_2 are isomorphic as abstract groups but X_1 and X_2 are not.

A full classification of the small finite subgroups of $\mathrm{GL}(2,\mathbb{C})$ up to conjugation was provided by Brieskorn [**2**] (also a complete list of dual graphs is given by Nikulin in [**31**]). The simplest ones are the finite cyclic groups, $G = \mathbb{Z}_d$. In this case any $0 < q < d$, $\gcd(d,q) = 1$ provides a possible linear representations on $\mathbb{C}^2$ given by the diagonal matrices

$$\xi \mapsto \begin{pmatrix} \xi & 0 \\ 0 & \xi^q \end{pmatrix},$$

where $\mathbb{Z}_d = \{\xi \in \mathbb{C} : \xi^d = 1\}$. The quotient space usually is denoted by $X_{d,q}$, or $\frac{1}{d}(1,q)$. The quotient spaces $\frac{1}{d}(1,q_1)$ and $\frac{1}{d}(1,q_2)$ are isomorphic if and only if $q_2 \in \{q_1, q_1'\}$, where $q_1 q_1' \equiv 1 \pmod{d}$.

For $d = 2$ there is only one quotient singularity, which can also be given by the germ $x^2 + y^2 + z^2 = 0$. However, for $d = 3$ there are two non–isomorphic ones: $\frac{1}{3}(1,1)$ is a complete intersection singularity in $\mathbb{C}^4$ whereas $\frac{1}{3}(1,2)$ is a hypersurface in $\mathbb{C}^3$ (moreover, their embedding dimensions are 4 and 3 respectively).

For more details (including the minimal resolution graphs of $\frac{1}{d}(1,q)$) see section 3.

From a different point of view, McKay [**23**] proved that there is a 1–1 correspondence between minimal graphs of quotient singularities and conjugacy classes of finite subgroups in $\mathrm{GL}(2,\mathbb{C})$. This is known as the McKay correspondence. (E.g., regarding the previous pair, the graph of $\frac{1}{3}(1,1)$ has one vertex, whereas $\frac{1}{3}(1,2)$ has two vertices.)

2.2. The notion of minimal generic curve germs of $(X,0)$. We define the *rational Lipman cone* by

$$\mathcal{S}_{\mathbb{Q}} := \{\ell' \in L_{\mathbb{Q}} \mid (\ell', E_v) \le 0 \text{ for all } v \in V\},$$

which is a cone generated over $\mathbb{Q}_{\ge 0}$ by E_v^*. By 2.1.3, if $s \in \mathcal{S}_{\mathbb{Q}} \setminus \{0\}$ then all the E_v–coordinates of s are strict positive.

Define $\mathcal{S}' := \mathcal{S}_{\mathbb{Q}} \cap L'$ as the semigroup (monoid) of anti-nef rational cycles of L'; it is generated over $\mathbb{Z}_{\ge 0}$ by the cycles E_v^*. As mentioned in section 2.1.3, any element of $\mathcal{S}'$ can be obtained as the exceptional part of the pull-back of an effective divisor of X.

Set also $\mathcal{S} := \mathcal{S}_{\mathbb{Q}} \cap L$.

2.2.1. *Artin's fundamental cycle and other minimal cycles.* The Lipman cone $\mathcal{S}'$ also admits a natural equivariant partition $\mathcal{S}'_h = \mathcal{S}' \cap L'_h$ indexed by H.

For $\ell'_1, \ell'_2 \in L_{\mathbb{Q}}$ with $\ell'_i = \sum_v l'_{iv} E_v$ $(i = \{1,2\})$ one considers an order relation $\ell'_1 \ge \ell'_2$ defined coordinatewise by $l'_{1v} \ge l'_{2v}$ for all $v \in V$. In particular, ℓ' is an effective rational cycle if $\ell' \ge 0$. We set also $\min\{\ell'_1, \ell'_2\} := \sum_v \min\{l'_{1v}, l'_{2v}\} E_v$ and analogously $\min\{F\}$ for a finite subset $F \subset L_{\mathbb{Q}}$.

First note that $\mathcal{S} = \mathcal{S}'_0$ has the following properties:

(1) if $Z = \sum n_v E_v \in \mathcal{S}$ and $Z \ne 0$, then $n_v > 0$ for all $v \in V$,
(2) if $Z_1, Z_2 \in \mathcal{S}$, then $Z_1 + Z_2 \in \mathcal{S}$,
(3) if $Z_1, Z_2 \in \mathcal{S}$, then $\min\{Z_1, Z_2\} \in \mathcal{S}$.

As a consequence, one can define the unique minimal element $Z_{\min}$ of $\mathcal{S} \setminus \{0\}$, called the *Artin's fundamental cycle* (*minimal cycle* or *numerical cycle*). It satisfy $Z_{\min} \geq E$.

Similarly, for any $h \in H$, $\mathcal{S}_h$ has the following properties:

(1) $s_1, s_2 \in \mathcal{S}'_h$ implies $s_2 - s_1 \in L$ and hence $\min\{s_1, s_2\} \in \mathcal{S}'_h$,
(2) for any $s \in L_{\mathbb{Q}}$ the set $\{s' \in \mathcal{S}'_h \mid s' \not\geq s\}$ is finite,
(3) for any h there exists a unique *minimal cycle* $s_h := \min\{\mathcal{S}'_h\}$ (see 2.2.2 below).

Note that for $h = 0$ one has $s_0 = \min\{\mathcal{S}'_0\} = 0$, while $Z_{\min} = \min\{\mathcal{S}'_0 \setminus 0\} \geq E$.

In general, for an arbitrary graph it is very difficult to determine the cycles s_h. For strings or star-shaped graphs see e.g. [**24**]. These will be revised in this note as well.

2.2.2. *Generalized Laufer's algorithm.* [**24**, Lemma 7.4] For any $\ell' \in L'$ there exists a unique minimal element $s(\ell')$ of the set $\{s \in \mathcal{S}' \mid s - \ell' \in L_{\geq 0}\}$. It can be obtained by the following algorithm. Set $x_0 := \ell'$. Then one constructs a *computation sequence* $\{x_i\}_i$ as follows. If x_i is already constructed and $x_i \notin \mathcal{S}'$ then there exits some $E_{v(i)}$ such that $(x_i, E_{v(i)}) > 0$. Then take $x_{i+1} := x_i + E_{v(i)}$. Then the procedure after finitely many steps stops, say at x_t, and necessarily $x_t = s(\ell')$.

Note that $s(r_h) = s_h$ and $r_h \leq s_h$, however, in general $r_h \neq s_h$. (This fact does not contradict the minimality of s_h in $\mathcal{S}'_h$ since r_h might not sit in $\mathcal{S}'_h$.)

2.2.3. *Laufer's criterion for rationality.* The classical algorithm of Laufer constructs a computation sequence from E (or, from any E_v) to $Z_{\min}$. According to [**22**], $(X, 0)$ is rational if and only if along this computation sequence $(x_i, E_{v(i)}) = 1$ always. (In particular, $\chi(Z_{\min}) = 1$ too, this identity is Artin's criterion for rationality.)

2.2.4. *Minimal generic h–curves on a surface singularity.* Consider a reduced curve (or an effective Weil divisor) $(C, 0)$ in $(X, 0)$. Let $\tilde{C}$ denote its strict transform by π. Then we define a cycle $\ell'_C \in L'$ in such a way that the divisor $\ell'_C + \tilde{C}$ is numerically trivial: $(\ell'_C + \tilde{C}, E_v) = 0$ for all v. We write this fact in the language of divisors as $\pi^*(C) = \ell'_C + \tilde{C}$.

DEFINITION 2.1. Fix some $h \in H$ and the minimal good resolution. We say that $(C, 0)$ is

(1) an *h–curve* if $[\ell'_C] = h$;
(2) a *minimal h*–curve if $\ell'_C = s_h$;
(3) a *minimal generic h*–curve if $\ell'_C = s_h$ and all the components of $\tilde{C}$ are smooth and do not intersect each other, and they intersect E transversally.

Also, a curve $(C, 0) \subset (X, 0)$ is called *minimal generic* if it is a minimal generic h–curve for some $h \in H$.

Note that minimal generic curves were introduced as *generic* curves in [**11**] and studied in the context of cyclic quotient singularities, however we think the term *minimal generic* is more appropriate.

From the definition it follows that for $h = 0$ the minimal h–curve is the empty curve, what we might include or we might eliminate (though when we compare this picture with some other classification statements usually it is convenient to have the empty curve also included).

Note that for fixed $h \in H$, all minimal generic h–curves are topologically equivalent: the number of components intersecting transversally E_v is exactly c_v, where

$s_h = \sum_v c_v E_v^*$. Analytically (in $\text{Pic}(\tilde{X})$) these curves might be different. However, if $(X,0)$ is rational (when $\text{Pic}(\tilde{X}) = L'$), for fixed h, all the minimal generic h–curves are linearly equivalent. In this case we will call such a curve 'the' minimal generic curve associated with h.

A minimal generic curve $(C,0)$ is not necessarily irreducible and if irreducible it is not necessarily smooth (though $\tilde{C}$ is smooth) (see Examples 6.3 and 6.3). Note also that if we take an irreducible smooth $\tilde{C}$ intersecting E_v transversally, then $C = \pi(\tilde{C})$ is a generic $[E_v^*]$–curve, but not necessarily minimal. Indeed, eg., since the E_8 graph is unimodular, the only minimal generic curve is the empty one.

In general, as s_h is hard to characterize, the position of the (strict transforms of the) minimal generic curves is also hard to characterize.

However, if $(X,0)$ is a cyclic quotient singularity, then any irreducible transversal smooth curve $\tilde{C}$ (intersecting any E_v) represents its minimal generic h–class (a fact proved in [**11**], see also the detailed discussion in section 3). This provides in the case of the minimal resolution of a cyclic quotient a bijection between irreducible minimal generic curves, the vertices of the dual graph, and the full special $\mathcal{O}_X$-modules in the spirit of the McKay correspondence.

2.3. Multivariable series. Let $\mathbb{Z}[[L']]$ be the $\mathbb{Z}$-module consisting of the $\mathbb{Z}$-linear combinations of the monomials $\mathbf{t}^{\ell'} := \prod_{v\in V} t_v^{l'_v}$, where $\ell' = \sum_v l'_v E_v \in L'$. Note that it is a $\mathbb{Z}$-submodule of the formal power series in variables $t_v^{1/d}, t_v^{-1/d}$, $v \in V$.

Consider a multivariable series $S(\mathbf{t}) = \sum_{\ell' \in L'} a(\ell')\mathbf{t}^{\ell'} \in \mathbb{Z}[[L']]$. Let us consider the support $\text{Supp}\, S(\mathbf{t}) := \{\ell' \in L' \mid a(\ell') \neq 0\}$ of the series and assume the following finiteness condition: for any $x \in L'$

$$\{\ell' \in \text{Supp}\, S(\mathbf{t}) \mid \ell' \not\geq x\} \text{ is finite.} \tag{5}$$

Throughout this paper we will use multivariable series in $\mathbb{Z}[[L']]$ as well as in $\mathbb{Z}[[L'_I]]$ for any $I \subset V$, where $L'_I = \text{pr}_I(L')$ is the projection of L' via $\text{pr}_I : L_{\mathbb{Q}} \to \oplus_{v\in I}\mathbb{Q}\langle E_v\rangle$. For example, if $S(\mathbf{t}) \in \mathbb{Z}[[L']]$ then $S(\mathbf{t}_I) := S(\mathbf{t})|_{t_v=1, v\notin I}$ is an element of $\mathbb{Z}[[L'_I]]$. In the sequel we use the notation $\ell'_I = \ell'|_I := \text{pr}_I(\ell')$ and $\mathbf{t}_I^{\ell'} := \mathbf{t}^{\ell'}|_{t_v=1, v\notin I}$ for any $\ell' \in L'$. Each coefficient $a_I(x)$ of $S(\mathbf{t}_I)$ is obtained as a summation of certain coefficients $a(y)$ of $S(\mathbf{t})$, where y runs over $\{\ell' \in \text{Supp}\, S(\mathbf{t}) \mid \ell'|_I = x\}$ (this is a finite sum by (5)). Moreover, $S(\mathbf{t}_I)$ satisfies a similar finiteness property as (5) in the variables $\mathbf{t}_I$.

Any $S(\mathbf{t}) \in \mathbb{Z}[[L']]$ decomposes in a unique way as $S(\mathbf{t}) = \sum_h S_h(\mathbf{t})$, where $S_h(\mathbf{t}) := \sum_{[\ell']=h} a(\ell')\mathbf{t}^{\ell'}$. The series $S_h(\mathbf{t})$ is called the h-part of $S(\mathbf{t})$. Note that the H-decomposition of the series $S(\mathbf{t}_I)$ is not well defined. That is, the restriction $S_h(\mathbf{t})|_{t_v=1, v\notin I}$ of the h-part $S_h(\mathbf{t})$ cannot be recovered from $S(\mathbf{t}_I)$ in general, since the class of ℓ' cannot be recovered from $\ell'|_I$. Hence, the notation $S_h(\mathbf{t}_I)$ (defined as $(S_h(\mathbf{t}))|_{t_v=1, v\notin I}$) is not ambiguous, but requires certain caution.

2.3.1. *The multivariable topological Poincaré series.* The *multivariable topological Poincaré series* (cf. [**4, 5, 27**]) is the Taylor expansion at the origin $Z(\mathbf{t}) = \sum_{\ell'} z(\ell')\mathbf{t}^{\ell'} \in \mathbb{Z}[[L']]$ of the rational *zeta function*

$$f(\mathbf{t}) = \prod_{v\in V} (1 - \mathbf{t}^{E_v^*})^{\text{val}_v - 2}. \tag{6}$$

One can consider its H-decomposition $Z(\mathbf{t}) = \sum_{h\in H} Z_h(\mathbf{t})$. The support of $Z(\mathbf{t})$ is in the Lipman cone $\mathcal{S}'$. Note that by 2.2.1(2) for any $x \in L'$ the set $\{\ell' \in \mathcal{S}' \mid \ell' \ngeq x\}$ is finite. Hence for any $h \in H$ one can consider the *counting function* $Q_h^\Gamma(\mathbf{t}) = Q_h(\mathbf{t})$ of $Z_h(\mathbf{t})$ defined as

$$(7)\qquad Q_h(\mathbf{t}) : L' \longrightarrow \mathbb{Z}, \qquad x \mapsto \sum_{\ell' \ngeq x,\ [\ell']=h} z(\ell').$$

Furthermore, for any $h \in H$ and $I \subset V$ one can also consider the *counting function* $Q_{h,I}^\Gamma(\mathbf{t}) = Q_{h,I}(\mathbf{t})$ of $Z_h(\mathbf{t}_I)$ defined as

$$(8)\qquad Q_{h,I}(\mathbf{t}) : L' \longrightarrow \mathbb{Z}, \qquad x \mapsto \sum_{\ell'|_I \ngeq x|_I,\ [\ell']=h} z(\ell').$$

2.3.2. *Surgery formulas for the counting function* $Q_h(\mathbf{t})$. Let us fix $I \subset V$. The set of vertices $V \setminus I$ determines the connected full subgraphs $\{\Gamma_k\}_k$, $\cup_k \Gamma_k = \Gamma \setminus I$. For each k we consider the inclusion operator $j_k : L(\Gamma_k) \to L(\Gamma)$, $E_v(\Gamma_i) \mapsto E_v(\Gamma)$, identifying naturally the corresponding E–base elements in the two graphs. This preserves the intersection forms. Let $j_k^* : L'(\Gamma) \to L'(\Gamma_k)$ be its dual, defined by $j_k^*(E_v^*(\Gamma)) = E_v^*(\Gamma_k)$ if $v \in V(\Gamma_k)$, and $j_k^*(E_v^*(\Gamma)) = 0$ otherwise. Then one has the projection formula $(j_k^*(\ell'), \ell)_{\Gamma_k} = (\ell', j_k(\ell))_\Gamma$ for any $\ell' \in L'(\Gamma)$ and $\ell \in L(\Gamma_k)$. This implies that

$$(9)\qquad j_k^*(Z_K) = Z_K(\Gamma_k),$$

where $Z_K = Z_K(\Gamma)$, respectively $Z_K(\Gamma_k)$, are the (anti)canonical cycles of Γ and Γ_k respectively. Note also that $j_k^*(E_v(\Gamma)) = E_v(\Gamma_k)$ for any $v \in V(\Gamma_k)$.

Furthermore, one has the following formula regarding the pull–back of the cycles s_h.

LEMMA 2.2 ([**17**, **21**]). *For any $h \in H$ one has $j_k^*(s_h) = s_{[j_k^*(s_h)]} \in L'(\Gamma_k)$.*

For any fixed $h \in H$ one has the following surgery formula for the counting functions:

THEOREM 2.3 ([**17**, Theorem 3.2.2],[**9**, Corollary 3.8]). *For any $\ell' = \sum_v a_v E_v^*$ with $a_v \gg 0$ and with the notation $[\ell'] = h$ one has the identity*

$$(10)\qquad Q_h^\Gamma(\ell') = Q_{h,I}^\Gamma(\ell') + \sum_k Q^{\Gamma_k}_{[j_k^*(\ell')]}(j_k^*(\ell')).$$

Moreover, if Γ is rational, then the equality (10) *holds on the entire cone $Z_K + \mathcal{S}'$, and $Q_{[Z_K]}(Z_K) = 0$.*

2.3.3. *The multivariable analytic Poincaré series.* Parallel to $Z(\mathbf{t})$ there is a series defined using the analytic structure of $(X, 0)$ (and a fixed resolution). Although in this note it will not be really used, for the convenience of the reader we include its definition here (together with the CDGZ–identity), since this correspondence was a leading motivation in the work of the authors, and initiated and resulted several of the statements, which are central in this discussion. (Eg., Theorem 2.3 was motivated by an analytic surgery formula.)

We fix a good resolution π of X. Consider $c : Y \to X$, the universal abelian covering of $(X, 0)$, let $\tilde{Y}$ be the normalized pull-back of π and c, and denote by

π_Y and $\tilde{c}$ the induced maps by the pull-back completing the following commutative diagram.

$$\begin{array}{ccc} \tilde{Y} & \xrightarrow{\tilde{c}} & \tilde{X} \\ {\scriptstyle \pi_Y}\downarrow & \# & \downarrow{\scriptstyle \pi} \\ Y & \xrightarrow[c]{} & X \end{array} \tag{11}$$

We define the following H-equivariant L'-indexed divisorial filtration of the local ring $\mathcal{O}_{Y,0}$: for any given $\ell' \in L'$ set

$$\mathcal{F}(\ell') := \{g \in \mathcal{O}_{Y,0} \mid \mathrm{div}(g \circ \pi_Y) \geq \tilde{c}^*(\ell')\}. \tag{12}$$

It is worth mentioning that the pull-back $\tilde{c}^*(\ell')$ is an integral cycle in $\tilde{Y}$ for any $\ell' \in L'$, cf. [**27**, Lemma 3.3]. The natural action of H on $(Y,0)$ induces an action on $\mathcal{O}_{Y,0}$ as follows: $h \cdot g(y) = g(h \cdot y)$, $g \in \mathcal{O}_{Y,0}$, $h \in H$. This action decomposes $\mathcal{O}_{Y,0}$ as $\oplus_{\lambda \in \hat{H}} (\mathcal{O}_{Y,0})_\lambda$ according to the characters $\lambda \in \hat{H} := \mathrm{Hom}(H, \mathbb{C}^*)$, where

$$(\mathcal{O}_{Y,0})_\lambda := \{g \in \mathcal{O}_{Y,0} \mid g(h \cdot y) = \lambda(h) g(y),\ \forall y \in Y, h \in H\}. \tag{13}$$

There exists a natural isomorphism $\theta : H \to \hat{H}$ given by $h \mapsto \exp(2\pi\sqrt{-1}(\ell', \cdot)) \in \mathrm{Hom}(H, \mathbb{C}^*)$, where ℓ' is any element of L' with $h = [\ell']$.

The subspace $\mathcal{F}(\ell')$ is invariant under this action and $\mathcal{F}(\ell')_{\theta(h)} = \mathcal{F}(\ell') \cap (\mathcal{O}_{Y,0})_{\theta(h)}$. Thus, one can define the *Hilbert function* $\mathfrak{h}(\ell')$ for any $\ell' \in L'$ as the dimension of the $\theta([\ell'])$-eigenspace $(\mathcal{O}_{Y,0}/\mathcal{F}(\ell'))_{\theta([\ell'])}$. The corresponding multivariable *Hilbert series* is

$$H(\mathbf{t}) = \sum_{\ell' \in L'} \mathfrak{h}(\ell') \mathbf{t}^{\ell'} \in \mathbb{Z}[[L']]. \tag{14}$$

The *multivariable analytic Poincaré series* $P(\mathbf{t}) = \sum_{\ell' \in L'} \mathfrak{p}(\ell') \mathbf{t}^{\ell'}$ can be defined by

$$P(\mathbf{t}) = -H(\mathbf{t}) \cdot \prod_{v \in V} (1 - t_v^{-1}). \tag{15}$$

2.3.4. *The CDGZ-identity.* The *CDGZ-identity,* named after Campillo, Delgado and Gusein-Zade thanks to their contributions in [**4**] and [**5**] relates the Poincaré series $P(\mathbf{t})$ and $Z(\mathbf{t})$ for rational surface singularities.

THEOREM 2.4 (CDGZ-identity [**4**]). *A rational surface singularity $(X, 0)$ and any of its resolution π satisfy the identity*

$$P(\mathbf{t}) = Z(\mathbf{t}). \tag{16}$$

The fourth author in [**26**], [**27**] extended the result for larger families of normal surface singularities with rational homology sphere link. The largest family for which the CDGZ-identity holds is the family of splice-quotient singularities with their special resolutions satisfying the end-curve conditions (see [**28**]).

2.4. Invariants of abstract curve germs. In this section we recall the most relevant local invariants of a curve germ considered in this paper.

2.4.1. *The delta invariant.* The delta invariant $\delta(C)$ of a reduced curve germ $(C,0)$ is defined as follows. Let $\gamma : (C,0)\tilde{} \to (C,0)$ be the normalization, where $(C,0)\tilde{}$ is the corresponding multigerm. Then $\delta(C) := \dim_{\mathbb{C}} \gamma_* \mathcal{O}_{(C,0)\tilde{}}/\mathcal{O}_{(C,0)}$.

Usually we also write r (or $r(C)$) for the number of irreducible components of $(C,0)$.

The delta invariant of a reduced curve can be determined inductively from the delta invariant of the components and the *Hironaka generalized intersection multiplicity*. Indeed, assume $(C,0)$ is embedded in some $(\mathbb{C}^n,0)$, and assume that $(C,0)$ is the union of two (not necessarily irreducible) germs $(C_1',0)$ and $(C_2',0)$ without common irreducible components. Assume that $(C_i',0)$ is defined by the ideal I_i in $\mathcal{O}_{(\mathbb{C}^n,0)}$ $(i=1,2)$. Then one can define *Hironaka's intersection multiplicity* by $(C_1',C_2')_{Hir} := \dim_{\mathbb{C}}(\mathcal{O}_{(\mathbb{C}^n,0)}/(I_1+I_2))$. Then, one has the following formula of Hironaka, see [**15**] or [**33**, 2.1] and [**3**],

$$\delta(C) = \delta(C_1') + \delta(C_2') + (C_1',C_2')_{Hir}. \tag{17}$$

In particular, if $(C,0)$ has r irreducible components, then using induction (and the obvious inequality $(C_1',C_2')_{Hir} \geq 1$) we get

$$\delta(C) \geq r-1. \tag{18}$$

2.4.2. *Ordinary r–tuples.* [**3, 13, 14, 33**] If a reduced (abstract) curve germ $(C,0)$ is (analytically equivalent with) the union of the coordinate axes of $(\mathbb{C}^r,0)$, then we call $(C,0)$ an *ordinary r–tuple.* Using e.g. Hironaka's formula we get that $\delta(C)=r-1$. This shows that the ordinary r–tuples realizes the optimal minimum of the bound (18). In fact, this property characterizes them: if for a curve $(C,0)$ with r components one has $\delta(C)=r-1$ then $(C,0)$ is necessarily an ordinary r–tuple. Their notation is R_r^r.

2.4.3. *R_r^{r-1} type r–tuples.* [**3,13,14,33**] Recall that a curve of type R_r^r consists of r lines in $(\mathbb{C}^r,0)$ in general position. We generalize this as follows. For $r\geq 3$, R_r^{r-1} denotes (the isomorphism class of) r lines in $(\mathbb{C}^{r-1},0)$ in general position. The plane curve singularity A_3 (resp. A_2) with equation $\{x^4+y^2=0\}$ (resp. $\{x^3+y^2=0\}$) will also be denoted by R_2^1 and R_1^0 respectively. Then, it turns out that a curve singularity $(C,0)$ with r components is R_r^{r-1} if and only if $\delta(C)=r$.

2.5. Invariants of embedded curve germs. We fix a normal surface singularity $(X,0)$ with a $\mathbb{Q}\mathrm{HS}^3$-link, and we consider a reduced curve germ $(C,0)\subset(X,0)$. Next we recall the definition and some properties of the *kappa invariant* $\kappa_X(C)$, an embedded invariant of the pair $(C,0)\subset(X,0)$. The interplay between $\delta(C)$ and $\kappa_X(C)$ has been studied in [**8, 9**]. This connection produced several explicit formulae for the delta invariant of minimal generic curves embedded into rational surface singularities. Furthermore, the definition (and several expressions) of $\kappa_X(C)$ explains and motivates several expressions intensively used in the present note.

2.5.1. *The κ–invariant of a cycle.* Fix a good resolution $\pi:\widetilde{X}\to X$. For any $\ell'\in L'$ define

$$\kappa_{\widetilde{X}}(\ell') := \mathfrak{h}(Z_K+\ell') = \dim_{\mathbb{C}}\left(\frac{\mathcal{O}_{Y,0}}{\mathcal{F}(Z_K+\ell')}\right)_{\theta([Z_K+\ell'])}, \tag{19}$$

where $(Y,0)$ is the universal abelian covering of $(X,0)$ as described in (11). Note that $\kappa_{\widetilde{X}}(\ell')$ in this way is defined via the analytic filtration $\ell'\mapsto\mathcal{F}(\ell')$, or equivalently, in terms of the Hilbert series $H(\mathbf{t})$, which can be identified with the counting function of the analytic Poincaré series $P(\mathbf{t})$. However, if $(X, 0)$ is rational, then by the CDGZ–identity 2.4 $P(\mathbf{t}) = Z(\mathbf{t})$, hence $\kappa_{\widetilde{X}}(\ell')$ is computable from the counting function of $Z(\mathbf{t})$ as $Q_{[Z_K+\ell']}(Z_K+\ell')$, cf. [**8**,**9**]. (In general, this expression is called the 'topological' κ–invariant.)

Next, assume that $(C, 0)$ is a reduced curve germ on $(X, 0)$ (with $\mathbb{Q}\mathrm{HS}^3$ link, but not necessarily rational) and choose a good embedded resolution $\pi : \widetilde{X} \to X$ of the pair $(C, 0) \subset (X, 0)$. Then consider the strict transform $\widetilde{C}$ and the total transform $\pi^*(C) = \ell'_C + \widetilde{C}$ as in 2.2.4. Note that $\ell'_C \in \mathcal{S}'$. It turns out that the invariant $\kappa_{\widetilde{X}}(\ell'_C)$ is independent of the resolution π (see [**8**, Corollary 3.4]).

This justifies the following definition of the κ–invariant of a reduced curve germ $(C, 0) \subset (X, 0)$, where $(X, 0)$ has a $\mathbb{Q}\mathrm{HS}^3$ link.

DEFINITION 2.5. The κ–invariant of the pair $(C, 0) \subset (X, 0)$ is defined as

$$\kappa_X(C) = \kappa_{\widetilde{X}}(\ell'_C) = \dim_{\mathbb{C}} \left(\frac{\mathcal{O}_Y}{\mathcal{F}(Z_K + \ell'_C)} \right)_{\theta([Z_K+\ell'_C])}. \tag{20}$$

One of the main results of [**8**] relates $\kappa_X(C)$ and the δ–invariant of $(C, 0)$ as follows.

THEOREM 2.6 ([**8**, Theorem 1.1 & 1.2]). *If $(X, 0)$ is a rational surface singularity and $(C, 0) \subset (X, 0)$ a reduced curve germ, then*

$$\kappa_X(C) = \delta(C) = \chi(-\ell'_C) - \chi(s_{-[\ell'_C]}). \tag{21}$$

In this paper we consider the problem of calculating the delta invariant of minimal generic curves embedded into rational surface singularities.

3. Old and new results for cyclic quotient singularities

3.1. Let $(X, 0)$ be a cyclic quotient surface singularity with the notation $(X, 0) := \frac{1}{d}(1, q)$, cf. 2.1.6. The dual graph Γ of the minimal resolution of X is the following string of vertices (see eg. [**1**])

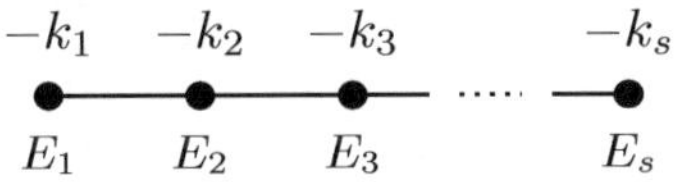

where the numerical data $\{k_i\}_{i=1}^s$ ($k_i \geq 2$) is encoded by the Hirzebruch–Jung (negative) continued fraction expansion

$$\frac{d}{q} = k_1 - \cfrac{1}{k_2 - \frac{1}{k_3 - \dots}} := [k_1, \dots, k_s].$$

We also set $0 < q' < d$ so that $qq' \equiv 1 \pmod d$. Moreover, for any $1 \leq i \leq j \leq s$ we write the continued fraction $[k_i, \dots, k_j]$ as a rational number d_{ij}/q_{ij} with $d_{ij} > 0$ and $\gcd(d_{ij}, q_{ij}) = 1$. Notice that d_{ij} is the determinant of the intersection matrix associated with the substring with vertices $v_i, \dots, v_j$. Then, $d_{1,s} = d$, $d_{2,s} = q$ and $d_{1,s-1} = q'$. We also set $d_{i,i-1} := 1$ and $d_{ij} := 0$ for $j < i - 1$.

Below we use the notation $\lfloor - \rfloor$ for the integral part, $\{-\}$ for the fractional part, and $\lceil - \rceil$ for the ceiling function.

We choose the class $[E_s^*]$ as a generator of $H = \mathbb{Z}_d$. For any class $a[E_s^*]$ with $0 \leq a < d$ we consider the minimal cycle $s_{a[E_s^*]}$ and we write $s_{a[E_s^*]} = \sum_{i=1}^s a_i E_i^*$. Then by [**24**, section 10.3], one has

$$a = d_{2,s} \cdot a_1 + d_{3,s} \cdot a_2 + \ldots + d_{s,s} \cdot a_{s-1} + a_s, \tag{22}$$

and the coefficients a_i are expressed from the integer a via the following recursive identity

(23)
$$a_i = \left\lfloor \frac{a - \sum_{j=1}^{i-1} d_{j+1,s} \cdot a_j}{d_{i+1,s}} \right\rfloor \quad (1 \le i \le s);$$

see also [**11**, Section 2.2]. In particular, $a_1 = \lfloor a/q \rfloor$.

For each fixed $h = [aE_s^*]$ the entries $\{a_i\}_{i=1}^s$ identify the minimal generic h–curve.

3.1.1. **Example.** Assume that $n/q = 15/11 = [2,2,3,2,2]$. Then the possible entries $(a_1, \dots, a_5)$ representing some $s_{a[E_s^*]}$ are the following 5–tuples

$$(1,0,1,0,0),\ (1,0,0,1,0),\ (1,0,0,0,1),\ (1,0,0,0,0),\ (0,1,1,0,0),$$

$$(0,1,0,1,0),\ (0,1,0,0,1),\ (0,1,0,0,0),\ (0,0,2,0,0),\ (0,0,1,1,0),$$

$$(0,0,1,0,1),\ (0,0,1,0,0),\ (0,0,0,1,0),\ (0,0,0,0,1),\ (0,0,0,0,0).$$

3.2. We consider a minimal generic curve $(C,0)$ on $(X,0)$ whose associated cycle is $\ell'_C = \sum_{i=1}^s a_i E_i^* = s_h$ for some $h = a[E_s^*]$, $0 < a < s$. If we denote by $r(C)$ the number of irreducible components of C, then we get $r(C) = \sum_i a_i$.

In this case one has the following simple formula for the delta invariant of $(C,0)$ in terms of the number of its branches.

THEOREM 3.1 ([**11**]). *For a minimal generic curve C on a cyclic quotient singularity $(X,0)$ one has $\delta(C) = r(C) - 1$. In particular, $(C,0)$ is an ordinary $r(C)$–tuple. (Cf.* 2.4.2.*)*

In the rest of the section, we summarize the existing proofs for the formula of Theorem 3.1.

3.2.1. The **first proof** was developed by the first and third authors in [**11**] using the methods of weighted blow-ups and an analytic type surgery formula for $\kappa_X(C)$. The authors used explicit equations for the irreducible minimal generic curves as $x^{q_i} - y^{\bar{q}_i}$, where q_i is a remainder appearing in the Hirzebruch-Jung decomposition of $\frac{d}{q}$ and $\bar{q}_i q = q_i \mod d$. The key of the proof is to show that $\kappa_X(C)$ can be obtained recursively as $\kappa_X(C) = \kappa_\pi + \kappa_{X'}(C')$ where $\pi' : X' \to X$ is a weighted blow-up and κ_π is either k_i or $k_i - 1$, where k_i is the number of transversal cuts on the exceptional divisor, depending on whether or not π is the last blow-up in the resolution. The invariant κ_π is obtained as the number of integer points in a triangle

(24)
$$\kappa_\pi := \#\{(i,j) \in \mathbb{Z}^2 \mid i,j \ge 1,\, k \ge i + qj \equiv k \mod d\} = \begin{cases} k_i - 1 & \text{if } k = k_i q \\ k_i & \text{otherwise.} \end{cases}$$

3.2.2. The **second proof** follows from one of the main result of [**8**] and [**9**] saying that for rational singularities $(X,0)$ and for $(C,0)$ as above one has $\delta(C) = \chi(-s_h) - \chi(s_{-h})$. The point is that when $(X,0)$ is cyclic quotient, the term $Exp := \chi(-s_h) - \chi(s_{-h})$ can be explicitly calculated. We follow [**8**, Example 5.9]. Indeed, we may write

(25)
$$Exp = \chi(s_h) - \chi(s_{-h}) - (s_h, Z_K)$$

since $\chi(-s_h) = \chi(s_h) - (s_h, Z_K)$. Then one can calculate

$$\chi(s_h) = \chi(s_{[aE_s^*]}) = \frac{a(1-d)}{2d} + \sum_{i=1}^{a} \left\{ \frac{iq'}{d} \right\}$$

using [**24**, 10.5.1]. Since $-h = [(d-a)E_s^*]$, $\chi(s_{-h})$ equals

$$\tag{26} \begin{aligned} \chi(s_{-h}) &= \frac{(d-a)(1-d)}{2d} + \sum_{j=1}^{d-a} \left\{ \frac{jq'}{d} \right\} \\ &= \frac{(d-a)(1-d)}{2d} + \sum_{i=1}^{d-1} \left(1 - \left\{ \frac{iq'}{d} \right\}\right) - \sum_{i=1}^{a-1} \left(1 - \left\{ \frac{iq'}{d} \right\}\right), \end{aligned}$$

which implies

$$\chi(s_h) - \chi(s_{-h}) = \frac{a}{d} - 1 - \left\{ \frac{aq'}{d} \right\}.$$

On the other hand $Z_K = E - E_1^* - E_s^*$ by (3), hence we get $(s_h, Z_K) = (s_h, E) - (s_h, E_1^* + E_s^*)$ where the first term is easily identified with $(s_h, E) = -r(C)$. For calculating the second term first we write $-(E_1^*, E_i^*) = d_{i+1,s}/d$ $(1 \le i \le s)$ following [**24**]. Then $(s_h, E_1^*) = -(\sum_i d_{i+1,s} \cdot a_i)/d = -a/d$ follows from (22). In order to compute (s_h, E_s^*) we create formally the symmetric situation. Since $[E_1^*] = [qE_s^*]$ and $[E_s^*] = [q'E_1^*]$, $h = [aE_s^*] = [aq'E_1^*]$. Write $a' \equiv aq' \pmod d$, $0 < a' < d$. Then for $h = [a'E_1^*]$ and $s_h = s_{[a'E_1^*]} = \sum_i a_i E_i^*$ the symmetric formula is

$$(s_h, E_s^*) = -\frac{a'}{d} = \left\lfloor \frac{aq'}{d} \right\rfloor - \frac{aq'}{d}.$$

Thus, by above calculation we get $(s_h, Z_K) = a/d - \lfloor aq'/d \rfloor + aq'/d - r(C)$. Finally, using also (25) and (26) we deduce $Exp = r(C) - 1$.

3.2.3. The **third proof** seems to reinterpret 3.2.1 in the language of Poincaré series. It is based on the surgery formula (10). It is more in the spirit of the topological approach developed in [**9**]. Here we use surgery formula techniques for the counting function of the coefficients of the topological Poincaré series, which will also serve as a base for further generalizations.

We proceed by induction on the number of vertices $|V|$ of Γ. First we assume that Γ contains only one vertex with self-intersection $-d$. Then $Z_K = (1 - 2/d)E$ and for any $0 < a < d$ the cycle of the generic curve is $\ell'_C = s_{[aE^*]} = aE^* = (a/d)E$ (cf. (23)). The topological Poincaré series associated with Γ is expressed as $Z(t) = \sum_{x \ge 0} (x+1) t^{x/d}$. By Theorem 2.6 we know that $\delta(C) = Q_{[Z_K + \ell'_C]}(Z_K + \ell'_C)$. In order to calculate it, we need to sum all the coefficients $(x+1)$ corresponding to $x \ge 0$, for which $x < d - 2 + a$ and $x \equiv d - 2 + a \pmod d$. Since there exists only one such x, namely $x = a - 2$, we get that $\delta(C) = a - 1$, which proves the statement for $|V| = 1$.

Now assume $|V| > 1$. Let v_1 be the first vertex of the string and we denote by Γ_2 the substring which we get from Γ by deleting v_1 and its adjacent edge. Denote the inclusion operator by $j_2 : L(\Gamma_2) \to L(\Gamma)$ and consider its dual operator j_2^* as in section 2.3.2. Then we apply the surgery formula in Theorem 2.3 for v_1 and the cycle $Z_K + \ell'_C \in Z_K + \mathcal{S}'$ in the rational surface context. Thus, we have

$$\tag{27} Q_{[Z_K + \ell'_C]}(Z_K + \ell'_C) = Q_{[Z_K + \ell'_C], v_1}(Z_K + \ell'_C) + Q^{\Gamma_2}_{[j_2^*(Z_K + \ell'_C)]}(j_2^*(Z_K + \ell'_C)).$$

First of all notice that by (9) one has $j_2^*(Z_K) = Z_K(\Gamma_2)$ where $Z_K(\Gamma_2)$ is the anti-canonical cycle associated with Γ_2. We set the notation $\ell'_{C,2} := j_2^*(\ell'_C) = \sum_{i=2}^s a_i E_i^*$ for the pull–back of the cycle $\ell'_C = \sum_{i=1}^s a_i E_i^*$.

Since C is a minimal generic curve one has $\ell'_C = s_h$ for $h = [\ell'_C] \in H$, hence by Lemma 2.2 we obtain that $\ell'_{C,2}$ is the cycle of a minimal generic curve C_2 on the cyclic quotient singularity whose minimal resolution graph is Γ_2. Note that $\ell'_{C,2} = 0$ exactly when $\sum_{i=2}^s a_i = 0$, in this case the corresponding restricted curve is empty. From (22) this happens exactly when $q|a$. Thus, by the inductive step, the second term of the right-hand side of (27) is

$$Q^{\Gamma_2}_{[Z_{K,2}+\ell'_{C,2}]}(Z_{K,2} + \ell'_{C,2}) = \delta(C_2) = \begin{cases} -1 + \sum_{i=2}^s a_i & \text{if } q \nmid a \\ 0 & \text{otherwise.} \end{cases} \tag{28}$$

It remains to calculate $Q_{[Z_K+s_h],v_1}(Z_K + s_h)$ from the topological Poincaré series $Z(\mathbf{t})$ of Γ. $Z(\mathbf{t})$, by definition, can be written in the following form

$$Z(\mathbf{t}) = T\Big[\frac{1}{(1-\mathbf{t}^{E_1^*})(1-\mathbf{t}^{E_s^*})}\Big] = \sum_{x,y\geq 0} \mathbf{t}^{xE_1^*+yE_s^*}.$$

Then we have to count all the cycles $\ell' = xE_1^* + yE_s^*$ satisfying the following properties:

$$\text{(i)} \quad \ell'|_{E_1} < (Z_K + s_h)|_{E_1}$$
$$\text{(ii)} \quad [\ell'] = [Z_K + s_h].$$

Recall that $h = [aE_s^*]$ for some $0 < a < d$. Then one has $s_h|_{E_1} = a/d$. Similarly $[Z_K] = [(d-q-1)E_s^*]$, hence $Z_K|_{E_1} = (d-q-1)/d$. Moreover, since $E_1^*|_{E_1} = q/d$ and $E_s^*|_{E_1} = 1/d$ the above counting problem with conditions (i) and (ii) can be transformed into the following lattice points counting: we count pairs of integers $x', y' \geq 1$ satisfying

$$\text{(i)} \ x'q + y' < d + a \quad \text{and} \quad \text{(ii)} \ x'q + y' \equiv a \pmod d.$$

This means that $x'q + y' = a$ with $x' \geq 1$ and $y' \geq 1$. The number of such pairs (x', y') is exactly $\lfloor a/q \rfloor$ if $q \nmid a$, and it is $\lfloor a/q \rfloor - 1$ if $q|a$. Finally note that from (23) $\lfloor a/q \rfloor = a_1$. Hence these facts combined with (28) ends the proof.

The connection between the counting function $Q_{[Z_K]+h,v}$ and the invariant κ_π from the first proof (24) remains to be better understood.

3.2.4. The **fourth proof**, again **new**, appears in [**10**]. Originally we planned to insert it in [**8**], or as an Appendix of the present manuscript, but during the reduction it evolved into an independent manuscript. It considers the canonical (minimal) embedding of the cyclic quotient singularity into $\mathbb{C}^e$ via its invariant monomials, and then one studies the tangent application of the parametrized curves $\widetilde{C} \to \mathbb{C}^e$. One of the advantages of this proof is that it can be naturally generalized for more general classes of singularities $(X, 0)$, and provides a direct method to compute the delta invariant of $(C, 0)$ for curves not satisfying the hypothesis of Theorem 3.1.

REMARK 3.2. The proofs in §3.2.1 and §3.2.3 end up with a certain lattice point counting. In fact, in the proof in §3.2.2 this aspect appears subtly in the arithmetical computation as well: behind these arithmetical formulae one can recognize either lattice point counting, or the corresponding Dedekind sum interpretations. The third proof makes connection with surgery formulae as well via explicit calculations of Ehrhart type quasi-polynomials. The proof §3.2.4 is completely different, it is more algebraic.

4. Delta invariant of minimal generic curves on rational singularities

4.1. The surgery formula. Let Γ be the minimal good dual resolution graph of a rational singularity $(X,0)$. Consider a minimal generic curve $(C,0)$ on $(X,0)$ with its associated cycle s_h for some $h\in H$, $h\neq 0$.

Let $\mathcal{N}$ denote the set of nodes of Γ, i.e. vertices with valency $\kappa_v\geq 3$. Assume that $\mathcal{N}\neq\emptyset$ (the $\mathcal{N}=\emptyset$ case has been already discussed in section 3). Then the other vertices $V\setminus\mathcal{N}$ determine the collection of connected strings $\{\Gamma_k\}_k$, which are minimal resolution graphs of cyclic quotient singularities. We also consider for any Γ_k the inclusion operator j_k and its dual j_k^* considered in section 2.3.2.

Fix $h\in H$, $h\neq 0$, and write

$$s_h=\sum_{v\in V}a_vE_v^*=\sum_{v\in\mathcal{N}}a_vE_v^*+\sum_k\sum_{v\in V(\Gamma_k)}a_{v,k}E_{v,k}^*,\qquad a_v\in\mathbb{Z}_{\geq 0}.$$

Then for every k we have the cycles $s_{h,k}:=j_k^*(s_h)=\sum_{v\in V(\Gamma_k)}a_{v,k}E_{v,k}^*$ on the minimal resolution of the corresponding cyclic quotient singularities $(X_k,0)$. Each $s_{h,k}$ is also an irreducible minimal cycle $s_{h,k}=s_{[j_k^*(s_h)]}$ on Γ_k by Lemma 2.2. It corresponds to the curve $(C_k,0)$, the union of those components of $(C,0)$, which intersect $E(\Gamma_k)=\cup_{v\in V(\Gamma_k)}E_v$. The curve $(C_k,0)$ is nonempty if and only if $r_k(C):=\sum_{v\in V(\Gamma_k)}a_{v,k}>0$. In such a case $r_k(C)$ is the the number of branches of $(C_k,0)$. Hence, whenever $s_{h,k}\neq 0$ (i.e. C_k is nonempty), by Theorem 3.1 one obtains that $\delta(C_k)=\sum_{v\in V(\Gamma_k)}a_{v,k}-1$.

Set also $r(C):=\sum_v a_v=\sum_k r_k(C)+\sum_{v\in\mathcal{N}}a_v$ for the number of components of C.

We apply the surgery formula from Theorem 2.3 in the rational singularity context for the set of nodes $\mathcal{N}$ and the cycle $Z_K+s_h\in Z_K+\mathcal{S}'$ in order to deduce the following identity

$$Q_{[Z_K+s_h]}(Z_K+s_h)=Q_{[Z_K+s_h],\mathcal{N}}(Z_K+s_h)+Q^{\Gamma_k}_{[j_k^*(Z_K+s_h)]}(j_k^*(Z_K+s_h)).\tag{29}$$

Then the above discussion implies the following surgery type formula for the delta invariant.

PROPOSITION 4.1. $\delta(C)=Q_{[Z_K+s_h],\mathcal{N}}(Z_K+s_h)+\sum_{k:r_k(C)>0}(r_k(C)-1)$.

4.2. The topology of star-shaped graphs.

4.2.1. In this section we focus on the topology of singularities $(X,0)$ whose minimal good resolution graph Γ is star-shaped. Let v_0 be the central vertex, i.e. the only node of Γ.

In this section will not assume that the graph is rational.

Note that the link of $(X,0)$ in this case is a Seifert rational homology sphere whose Seifert structure is determined by its normalized Seifert invariants. Our aim is to express arithmetically the value $Q_{[Z_K+s_h],v_0}(Z_K+s_h)$ of the one-variable counting function $Q_{[Z_K+s_h],v_0}(t)$ in terms of the Seifert invariants.

4.2.2. *Seifert invariants, lifts and equivariant decomposition of* $Z(\mathbf{t})$ *(cf.* [**19**, Section 6]*).* Assume that the star-shaped graph Γ has ν legs. Each leg is a string with normalized Seifert invariants (d_i,q_i) where $0<q_i<d_i$ and $\gcd(d_i,q_i)=1$, $i=1,\dots,\nu$. That is, if the ith leg has s_i vertices, say $v_{i1},\dots,v_{is_i}$ with self-intersection

numbers $-k_{i1},\dots,-k_{is_i}$ such that v_{i1} is connected by the central vertex, then they are expressed by the Hirzebruch–Jung continued fraction $d_i/q_i = [k_{i1},\dots,k_{is_i}]$ ($k_{ij}\geq 2$), see section 3. The corresponding base elements in L will be denoted by $\{E_{ij}\}_{j=1}^{s_i}$. The central vertex v_0 with decoration $-k$ defines the higher-valency element E_0 in L.

We use the notation from section 3, for example we will write $d^i_{j_1j_2}$ for the determinant of the subchain of the ith leg connecting the vertices v_{ij_1} and v_{ij_2}.

The plumbed 3-manifold associated with such a star-shaped graph Γ, or, equivalently, the link of $(X,0)$, has a Seifert structure and the collection of normalized Seifert invariants is denoted by $Sf=(-k;(d_i,q_i)_{i=1}^{\nu})$. The orbifold Euler number is defined as

$$e=-k+\sum_i q_i/d_i.$$

The negative definiteness of the intersection form is equivalent to $e<0$.

Fix some $h\in H$. We say that $\ell\in L'$ is a *lift* of h if $[\ell'_h]=h$; ℓ'_h is called a *reduced lift* of h if $[\ell'_h]=h$ and ℓ'_h has the form $\sum_{i=0}^{\nu}c_iE^*_i$, where we abbreviate $E_i:=E_{is_i}$ for simplicity. Note that in general any coefficient of a lift might depend on the choice of the lift.

Next we list some useful identities regarding the intersection of the dual base elements: $(E^*_0,E^*_0)=1/e$, $(E^*_0,E^*_i)=1/d_ie$, $(E^*_{ij},E^*_0)=(d^i_{j+1,s_i}E^*_{is_i},E^*_0)$, and $[E^*_{ij}]=[d^i_{j+1,s_i}E^*_{is_i}]$ for any $1\leq i\leq\nu, 1\leq j\leq s_i$ (see eg. [**24**, 11.1] or [**19**, Section 6.1]).

For any cycle $\ell'\in L'$ the E_0-coefficient of ℓ' will be denoted by $\ell'_0:=(\ell',-E^*_0)$. Furthermore, if $\ell'=c_0E^*_0+\sum_{i,j}c_{ij}E^*_{ij}\in L'$, then we define its *reduced transform* by

$$\ell'_{red}:=c_0E^*_0+\sum_{i,j}c_{ij}\cdot d^i_{j+1,s_i}E^*_i. \tag{30}$$

Write $\ell'_{red}=\sum_{i=0}^{\nu}c_iE^*_i$. The above identities show that $[\ell']=[\ell'_{red}]$ in H, $\ell'_0=\ell'_{red,0}$, and this coefficient can be expressed as

$$\ell'_{red,0}=\frac{1}{|e|}(c_0+\sum_{i=1}^{\nu}\frac{c_i}{d_i}). \tag{31}$$

Thus, for any fixed $h\in H$, if $\ell'_h\in L'$ is a lift of h then its reduced transform is a reduced lift of h with the same E_0-coefficient.

4.2.3. *The coefficients of s_h.* In this subsection we characterize those elements $\ell'=a_0E^*_0+\sum_{i,j}a_{ij}E^*_{ij}\in L'$, which are of type s_h for some $h\in H$. This means that if we take $\{a_0,\{a_{ij}\}_{ij}\}$ generic transversal discs along each irreducible exceptional divisor $\{E_0,\{E_{ij}\}_{ij}\}$, then their collection constitutes the strict transform of a minimal generic curve. We follow [**24**, Prop. 11.5].

Let us consider the reduced transform $\ell'_{red}=\sum_{i=0}^{\nu}a_iE^*_i$ of ℓ' as well.

Clearly ℓ' determines ℓ'_{red} canonically by (30). Conversely, we claim that whenever ℓ' equals some s_h then from its reduced transform ℓ'_{red} one can recover ℓ'. First recall that in the cyclic quotient case, for any $0\leq a<d$, the number a (that is, the class $a[E^*_s]$) determines $s_{a[E^*_s]}=\sum_{i=1}^{s}a_iE^*_i$ via the algorithms described in section 3, see eg. (23).

In a very same way, each a_i ($i=1,\dots,\nu$) determines $\{a_{ij}\}_{j=1}^{s_i}$ by the same algorithms, eg. by (23).

Next we wish to characterize the coefficients of ℓ'_{red}. By [**24**], $\ell'_{red} = \sum_{i=0}^{\nu} a_i E_i^*$ is the reduced transform of some s_h if and only if

$$\text{(32)} \qquad \begin{cases} a_0 \geq 0, \ d_i > a_i \geq 0 & (1 \leq i \leq \nu) \\ 1 + a_0 - kt + \sum_{i=1}^{\nu} \left\lfloor \frac{q_i t + a_i}{d_i} \right\rfloor \leq 0 & \text{for any } t > 0. \end{cases}$$

Note that the last restriction for $t = 1$ gives $a_0 \leq k - 1$ as well.

4.2.4. *The topological Poincaré series reduced to the variable v_0.* In the sequel, associated with any $h \in H$ we fix a reduced lift $\ell' = \sum_{i=0}^{\nu} c_i E_i^*$ and we define the following quasi-linear function

$$\text{(33)} \qquad \mathfrak{N}_{\mathbf{c}} : \mathbb{Z} \to \mathbb{R}, \quad \mathfrak{N}_{\mathbf{c}}(n) := 1 + c_0 + kn - \sum_{i=1}^{\nu} \left\lceil \frac{q_i n - c_i}{d_i} \right\rceil, \quad n \in \mathbb{Z},$$

where the index vector $\mathbf{c} := (c_0, \ldots, c_\nu)$ reflects for the dependence of $\mathfrak{N}_{\mathbf{c}}$ on the chosen reduced lift. Then, by [**19**, (6.2.3)] (see also [**20**] for the general case) the h-equivariant part of the topological Poincaré series $Z(\mathbf{t})$, reduced to the node v_0, can be expressed as follows

$$\text{(34)} \qquad Z_h(t) = \sum_{n \geq -\ell'_0} \max\{0, \mathfrak{N}_{\mathbf{c}}(n)\}\ t^{n+\ell'_0}.$$

REMARK 4.2. Consider the trivial class $h = 0$ with the obvious reduced lift $\ell' = 0$. Then the function from (33) $\mathfrak{N}_{\mathbf{0}}(n) := 1 + kn - \sum_{i=1}^{m} \lceil q_i n / d_i \rceil$ is the *Dolgachev–Pinkham–Demazure* function (see eg. [**32**]) targeting the dimension of the n-graded piece of the local algebra of the weighted homogeneous singularity whose resolution graph is Γ.

4.2.5. *Interpretation of $Q_{[Z_K+s_h],v_0}(Z_K + s_h)$.* Now we can apply the previous section to express the counting function $Q_{[Z_K+s_h],v_0}(Z_K + s_h)$. Fix $h \in H$ and write $s_h = a_0 E_0^* + \sum_{i,j} a_{ij} E_{ij}^*$ as above. We consider its reduced transform $s_{h,red} = a_0 E_0^* + \sum_i a_i E_i^*$ where $a_i := \sum_j a_{ij} d^i_{j+1,s_i}$.

The key here is to choose the special reduced lift $Z_K - E + s_{h,red}$ associated with the class $[Z_K + s_h]$, which by using (3) reads as

$$Z_K - E + s_{h,red} = (a_0 + \nu - 2)E_0^* + \sum_i (a_i - 1)E_i^*.$$

Its E_0-coefficient can be computed by the above intersection identities, it is $\gamma + s_{h,0} = (1/|e|)(a_0 + \nu - 2 + \sum_i (a_i - 1)/d_i)$, with the notation $\gamma := Z_{K,0} - 1$.

REMARK 4.3. Note that γ is a key numerical invariant which can be expressed as $\gamma = (\nu - 2 - \sum_i 1/d_i)/|e| \in \mathbb{Q}$. It has several names in the literature: the 'exponent' of the weighted homogeneous germ $(X, 0)$, or $-\gamma$ is called the *log discrepancy* of E_0. $\mathfrak{o}\gamma$ (where $\mathfrak{o}$ is the order of the class of E_0^* in H) also appears as the Goto–Watanabe a–invariant (cf. [**12**, (3.1.4)]) of the universal abelian covering of $(X, 0)$. In [**30**] $e\gamma$ appears as the orbifold Euler characteristic (see also [**29**, 3.3.6]).

By (34) the corresponding $[Z_K + s_h]$-part of the topological Poincaré series is

$$\text{(35)} \qquad \begin{aligned} Z_{[Z_K+s_h]}(t) &= \sum_{n \geq -\gamma - s_{h,0}} \max\{0, \mathfrak{N}_{\tilde{\mathbf{a}}}(n)\}\ t^{n+\gamma+s_{h,0}}, \\ \mathfrak{N}_{\tilde{\mathbf{a}}}(n) &= a_0 + \nu - 1 + kn - \sum_{i=1}^{\nu} \left\lceil \frac{q_i n - a_i + 1}{d_i} \right\rceil \end{aligned}$$

associated with the index vector $\widetilde{\mathbf{a}} = (a_0 + \nu - 2, a_1 - 1, \dots, a_\nu - 1)$.

Note that the E_0–coefficient of $Z_k + s_h$ is one larger than $\gamma + s_{h,0}$. Therefore, using the definition of the counting function one gets

$$Q_{[Z_K+s_h],v_0}(Z_K + s_h) = \sum_{-\gamma - s_{h,0} \leq n \leq 0} \max\{0, \mathfrak{N}_{\widetilde{\mathbf{a}}}(n)\}. \tag{36}$$

4.2.6. Let us mention immediately that the summation interval is not empty, that is, $-\gamma - s_{h,0} \leq 0$ whenever $h \neq 0$. Indeed, $s_{h,0} = (a_0 + \sum_i a_i/d_i)/|e|$, cf. (31). On the other hand, by [**24**], $Z_{K,0} = 1 + \chi/e$, where χ is the orbifold Euler characteristic $2 - \nu + \sum_i 1/d_i$. Therefore we have to check the inequality

$$a_0 + \sum_i a_i/d_i \geq \chi = 2 - \nu + \sum_i 1/d_i,$$

which follows by an elementary computation, since each $d_i \geq 2$ and at least one of the member of $\{a_0, a_1, \dots, a_\nu\}$ is non–zero.

4.2.7. Next, we relate the above formula (36) with another expression associated with the class $h \in H$. By the 'combinatorial duality' Theorem 4.4.1 of [**18**] we have that $Q_{[Z_K]+h,v_0}(Z_K - r_{-h}) = \mathrm{pc}(Z_{-h}(t))$, where $\mathrm{pc}(Z_{-h}(t))$ denotes the *periodic constant* of the series Z_{-h}. Note that $Q_{[Z_K]+h,v_0}(Z_K - r_{-h})$ counts the coefficients in $Z_{[Z_K+s_h]}(t)$ with the condition that the exponent $n+\gamma+s_{h,0}$ satisfies $0 \leq n+\gamma+s_{h,0} < Z_{K,0} - r_{-h,0} = \gamma+1-r_{-h,0}$, or, $-\gamma - s_{h,0} \leq n < 1 - r_{-h,0} - s_{h,0}$.

Write $s_{h,0}$ as $r_{h,0} + \Delta_h$ for some $\Delta_h \in \mathbb{Z}_{\geq 0}$. Note also that $r_{-h,0} = 1 - r_{h,0}$ if $r_{h,0} \neq 0$, otherwise it is zero. Hence $1 - r_{-h,0} - s_{h,0} = 1 - r_{-h,0} - r_{h,0} - \Delta_h$ equals $-\Delta_h$ if $r_{h,0} \neq 0$, and equals $-\Delta_h + 1$ otherwise. It can also be written as $1 + \lfloor -s_{h,0} \rfloor$. Hence

$$\mathrm{pc}(Z_{-h}(t)) = \sum_{-\gamma - s_{h,0} \leq n \leq \lfloor -s_{h,0} \rfloor} \max\{0, \mathfrak{N}_{\widetilde{\mathbf{a}}}(n)\}. \tag{37}$$

Note that the summation in the right-hand side is non–empty only if $-\gamma - s_{h,0} \leq \lfloor -s_{h,0} \rfloor$. This in the case $r_{h,0} = 0$ means $\gamma \geq 0$, and in the case $r_{h,0} \neq 0$ means $\gamma + r_h \geq 1$. (Otherwise $\mathrm{pc}(Z_{-h}(t)) = 0$ automatically.)

Note that in general we have the following facts. If $(X, 0)$ is ADE then $Z_K = 0$, hence $\gamma = -1$. In all other cases $Z_K > 0$, hence $\gamma > -1$. If $(X, 0)$ is quotient singularity then all the coefficients of Z_k are less than 1, hence $\gamma < 0$.

Summarized, if $-\gamma - s_{h,0} \leq \lfloor -s_{h,0} \rfloor$ then

$$Q_{[Z_K+s_h],v_0}(Z_K + s_h) = \mathrm{pc}(Z_{-h}(t)) + \sum_{1+\lfloor -s_{h,0} \rfloor \leq n \leq 0} \max\{0, \mathfrak{N}_{\widetilde{\mathbf{a}}}(n)\}. \tag{38}$$

The sum on the right-hand side can also be empty in some cases.

Finally let us comment the expression $\mathrm{pc}(Z_{-h}(t))$. For any star-shaped graph $\mathrm{pc}(Z_{-h}(t))$ can be identified with the (normalized) Seiberg–Witten invariant associated with the link (and spin^c structure corresponding to $-h$). For this identity and expression see eg. [**27**] or [**19**]. If the CDGZ identity holds, then it equals the equivariant geometric genus $p_g(X, 0)_{-h}$ (the $(-h)$–equivariant part of the geometric genus of the universal abelian covering) [**27**]. This is true even if the CDGZ identity does not hold, but the equivariant Seiberg–Witten Conjecture holds (see again [**27**]). If $(X, 0)$ is rational then it can be replaced by a much simpler combinatorial expression, see below.

We emphasize again that the identities of this subsection 4.2 hold for any star-shaped graph whose plumbing (Seifert) 3–manifold is a rational homology sphere, and the rationality of Γ is not required.

REMARK 4.4. Using the criterion (32) one verifies that $\mathfrak{N}_{\widetilde{\mathbf{a}}}(n) \leq \nu - 2$ for any $n < 0$.

4.3. The case of rational star-shaped graphs. Next, we wish to add some comments regarding the formulae of subsection 4.2 in the case when Γ is rational. By 2.2.3, if Γ is rational then necessarily $-e_{v_0} \geq \nu - 1$, where e_{v_0} denotes the self-intersection E_0^2 of E_0.

In the rational case, the CDGZ identity, or the equivariant Seiberg–Witten Conjecture hold, hence $\mathrm{pc}(Z_{-h}(t)) = p_g(X,0)_{-h}$. Furthermore, it can be reinterpreted combinatorially (see [**25**, Cor. 4.5.3] or [**19**, 6.4]) as

$$\mathrm{pc}(Z_{-h}(t)) = \chi(r_{-h}) - \chi(s_{-h}).$$

The identity $\mathrm{pc}(Z_{-h}(t)) = p_g(X,0)_{-h}$ has the following interpretation too. Note that $\sum_h p_g(X,0)_{-h}$ is exactly the geometric genus of the universal abelian covering. $p_g(X,0)_0 = p_g(X,0) = 0$ by the rationality of $(X,0)$, but the vanishing of the other terms tests whether the universal abelian covering is rational or not. (Eg., the rational graph with Seifert invariants $(-2;(3,1)_{i=1}^3)$ has non–rational universal abelian covering, cf. [**25**, Ex. 4.5.4].)

If $(X,0)$ is a quotient singularity then the universal abelian covering is an ADE germ, hence in this case $\mathrm{pc}(Z_{-h}(t)) = \chi(r_{-h}) - \chi(s_{-h}) = p_g(X,0)_{-h} = 0$ for any $h \in H$.

4.4. The delta invariant formula revisited. Assume that Γ is star-shaped and rational, and we adopt the notations and results of the previous subsections.

Let us compute the contribution $\mathfrak{N}_{\widetilde{\mathbf{a}}}(0)$. Note that $a_i = 0$ if and only if $\sum_j a_{ij} = 0$ if and only if $r_i(C) = 0$. Therefore,

$$\mathfrak{N}_{\widetilde{\mathbf{a}}}(0) = a_0 + \nu - 1 - \sum_{i: r_i(C) \neq 0} \left\lceil \frac{1 - a_i}{d_i} \right\rceil - \sum_{i: r_i(C) = 0} \left\lceil \frac{1}{d_i} \right\rceil.$$

Since $a_i < d_i$ (cf. (32)), we get $\mathfrak{N}_{\widetilde{\mathbf{a}}}(0) = a_0 + \nu - 1 - \#\{i : r_i(C) = 0\}$. This identity combined with Proposition 4.1 and (36) provides the following.

THEOREM 4.5. *If $(C,0)$ is a minimal generic curve germ on a rational singularity $(X,0)$ with a star-shaped graph, $r(C)$ is the number of components of $(C,0)$, and $\sum_{i=0}^{\nu} a_i E_i^*$ is the reduced transform of $s_h \neq 0$, then*

$$\delta(C) = r(C) - 1 + \sum_{-\gamma - s_{h,0} \leq n \leq -1} \max\{0, \mathfrak{N}_{\widetilde{\mathbf{a}}}(n)\}. \tag{39}$$

In particular, $(C,0)$ is an ordinary $r(C)$–tuple if and only if the above sum vanishes.

Furthermore, if $-\gamma - s_{h,0} \leq \lfloor -s_{h,0} \rfloor$ then

$$\delta(C) = r(C) - 1 + \chi(r_{-h}) - \chi(s_{-h}) + \sum_{-\lfloor s_{h,0} \rfloor \leq n \leq -1} \max\{0, \mathfrak{N}_{\widetilde{\mathbf{a}}}(n)\}. \tag{40}$$

Recall that $\chi(r_{-h}) \geq \chi(s_{-h})$. Hence the above expressions show exactly how far the minimal generic curve numerically from an ordinary $r(C)$-tuple is.

4.4.1. *Further topological connections.* Consider the series $Z_{[Z_K+s_h]}(t)$ from (35). The following is called the *polynomial part of* $Z_{[Z_K+s_h]}(t)$ in [**19**, (6.2.4)]:

$$Z^+_{[Z_K+s_h]}(t) = \sum_{n\geq -\gamma - s_{h,0}} \max\{0, -\mathfrak{N}_{\widetilde{\mathbf{a}}}(n)\}\ t^{n+\gamma+s_{h,0}}. \tag{41}$$

According to [**19, 25**], since $(X,0)$ is rational, $Z^+_{[Z_K+s_h]}(1) = p_g(X,0)_{[Z_K]+h} = \chi(r_{[Z_K]+h})) - \chi(s_{[Z_K]+h})$. Therefore, if the universal abelian covering of $(X,0)$ is rational then $Z^+_{[Z_K+s_h]}(1) = p_g(X,0)_{[Z_K]+h} = 0$, hence in this case $\mathfrak{N}_{\widetilde{\mathbf{a}}}(n) \geq 0$ for all $n \geq -\gamma - s_{h,0}$.

This fact, for $(X,0)$ quotient singularity, and for the relevant interval $-\gamma - s_{h,0} \leq n \leq 0$, will be verified combinatorially in 5.2.2.

5. Minimal generic curves on quotient singularities

5.1. The general formula. Let $(X,0)$ be a non-cyclic quotient singularity.

It is well–known that $(X,0)$ is a quotient singularity if and only if Γ is *numerically log terminal*, that is, all the coefficients of Z_K are < 1.

If Γ is minimal and star-shaped (but not a string) then $(X,0)$ is quotient if and only if $\nu = 3$ and $\sum_i 1/d_i > 1$. (Recall that from Laufer algorithm we also have $k \geq 2$.)

In particular, $\gamma \in [-1,0)$ and the inequality $-\gamma - s_{h,0} \leq \lfloor -s_{h,0} \rfloor$ never happens.

Note that $\nu = 3$ and Remark 4.4 show that $\max\{0, \mathfrak{N}_{\widetilde{\mathbf{a}}}(n)\} \in \{0,1\}$ for any $n < 0$. In fact, by 4.4.1, $\max\{0, \mathfrak{N}_{\widetilde{\mathbf{a}}}(n)\} = \mathfrak{N}_{\widetilde{\mathbf{a}}}(n)$ for $n < 0$, a fact which will be reproved in 5.2.2.

Then from the previous section we have the delta invariant formula (39), subject to the new conditions imposed by the fact that $(X,0)$ is a non–cyclic quotient singularity.

5.2. Let us concentrate first one the summation interval in the formula of (39) $(h \neq 0)$.

Note that if $-\gamma - s_{h,0} \leq -1$, then $s_{h,0} \geq 1 - \gamma$. Since $\gamma < 0$ (because $(X,0)$ is a quotient) we get that necessarily $s_{h,0} > 1$. This proves (via (39)) the following.

5.2.1.

CLAIM 1. If $s_{h,0} \leq 1$ then $\delta(C) = r(C) - 1$, hence $(C,0)$ is an ordinary $r(C)$–tuple.

Next, we wish to eliminate 'max' from the expression $\max\{0, \mathfrak{N}_{\widetilde{\mathbf{a}}}(n)\}$ for some n.

5.2.2.

CLAIM 2. If $-\gamma - s_{h,0} \leq -t \leq 0$, then $\mathfrak{N}_{\widetilde{\mathbf{a}}}(-t) \geq 0$.

Indeed, using the same identities as in 4.2.6, we see that $-\gamma - s_{h,0} \leq -t$ reads as

$$a_0 + \sum_i a_i/d_i \geq t|e| + \chi = t(k - \sum_i q_i/d_i) - 1 + \sum_i 1/d_i,$$

or

$$\mathfrak{P}_{\widetilde{\mathbf{a}}}(t) \geq 0, \text{ where } \mathfrak{P}_{\widetilde{\mathbf{a}}}(t) := 1 + a_0 - kt + \sum_i \frac{q_i t + a_i - 1}{d_i}. \tag{42}$$

This can/should be compared with the following two expressions. The first one is

$$\mathfrak{N}_{\widetilde{\mathbf{a}}}(-t) = 2 + a_0 - kt + \sum_{i=1}^{\nu} \Big\lfloor \frac{q_i t + a_i - 1}{d_i} \Big\rfloor. \tag{43}$$

The second one is the expression from (32)

$$\mathfrak{R}_{\widetilde{\mathbf{a}}}(t) = 1 + a_0 - kt + \sum_{i=1}^{\nu} \Big\lfloor \frac{q_i t + a_i}{d_i} \Big\rfloor. \tag{44}$$

By the minimality assumption and (32), if $\widetilde{\mathbf{a}}$ represents some s_h, then $\mathfrak{R}_{\widetilde{\mathbf{a}}}(t) \leq 0$ for any $t > 0$. Also, as we already mentioned (and it can be seen easily), $\mathfrak{N}_{\widetilde{\mathbf{a}}}(-t) \leq \mathfrak{R}_{\widetilde{\mathbf{a}}}(t) + 1$.

We wish to prove that (42) implies $\mathfrak{N}_{\widetilde{\mathbf{a}}}(-t) \geq 0$. Write $(q_i t + a_i)/d_i$ as $\lfloor (q_i t + a_i)/d_i \rfloor + r_i/d_i$ for some $0 \leq r_i < d_i$. Then

$$\begin{aligned} \mathfrak{N}_{\widetilde{\mathbf{a}}}(-t) - \mathfrak{P}_{\widetilde{\mathbf{a}}}(t) &= 1 + \sum_{i:r_i=0} \Big(-1 + \frac{1}{d_i} \Big) + \sum_{i:r_i \neq 0} \Big(-\frac{r_i - 1}{d_i} \Big) \\ &= 1 + \sum_{i} \frac{1}{d_i} + \sum_{i:r_i=0} (-1) + \sum_{i:r_i \neq 0} \frac{-r_i}{d_i}. \end{aligned}$$

Since $\sum_i 1/d_i > 1$ (and we have three legs), the right-hand side is > -1, hence $\mathfrak{N}_{\widetilde{\mathbf{a}}}(-t) - \mathfrak{P}_{\widetilde{\mathbf{a}}}(t) > -1$, or $\mathfrak{N}_{\widetilde{\mathbf{a}}}(-t) > \mathfrak{P}_{\widetilde{\mathbf{a}}}(t) - 1 \geq -1$. Since it is an integer, we obtain the claim.

Note also that $\mathfrak{N}_{\widetilde{\mathbf{a}}}(-1) - \mathfrak{R}_{\widetilde{\mathbf{a}}}(1) = 1 - \#\{i : r_i = 0\}$. Hence we get
5.2.3.

CLAIM 3. $\mathfrak{N}_{\widetilde{\mathbf{a}}}(-1) = 1$ then necessarily $\mathfrak{R}_{\widetilde{\mathbf{a}}}(1) = 0$ and $r_i > 0$ for all i.

Maybe it is worth to notice that for $\mathfrak{N}_{\widetilde{\mathbf{a}}}(-1)$ we have the upper bound 1, and we also proved that $\mathfrak{N}_{\widetilde{\mathbf{a}}}(-1) \geq 0$ whenever $-\gamma - s_{h,0} \leq -1$. But, in general, there exists no lower bound, for it, it can be a very negative number (if $k \ll 0$).

5.3. We can now restate Theorem 4.5 for the particular case of quotient surface singularities providing a very explicit formula for the delta invariant of minimal generic curves. The two exceptional cases where the general formula does not apply are described by their Seifert invariants (see 4.2.2) and the description of the minimal generic cycle s_h follows the notation in Figures 1 and 2.

THEOREM 5.1. *If $(C,0)$ is a minimal generic curve germ on a quotient singularity $(X,0)$, then*

$$\delta(C) = r(C) - 1 + \varepsilon(C), \qquad \varepsilon(C) \in \{0,1\}.$$

The explicit value of $\varepsilon(C)$ is given as:

$$\varepsilon(C) = \begin{cases} 0 & \text{if } (X,0) \text{ is cyclic, } \mathbb{E}_6,\ \mathbb{E}_7, \text{ or } s_{h,0} \leq 1, \\ 1 & \text{if } (X,0) \text{ is } (-2;(2,1),(3,2),(5,2)) \text{ and } s_h = E^*_{3,1} + E^*_{1,1}, \\ 1 & \text{if } (X,0) \text{ is } (-2;(2,1),(3,2),(5,3)) \text{ and } s_h = E^*_{1,1}, \\ \mathfrak{N}_{\widetilde{\mathbf{a}}}(-1) & \text{otherwise,} \end{cases}$$

where

$$\mathfrak{N}_{\widetilde{\mathbf{a}}}(-1) = 2 + a_0 - k + \sum_{i=1}^{\nu} \Big\lfloor \frac{q_i + a_i - 1}{d_i} \Big\rfloor. \tag{45}$$

Moreover, $(C,0)$ is of type R_r^r if $\varepsilon(C)=0$, and of type R_r^{r-1} if $\varepsilon(C)=1$ (cf. Examples 2.4.2 and 2.4.3).

REMARK 5.2.

(1) Note that the exceptional cases in Theorem 5.1 are special in the sense that $\mathfrak{N}_{\tilde{\mathbf{a}}}(-1)=0$ and $\mathfrak{N}_{\tilde{\mathbf{a}}}(-2)=1$ (see Examples 6.3(5) and 6.4(5)). These are the only exceptions in the quotient singularity case where there is a contribution from $\mathfrak{N}_{\tilde{\mathbf{a}}}(n)=0$ for $n<-1$ (compare with equation (39)).

(2) In (45) the term $\mathfrak{N}_{\tilde{\mathbf{a}}}(-1)$ is in general non-trivial, as we will see several examples in the sequel. This corresponds in particular to the fact that not any minimal general curve is an ordinary $r(C)$–tuple.

(3) On the other hand, the above Theorem has the following consequence: If $(C,0)$ is a minimal generic curve on $(X,0)$ then any sub-collection of $(C,0)$ with $r(C)-1$ components is an ordinary $(r(C)-1)$–tuple (since any proper subsets of any R_r^r or R_r^{r-1} is an ordinary r'–tuple). This means that a nontrivial contribution of type $\mathfrak{N}_{\tilde{\mathbf{a}}}(-1)$ can appear only for the 'maximal elements' of type s_h (and not for all of them). (Compare also with 5.2.3.) Maximality here means the following: s_h is maximal if $s_h+E_v^*$ does not equal $s_{h+[E_v^*]}$ for any v.

This fact is hard to see directly. However, we can exemplify it in the following two cases.

First assume that $\tilde{\mathbf{a}}$ represents s_h and $s_h+E_0^*=s_{h+[E_0^*]}$ too, that is, $\tilde{\mathbf{a}}+E_0^*$ is also a 'minimal system'. Then $\mathfrak{N}_{\tilde{\mathbf{a}}}(-t)\leq \mathfrak{R}_{\tilde{\mathbf{a}}}(t)+1=\mathfrak{R}_{\tilde{\mathbf{a}}+E_0^*}(t)\leq 0$ for any $t>0$.

Similarly, if $d_1=2$, and E_1 is the (-2)–exceptional curve of this leg, we have a similar statement (with a slightly different proof). If $\tilde{\mathbf{a}}$ represents s_h and $s_h+E_1^*=s_{h+[E_1^*]}$, then $\mathfrak{N}_{\tilde{\mathbf{a}}}(-t)=\mathfrak{R}_{\tilde{\mathbf{a}}+E_1^*}(t)\leq 0$ for any $t>0$.

EXAMPLE 5.3. **A non-quotient rational singularity.** We end this section with an example, which shows that if $(X,0)$ is a non-quotient singularity then $\delta(C)-r(C)+1$ can be strict larger than one (hence $(C,0)$ is not of type R_r^r or R_r^{r-1}).

Assume that the graph of $(X,0)$ is the following.

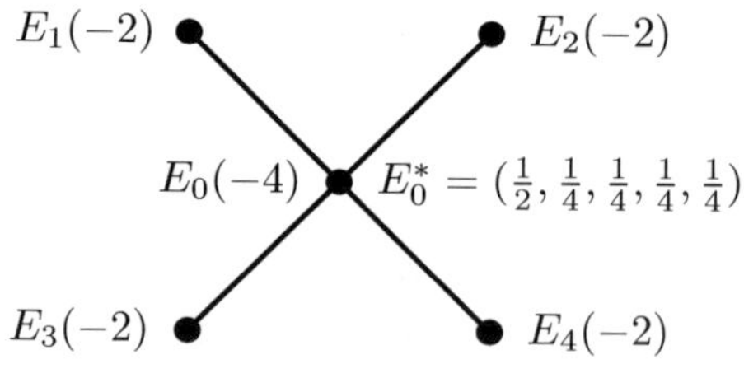

One verifies that $Z_K=2E_0^*$ and $Z_{\min}=E$, hence $(X,0)$ is minimal rational.

Choose $\ell'=3E_0^*\in\mathcal{S}'$ and $h=[\ell']\in H$. The cycle r_h is $\ell'-E_0$ and it does not belong to $\mathcal{S}'$, so $s_h=\ell'$. Let C be the reduced curve associated with s_h. Clearly $r(C)=3$.

Note that $\gamma=Z_{K,0}-1=0$, hence in (39) $-\gamma-s_{h,0}=-3/2$ and $\delta(C)=r(C)-1+\max\{0,\mathfrak{N}_{\tilde{\mathbf{a}}}(-1)\}$. But a direct verification shows that $\mathfrak{N}_{\tilde{\mathbf{a}}}(-1)=2$.

This example shows that $\nu = 3$ is a necessary condition in Theorem 5.1. In fact, increasing $\nu = k$, one can construct rational star-shaped graphs as above with (-2)-legs and $\delta(C) - r(C) = k - 3$.

6. Proof of Theorem 5.1

Recall that Theorem 5.1 in the cyclic case has already been discussed in section 3. For the non-cyclic quotient singularities the proof is based on (39), after showing that, except for the exceptional cases, only $n = -1$ might give a non-zero contribution from $\mathfrak{N}_{\tilde{\mathbf{a}}}(n)$. This will be the purpose of the coming section.

6.1. Preliminary results. First we will prove two combinatorial lemmas covering different families (with some overlaps) of non-cyclic quotient singularities.

LEMMA 6.1. *If $(X, 0)$ is a non-cyclic quotient singularity, then for any $h \in H$ the following facts hold:*

$$\begin{cases} \text{if } k \geq 3 \text{ then } s_{h,0} < 2 \\ \text{if } k = 2 \text{ and } q_1 = q_2 = q_3 = 1 \text{ then } s_{h,0} \leq 2. \end{cases}$$

PROOF. Assume that $s_h = a_0 E_0^* + \sum_{ij} a_{ij} E_{ij}^*$. Then by the discussion from 4.2.2

$$s_{h,0} = (s_{h,red})_0 = \big(a_0 + \sum_i a_i/d_i\big)/|e|. \tag{46}$$

In the first case we need to show that this expression is < 2. Formula (33) for $n = -1$ becomes

$$1 + a_0 - k + \sum_i \lfloor (q_i + a_i)/d_i \rfloor \leq 0.$$

Note that by (32) $a_i < d_i$. Let I be the set of indeces when $q_i + a_i \geq d_i$. Hence the above ineguality reads as $1 + a_0 - k + \#I \leq 0$. Using this we have

$$\begin{aligned} a_0 + \sum_i \frac{a_i}{d_i} &\leq k - 1 - \#I + \sum_{i \notin I} \frac{d_i - 1 - q_i}{d_i} + \Big(\sum_{i \in I} \frac{d_i - 1 - q_i}{d_i} + \sum_{i \in I} \frac{q_i}{d_i}\Big) \\ &\leq k - 1 + \sum_i \frac{d_i - 1 - q_i}{d_i}. \end{aligned}$$

Hence it is enough to prove that

$$k - 1 + \sum_i \frac{d_i - 1 - q_i}{d_i} < 2|e|, \tag{47}$$

or $|e| + 2 - \sum_i 1/d_i < 2|e|$. This transforms into $2 - \sum_i 1/d_i < |e| = k - \sum_i q_i/d_i$. But $k - \sum_i q_i/d_i \geq k - \sum_i (d_1 - 1)/d_i = k - 3 + \sum_i 1/d_i$, hence it is enough to prove that $k - 3 + \sum_i 1/d_i > 2 - \sum_i 1/d_i$. This is $k + 2\sum_i 1/d_i > 5$, which is true for $k \geq 3$ since $\sum_i 1/d_i > 1$. In the second case we repeat the argument and we stop at (47), where we have equallity. $\square$

EXAMPLE 6.2. For $k = 2$, in general, there exists no such bound for $s_{h,0}$, in fact there exist no universal bound at all for $s_{h,0}$. Eg., if we take the D_{d+2} graph (with all decorations -2), with Seifert invariants $(2,1), (2,1), (d, d-1)$, and if E_1 denotes the end vertex of a $(2,1)$ leg, then $E_1^* = s_{[E_1^*]}$ and $s_{h,0} = d/2$, a number which can be arbitrarily large.

LEMMA 6.3. *If $(X,0)$ is a non-cyclic quotient singularity with $k=2$ and $q_1 = q_2 = 1$ then $\mathfrak{N}_{\tilde{\mathbf{a}}}(n) \leq 0$ for any $n < -1$.*

PROOF. We compare $\mathfrak{N}_{\tilde{\mathbf{a}}}(-(t+1))$ with $\mathfrak{R}_{\tilde{\mathbf{a}}}(t)$. Note that if we use for the third floor expression $\lfloor x + (q_3-1)/d_3 \rfloor \leq \lfloor x \rfloor + 1$ we get $\mathfrak{N}_{\tilde{\mathbf{a}}}(-(t+1)) \leq \mathfrak{R}_{\tilde{\mathbf{a}}}(t) \leq 0$. □

6.2. The non-cyclic case. To end the proof, first assume that $k \geq 3$. Then by Lemma 6.1 $s_{h,0} < 2$, hence $-\gamma - s_{h,0} > -2$, hence the summation in (39) reduces to the term corresponding to $n=-1$. Then, by Claim 2 from 5.2.2, we have $\mathfrak{N}_{\tilde{\mathbf{a}}}(-1) \geq 0$ as well.

By Laufer's criterion $k \geq \nu - 1 = 2$, hence the only remaining case is $k=2$, what we will assume next.

If $q_1 = q_2 = 1$, then by Lemma 6.3 we conclude again that only the term $n=-1$ contributes. Note that this case includes all the infinite family of Seifert invariants $(-2;(2,1),(2,1),(d_3,q_3))$.

Next assume that Γ is one of the classical $\mathbb{E}_6$, $\mathbb{E}_7$, and $\mathbb{E}_8$ graphs. Their groups H are $\mathbb{Z}_3$, $\mathbb{Z}_2$, and $\mathbb{Z}_1$ respectively (hence the case $\mathbb{E}_8$ contains no non–zero s_h). In the first two cases one verifies easily that each non–zero s_h is represented by an irreducible curve supported in an end exceptional divisor, along which the multiplicity of the minimal cycle $Z_{\min}$ is one. This means that the irreducible curve $(C,0)$ (embedded in $(X,0) \subset (\mathbb{C}^e,0)$, where $e=3$ is the embedded dimension of $(X,0)$) is intersected by the generic hyperplane with intersection multiplicity 1. This means that $(C,0)$ is smooth, hence R_1^1. (The statements regarding these examples can also be verified directly.)

In this way we covered all cases, except for the singularities with Seifert invariants $(-2;(2,1),(3,2),(5,3))$ and $(-2;(2,1),(3,2),(5,2))$. There are verified below.

6.3. The exceptional singularity $(-2;(2,1),(3,2),(5,2))$. In this case one has $H = \mathbb{Z}/13\mathbb{Z}[E_0^*]$. The canonical cycle is $Z_K = E_{3,1}^*$ with $Z_{K,0} = 12/13$.

Consider any representative of a class $h \in H$, say hE_0^*. Using the generalized Laufer's algorithm one obtains the minimal cycle s_h.

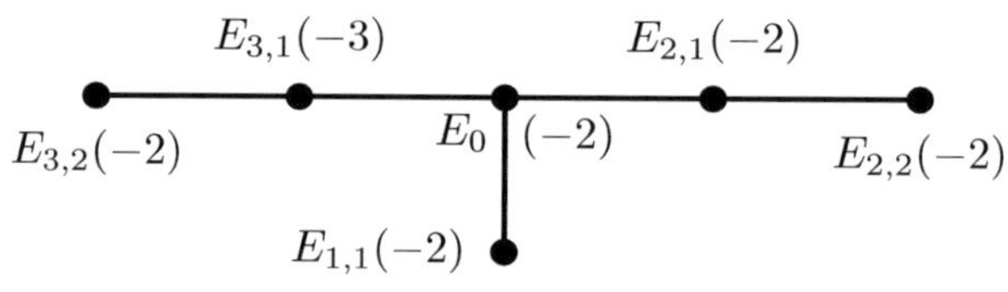

FIGURE 1

Note that if $s_h = a_0 E_0^* + \sum_{i,j} a_{i,j} E_{i,j}^*$, then

$$s_{h,red} = \sum_{i=0}^{3} a_i E_i^* = a_0 E_0^* + a_{1,1} E_1^* + (2a_{2,1} + a_{2,2}) E_2^* + (2a_{3,1} + a_{3,2}) E_3^*,$$

where $E_1^* = E_{1,1}^*$, $E_2^* = E_{2,2}^*$, $E_3^* = E_{3,2}^*$ (see 4.2.3) and hence $\mathfrak{N}_{\tilde{\mathbf{a}}}(-t)$ equals

$$(48)\quad 2 + a_0 - 2t + \left\lfloor \frac{t + a_{1,1} - 1}{2} \right\rfloor + \left\lfloor \frac{2t + 2a_{2,1} + a_{2,2} - 1}{3} \right\rfloor + \left\lfloor \frac{2t + 2a_{3,1} + a_{3,2} - 1}{5} \right\rfloor.$$

The values of the delta invariant for the minimal generic h-curves C_h for $h \neq 0$ are given below. They are obtained from Theorem 4.5. Each $\kappa(s_h)$ is obtained using Theorem 2.6. The values $\mathfrak{N}_{\tilde{\mathbf{a}}}(n)$ for $-\lfloor s_{h,0}+\gamma\rfloor \leq n \leq -1$ are obtained explicitly from $s_{h,red}$ and (48).

(1) $s_{h,0} \leq 1$, $\lfloor s_{h,0}+\gamma\rfloor = 0$ for:

$$\delta(C_3)=\kappa(E^*_{3,1})=0,\quad \delta(C_8)=\kappa(E^*_{3,2})=0,\quad \delta(C_9)=\kappa(E^*_{2,2})=0.$$

(2) $s_{h,0} > 1$, $\lfloor s_{h,0}+\gamma\rfloor = 1$, $\mathfrak{N}_{\tilde{\mathbf{a}}}(-1)=0$ for:

$$\begin{aligned} &\delta(C_2)=\kappa(E^*_{3,2}+E^*_{1,1})=1, && \delta(C_4)=\kappa(E^*_{3,2}+E^*_{2,2})=1,\\ &\delta(C_7)=\kappa(E^*_{1,1})=0, && \delta(C_{11})=\kappa(E^*_{3,1}+E^*_{3,2})=1,\\ &\delta(C_{12})=\kappa(E^*_{3,1}+E^*_{2,2})=1. \end{aligned}$$

(3) $s_{h,0} > 1$, $\lfloor s_{h,0}+\gamma\rfloor = 1$, $\mathfrak{N}_{\tilde{\mathbf{a}}}(-1)=1$ for:

$$\delta(C_5)=\kappa(E^*_{2,1})=1,\quad \delta(C_6)=\kappa(2E^*_{3,1})=2.$$

(4) $s_{h,0} > 1$, $\lfloor s_{h,0}+\gamma\rfloor = 2$, $\mathfrak{N}_{\tilde{\mathbf{a}}}(-1)=1$, $\mathfrak{N}_{\tilde{\mathbf{a}}}(-2)=0$ for:

$$\delta(C_1)=\kappa(E^*_0)=1.$$

(5) $s_{h,0} > 1$, $\lfloor s_{h,0}+\gamma\rfloor = 2$, $\mathfrak{N}_{\tilde{\mathbf{a}}}(-1)=0$, $\mathfrak{N}_{\tilde{\mathbf{a}}}(-2)=1$ for:

$$\delta(C_{10})=\kappa(E^*_{3,1}+E^*_{1,1})=2.$$

Note that the classes of the irreducible cuts are the following: $[E^*_0]=1$, $[E^*_{1,1}]=7$, $[E^*_{2,1}]=5$, and $[E^*_{2,2}]=9$, $[E^*_{3,1}]=3$, $[E^*_{3,2}]=8$.

The curves C_3, C_7, C_8, and C_9 are smooth irreducible minimal generic, that is, of type R^1_1. The curve C_1 and C_5 are singular irreducible minimal generic h-curves, that is, of type R^0_1, As for the rest, C_6 and C_{10} are not irreducible with smooth tangent components, that is, of type R^1_2, see 2.4.3 and the remaining ones are nodal, that is, of type R^2_2.

6.4. The exceptional singularity $(-2;(2,1),(3,2),(5,3))$. Similarly, one has $H=\mathbb{Z}/7\mathbb{Z}[E^*_0]$. The canonical cycle is $Z_K=E^*_{3,2}$ with $Z_{K,0}=6/7$.

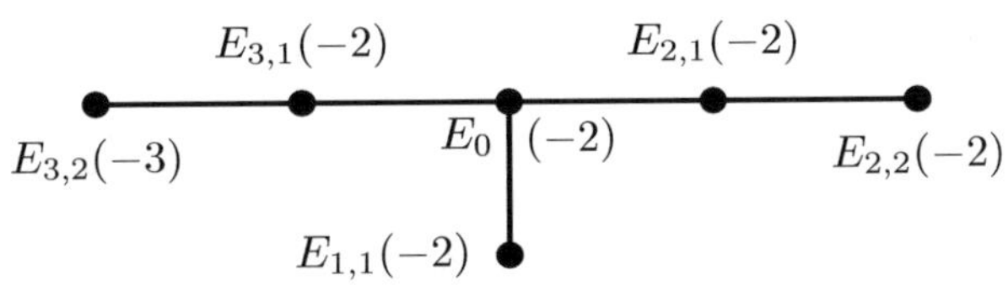

FIGURE 2

The values of the delta invariant for the minimal generic h-curves C_h for $h \neq 0$ are given below. Note that here the expression of $\mathfrak{N}_{\tilde{\mathbf{a}}}(-t)$ becomes:

$$2+a_0-2t+\left\lfloor\frac{t+a_{1,1}-1}{2}\right\rfloor+\left\lfloor\frac{2t+2a_{2,1}+a_{2,2}-1}{3}\right\rfloor+\left\lfloor\frac{3t+3a_{3,1}+a_{3,2}-1}{5}\right\rfloor.$$

(1) $s_{h,0} \leq 1$, $\lfloor s_{h,0}+\gamma\rfloor = 0$ for:

$$\delta(C_3)=\kappa(E^*_{3,2})=0.$$

(2) $s_{h,0} > 1$, $\lfloor s_{h,0}+\gamma\rfloor = 1$, $\mathfrak{N}_{\tilde{\mathbf{a}}}(-1)=0$ for:

$$\delta(C_5)=\kappa(E^*_{2,2})=0,\qquad \delta(C_6)=\kappa(2E^*_{3,2})=1.$$

(3) $s_{h,0} > 1$, $\lfloor s_{h,0} + \gamma \rfloor = 2$, $\mathfrak{N}_{\tilde{\mathbf{a}}}(-1) = 0$, $\mathfrak{N}_{\tilde{\mathbf{a}}}(-2) = 0$ for:

$$\delta(C_1) = \kappa(E^*_{3,2} + E^*_{2,2}) = 1.$$

(4) $s_{h,0} > 1$, $\lfloor s_{h,0} + \gamma \rfloor = 2$, $\mathfrak{N}_{\tilde{\mathbf{a}}}(-1) = 1$, $\mathfrak{N}_{\tilde{\mathbf{a}}}(-2) = 0$ for:

$$\delta(C_2) = \kappa(E^*_{3,1}) = 1.$$

(5) $s_{h,0} > 1$, $\lfloor s_{h,0} + \gamma \rfloor = 2$, $\mathfrak{N}_{\tilde{\mathbf{a}}}(-1) = 0$, $\mathfrak{N}_{\tilde{\mathbf{a}}}(-2) = 1$ for:

$$\delta(C_4) = \kappa(E^*_{1,1}) = 1.$$

The classes of the irreducible cuts are the following: $[E^*_0] = 1$, $[E^*_{1,1}] = 4$, $[E^*_{2,1}] = [E^*_{3,2}] = 3$, $[E^*_{2,2}] = 5$, and $[E^*_{3,1}] = 2$.

This ends the proof of the theorem.

References

[1] W. Barth, C. Peters, and A. Van de Ven, *Compact complex surfaces*, Ergebnisse der Mathematik und ihrer Grenzgebiete, 3. Folge, Band **4**, A Series of Modern Surveys in Mathematics, Springer-Verlag 1984.

[2] E. Brieskorn, *Rationale Singularitäten komplexer Flächen* (German), Invent. Math. **4** (1967/68), 336–358, DOI 10.1007/BF01425318. MR222084

[3] R.-O. Buchweitz and G.-M. Greuel, *The Milnor number and deformations of complex curve singularities*, Invent. Math. **58** (1980), no. 3, 241–281, DOI 10.1007/BF01390254. MR571575

[4] A. Campillo, F. Delgado, and S. M. Gusein-Zade, *Poincaré series of a rational surface singularity*, Invent. Math. **155** (2004), no. 1, 41–53, DOI 10.1007/s00222-003-0310-y. MR2025300

[5] A. Campillo, F. Delgado, and S. M. Gusein-Zade, *Universal abelian covers and multi-index filtration*, Func. Anal. and its Appl. **42** (2008), no. 2, 83–88.

[6] A. Campillo, F. Delgado, and S. M. Gusein-Zade, *Hilbert function, generalized Poincaré series and topology of plane valuations*, Monatsh. Math. **174** (2014), no. 3, 403–412, DOI 10.1007/s00605-013-0547-5. MR3223495

[7] J. I. Cogolludo-Agustín, *Topological invariants of the complement to arrangements of rational plane curves*, Mem. Amer. Math. Soc. **159** (2002), no. 756, xiv+75, DOI 10.1090/memo/0756. MR1921584

[8] J.I. Cogolludo-Agustín, T. László, J. Martín-Morales, and A. Némethi, *Delta invariant of curves on rational surfaces I. An analytic approach*, `https://doi.org/10.1142/S0219199721500528`

[9] J.I. Cogolludo-Agustín, T. László, J. Martín-Morales, and A. Némethi, *The delta invariant of curves on rational surfaces II. Poincaré series and topological aspects*, `arXiv:2003.07110 [math.GT]`.

[10] J.I. Cogolludo-Agustín, T. László, J. Martín-Morales, and A. Némethi, manuscript in preparation.

[11] J. I. Cogolludo-Agustín and J. Martín-Morales, *The correction term for the Riemann-Roch formula of cyclic quotient singularities and associated invariants*, Rev. Mat. Complut. **32** (2019), no. 2, 419–450, DOI 10.1007/s13163-018-0280-7. MR3942923

[12] S. Goto and K. Watanabe, *On graded rings. I*, J. Math. Soc. Japan **30** (1978), no. 2, 179–213, DOI 10.2969/jmsj/03020179. MR494707

[13] G.M. Greuel, *Deformationen spezieller Kurvensingularitäten und eine Formel von Deligne*, Teil II der Habilitationschrift Bonn 1979.

[14] G.-M. Greuel, *On deformation of curves and a formula of Deligne*, Algebraic geometry (La Rábida, 1981), Lecture Notes in Math., vol. 961, Springer, Berlin, 1982, pp. 141–168, DOI 10.1007/BFb0071281. MR708332

[15] H. Hironaka, *On the arithmetic genera and the effective genera of algebraic curves*, Mem. Coll. Sci. Univ. Kyoto Ser. A. Math. **30** (1957), 177–195, DOI 10.1215/kjm/1250777055. MR90850

[16] J. Kollár and N. I. Shepherd-Barron, *Threefolds and deformations of surface singularities*, Invent. Math. **91** (1988), no. 2, 299–338, DOI 10.1007/BF01389370. MR922803

[17] T. László, J. Nagy, and A. Némethi, *Surgery formulae for the Seiberg-Witten invariant of plumbed 3-manifolds*, Rev. Mat. Complut. **33** (2020), no. 1, 197–230, DOI 10.1007/s13163-019-00297-z. MR4047693

[18] T. László, J. Nagy, and A. Némethi, *Combinatorial duality for Poincaré series, polytopes and invariants of plumbed 3-manifolds*, Selecta Math. (N.S.) **25** (2019), no. 2, Paper No. 21, 31.

[19] T. László and A. Némethi, *Ehrhart theory of polytopes and Seiberg-Witten invariants of plumbed 3-manifolds*, Geom. Topol. **18** (2014), no. 2, 717–778, DOI 10.2140/gt.2014.18.717. MR3180484

[20] T. László and Zs. Szilágyi, *Non-normal affine monoids, modules and Poincaré series of plumbed 3-manifolds*, Acta Math. Hungar. **152** (2017), no. 2, 421–452, DOI 10.1007/s10474-017-0726-2. MR3682892

[21] T. László and Z. Szilágyi, *On Poincaré series associated with links of normal surface singularities*, Trans. Amer. Math. Soc. **372** (2019), no. 9, 6403–6436, DOI 10.1090/tran/7802. MR4024526

[22] H. B. Laufer, *On rational singularities*, Amer. J. Math. **94** (1972), 597–608, DOI 10.2307/2374639. MR330500

[23] J. McKay, *Graphs, singularities, and finite groups*, The Santa Cruz Conference on Finite Groups (Univ. California, Santa Cruz, Calif., 1979), Proc. Sympos. Pure Math., vol. 37, Amer. Math. Soc., Providence, R.I., 1980, pp. 183–186. MR604577

[24] A. Némethi, *On the Ozsváth-Szabó invariant of negative definite plumbed 3-manifolds*, Geom. Topol. **9** (2005), 991–1042, DOI 10.2140/gt.2005.9.991. MR2140997

[25] A. Némethi, *Graded roots and singularities*, Proc. *Advanced School and Workshop* on *Singularities in Geometry and Topology* ICTP (Trieste, Italy), World Sci. Publ., Hackensack, NJ, 2007, 394–463.

[26] A. Némethi, *Poincaré series associated with surface singularities*, in: Singularities I: Algebraic and Analytic Aspects, International Conference in honor of the 60th birthday of Lê Dũng Tráng, Cuernavaca, Mexico, 2007, Contemp. Math. **474** (2008), 271–299.

[27] A. Némethi, *The cohomology of line bundles of splice-quotient singularities*, Adv. Math. **229** (2012), no. 4, 2503–2524, DOI 10.1016/j.aim.2012.01.003. MR2880230

[28] A. Némethi, *Pairs of invariants of surface singularities*, Proc. Int. Cong. of Math. 2018 Rio de Janeiro, vol. 1, 2018, pp. 745–776.

[29] A. Némethi and T. Okuma, *The embedding dimension of weighted homogeneous surface singularities*, J. Topol. **3** (2010), no. 3, 643–667, DOI 10.1112/jtopol/jtq019. MR2684515

[30] W. D. Neumann, *Abelian covers of quasihomogeneous surface singularities*, Singularities, Part 2 (Arcata, Calif., 1981), Proc. Sympos. Pure Math., vol. 40, Amer. Math. Soc., Providence, RI, 1983, pp. 233–243. MR713252

[31] V. V. Nikulin, *del Pezzo surfaces with log-terminal singularities* (Russian), Mat. Sb. **180** (1989), no. 2, 226–243, 304, DOI 10.1070/SM1990v066n01ABEH001314; English transl., Math. USSR-Sb. **66** (1990), no. 1, 231–248. MR993456

[32] H. Pinkham, *Normal surface singularities with C^* action*, Math. Ann. **227** (1977), no. 2, 183–193, DOI 10.1007/BF01350195. MR432636

[33] J. Stevens, *Kulikov singularities, a study of a class of complex surface singularities with their hypeplane section*, Ph.D. thesis, Leiden, 1985.

Departamento de Matemáticas, IUMA, Universidad de Zaragoza, C. Pedro Cerbuna 12, 50009 Zaragoza, Spain
Email address: jicogo@unizar.es

Babeş-Bolyai University, Faculty of Mathematics and Computer Science, Str. Mihail Kogălniceanu nr. 1, 400084 Cluj-Napoca, Romania
Email address: laszlo.tamas@math.ubbcluj.ro

Departamento de Matemáticas, IUMA, Universidad de Zaragoza, C. Pedro Cerbuna 12, 50009 Zaragoza, Spain
Email address: jorge@unizar.es

Alfréd Rényi Institute of Mathematics, Hungarian Academy of Sciences, Reáltanoda utca 13-15, H-1053, Budapest, Hungary; ELTE - University of Budapest, Dept. of Geometry, Budapest, Hungary; and BCAM - Basque Center for Applied Math., Mazarredo, 14 E48009 Bilbao, Basque Country, Spain
Email address: nemethi.andras@renyi.mta.hu

Contemporary Mathematics
Volume **778**, 2022
https://doi.org/10.1090/conm/778/15661

Motivic Poincaré series of cusp surface singularities

János Nagy and András Némethi

Abstract. We target multivariable series associated with resolutions of complex analytic normal surface singularities. In general, the equivariant multivariable analytical and topological Poincaré series are well–defined and have good properties only if the link is a rational homology sphere. We wish to create a model when this assumption is not valid: we analyse the case of cusps. For such germs we define even the motivic versions of these two series, we prove that they are equal, and we provide explicit combinatorial expression for them. This is done via a motivic multivariable series associated with the space of effective Cartier divisors of the reduced exceptional curve.

1. Introduction

1.1. Let us consider a complex analytic normal surface singularity and fix one of its resolutions and the divisorial filtration of the local algebra associated with the irreducible exceptional divisors. The multivariable Poincaré series $\mathcal{P}_0(\mathbf{t})$ associated with this filtration is one of the strongest analytic invariants of the germ. For definition, particular examples and several properties see e.g. [**3, 5, 7, 23, 24, 26–28**].

In several special cases (even if the germ is not taut, but the analytic structure is 'nice'), $\mathcal{P}_0(\mathbf{t})$ can be recovered from the topology of the link. These cases include e.g. the rational or minimally elliptic singularities (with rational homology sphere link) [**26**]. The most general case when such a characterization was established is the family of splice quotient singularities [**28**]. (For such germs the analytic type is defined canonically from the graph [**32, 34**].)

However, in all the cases when such a characterization was established the link of the corresponding germs are rational homology spheres, $\mathbb{Q}HS^3$, (that is, the dual graph is a tree of $\mathbb{P}^1$'s). In this note we wish to step over this obstruction by analysing the case of cusps, when the dual graph consists of a loop. (The rationality of the irreducible exceptional divisors, due to the complexity of the moduli space of algebraic curves, presumably cannot be dropped in such topological comparison without some other essential analytic assumption.)

In fact, there exists even a more general series, the equivariant multivariable Poincaré series $\mathcal{P}(\mathbf{t})$ associated with the divisorial filtration of the local algebra of the universal abelian covering of the germ. It behaves in a more uniform and conceptual way in several geometric construction, e.g. in the context of abelian

2020 *Mathematics Subject Classification.* Primary 32S05, 32S25, 32S50; Secondary 14Bxx, 14J80.

The authors were partially supported by NKFIH Grant "Élvonal (Frontier)" KKP 126683.

coverings and associated bundles. However, its definition is also obstructed: it is defined in terms of the universal abelian covering, which is well–defined only when the link is a rational homology sphere. Again, we will step over this obstruction as well in the case of cusps.

For definition and certain properties of cusp surface singularities see [**15, 23, 25, 38**].

When the link is a rational homology sphere there is a concrete 'topological candidate' for $\mathcal{P}(\mathbf{t})$, denoted by $\mathcal{Z}(\mathbf{t})$, a multivariable series defined from the combinatorics of the resolution graph (well–defined in any case, even if $\mathcal{P}(\mathbf{t}) \neq \mathcal{Z}(\mathbf{t})$). In the cases mentioned above (rational, splice quotient) in fact one has $\mathcal{P}(\mathbf{t}) = \mathcal{Z}(\mathbf{t})$. This series $\mathcal{Z}(\mathbf{t})$ has several other pleasant properties, it has deep connections with several topological invariants of the link (e.g. with the Seiberg–Witten invariant, lattice cohomology, etc) [**27,29,30**]. In this way it created a bridge between analytic and topological invariants.

However, usually (e.g. for a non-rational or non–elliptic topological types) for a 'non–nice' analytic type (e.g., for a generic one) $\mathcal{P}(\mathbf{t}) \neq \mathcal{Z}(\mathbf{t})$). In such cases the challenge is to find new topological candidate series, which might recover $\mathcal{P}(\mathbf{t})$ for several 'bad' (but interesting) families of analytic types. Hence different geometric realizability (equivalent definitions) even of 'classical' $\mathcal{Z}(\mathbf{t})$ is crucial: they might suggets/impose different generalizations/versions, which might fit with different analytic structures. (This partly motivated that in the body of the article we list several parallel realizabilities.)

1.2. This article has several goals. Firstly, we wish to provide a model for the definition of $\mathcal{P}(\mathbf{t})$ (in terms of analytic data) when the link is not a rational homology sphere (hence the universal abelian covering does not exists). We will do this for the cusp normal surface singularities. (Maybe here is appropriate to eliminate a possible confusion from the start: Though cusp singularities are taut [**14**], hence there is no analytic moduli of their analytic structures, the construction of $\mathcal{P}(\mathbf{t})$ involves — must involve — certain line bundles associated with fixed Chern classes; since the corresponding Picard groups are non–trivial, these choices have an analytic moduli.)

Second, we wish to provide a good topological candidate $\mathcal{Z}(\mathbf{t})$ for $\mathcal{P}(\mathbf{t})$. (Note that the 'old' definition of $\mathcal{Z}(\mathbf{t})$, though well–defined, for cusps gives the meaningless constant series 1, which definitely should be modified.)

In order to provide these two definitions we rely on two collections of linear subspace arrangements, one of them defined analytically, the other one topologically. Both are indexed by the possible first Chern classes of line bundles of the resolution. Both series in the new situation are defined via these arrangements by considering the topological Euler characteristic of the projectivised arrangement complements. For details regarding these arrangements see [**29, 30**]. (Since the cusp singularities are minimally elliptic, in the introductory part we emphasize more the known facts regarding the rational and minimally elliptic cases, the second one under the $\mathbb{Q}HS^3$–link assumption. This serves as a good comparison for the newly established methods and formulae.)

Once the definitions are settled, we prove that in the case of cusps one has $\mathcal{P}(\mathbf{t}) = \mathcal{Z}(\mathbf{t})$ indeed. Hence, $\mathcal{Z}(\mathbf{t})$ is a good topological candidate for $\mathcal{P}(\mathbf{t})$ even if the graph has 1–cycles.

In fact, the new definition (based on the complements of subspace arrangements) allows us to extend both series to their motivic versions (with coefficient in the Grothendieck ring of varieties), denoted by $\mathcal{P}(\mathbb{L},\mathbf{t})$ and $\mathcal{Z}(\mathbb{L},\mathbf{t})$. For these extended versions we also establish the identity $\mathcal{P}(\mathbb{L},\mathbf{t})=\mathcal{Z}(\mathbb{L},\mathbf{t})$.

Surprisingly, this new extensions direct us to an unexpected new territory. It turns out that these objects can be related with a multivariable motivic series associated with the effective Cartier divisors (supported by the reduced exceptional curve and indexed by the possible Chern classes), denoted by $\mathcal{Z}^{\mathrm{ECa}}(\mathbb{L},\mathbf{t})$. (Maybe is worth to mention that for minimal resolution of cusps the reduced exceptional divisor equals the Artin fundamental cycle, the minial elliptic cyle and the anticanonical cycle as well [**15**].) This connection was suggested and imposed by the recent manuscripts of the authors regarding the Abel maps associated with a resolution of a normal surface singularity [**19**–**22**]. The point is that the series $\mathcal{Z}^{\mathrm{ECa}}(\mathbb{L},\mathbf{t})$ has a very natural form in terms of the graph (for notations see 2.1, $\mathcal{V}$ and $\mathcal{E}$ are the set of vertices and edges, while E_v^* stay for the dual base elements):

$$\mathcal{Z}^{\mathrm{ECa}}(\mathbb{L},\mathbf{t})=\frac{\prod_{(u,v)\in\mathcal{E}}(1-\mathbf{t}^{E_u^*}-\mathbf{t}^{E_v^*}+\mathbb{L}\mathbf{t}^{E_u^*+E_v^*})}{\prod_{v\in\mathcal{V}}(1-\mathbf{t}^{E_v^*})(1-\mathbb{L}\mathbf{t}^{E_v^*})}.$$

(This is valid for any singularity — not necessarily for a cusp —, whenever all the irreducible exceptional curves are rational.) Then, in the case of cusp, the following phenomenon happens. For relevant Chern classes l', the motivic information of any fiber of the Abel map $\mathrm{ECa}^{-l'}(E)\to\mathrm{Pic}^{-l'}(E)=\mathbb{C}^*$ can be related with $\mathcal{P}(\mathbf{t})$. It turns out that for a cusp and its minimal resolution

$$\mathcal{Z}(\mathbb{L},\mathbf{t})=\mathcal{P}(\mathbb{L},\mathbf{t})=1+\frac{\mathcal{Z}^{\mathrm{ECa}}(\mathbb{L},\mathbf{t})-1}{\mathbb{L}-1}.$$

Furthermore (when the exceptional curve has at least two components) then

$$\begin{aligned}\mathcal{P}(\mathbf{t})=\mathcal{Z}(\mathbf{t})&=\Big(1+\frac{\mathcal{Z}^{\mathrm{ECa}}(\mathbb{L},\mathbf{t})-1}{\mathbb{L}-1}\Big)\Big|_{\mathbb{L}=1}\\&=1+\sum_{v\in\mathcal{V}}\frac{\mathbf{t}^{E_v^*}}{1-\mathbf{t}^{E_v^*}}+\sum_{(u,v)\in\mathcal{E}}\frac{\mathbf{t}^{E_v^*}}{1-\mathbf{t}^{E_v^*}}\cdot\frac{\mathbf{t}^{E_u^*}}{1-\mathbf{t}^{E_u^*}}.\end{aligned}$$

For the remaining case see Theorem 3.2.

2. Preliminaries regarding normal surface singularities

2.1. Definitions, notations. Let (X,o) be a complex normal surface singularity. Let $\pi:\widetilde{X}\to X$ be a good resolution with dual graph Γ whose vertices are denoted by $\mathcal{V}$ and edges by $\mathcal{E}$. Set $E:=\pi^{-1}(o)$. Let M be the link of (X,o).

Set $L:=H_2(\widetilde{X},\mathbb{Z})$. It is freely generated by the classes of the irreducible exceptional curves. If L' denotes $H^2(\widetilde{X},\mathbb{Z})$, then the intersection form $(\,,\,)$ on L provides an embedding $L\hookrightarrow L'$ with factor the torsion part of the first homology group H of the link. (In fact, L' is the dual lattice of $(L,(\,,\,))$, it can also be identified with $H_2(\widetilde{X},\partial\widetilde{X},\mathbb{Z})$.) Moreover, $(\,,\,)$ extends to L'. L' is freely generated by the duals E_v^*, where $(E_v^*,E_w)=-1$ for $v=w$ and $=0$ else.

Effective classes $l=\sum r_vE_v\in L'$ with all $r_v\in\mathbb{Q}_{\geq 0}$ are denoted by $L'_{\geq 0}$ and $L_{\geq 0}:=L'_{\geq 0}\cap L$. Denote by $\mathcal{S}'$ the (Lipman's) anti-nef cone $\{l'\in L'\,:\,(l',E_v)\leq 0$ for all $v\}$. It is generated over $\mathbb{Z}_{\geq 0}$ by the base-elements E_v^*. Since all the entries of E_v^* are strict positive, $\mathcal{S}'$ is a sub-cone of $L'_{\geq 0}$, and for any fixed $a \in L'$ the set $\{l' \in \mathcal{S}' : l' \not\geq a\}$ is finite. Set $\mathfrak{C} := \{\sum l'_v E_v \in L',\ 0 \leq l'_v < 1\}$. For any $l' \in L'$ write its class in H by $[l']$. Denote by $\theta : H \to \widehat{H}$ the isomorphism $[l'] \mapsto e^{2\pi i (l', \cdot)}$ of H with its Pontrjagin dual $\widehat{H}$.

Let $K \in L'$ be the *canonical class* satisfying $(K + E_v, E_v) = -2 + 2g_v$ for all $v \in \mathcal{V}$, where g_v is the genus of E_v. Set $\chi(l') = -(l', l' + K)/2$. By Riemann-Roch theorem $\chi(l) = \chi(\mathcal{O}_l)$ for $l \in L_{>0}$.

In this preliminary section we will assume that M is a rational homology sphere. (This is exactly that assumption what we wish to drop later.) This happens if and only if Γ is a tree and $g_v = 0$ for all v.

In subsections 2.2 and 2.3 we list certain analytic invariants, then we continue with the topological ones. For more details regarding this part see e.g. [**3, 5, 7, 23, 24, 26–28**]. These results are present in the literature (though slightly scattered), nevertheless here we collect the relevant ones. The reason is that this list of constructions and results will serve as prototypes for further generalizations, in some cases they already suggest the need of modifications as well.

2.2. Natural line bundles. Natural line bundles are provided by the splitting of the cohomological exponential exact sequence [**24**, §4.2]:

$$0 \to H^1(\widetilde{X}, \mathcal{O}_{\widetilde{X}}) \to \mathrm{Pic}(\widetilde{X}) \xrightarrow{c_1} L' \to 0.$$

The first Chern class c_1 has an obvious section on the subgroup L, namely $l \mapsto \mathcal{O}_{\widetilde{X}}(l)$. This section has a unique extension $\mathcal{O}(\cdot)$ to L'. We call a line bundle *natural* if it is in the image of this section. By this definition, a line bundle $\mathcal{L}$ is natural if and only if there exists a positive integer n such that $\mathcal{L}^{\otimes n}$ has the form $\mathcal{O}_{\widetilde{X}}(l)$ for some $l \in L$.

One can recover the natural line bundles via coverings as follows. Let $c : (X_a, o) \to (X, o)$ be the universal abelian covering of (X, o). (Note that the existence of the universal abelian covering is guaranteed by the fact that the link is a rational homology sphere. Indeed c is a finite regular covering over $X \setminus \{o\}$ and this regular covering has a unique non-regular extension to the level of germs of normal surface singularities. The regular covering is associated with the representation $\pi_1(M) \to H_1(M, \mathbb{Z}) = \mathrm{Tors}(H_1(M, \mathbb{Z})) = H$.) Furthermore, let $\pi_a : \widetilde{X}_a \to X_a$ the normalized pullback of π by c, and $\widetilde{c} : \widetilde{X}_a \to \widetilde{X}$ the morphism which covers c. Then the action of H on (X_a, o) lifts to an action on $\widetilde{X}_a$ and one has an H–eigensheaf decomposition ([**24**, 4.2.9] or [**35**, (3.5)]):

$$\widetilde{c}_* \mathcal{O}_{\widetilde{X}_a} = \bigoplus_{l' \in \mathfrak{C}} \mathcal{O}_{\widetilde{X}}(-l') \quad (\mathcal{O}_{\widetilde{X}}(-l') \text{ being the } \theta([l'])\text{-eigenspace of } \widetilde{c}_* \mathcal{O}_{\widetilde{X}_a}). \tag{2.1}$$

Then write any $l' \in L'$ as $l'_0 + l$ with $l'_0 \in \mathfrak{C}$ and $l \in L$, and set $\mathcal{O}_{\widetilde{X}}(-l') := \mathcal{O}_{\widetilde{X}}(-l'_0) \otimes \mathcal{O}_{\widetilde{X}}(-l)$.

2.3. Series associated with the divisorial filtration. Once a resolution π is fixed, $\mathcal{O}_{X_a,o}$ inherits the *divisorial multi-filtration* (cf. [**26**, (4.1.1)]):

$$\mathcal{F}(l') := \{f \in \mathcal{O}_{X_a,o} \,|\, \mathrm{div}(f \circ \pi_a) \geq \widetilde{c}^*(l')\}. \tag{2.2}$$

Let $\mathfrak{h}(l')$ be the dimension of the $\theta([l'])$-eigenspace of $\mathcal{O}_{X_a,o}/\mathcal{F}(l')$. Then, one defines the *equivariant divisorial Hilbert series* by

$$\mathcal{H}(\mathbf{t}) = \sum_{l'\in L'} \mathfrak{h}(l')t_1^{l'_1}\cdots t_s^{l'_s} = \sum_{l'\in L'} \mathfrak{h}(l')\mathbf{t}^{l'} \in \mathbb{Z}[[L']] \quad (l' = \sum_i l'_i E_i). \tag{2.3}$$

Notice that the terms of the sum reflect the H-eigenspace decomposition too: $\mathfrak{h}(l')\mathbf{t}^{l'}$ contributes to the $\theta([l'])$-eigenspace. For example, $\sum_{l\in L}\mathfrak{h}(l)\mathbf{t}^l$ corresponds to the H-invariants, hence it is the *Hilbert series* of $\mathcal{O}_{X,o}$ associated with the $\pi^{-1}(o)$-divisorial multi-filtration.

The 'graded version' associated with the Hilbert series is defined (cf. [**5, 7**]) as

$$\mathcal{P}(\mathbf{t}) = -\mathcal{H}(\mathbf{t})\cdot\prod_v(1-t_v^{-1}) \in \mathbb{Z}[[L']]. \tag{2.4}$$

Usually this is called the equivarient multivariable analytic Poincaré series of (X, o).

If we write the series $\mathcal{P}(\mathbf{t})$ as $\sum_{l'}\mathfrak{p}(l')\mathbf{t}^{l'}$, then

$$\mathfrak{p}(l') = \sum_{I\subseteq\mathcal{V}}(-1)^{|I|+1}\dim\frac{H^0(\widetilde{X},\mathcal{O}_{\widetilde{X}}(-l'))}{H^0(\widetilde{X},\mathcal{O}_{\widetilde{X}}(-l'-E_I))} \tag{2.5}$$

and $\mathcal{P}$ is supported in the cone $\mathcal{S}'$.

Although the multiplication by $\prod_v(1-t_v^{-1})$ in $\mathbb{Z}[[L']]$ is not injective, hence apparently $\mathcal{P}$ contains less information then $\mathcal{H}$, they, in fact, determine each other. Indeed, for any $l'\in L'$ one has

$$\mathfrak{h}(l') = \sum_{l\in L,\, l\ngeq 0}\mathfrak{p}(l'+l). \tag{2.6}$$

2.4. Linear subspace arrangements associated with the filtration [29–31]. Fix a normal surface singularity, one of its resolutions and the filtration $\{\mathcal{F}(l')\}_{l'\in L'}$. For any $l'\in L'$, the linear space

$$(\mathcal{F}(l')/\mathcal{F}(l'+E))_{\theta([l'])} = H^0(\mathcal{O}_{\widetilde{X}}(-l'))/H^0(\mathcal{O}_{\widetilde{X}}(-l'-E))$$

naturally embeds into

$$T(l') := H^0(\mathcal{O}_E(-l')).$$

Let its image be denoted by $A(l')$. Furthermore, for every $v\in\mathcal{V}$, consider the linear subspace $T_v(l')$ of $T(l')$ given by

$$T_v(l') := H^0(\mathcal{O}_{E-E_v}(-l'-E_v)) = \ker\big(H^0(\mathcal{O}_E(-l'))\to H^0(\mathcal{O}_{E_v}(-l'))\big)\subset T(l').$$

Then the image $A_v(l')$ of $H^0(\mathcal{O}_{\widetilde{X}}(-l'-E_v)/H^0(\mathcal{O}_{\widetilde{X}}(-l'-E))$ in $T(l')$ satisfies $A_v(l') = A(l')\cap T_v(l')$.

DEFINITION 2.1. The (finite dimensional) arrangement of linear subspaces $\mathcal{A}_{\text{top}}(l') = \{T_v(l')\}_v$ in $T(l')$ is called the 'topological arrangement' at $l'\in L'$. The arrangement of linear subspaces $\mathcal{A}_{\text{an}}(l') = \{A_v(l') = T_v(l')\cap A(l')\}_v$ in $A(l')$ is called the 'analytic arrangement' at $l'\in L'$. The corresponding projectivized arrangement complements will be denoted by $\mathbb{P}(T(l')\setminus\cup_v T_v(l'))$ and $\mathbb{P}(A(l')\setminus\cup_v A_v(l'))$ respectively.

If $l'\notin\mathcal{S}'$ then there exists v such that $(E_v, l') > 0$, that is $h^0(\mathcal{O}_{E_v}(-l')) = 0$, proving that $T_v(l') = T(l')$. Hence $A_v(l') = A(l')$ too. In particular, both arrangement complements are empty.

The inclusion–exclusion principle and $\dim V_I = \chi_{\rm top}(\mathbb{P}V_I)$ shows the following fact.

LEMMA 2.2. *Assume that $\{V_\alpha\}_{\alpha\in\Lambda}$ is a finite family of linear subspaces of a finite dimensional linear space V. For $I\subset\Lambda$ set $V_I := \cap_{\alpha\in I}V_\alpha$ (where $V_\emptyset = V$). Then*

$$\chi_{\rm top}(\mathbb{P}(V\setminus\cup_\alpha V_\alpha)) = \sum_{I\subset\Lambda}(-1)^{|I|}\dim V_I.$$

If $\Lambda\neq\emptyset$, then this also equals $\sum_I(-1)^{|I|+1}\mathrm{codim}(V_I\subset V)$.

In particular, this lemma and identity (2.5) imply the following.

COROLLARY 2.1. *For any $l'\in\mathcal{S}'$ one has*

$$\mathfrak{p}(l') = \chi_{\rm top}(\,\mathbb{P}(A(l')\setminus\cup_v A_v(l'))\,).$$

Then for any $l'\in L'$ and $I\subset\mathcal{V}$ one has:

$$(2.7)\quad \dim A(l') = \mathfrak{h}(l'+E)-\mathfrak{h}(l'),\quad \dim\cap_{v\in I}A_v(l') = \mathfrak{h}(l'+E)-\mathfrak{h}(l'+E_I).$$

Thus, we can expect that the analytic arrangement is rather sensitive to the modification of the analytic structure, and in general, does not coincide with the topological arrangement.

In the next paragraphs we will show that whenever the link of the singularity is a rational homology sphere the *topological arrangement* $\mathcal{A}_{\rm top}$ is indeed topological, it depends only on the combinatorics of the resolution graph. We will need the following technical definition.

LEMMA 2.3. **[26]**

(1) *For any $l'\in L'$ and subset $I\subset\mathcal{V}$ there exists a unique minimal subset $J(l',I)\subset\mathcal{V}$ which contains I, and has the following property:*

$$(2.8)\qquad \textit{there is no } v\in\mathcal{V}\setminus J(l',I) \textit{ with } (E_v, l'+E_{J(l',I)})>0.$$

(2) *$J(l',I)$ can be found by the next algorithm: one constructs a sequence $\{I_m\}_{m=0}^k$ of subsets of $\mathcal{V}$, with $I_0=I$, $I_{m+1}=I_m\cup\{v(m)\}$, where the index $v(m)$ is determined as follows. Assume that I_m is already constructed. If I_m satisfies (2.8) we stop and $m=k$. Otherwise, there exists at least one v with $(E_v,l'+E_{I_m})>0$. Take $v(m)$ one of them and continue the algorithm with I_{m+1}. Then $I_k=J(l',I)$.*

PROPOSITION 2.1. **[26]** *Assume that the resolution graph is a tree of rational curves. For any $l'\in L'$ and $I\subset\mathcal{V}$ write $J(I):=J(l',I)$. Then the following facts hold.*

(a) *One has the following commutative diagram with exact rows*

$$\begin{array}{ccccccccc}
0 & \longrightarrow & H^0(\mathcal{O}_{E-E_{J(I)}}(-l'-E_{J(I)})) & \longrightarrow & H^0(\mathcal{O}_E(-l')) & \stackrel{k}{\twoheadrightarrow} & H^0(\mathcal{O}_{E_{J(I)}}(-l')) & \longrightarrow & 0\\
 & & \simeq\downarrow j & & \| & & \downarrow i & & \\
0 & \longrightarrow & H^0(\mathcal{O}_{E-E_I}(-l'-E_I)) & \longrightarrow & H^0(\mathcal{O}_E(-l')) & \longrightarrow & H^0(\mathcal{O}_{E_I}(-l')) & & \\
 & & \| & & \| & & & & \\
 & & \cap_{v\in I}T_v(l') & \hookrightarrow & T(l') & & & &
\end{array}$$

where j is an isomorphism (hence $\cap_{v\in I}T_v(l')=\cap_{v\in J(I)}T_v(l')$), i is injective and k is surjective.

(b) $\dim \cap_{v\in J(I)}T_v(l') = \chi(\mathcal{O}_{E-E_{J(I)}}(-l'-E_{J(I)})) = \chi(l'+E)-\chi(l'+E_{J(I)})$.
(c) *In particular, if* $J(I_1) = J(I_2)$ *then* $\cap_{v\in I_1}T_v(l') = \cap_{v\in I_2}T_v(l')$*, and if* $J(I_1) \subsetneq J(I_2)$ *then* $\cap_{v\in I_1}T_v(l') \supsetneq \cap_{v\in I_2}T_v(l')$*. Therefore,* $J(I)$ *is the unique maximal subset* $I_{max} \subset \mathcal{V}$*, such that* $I \subset I_{max}$*, and* $\cap_{v\in I}T_v(l') = \cap_{v\in I_{max}}T_v(l')$.
(d) *Part (b) for* $I = \emptyset$ *reads as follows:* $\dim T(l') = \dim \cap_{v\in J(\emptyset)}T_v(l') = \chi(l'+E)-\chi(l'+E_{J(\emptyset)})$*. Hence, if* $l' \in \mathcal{S}'$ *then* $\dim T(l') = -(l', E)+1$.
(e) $\operatorname{codim}(\cap_{v\in I}T_v(l') \hookrightarrow T(l')) = \chi(l'+E_{J(I)}) - \chi(l'+E_{J(\emptyset)})$.
(f) *In particular, the arrangement complement is non–empty if and only if* $J(\emptyset) = \emptyset$ *(if and only if* $l' \in \mathcal{S}'$*).*

Therefore, if the graph is a tree of rational curves then the isotopy type of the arrangement $\mathcal{A}_{\rm top}$ depends only on the combinatorial data of the graph. Note also that $J(l', I)$, and the topological linear subspace arrangement $\mathcal{A}_{\rm top}$ too, depend only on the E^*–coefficients of l' and on the *shape of the graph* Γ, that is, on the valencies $\{\kappa_v\}_{v\in\mathcal{V}}$ but not on the Euler numbers $\{E_v^2\}_v$.

At topological Euler characteristic level one has:

COROLLARY 2.2. *If the graph is a tree of rational curves and* $l' \in \mathcal{S}'$ *then*

$$\chi_{\rm top}(\,\mathbb{P}(T(l')\setminus \cup_v T_v(l'))\,) = \sum_{I\subset\mathcal{V}}(-1)^{|I|+1}\chi(l'+E_{J(l',I)}).$$

PROOF. Use Lemma 2.2 and Proposition 2.1(b). □

EXAMPLE 2.4. Using special vanishing theorems and computation sequences of rational and elliptic singularities (cf. [**23,25**]) one can prove the following results as well (see e.g. [**30,31**]).

(I) Assume the following situations:

(a) either (X, o) is rational, π is arbitrary resolution, and $l' \in \mathcal{S}'$ is arbitrary,
(b) or (X, o) is minimally elliptic singularity with $H^1(\widetilde{X},\mathbb{Z}) = 0$, π is a resolution such that the support of the elliptic cycle equals E, and we also assume that for the fixed $l' \in \mathcal{S}'$ there exists a computation sequence $\{x_i\}_i$ for the fundamental cycle Z_{min} (in the sense of Laufer [**13**]), which contains E as one of its terms, and it jumps (that is, $(x_i, E_1) = 2$) at some E_1 with $(E_1, l') < 0$.

Then the topological and analytic arrangements at l' agree, $\mathcal{A}_{\rm top}(l') = \mathcal{A}_{\rm an}(l')$.

(II) For minimally elliptic singularities it can happen that $\mathcal{A}_{\rm top}(l') \neq \mathcal{A}_{\rm an}(l')$, even for the minimal resolution. E.g., in the case of the minimal good resolution of $\{x^2+y^3+z^7=0\}$, or in the case of minimal resolution of $\{x^2+y^3+z^{11}=0\}$ (which is good), for $l = Z_{min}$ one has $\dim(T(Z_{min})) = 2$ and $\dim(A(Z_{min})) = 1$.

(III) For any $l' \in \mathcal{S}'$ one has the exact sequence

$$0 \to A(l') \to T(l') \to H^1(\mathcal{O}_{\widetilde{X}}(-l'-E)) \to H^1(\mathcal{O}_{\widetilde{X}}(-l'))$$

Hence, $\mathcal{A}_{\rm an}(l') = \mathcal{A}_{\rm top}(l')$ whenever $H^1(\mathcal{O}_{\widetilde{X}}(-l'-E)) = 0$. This happens e.g. if $l' = \sum_v a_vE_v^*$ with $a_v \gg 0$, in which case $H^1(\mathcal{O}_{\widetilde{X}}(-l'-E)) = 0$ by the Grauert–Riemenschneider Vanishing Theorem.

2.5. The topological series $\mathcal{Z}(\mathbf{t})$ [5,7,23,29–31]. The series $\mathcal{Z}(\mathbf{t}) \in \mathbb{Z}[[\mathcal{S}']]$ is defined by the rational function $z(\mathbf{x})$ in variables $x_v = \mathbf{t}^{E_v^*}$, or by its Taylor expansion at the origin, where

$$z(\mathbf{x}) := \prod_{v\in\mathcal{V}} (1-x_v)^{\kappa_v-2}, \tag{2.9}$$

and κ_v is the valency of the vertex v. Hence it is the expansion of $\mathcal{Z}(\mathbf{t}) = \prod_v (1 - \mathbf{t}^{E_v^*})^{\kappa_v-2}$.

We start to list some other appearances of $\mathcal{Z}(\mathbf{t})$.

If Σ is a topological space, let $S^a\Sigma$ ($a \geq 0$) denote its symmetric product $\Sigma^a/\mathfrak{S}_a$. For $a = 0$, by convention, $S^0\Sigma$ is a point. Then, by Macdonald formula [**17**],

$$\sum_{a\geq 0} \chi_{\text{top}}(S^a\Sigma)\, x^a = (1-x)^{-\chi(\Sigma)}. \tag{2.10}$$

Since $E_v \simeq \mathbb{P}^1$, and E_v° is the regular part of E_v, then $\chi_{\text{top}}(E_v^\circ) = 2 - \kappa_v$.

The first formula of $Z(\mathbf{t})$ [5, 7]. With the notation $\mathbf{x}^{\mathbf{a}} = x_1^{a_1}\cdots x_s^{a_s}$,

$$\begin{aligned} z(\mathbf{x}) &= \prod_v \sum_{a_v\geq 0} \chi_{\text{top}}(S^{a_v}E_v^\circ)\, x_v^{a_v} = \sum_{\mathbf{a}\geq 0}\prod_v \chi_{\text{top}}(S^{a_v}E_v^\circ)\mathbf{x}^{\mathbf{a}} \\ &= \sum_{\mathbf{a}\geq 0}\Big(\prod_v (-1)^{a_v}\binom{\kappa_v-2}{a_v}\Big)\,\mathbf{x}^{\mathbf{a}}, \end{aligned} \tag{2.11}$$

where, for *any* integer b, $\binom{b}{0} = 1$ and $\binom{b}{a} = b(b-1)\cdots(b-a+1)/a!$ for $a \in \mathbb{Z}_{>0}$ as usual.

The next interpretation of $\mathcal{Z}(\mathbf{t})$ is in terms of $J(l', I)$, cf. 2.3.

The second formula of $\mathcal{Z}(\mathbf{t})$ [26].

$$\mathcal{Z}(\mathbf{t}) = \sum_{l'\in\mathcal{S}'}\sum_{I\subset\mathcal{V}} (-1)^{|I|+1}\chi(l' + E_{J(l',I)})\mathbf{t}^{l'}. \tag{2.12}$$

REMARK 2.5. The above formula can be compared with

$$\mathcal{P}(\mathbf{t}) = \sum_{l'\in\mathcal{S}'}\sum_{I\subset\mathcal{V}} (-1)^{|I|+1}\Big(\ \chi(l' + E_{J(l',I)}) - h^1(\mathcal{O}_{\widetilde{X}}(-l' - E_{J(l',I)}))\Big)\mathbf{t}^{l'}.$$

This combined with (2.12) gives

$$\mathcal{Z}(\mathbf{t}) - \mathcal{P}(\mathbf{t}) = \sum_{l'\in\mathcal{S}'}\sum_{I\subset\mathcal{V}} (-1)^{|I|+1}\ h^1(\mathcal{O}_{\widetilde{X}}(-l' - E_{J(l',I)}))\ \mathbf{t}^{l'}.$$

The third formula of $\mathcal{Z}(\mathbf{t})$.

$$\mathcal{Z}(\mathbf{t}) = \sum_{l'\in\mathcal{S}'} \chi_{\text{top}}(\,\mathbb{P}(T(l')\setminus \cup_v T_v(l'))\,)\cdot \mathbf{t}^{l'}. \tag{2.13}$$

This follows from the combination of Corollary 2.2 and (2.12). Then, by 2.4 and Corollaries 2.1 and 2.13 one also has:

COROLLARY 2.3. $\mathcal{P}(\mathbf{t}) = \mathcal{Z}(\mathbf{t})$ *in the following cases:*

(a) (X, o) *is rational, and* π *is arbitrary resolution,*
(b) *or* (X, o) *is minimally elliptic singularity, and it satisfies the assumptions of 2.4(I).*

In fact, under the condition of Corollary 2.3, in [**26**, p. 280-281] it is proved that

$$\mathcal{P}(\mathbf{t}) = \mathbf{t}^0 + \sum_{l' \in \mathcal{S}' \setminus \{0\}} \sum_I (-1)^{|I|+1} \chi(l' + E_{J(l',I)}) \cdot \mathbf{t}^{l'}. \tag{2.14}$$

REMARK 2.6. Part *(b)* can be improved by adding some additional cases when $\mathcal{A}_{\mathrm{an}}(l') \neq \mathcal{A}_{\mathrm{top}}(l')$, but the Euler characteristics of the two arrangement complements agree. E.g., if (X,o) is minimally elliptic singularity whose minimal resolution is good, and if π is this minimal resolution, then $\mathcal{P}(\mathbf{t}) = \mathcal{Z}(\mathbf{t})$. In general, $\mathcal{P}(\mathbf{t}) = \mathcal{Z}(\mathbf{t})$ if and only if (X,o) is a splice quotient singularity [**28, 29**].

2.6. $\mathbb{P}(\mathbf{T}(\mathbf{l}') \setminus \cup_{\mathbf{v}} \mathbf{T}_{\mathbf{v}}(\mathbf{l}'))$ as a space of effective Cartier divisors [19, 31]. For any cycle $Z \in L$, $Z \geq E$, let $\mathrm{ECa}(Z)$ be the set of effective Cartier divisors on Z. Their supports are zero–dimensional in E. Taking the class of a Cartier divisor provides the *Abel map* $\mathrm{ECa}(Z) \to \mathrm{Pic}(Z)$. Let $\mathrm{ECa}^{l'}(Z)$ be the subset of $\mathrm{ECa}(Z)$, which consists of divisors whose associated line bundles have Chern class $l' \in L'$. Set $c^{l'} : \mathrm{ECa}^{l'}(Z) \to \mathrm{Pic}^{l'}(Z)$ for the restricted Abel map. Regarding the existence of $\mathrm{ECa}^{l'}(Z)$ as an algebraic variety we make the following comment. First, by a theorem of Artin [**1**, 3.8], there exists an affine algebraic variety Y and a point $y \in Y$ such that (Y, y) and (X, o) have isomorphic formal completions. Then, according to Hironaka [**9**], (Y, y) and (X, o) are analytically isomorphic. In particular, we can regard Z as a projective algebraic scheme, in which case $\mathrm{ECa}^{l'}(Z)$ together with the algebraic Abel map, as part of the general theory, was constructed by Grothendieck [**8**], see also the article of Kleiman [**11**]. In particular, $c^{l'} : \mathrm{ECa}^{l'}(Z) \to \mathrm{Pic}^{l'}(Z)$ is algebraic. For an explicit description and several properties see [**19–21**].

From definition, $\mathrm{ECa}^{l'}(Z) \neq \emptyset$ if and only if there exists $\mathcal{L} \in \mathrm{Pic}^{l'}(Z)$, which has a global section vanishing somewhere, but it has no fixed components. If this happens then $l' \neq 0$ and $H^0(\mathcal{L}|_{E_v}) \neq 0$ for any v, hence $-l' \in \mathcal{S}' \setminus \{0\}$. Conversely, if $-l' \in \mathcal{S}' \setminus \{0\}$ then one constructs elements of $\mathrm{ECa}^{l'}(Z)$ by some generic cuts of E enumerated by l'. Hence, it is natural to modify the definition, and redefine formally $\mathrm{ECa}^0(Z)$ as a point, the space of the 'empty divisor' $\{\emptyset\}$. It is sent by the Abel map to $\mathcal{O}_Z$. Hence, finally, $\mathrm{ECa}^{l'}(Z) \neq \emptyset$ if and only if $-l' \in \mathcal{S}'$.

In [**19**] is proved that for any $-l' \in \mathcal{S}'$, $\mathrm{ECa}^{l'}(Z)$ is irreducible, quasiprojective, smooth and of dimension $-(l', Z)$.

For any $\mathcal{L} \in \mathrm{Pic}^{l'}(Z)$ define $H^0(Z, \mathcal{L})_{\mathrm{reg}}$, the set of regular sections (or, section without fixed components) by $H^0(Z, \mathcal{L}) \setminus \cup_v H^0(Z - E_v, \mathcal{L}(-E_v))$. Then the preimage of $\mathcal{L}$ by the Abel map is $H^0(Z, \mathcal{L})_{\mathrm{reg}} / H^0(\mathcal{O}_Z^*)$ [**10**, §3].

Next, assume that $Z = E$ and $l' \in \mathcal{S}'$. In this case $h^1(\mathcal{O}_E) = 0$, $\mathrm{Pic}^{-l'}(E)$ consists of a point, say $\{\mathcal{O}_E(-l')\}$, and $H^0(\mathcal{O}_E^*) = \mathbb{C}^*$. Hence, by the above discussion,

$$\mathrm{ECa}^{-l'}(E) = H^0(\mathcal{O}_E(-l'))_{\mathrm{reg}} / \mathbb{C}^* = \mathbb{P}(T(l') \setminus \cup_v T_v(l')). \tag{2.15}$$

The forth formula of $\mathcal{Z}(\mathbf{t})$. If the link is a rational homology sphere then

$$\mathcal{Z}(\mathbf{t}) = \sum_{l' \in L'} \chi_{\mathrm{top}}(\mathrm{ECa}^{-l'}(E)) \cdot \mathbf{t}^{l'}.$$

2.7. The extension of $\mathcal{Z}(\mathbf{t})$ to the Grothendieck ring. The information contained in $\mathcal{Z}(\mathbf{t})$ can be improved if we modify the 'third formula' $\mathcal{Z}(\mathbf{t}) = \sum_{l'\in\mathcal{S}'}\chi_{\mathrm{top}}(\mathbb{P}(T(l')\setminus\cup_v T_v(l'))\mathbf{t}^{l'}$. Namely, we replace the topological Euler characteristic of $\mathbb{P}(T(l')\setminus\cup_v T_v(l'))$ with the class of this space in the Grothendieck ring of complex quasi–projective varieties.

$$\mathcal{Z}(\mathbb{L},\mathbf{t}) = \sum_{l'\in\mathcal{S}'} [\mathbb{P}(T(l')\setminus\cup_v T_v(l'))]\,\mathbf{t}^{l'}. \tag{2.16}$$

Let $\mathbb{L}$ be the class of the 1–dimensional affine space. Then, by inclusion–exclusion principle (as the analogue of 2.2) one has the following. If $\{V_\alpha\}_{\alpha\in\Lambda}$ is a finite family of linear subspaces of a finite dimensional linear space V, and for $I\subset\Lambda$ one writes $V_I := \cap_{\alpha\in I}V_\alpha$, then

$$[V\setminus\cup_\alpha V_\alpha] = \sum_I(-1)^{|I|}\mathbb{L}^{\dim(V_I)}, \quad [\mathbb{P}(V\setminus\cup_\alpha V_\alpha)] = \big(\sum_I(-1)^{|I|}\mathbb{L}^{\dim(V_I)}\big)/(\mathbb{L}-1).$$

Hence, using 2.1, (2.16) reads as

$$\begin{aligned}\mathcal{Z}(\mathbb{L},\mathbf{t}) &= \frac{1}{\mathbb{L}-1}\cdot\sum_{l'\in\mathcal{S}'}\sum_{I\subset\mathcal{V}}(-1)^{|I|}\,\mathbb{L}^{\chi(l'+E)-\chi(l'+E_{J(l',I)})}\,\mathbf{t}^{l'}\\ &= \sum_{l'\in\mathcal{S}'}\sum_{I\subset\mathcal{V}}(-1)^{|I|}\cdot\frac{\mathbb{L}^{\chi(l'+E)-\chi(l'+E_{J(l',I)})}-1}{\mathbb{L}-1}\,\mathbf{t}^{l'}.\end{aligned} \tag{2.17}$$

Note that $\lim_{\mathbb{L}\to 1}\mathcal{Z}(\mathbb{L},\mathbf{t}) = \mathcal{Z}(\mathbf{t})$. The analogue of the topological/combinatorial identity (2.12) is:

THEOREM 2.7. **[18, 29, 30]**

$$\mathcal{Z}(\mathbb{L},\mathbf{t}) = \frac{\prod_{(u,v)\in\mathcal{E}}(1-\mathbf{t}^{E^*_u}-\mathbf{t}^{E^*_v}+\mathbb{L}\mathbf{t}^{E^*_u+E^*_v})}{\prod_{v\in\mathcal{V}}(1-\mathbf{t}^{E^*_v})(1-\mathbb{L}\mathbf{t}^{E^*_v})}. \tag{2.18}$$

One defines similarly the analytic version as well: $\mathcal{P}(\mathbb{L},\mathbf{t}) = \sum_{l'\in\mathcal{S}'}[\mathbb{P}(A(l')\setminus\cup_v A_v(l'))]\,\mathbf{t}^{l'}$ **[6]**. (This will be improved in the next section, cf. Example 3.4.)

3. The extension of the series to cusp singularities

3.1. Notations and preliminaries regarding cusps. Assume that (X,o) is an arbitrary normal surface singularity with $b_1(M) = \dim H_1(M,\mathbb{Q})$ not necessarily zero. Then the long exact sequence of the pair $(\widetilde{X},\partial\widetilde{X})$ with $L = H_2(\widetilde{X},\mathbb{Z})$ and $L' = H_2(\widetilde{X},\partial\widetilde{X},\mathbb{Z})$ gives

$$0\to L\to L'\to H_1(M,\mathbb{Z})\to H_1(\widetilde{X},\mathbb{Z}) = \mathbb{Z}^{b_1(M)}\to 0,$$

where $b_1(M) = 2\sum_v g_v + c(\Gamma)$, $c(\Gamma)$ being the number of independent cycles in the dual graph Γ. Hence, in this case $H := L'/L$ is identified with the torsion part $\mathrm{Tors}H_1(M,\mathbb{Z})$.

The point is that in general there exists no canonical splitting of the exact sequence $0\to H\to H_1(M,\mathbb{Z})\to\mathbb{Z}^{b_1(M)}\to 0$. Different choices of splittings $H_1(M,\mathbb{Z})\to H$ composed with $\pi_1(M)\to H_1(M,\mathbb{Z})$ provide essentially different representations $\pi_1(M)\to H$, hence different H–coverings. In particular, via such representations the definition of the natural line bundles (following the method of 2.2) is not well–defined. In fact, in general, any other definition of the natural line bundles fails (either by this ambiguity, or by the fact that $\mathrm{Pic}^0(\widetilde{X}) = H^1(\widetilde{X}, \mathcal{O}_{\widetilde{X}})/H^1(\widetilde{X}, Z)$ is not torsion free).

On the other hand, all other combinatorial invariants, e.g. $\mathcal{S}'$, are defined similarly, see e.g. [**23**, **25**].

In the case of cusp singularities, $g_v = 0$ for all $v \in \mathcal{V}$, and $b_1(M) = c(\Gamma) = 1$. Furthermore, since $p_g = 1$ we also have $\mathrm{Pic}^0(\widetilde{X}) = \mathbb{C}/\mathbb{Z}$.

In the previous section, in the definition of the series $\mathcal{P}(\mathbf{t})$, we used the assumption that the link is rational homology sphere. This was really necessary, since the definition was based on the existence of the natural line bundles (defined via the universal abelian covering). The point is that usually the cohomological properties of a natural line bundle and of a line bundle with the same Chern class differ, hence the identification of the natural bundles is crucial. (On the other hand, if we wish to define only the H–equivariant part of $\mathcal{P}(\mathbf{t})$, namely $P_0(\mathbf{t}) := \sum_{l \in L} \mathfrak{p}(l)\mathbf{t}^l$, then we do not need any covering, and it can be defined as the Poincaré series of the divisorial filtration of $\mathcal{O}_{X,o}$ associated with the irreducible components of $\pi^{-1}(0) = E$. Nevertheless, in this way we loose essential part of the theory, namely all the geometry related with not integral Chern classes.)

Let us look also at the 'old' definition of $\mathcal{Z}(\mathbf{t})$ too, cf. 2.5. E.g., for the minimal resolution of cusps (when $\kappa_v = 2$ for all v), (2.9) gives $\mathcal{Z}(\mathbf{t}) \equiv 1$, an object which definitely carries no information.

Hence, in general, the possible extensions of the series $\mathcal{P}(\mathbf{t})$ and $\mathcal{Z}(\mathbf{t})$ are seriously obstructed.

However, in this section we explain that an extension can be done even if the link is not rational homology sphere, at least in the case of cusps. This might serve as a model for further generalizations (at least for the cases when all the exceptional curves are rational). In the case of cusps several analytic vanishing statements are present, facts which make the definition and results work.

3.2. The series $\mathcal{P}(\mathbf{t})$ for cusps. Let us assume that (X, o) is a cusp singularity, and we fix its minimal resolution $\widetilde{X}$. The definition of $\mathcal{P}(\mathbf{t})$ can be done in two different ways. The first one is a 'naive' one: one defines $\mathcal{P}(\mathbf{t})$ by the identity (2.5),

$$\mathfrak{p}(l') = \sum_{I \subset \mathcal{V}} (-1)^{|I|} \dim \frac{H^0(\widetilde{X}, \mathcal{O}_{\widetilde{X}}(-l' - E_I))}{H^0(\widetilde{X}, \mathcal{O}_{\widetilde{X}}(-l' - E))}. \tag{3.1}$$

once we clarify the meaning of $\mathcal{O}_{\widetilde{X}}(-l') \in \mathrm{Pic}(\widetilde{X})$ for any $l' \in \mathcal{S}'$.

Before we make the choice of $\mathcal{O}_{\widetilde{X}}(-l') \in \mathrm{Pic}(\widetilde{X})$ we make two remarks. Let us fix any line bundle $\mathcal{L} \in \mathrm{Pic}(\widetilde{X})$ with $c_1(\mathcal{L}) = l'$. Then, for any effective $l \in L$, from the cohomological exact sequence of $0 \to \mathcal{L}(-l) \to \mathcal{L} \to \mathcal{L}|_l \to 0$, we have

$$\dim H^0(\mathcal{L})/H^0(\mathcal{L}(-l)) - \chi(l) - (l, l') + h^1(\mathcal{L}(-l)) - h^1(\mathcal{L}) = 0.$$

On the other hand, by a Laufer type algorithm, for any $\mathcal{L} \in \mathrm{Pic}(\widetilde{X})$, there exists $l \in L_{\geq 0}$ such that $c_1(\mathcal{L} \otimes \mathcal{O}_{\widetilde{X}}(-l)) \in -\mathcal{S}'$ and $h^1(\mathcal{L}) - h^1(\mathcal{L} \otimes \mathcal{O}_{\widetilde{X}}(-l))$ is topological (see e.g. [**24**, Prop. 4.3.3]. Next, for $\mathcal{L} \in \mathrm{Pic}(\widetilde{X})$ with $c_1(\mathcal{L}) \in -\mathcal{S}'$ one has (cf. [**15**], [**23**, p. 333], [**36**, 1.7], [**31**])

$$h^1(\widetilde{X}, \mathcal{L}) = 0 \ \text{ unless } \mathcal{L} = \mathcal{O}_{\widetilde{X}}, \text{ when } h^1(\widetilde{X}, \mathcal{O}_{\widetilde{X}}) = 1. \tag{3.2}$$

In particular, in all our relevant cases in the computation of the $\mathfrak{p}(l')$ in (3.1) for $l' \in \mathcal{S}'$, the expression from the right hand side is independent on the choice of $\mathcal{O}_{\widetilde{X}}(-l') \in \mathrm{Pic}(\widetilde{X})$, basically it depends (topologically/combinatorially) only on l'.

The second definition identifies precisely the bundles $\mathcal{O}_{\widetilde{X}}(-l') \in \mathrm{Pic}(\widetilde{X})$, and even an H–covering, which replaces the universal abelian covering. Once the covering is fixed, the analogues of the natural line bundles are defined via an eigensheaf decomposition as in (2.1) (and the line after it), and all the analytic filtrations $\{\mathcal{F}(l')\}_{l'}$ can be defined as in (2.2), and all the basic statements of that subsection can be reproved. Again, here in the case of cusps, such a natural covering exists, it is called the 'discriminant covering' of the cusp. The point is that any two splittings $H_1(M,\mathbb{Z}) \to H$ can be identified by an automorphism of $\pi_1(M)$ **[39]**. See also **[33]** for the definition of this covering and several other properties of it. Summed up, in the special case of cusps, any splitting gives basically the same covering, on which we can rely.

As we already said, in the case of cusps, the cohomology of line bundles are topological (in the sense of (3.2)), hence $\mathcal{P}(\mathbf{t})$ also should be topological. Let us rewrite its coefficients in terms of χ. Let us fix some $l' \in \mathcal{S}'$ and write $\overline{I} := \mathcal{V} \setminus I$. Then from the exact sequence

$$0 \to \mathcal{O}_{\widetilde{X}}(-l'-E) \to \mathcal{O}_{\widetilde{X}}(-l'-E_I) \to \mathcal{O}_{E_{\overline{I}}}(-l'-E_I) \to 0, \tag{3.3}$$

and from the vanishing $h^1(\mathcal{O}_{\widetilde{X}}(-l'-E)) = 0$ (since $Z_K = E$ and $l' \in \mathcal{S}'$, hence Grauert–Riemenschneider vanishing holds), we get that

$$\mathfrak{p}(l') = \sum_I (-1)^{|I|} h^0(\mathcal{O}_{E_{\overline{I}}}(-l'-E_I)).$$

Let $J := J(l', I)$ be as in Lemma 2.3. Then $h^0(\widetilde{X}, \mathcal{O}_{\widetilde{X}}(-l'-E_I)) = h^0(\widetilde{X}, \mathcal{O}_{\widetilde{X}}(-l'-E_{J(l',I)}))$, hence again by (3.3) we get that one also has

$$\mathfrak{p}(l') = \sum_I (-1)^{|I|} h^0(\mathcal{O}_{E_{\overline{J(l',I)}}}(-l'-E_{J(l',I)})) = \sum_I (-1)^{|I|} h^0(\mathcal{O}_{E_{\overline{J}}}(-l'-E_J)).$$

We will separate the case $l'=0$. Note that $\mathfrak{p}(0)=1$ for *any* singularity. Here this can be seen as follows. For $I=\emptyset$ one has $\overline{J} = \mathcal{V}$ hence $h^0(\mathcal{O}_{E_{\overline{J}}}(-E_J)) = 1$. But for $I \neq \emptyset$, $h^0(\mathcal{O}_{E_{\overline{J}}}(-E_J)) = 0$ (since $J = \mathcal{V}$). Hence $\mathfrak{p}(0)=1$.

If $l' \neq 0$ then $h^1(\mathcal{O}_{E_{\overline{J}}}(-l'-E_J)) = 0$. This for $I=\emptyset$ reads as $h^1(\mathcal{O}_E(-l'))=0$. But by Grauert–Riemenschneider vanishing $h^1(\mathcal{O}_E(-l')) = h^1(\mathcal{O}_{\widetilde{X}}(-l'))$, which vanishes by (3.2).

For $I \neq \emptyset$ it follows from the fact that each component of $\overline{J}$ is a string and all the Chern degrees are ≥ 0. In particular, $h^0(\mathcal{O}_{E_{\overline{J}}}(-l'-E_J)) = \chi(\mathcal{O}_{E_{\overline{J}}}(-l'-E_J)) = \chi(l'+E) - \chi(l'+E_J)$. Hence

$$\mathcal{P}(\mathbf{t}) = \mathbf{t}^0 + \sum_{l' \in \mathcal{S}' \setminus \{0\}} \sum_I (-1)^{|I|+1} \chi(l'+E_{J(l',I)}) \cdot \mathbf{t}^{l'}. \tag{3.4}$$

This can be compared with the expression from (2.14), valid in the rational homology sphere link case for certain minimally elliptic singularities (with rational homology sphere link).

3.3. The series $\mathcal{Z}(\mathbf{t})$ for cusps. The topological linear subspace arrangement can be defined analogously as suggested by (2.13). Namely, one sets $T(l') := H^0(\mathcal{O}_E(-l'))$ and $T_v(l') := H^0(\mathcal{O}_{E-E_v}(-l'-E_v))$ for each $v \in \mathcal{V}$. Then one defines

$$\mathcal{Z}(\mathbf{t}) = \sum_{l' \in \mathcal{S}'} \chi_{\text{top}}(\,\mathbb{P}(T(l') \setminus \cup_v T_v(l'))\,) \cdot \mathbf{t}^{l'}.$$

Since we use the reduced structure of E, one can argue (similarly as in the rational homology sphere case) that $\mathcal{Z}(\mathbf{t})$ depends only on the combinatorics of the graph. (This will follow from the next lemma and its proof as well.)

LEMMA 3.1. *$\mathcal{Z}(\mathbf{t}) = \mathcal{P}(\mathbf{t})$, or,*

$$\mathcal{Z}(\mathbf{t}) = \mathbf{t}^0 + \sum_{l' \in \mathcal{S}' \setminus \{0\}} \sum_I (-1)^{|I|+1} \chi(l' + E_{J(l',I)}) \cdot \mathbf{t}^{l'}. \tag{3.5}$$

PROOF. One can proceed in two different ways. The first version is that one notice that $H^0(\mathcal{O}_{\widetilde{X}}(-l')) \to H^0(\mathcal{O}_E(-l'))$ is onto (since $h^1(\mathcal{O}_{\widetilde{X}}(-l'-E)) = 0$). Hence, analytic and the topological linear subspace arrangements are the same. The analytic one can be connected with $\mathcal{P}(\mathbf{t})$ similarly as in Corollary 2.1, hence $\mathcal{P} = \mathcal{Z}$ follows.

In a different way, one can analyse the validity of Proposition 2.1 as well. One sees that for $l' \in \mathcal{S}' \setminus \{0\}$ *(a)-(b)-(c)* are valid. *(d)* should be modified as follows: $\dim T(l') = h^0(\mathcal{O}_E(-l')) = \chi(\mathcal{O}_E(-l')) = -(E, l')$. Moreover, *(e)* remain valid as well. In particular, as in 2.2 one proves (3.5). $\square$

3.4. The computation of $\mathcal{P}(\mathbf{t}) = \mathcal{Z}(\mathbf{t})$. Assume first that $|\mathcal{V}| \geq 3$.

(a) Assume that $l' = E_v^*$ for some v. Then $J = \emptyset$ for $I = \emptyset$ and $J = \mathcal{V}$ otherwise. Hence, by (3.4) $\mathfrak{p}(l') = 1$. Similarly, if $l' = kE_v^*$ for some $v \in \mathcal{V}$ and $k \geq 2$ then $J = \emptyset$ for $I = \emptyset$, and $J = \mathcal{V}$ if $v \in I$, and $J = \mathcal{V} \setminus v$ if $I \neq \emptyset$ and $v \notin I$. Hence, again by (3.4) $\mathfrak{p}(l') = 1$.

(b) Assume that v and v' are adjacent vertices, and $l' = kE_v^* + k'E_{v'}^*$ with $k, k' > 0$. Let $w \neq v'$ adjacent vertex with v. Let $\mathcal{P}_v := \{I \subset \mathcal{V}, v \in I\}$. Then its elements can be put in pairs $(I, I \cup w)$ with $w \notin I$, and with $J(l', I) = J(l', I \cup w)$. Hence the contribution in (3.4) corresponding to the sum over $\mathcal{P}_v$ is zero. The same is true, by similar argument, for the subset $\mathcal{P}_{v,v'} := \{I \subset \mathcal{V}, v \notin I, v' \in I\}$. Hence we remain with subsets I with $I \cap \{v, v'\} = \emptyset$. For them $J = \emptyset$ if $I = \emptyset$ (with contribution $\chi(l')$), and $J = \mathcal{V} \setminus \{v, v'\}$ (with contribution $\chi(l') + 1$) else. Hence $\mathfrak{p}(l') = 1$ again.

(c) We claim that $\mathfrak{p}(l') = 0$ in all other cases. Write l' as $\sum_{v \in S} a_v E_v^*$ with all $a_v > 0$. Assume that the E^* support S of l' is not $\mathcal{V}$. Fix a maximal connected string Γ_1 in $\mathcal{V} \setminus S$. Let $v, v' \in S$ be the two adjacent vertices of Γ_1. Then, by excluding the cases already discussed, we know that $v \neq v'$ and $\Gamma_1 \cup \{v, v'\} \neq \mathcal{V}$. Then we compute $\mathfrak{p}(l')$ via (3.4). Similarly as in case (b), the contribution in the sum corresponding to the subset I with $I \cap \{v, v'\} \neq \emptyset$ is zero (choose the adjacent w in Γ_1). Hence, in the sequel we consider sets I with $I \cap \{v, v'\} = \emptyset$. For each such I, note that $\tilde{J} := J \setminus \{v, v'\}$ has the property that $\chi(l' + E_J) = \chi(l' + E_{\tilde{J}})$ (†). Indeed, if at some step of the algorithm from Lemma 2.3(2) we have $I_m = \tilde{J}$, and by the algorithm we have to add v (or v'), then $(E_v, l' + E_{I_m}) = 1$, hence (†) holds. Write I as $I_1 \cup I_2$ with $I_1 \subset \Gamma_1$, while $I_2 \cap \Gamma_1 = \emptyset$. Then $\tilde{J}$ also decomposes as $J_1 \cup J_2$ with similar properties. Note also that $(l' + E_{J_2}, E_{J_1}) = 0$

Hence $\sum_{I_1\cup I_2}(-1)^{|I_1|+|I_2|}\chi(l'+E_{\bar J})=\sum_{I_1\cup I_2}(-1)^{|I_1|+|I_2|}(\chi(l'+E_{J_2})+\chi(E_{J_1}))=\sum_{I_2}(-1)^{|I_2|}\chi(l'+E_{J_2})\sum_{I_1}(-1)^{|I_1|}+\sum_{I_1}(-1)^{|I_1|}\chi(E_{J_1})\sum_{I_2}(-1)^{|I_2|}=0$.

Finally assume that $S=\mathcal{V}$. Similarly as in (†) above, we have $\chi(l'+E_J)=\chi(l'+E_I)$, which equals $\chi(l'+E)-\chi(\mathcal{O}_{E_{\bar I}}(-l'-E_I))$. (This identity can be proved via $h^1(\mathcal{O}_{E_{\bar I}}(-l'-E_I))=0$ as well.) Now the vanishing $\sum_I(-1)^{|I|}\chi(\mathcal{O}_{E_{\bar I}}(-l'-E_I))=\sum_I(-1)^{|I|}(\chi(E_{\bar I})-(E_{\bar I},l'+E_I))=0$ follows from combinatorial arguments.

Hence we proved that whenever $|\mathcal{V}|\geq 3$ then the following holds.

THEOREM 3.2.

$$\mathcal{P}(\mathbf{t})=\mathcal{Z}(\mathbf{t})=1+\sum_{v\in\mathcal{V}}\frac{\mathbf{t}^{E^*_v}}{1-\mathbf{t}^{E^*_v}}+\sum_{(u,v)\in\mathcal{E}}\frac{\mathbf{t}^{E^*_v}}{1-\mathbf{t}^{E^*_v}}\cdot\frac{\mathbf{t}^{E^*_u}}{1-\mathbf{t}^{E^*_u}}.$$

By direct verification, the same formula holds for $\mathcal{V}=\{v,u\}$ as well. Namely

$$\mathcal{P}(\mathbf{t})=\mathcal{Z}(\mathbf{t})=1+\frac{\mathbf{t}^{E^*_v}}{1-\mathbf{t}^{E^*_v}}+\frac{\mathbf{t}^{E^*_u}}{1-\mathbf{t}^{E^*_u}}+2\cdot\frac{\mathbf{t}^{E^*_v}}{1-\mathbf{t}^{E^*_v}}\cdot\frac{\mathbf{t}^{E^*_u}}{1-\mathbf{t}^{E^*_u}}.$$

If $\mathcal{V}=\{v\}$ then

$$\mathcal{P}(t)=\mathcal{Z}(t)=1+\mathbf{t}^{E^*}/(1-\mathbf{t}^{E^*})^2.$$

3.5. The motivic series associated with $\{\mathrm{ECa}^{l'}(E)\}_{l'}$ (general (X,o)). Let us assume that $\widetilde{X}$ is a good resolution of a normal surface singularity (X,o) such that each E_v is rational, however, we allow cycles in the graph (hence in this subsection (X,o) is not necessarily a cusp). For any $l'\in\mathcal{S}'$ let $\mathrm{ECa}^{-l'}(E)$ be the space of effective Cartier divisors of E, cf. 2.6 or [**19**]. It is a quasiprojective variety, cf. [**8, 10, 11, 19**], non–empty if and only if $l'\in\mathcal{S}'$. (See also 2.6.) Let us define the generating function of their classes in the Grothendieck ring

$$\mathcal{Z}^{\mathrm{ECa}}(\mathbb{L},\mathbf{t}):=\sum_{l'\in\mathcal{S}'}[\mathrm{ECa}^{-l'}(E)]\cdot\mathbf{t}^{l'}. \tag{3.6}$$

THEOREM 3.3. *With the above notations*

$$\mathcal{Z}^{\mathrm{ECa}}(\mathbb{L},\mathbf{t})=\frac{\prod_{(u,v)\in\mathcal{E}}(1-\mathbf{t}^{E^*_u}-\mathbf{t}^{E^*_v}+\mathbb{L}\mathbf{t}^{E^*_u+E^*_v})}{\prod_{v\in\mathcal{V}}(1-\mathbf{t}^{E^*_v})(1-\mathbb{L}\mathbf{t}^{E^*_v})}.$$

PROOF. Write $l'=\sum_v a_vE^*_v$ with $a_v\in\mathbb{Z}_{\geq 0}$. Recall that effective Cartier divisors are local nonzero sections of the sheaf $\mathcal{O}_E$ up to local invertible elements of $\mathcal{O}^*_E$. At any intersection point $p\in E_v\cap E_u$, $(v,u)=e\in\mathcal{E}$, with local coordinates $(x,y)\in U$, $\{x=0\}=E_v\cap U$, $\{y=0\}=E_u\cap U$, an effective Cartier divisor supported at p has the form $D_{k_v,k_u}=c^e_vx^{k^e_v}+c^e_uy^{k^e_u}$ (in $\mathbb{C}\{x,y\}/(xy)$, up to an invertible element), where c^e_v, $c^e_u\in\mathbb{C}^*$. It is nonempty if and only if $k^e_v\geq 1$ and $k^e_u\geq 1$. $\mathrm{ECa}^{-l'}(E)$ has a natural stratification according to the support of the divisors. The sum of the degrees of the divisors with support on E_v should be a_v, out of this, say a°_v, is provided by those supported on E°_v, and the others by divisors supported on intersection points of type $E_v\cap E_u$. Such a divisor contribute in the degree with k^e_u. Hence $a_v=a^\circ_v+\sum_{(v,u)=e\in\mathcal{E}}k^e_u$ $(\dagger_v)$. Hence, with fixed $\{a_v\}_v$, we consider the stratification of $\mathrm{ECa}^{l'}(E)$ according to the system $S(l'):=\{\{a^\circ_v\}_v,(k^e_v,k^e_u)_{(v,u)\in\mathcal{E}}\}$ satisfying $(\dagger_v)$ for any v and $k^e_u,k^e_v\geq 1$ for any such edge e which has a contribution. For $s=\{\{a^\circ_v\}_v,(k^e_v,k^e_u)_{(v,u)\in\mathcal{E}_s}\}\in S(l')$ (where $\mathcal{E}_s$

is the set of edges which contributes in this stratum) the corresponding stratum, $\mathrm{ECa}^{l'}_s(E)$, satisfies

$$[\mathrm{ECa}^{l'}_s(E)] = (\mathbb{L}-1)^{|\mathcal{E}_s|}\cdot \prod_v [S^{a_v^\circ}E_v^\circ].$$

First we compute the class $[S^{a_v^\circ}E_v^\circ]$ for any v. Let E_v^c be a set consisting of κ_v-1 distinct point, and x a formal variable. Then $\sum_{i\geq 0}x^i[S^iE_v^\circ]\cdot\sum_{i\geq 0}x^i[S^iE_v^c] = \sum_{i\geq 0}x^i[S^i\mathbb{C}] = \sum_{i\geq 0}x^i\mathbb{L}^i$. Hence

$$\sum_{i\geq 0}x^i[S^iE_v^\circ] = (1-x)^{\kappa_v-1}/(1-\mathbb{L}x).$$

Next, let S be the set of all the systems when we vary l', that is, $S=\cup_{l'\in\mathcal{S}'}S(l')$. Then, in S, $\{a_v\}_v$ is not fixed anymore, and the integers $a_v^\circ\geq 0$, $k_v^e, k_u^e\geq 1$ run independently. Therefore, with the usual substitution $x_v=\mathbf{t}^{E_v^*}$,

$$\begin{aligned}
Z^{\mathrm{ECa}}(\mathbb{L},\mathbf{x}) &= \sum_{l'}\sum_{s\in S(l')}\Big((\mathbb{L}-1)^{|\mathcal{E}_s|}\cdot\prod_v[S^{a_v^\circ}E_v^\circ]\cdot\prod_v x_v^{a_v^\circ+\sum_{(v,u)=e\in\mathcal{E}_s}k_u^e}\Big)\\
&= \sum_{s\in S}\Big(\prod_v x_v^{a_v^\circ}[S^{a_v^\circ}E_v^\circ]\cdot(\mathbb{L}-1)^{|\mathcal{E}_s|}\cdot\prod_v x_v^{\sum_{(v,u)=e\in\mathcal{E}_s}k_u^e}\Big)\\
&= \prod_v\frac{(1-x_v)^{\kappa_v-1}}{1-\mathbb{L}x_v}\cdot\sum_{\mathcal{E}'\subset\mathcal{E}}\ \prod_{(v,u)=e\in\mathcal{E}'}(\mathbb{L}-1)\cdot\frac{x_v}{1-x_v}\cdot\frac{x_u}{1-x_u}\\
&= \prod_v\frac{(1-x_v)^{\kappa_v-1}}{1-\mathbb{L}x_v}\cdot\prod_{(v,u)=e\in\mathcal{E}}\Big(1+(\mathbb{L}-1)\cdot\frac{x_v}{1-x_v}\cdot\frac{x_u}{1-x_u}\Big)\\
&= \prod_v\frac{(1-x_v)^{\kappa_v-1}}{1-\mathbb{L}x_v}\cdot\prod_{(v,u)=e\in\mathcal{E}}\frac{1-x_v-x_u+\mathbb{L}x_vx_u}{(1-x_v)(1-x_u)},
\end{aligned}$$

which is equivalent with the needed expression. □

EXAMPLE 3.4. If the graph is a tree then $\mathrm{ECa}^{-l'}(E)=\mathbb{P}(T(l')\setminus\cup_vT_v(l'))$ and $\mathcal{Z}^{\mathrm{ECa}}(\mathbb{L},\mathbf{t})=\mathcal{Z}(\mathbb{L},\mathbf{t})$, hence we recover Theorem 2.7.

REMARK 3.5. Since $\mathcal{Z}_K-E=\sum_v(\kappa_v-2)E_v^*$ we have $\mathbf{t}^{\mathcal{Z}_K-E}\cdot\mathcal{Z}(\frac{1}{\mathbf{t}})=\mathcal{Z}(\mathbf{t})$. Similarly, the motivic expression satisfies the functional equation

$$\mathbb{L}^{c(\Gamma)-1}\cdot\mathbf{t}^{\mathcal{Z}_K-E}\cdot\mathcal{Z}(\mathbb{L},\mathbf{t})\big|_{t_v\mapsto(\mathbb{L}t_v)^{-1}}=\mathcal{Z}(\mathbb{L},\mathbf{t}).$$

3.6. The motivic $\mathcal{P}(\mathbb{L},\mathbf{t})$ and $\mathcal{Z}(\mathbb{L},\mathbf{t})$ for a cusp singularity. Let us return to a cusp singularity (X,o). Recall that the vanishing $h^1(\mathcal{O}_{\widetilde{X}}(-l'-E))=0$ guarantees that the analytic and topological arrangements are the same. In particular, $\mathcal{P}(\mathbb{L},\mathbf{t})=\mathcal{Z}(\mathbb{L},\mathbf{t})$. On the other hand, for a cusp $\mathcal{Z}^{\mathrm{ECa}}(\mathbb{L}=1,\mathbf{t})=\prod_v(1-\mathbf{t}^{E_v^*})^{\kappa_v-2}\equiv 1$. However, this '1' is not the right substitute for $Z(\mathbf{t})$!

Indeed, for any $l'\in\mathcal{S}'$ we have to consider the Abel map $c^{-l'}:\mathrm{ECa}^{-l'}(E)\to\mathrm{Pic}^{-l'}(E)$, and for any $\mathcal{L}\in\mathrm{Pic}^{-l'}(E)$ the class $[\mathbb{P}(H^0(E,\mathcal{L})_{\mathrm{reg}})]=[(c^{-l'})^{-1}(\mathcal{L})]$ instead of the whole $[\mathrm{ECa}^{-l'}(E)]$. Let us denote this dependence by $\mathcal{L}=\mathcal{O}_E(-l')$ (though it can be an arbitrary choice when $l'\neq 0$ and $\mathcal{O}_E(-l')=\mathcal{O}_E$ for $l'=0$). Hence, we define

$$\mathcal{Z}(\mathbb{L},\mathbf{t}):=\sum_{l'\in\mathcal{S}'}[\mathbb{P}(H^0(E,\mathcal{O}_E(-l'))_{\mathrm{reg}})]\cdot\mathbf{t}^{l'}=\sum_{l'\in\mathcal{S}'}[\mathbb{P}(T(l')\setminus\cup_vT_v(l'))]\cdot\mathbf{t}^{l'}.$$

The second identity follows similarly as in (2.15).

The substitution $\mathbb{L} = 1$ replaces the element $[\mathbb{P}(H^0(E, \mathcal{O}_E(-l'))_{\mathrm{reg}})]$ from the Grothendieck ring by $\chi_{top}(\mathbb{P}(H^0(E, \mathcal{O}_E(-l'))_{\mathrm{reg}}))$, hence $\mathcal{Z}(\mathbb{L} = 1, \mathbf{t})$ is exactly $\mathcal{Z}(\mathbf{t})$ considered in 3.3.

THEOREM 3.6. $\mathcal{Z}(\mathbb{L}, 0) = 1$. *Furthermore, for* $l' \in \mathcal{S} \setminus \{0\}$ *one has*

$$[\mathbb{P}(H^0(E, \mathcal{O}_E(-l'))_{\mathrm{reg}})] = [\mathrm{ECa}^{-l'}(E)]/(\mathbb{L} - 1).$$

PROOF. First we claim that $\mathrm{Pic}^0(E)$, regarded as the kernel of $c_1 : H^1(\mathcal{O}_E^*) \to L'$ in $H^1(\mathcal{O}_E^*)$, is the multiplicative group $\mathbb{C}^*$. Indeed, the exponential map $exp(2\pi i \cdot) : \mathcal{O}_E \mapsto \mathcal{O}_E^*$ induces an isomorphism $H^1(\mathcal{O}_E)/H^1(E, \mathbb{Z}) = \mathbb{C}/\mathbb{Z} \overset{exp(2\pi i \cdot)}{\longrightarrow} \mathbb{C}^*$.

Note that if $f : X \to Y$ is an algebraic locally trivial fibration with fiber F, then $[X] = [Y][F]$. In the proof, instead of the algebraic locally triviality of the Abel map we will prove the locally triviality of its restrictions on the strata of $\mathrm{ECa}^{-l'}(E)$ consider in the proof of Theorem 3.3.

First, consider the Abel map $c^{-E_v^*} : \mathrm{ECa}^{-E_v^*}(E) \to \mathrm{Pic}^{-E_v^*}(E)$ for a fixed $v \in \mathcal{V}$ (for more details see also [**19**]). The source consists of divisors of degree one supported on E_v°, hence it can be identified with E_v°. The Abel map can be described by Laufer integration. Recall that $p_g = 1$ and (X, o) is Gorenstein. Then we fix the Gorenstein differential form in $\widetilde{X} \setminus E$ with pole E, and we apply the Laufer integration procedure [**13**] [**15**, p. 1281], or [**19**, 7.1]. We obtain that $c^{-E_v^*} : E_v^\circ \to \mathbb{C}^*$ is a tautological bijection.

Next, consider for any $k \geq 1$ the Abel map $c^{-kE_v^*} : \mathrm{ECa}^{-kE_v^*}(E) \to \mathrm{Pic}^{-kE_v^*}(E)$. One has a multiplicative structure $\prod_{i=1}^k \mathrm{ECa}^{-E_v^*}(E) \to \mathrm{ECa}^{-kE_v^*}(E)$ given by union (sum) of divisors, while $\prod_{i=1}^k \mathrm{Pic}^{-E_v^*}(E) \to \mathrm{Pic}^{-kE_v^*}(E)$ given by the tensor product of line bundles, that is, multiplication in $\mathbb{C}^*$. These operators commute with the Abel maps. Hence $c^{-kE_v^*} : S^k E_v^\circ = S^k\mathbb{C}^* \to \mathbb{C}^*$ is given by $c^{-kE_v^*}(\sum_{i=1}^k p_i) = \prod_i c^{-E_v^*}(p_i)$ (product in $\mathbb{C}^*$). This is surjective and it is an algebraic fibration over $\mathbb{C}^*$.

Finally, consider a stratum corresponding to $e = (v, u) \in \mathcal{E}$ consisting of divisors of type D_{k_v,k_u}. As a space, it is isomorphic to $\mathbb{C}^*$. We claim that the Abel map restricted to it is an isomorphism. Indeed, if we blow up $p = E_u \cap E_v$ several times conveniently at its infinitesimal close points, this strata can be identified with the divisors on some newly created exceptional divisor $E_{new}^\circ \simeq \mathbb{C}^*$ given by $x^{d_v} + a = 0$, where $a \in \mathbb{C}^*$ is the parameter of the stratum, and $d_v = \gcd\{k_v, k_u\}$. By the discussion from the previous case, applied for $S^{d_v}E_{new}^\circ \to \mathbb{C}^*$, we get that the subset given by $\{x^{d_v} + a = 0\}_{a\in\mathbb{C}^*}$ maps isomorphically to $\mathbb{C}^*$.

Since the stratification is algebraic, and the Abel map is also algebraic, we are done. $\square$

COROLLARY 3.1. *For a cusp singularity* $\mathcal{Z}(\mathbb{L}, \mathbf{t}) = \mathcal{P}(\mathbb{L}, \mathbf{t})$ *is given by*

$$\mathcal{Z}(\mathbb{L}, \mathbf{t}) = 1 + \frac{\mathcal{Z}^{\mathrm{ECa}}(\mathbb{L}, \mathbf{t}) - 1}{\mathbb{L} - 1},$$

where $\mathcal{Z}^{\mathrm{ECa}}(\mathbb{L}, \mathbf{t})$ *is given in Theorem* 3.3.

REMARK 3.7. We invite the reader to verify that

$$\left(1+\frac{\mathcal{Z}^{\mathrm{ECa}}(\mathbb{L},\mathbf{t})-1}{\mathbb{L}-1}\right)\Big|_{\mathbb{L}=1}$$

provides exactly the expressions from 3.2 for $\mathcal{Z}(\mathbf{t})$ determined by a direct verification. (For this, one has to determine the first terms of the Taylor expansion of $\mathbb{L}\mapsto\mathcal{Z}^{\mathrm{ECA}}$ at $\mathbb{L}=1$.) In this way we provide a second independent alternative proof for the expression from Theorem 3.2.

EXAMPLE 3.8. For a cusp with one vertex one has $\mathcal{Z}(\mathbb{L},\mathbf{t})=1+\sum_{k\geq 1}[\mathbb{P}^{k-1}]\cdot t^{kE^*}$.

REMARK 3.9. The expression from the right hand side of Theorem 3.2 is rather surprising: the series

$$z(\mathbf{x})=1+\sum_{v\in\mathcal{V}}\frac{x_v}{1-x_v}+\sum_{(u,v)\in\mathcal{E}}\frac{x_v}{1-x_v}\cdot\frac{x_u}{1-x_u}\in\mathbb{Z}[[\mathbf{x}]]$$

appears in a natural way in a rather different context in the literature. Indeed, identify the minimal good resolution graph Γ with a simplicial complex with vertex set $\mathcal{V}$ and having only 1–simplices, namely the edges. Then the Hilbert series of the graded Stanley–Reisner ring associated with Γ is exactly $z(\mathbf{x})$ [**2**, **16**]. This might provide some new starting point in the direction of the explicit determination of the equation of the local ring of (X,o) as well (compare e.g. with [**37**]).

References

[1] M. Artin, *Algebraic approximation of structures over complete local rings*, Inst. Hautes Études Sci. Publ. Math. **36** (1969), 23–58. MR268188

[2] Winfried Bruns and Jürgen Herzog, *Cohen-Macaulay rings*, Cambridge Studies in Advanced Mathematics, vol. 39, Cambridge University Press, Cambridge, 1993. MR1251956

[3] Steven Dale Cutkosky, Jürgen Herzog, and Ana Reguera, *Poincaré series of resolutions of surface singularities*, Trans. Amer. Math. Soc. **356** (2004), no. 5, 1833–1874, DOI 10.1090/S0002-9947-03-03346-4. MR2031043

[4] S. M. Guseĭn-Zade, F. Del′gado, and A. Kampil′o, *On the monodromy of a plane curve singularity and the Poincaré series of its ring of functions* (Russian), Funktsional. Anal. i Prilozhen. **33** (1999), no. 1, 66–68, DOI 10.1007/BF02465144; English transl., Funct. Anal. Appl. **33** (1999), no. 1, 56–57. MR1711890

[5] A. Campillo, F. Delgado, and S. M. Gusein-Zade, *Poincaré series of a rational surface singularity*, Invent. Math. **155** (2004), no. 1, 41–53, DOI 10.1007/s00222-003-0310-y. MR2025300

[6] A. Campillo, F. Delgado, and S. M. Gusein-Zade, *Multi-index filtrations and generalized Poincaré series*, Monatsh. Math. **150** (2007), no. 3, 193–209, DOI 10.1007/s00605-006-0437-1. MR2308548

[7] S. M. Guseĭn-Zade, F. Del′gado, and A. Kampil′o, *Universal abelian covers of rational surface singularities, and multi-index filtrations* (Russian, with Russian summary), Funktsional. Anal. i Prilozhen. **42** (2008), no. 2, 3–10, 95, DOI 10.1007/s10688-008-0013-7; English transl., Funct. Anal. Appl. **42** (2008), no. 2, 83–88. MR2438013

[8] Alexander Grothendieck, *Fondements de la géométrie algébrique. [Extraits du Séminaire Bourbaki, 1957–1962.]* (French), Secrétariat mathématique, Paris, 1962. MR0146040

[9] Heisuke Hironaka, *On the equivalence of singularities. I*, Arithmetical Algebraic Geometry (Proc. Conf. Purdue Univ., 1963), Harper & Row, New York, 1965, pp. 153–200. MR0201433

[10] Kleiman, St. L.: The Picard scheme, in 'Fundamental Algebraic Geometry: Grothendieck's FGA Explained', Mathematical Surveys and Monographs Volume: 123; 2005, 248–333.

[11] Steven L. Kleiman, *The Picard scheme*, Alexandre Grothendieck: a mathematical portrait, Int. Press, Somerville, MA, 2014, pp. 35–74. MR3287693

[12] Henry B. Laufer, *Normal two-dimensional singularities*, Annals of Mathematics Studies, No. 71, Princeton University Press, Princeton, N.J.; University of Tokyo Press, Tokyo, 1971. MR0320365

[13] Henry B. Laufer, *On rational singularities*, Amer. J. Math. **94** (1972), 597–608, DOI 10.2307/2374639. MR330500

[14] Henry B. Laufer, *Taut two-dimensional singularities*, Math. Ann. **205** (1973), 131–164, DOI 10.1007/BF01350842. MR333238

[15] Henry B. Laufer, *On minimally elliptic singularities*, Amer. J. Math. **99** (1977), no. 6, 1257–1295, DOI 10.2307/2374025. MR568898

[16] Ezra Miller and Bernd Sturmfels, *Combinatorial commutative algebra*, Graduate Texts in Mathematics, vol. 227, Springer-Verlag, New York, 2005. MR2110098

[17] I. G. Macdonald, *The Poincaré polynomial of a symmetric product*, Proc. Cambridge Philos. Soc. **58** (1962), 563–568. MR143204

[18] J. Nagy, BSc Thesis, ELTE, Budapest, 2016 (in Hungarian).

[19] János Nagy and András Némethi, *The Abel map for surface singularities I: generalities and examples*, Math. Ann. **375** (2019), no. 3-4, 1427–1487, DOI 10.1007/s00208-019-01873-w. MR4023382

[20] János Nagy and András Némethi, *The Abel map for surface singularities II. Generic analytic structure*, Adv. Math. **371** (2020), 107268, 38, DOI 10.1016/j.aim.2020.107268. MR4118773

[21] Nagy, J. and Némethi, A.: The Abel map for surface singularities III. Elliptic germs. to appear in *Math. Zeitschrift*; DOI 10.1007/s00209-021-02830-7.

[22] János Nagy and András Némethi, *On the topology of elliptic singularities*, Pure Appl. Math. Q. **16** (2020), no. 4, 1123–1146, DOI 10.4310/PAMQ.2020.v16.n4.a11. MR4180244

[23] A. Némethi, *Five lectures on normal surface singularities*, Low dimensional topology (Eger, 1996/Budapest, 1998), Bolyai Soc. Math. Stud., vol. 8, János Bolyai Math. Soc., Budapest, 1999, pp. 269–351. With the assistance of Ágnes Szilárd and Sándor Kovács. MR1747271

[24] András Némethi, *Graded roots and singularities*, Singularities in geometry and topology, World Sci. Publ., Hackensack, NJ, 2007, pp. 394–463, DOI 10.1142/9789812706812_0013. MR2311495

[25] András Némethi, *"Weakly" elliptic Gorenstein singularities of surfaces*, Invent. Math. **137** (1999), no. 1, 145–167, DOI 10.1007/s002220050327. MR1703331

[26] András Némethi, *Poincaré series associated with surface singularities*, Singularities I, Contemp. Math., vol. 474, Amer. Math. Soc., Providence, RI, 2008, pp. 271–297, DOI 10.1090/conm/474/09260. MR2454352

[27] András Némethi, *The Seiberg-Witten invariants of negative definite plumbed 3-manifolds*, J. Eur. Math. Soc. (JEMS) **13** (2011), no. 4, 959–974, DOI 10.4171/JEMS/272. MR2800481

[28] András Némethi, *The cohomology of line bundles of splice-quotient singularities*, Adv. Math. **229** (2012), no. 4, 2503–2524, DOI 10.1016/j.aim.2012.01.003. MR2880230

[29] András Némethi, *Pairs of invariants of surface singularities*, Proceedings of the International Congress of Mathematicians—Rio de Janeiro 2018. Vol. II. Invited lectures, World Sci. Publ., Hackensack, NJ, 2018, pp. 749–779. MR3966788

[30] A. Némethi, *Linear subspace arrangements associated with normal surface singularities*, Journal of Singularities, **18** (2018), 464–476; volume dedicated to E. Brieskorn.

[31] A. Némethi, *Normal surface singularities*, book in preparation, to be published in 2022 by Ergebnisse der Mathematik und ihrer Grenzgebiete. 3. Folge, Springer.

[32] Walter D. Neumann and Jonathan Wahl, *Universal abelian covers of surface singularities*, Trends in singularities, Trends Math., Birkhäuser, Basel, 2002, pp. 181–190. MR1900786

[33] Walter D. Neumann and Jonathan Wahl, *Universal abelian covers of quotient-cusps*, Math. Ann. **326** (2003), no. 1, 75–93, DOI 10.1007/s00208-002-0405-6. MR1981612

[34] Walter D. Neumann and Jonathan Wahl, *Complete intersection singularities of splice type as universal abelian covers*, Geom. Topol. **9** (2005), 699–755, DOI 10.2140/gt.2005.9.699. MR2140991

[35] Tomohiro Okuma, *The geometric genus of splice-quotient singularities*, Trans. Amer. Math. Soc. **360** (2008), no. 12, 6643–6659, DOI 10.1090/S0002-9947-08-04559-5. MR2434304

[36] A. Röhr, *A vanishing theorem for line bundles on resolutions of surface singularities*, Abh. Math. Sem. Univ. Hamburg **65** (1995), 215–223, DOI 10.1007/BF02953328. MR1359130

[37] Jan Stevens, *Degenerations of elliptic curves and equations for cusp singularities*, Math. Ann. **311** (1998), no. 2, 199–222, DOI 10.1007/s002080050184. MR1625766

[38] Philip Wagreich, *Elliptic singularities of surfaces*, Amer. J. Math. **92** (1970), 419–454, DOI 10.2307/2373333. MR291170

[39] C. T. C. Wall, *Real forms of cusp singularities. II*, Math. Proc. Cambridge Philos. Soc. **102** (1987), no. 2, 193–201, DOI 10.1017/S0305004100067220. MR898140

Alfréd Rényi Institute of Mathematics, Reáltanoda utca 13-15, H-1053, Budapest, Hungary

Email address: nagy.janos@renyi.hu

Alfréd Rényi Institute of Mathematics, Reáltanoda utca 13-15, H-1053, Budapest, Hungary; ELTE - University of Budapest, Dept. of Geometry, Budapest, Hungary; and BCAM - Basque Center for Applied Math., Mazarredo, 14 E48009 Bilbao, Basque Country, Spain

Email address: nemethi.andras@renyi.hu

Contemporary Mathematics
Volume **778**, 2022
https://doi.org/10.1090/conm/778/15662

Non-Archimedean electrostatics

Christopher D. Sinclair

ABSTRACT. We introduce ensembles of repelling charged particles restricted to a ball in a non-archimedean field (such as the p-adic rational numbers) with interaction energy between pairs of particles proportional to the logarithm of the (p-adic) distance between them. In the *canonical ensemble*, a system of N particles is put in contact with a heat bath at fixed inverse temperature β and energy is allowed to flow between the system and the heat bath. Using standard axioms of statistical physics, the relative density of states is given by the β power of the (p-adic) absolute value of the Vandermonde determinant in the locations of the particles. The partition function is the normalizing constant (as a function of β) of this ensemble, and we identify a recursion that allows this to be computed explicitly in finite time. Probabilities of interest, including the probabilities that fixed subsets will have a prescribed number of particles, and the conditional distribution of particles within a subset given a prescribed occupation number, are given explicitly in terms of the partition function. We then turn to the *grand canonical ensemble* where both the energy and number of particles are variable. We compute similar probabilities to those in the canonical ensemble and show how these probabilities can be given in terms the canonical and grand canonical partition functions. Finally, we briefly consider the multi-component ensemble where particles are allowed to take different integer charges, and we connect basic properties of this ensemble to the canonical and grand canonical ensembles.

1. Introduction

This paper lies at the intersection of number theory, probability and mathematical statistical physics.

We consider a collection of charged particles confined to a compact region of a complete non-archimedean field (*e.g.* $\mathbb{Q}_p$) at a fixed temperature. We may think of this as a non-archimedean (p-adic) *plasma*, and since the particles have identical charges, they have a tendency to repel. What might we want to know about this plasma? For starters we might want to know how many particles there are. This actually introduces two models: the *canonical ensemble* with a fixed number of particles, and the *grand canonical ensemble* where the number of particles is variable (but in a specific way suggested by physical laws). In either of these settings we might want to know the probability of finding a specified number of particles in a specified subregion of our domain. More specifically: Given a disjoint union of subregions, and an occupation number for each of these regions, what is

2020 *Mathematics Subject Classification.* Primary 60B20, 60G55, 82B23, 11C08, 11R42, 11R04.

the probability of each set having the specified number of particles? A specific, but particularly salient example follows when we ask for the probability that a given subregion contains no particles. This latter probability is called a *gap probability* and for now we focus on this quantity as a proxy for more nuanced statistical information about counts of particles.

What might be a reasonable answer? Certainly a formula for this probability would be ideal. Moreover, if the subregion can be described using a finite amount of data, the ideal formula would require only a finite number of maneuvers to calculate this probability. Another possible solution would be to describe a recurrence for gap probabilities that terminate in a finite number of steps, given a finitely-described subregion. We will be aiming for the latter, and in the canonical ensemble, we will provide such recursions for gap probabilities and other common statistical quantities (like the free energy, partition function, etc).

Some of the results here already appear in the literature. Non-archimedean Coulomb gases appear first in [**23**] and [**24**] as well as more recently in [**22**] (the latter gives a massive generalization of Theorem 4.5). From the number theoretic veiwpoint, it is as natural to study p-adic electrostatics as ordinary (real or complex) electrostatics under the arithmetic ethos that all completions of $\mathbb{Q}$ be investigated equally. There are also purely physical reasons for studying p-adic electrostatics (and p-adic physics more generally). The physical fact that the topology of space time at the finest (Planck) scale is not that of a continuum has lead to conjectures of a non-archimedean space-time at the Planck scale [**20, 21**].

The author's of [**5**] investigate the probability that a polynomial with p-adic coefficients splits completely (has all roots) in the p-adic integers. The roots of such polynomials behave like our p-adic electrons at a very specific temperature. One of their main results gives a functional equation for the generating function (over degree of polynomials) of these probabilities. Here we report a similar functional equation for the grand canonical partition function, generalized to all temperatures (and with a new, different proof).

In another direction, Igusa studies local zeta functions [**13**]of which our canonical partition function (as a function of temperature) is a very specific examples. The study of Igusa local zeta functions is deep and connects many different areas of mathematics and physics. The proceedings in which this paper appear are a testament to the importance and wealth of applications of Igusa's work.

Alternate titles for this paper include "The p-adic Selberg Integral" or "p-adic Random Matrix Theory" due to the appearance of a non-archimedean version of the Selberg Integral which appears as the partition function of the canonical ensemble. The Selberg Integral is an important special function [**18**] and the fact that our partition function is a p-adic analog is reason enough to study it. See [**11**] for a more comprehensive look at the importnce of the Selberg integral. Amazingly, the p-adic Selberg Integral appears in string theory as the partition function of the (non-archimedean) Koba-Nielsen string open amplitude, providing yet another deep connection between number theory and mathematical physics [**2–4, 14, 15**].

The connection to random matrix theory is (currently) more tentative, since there are no random matrices introduced in this paper. However, in Hermitian random matrix theory, a Selberg-like integral appears as the normalization constant for certain ensembles of matrices [**16**]. In some instances, determining a closed-form for those Selberg-like integrals leads to the solvability of the related ensemble of random matrices. The results we present here suggest that if this analogy holds for p-adic random matrices, then those ensembles are solvable in the sense that we can determine probabilities of interest about the locations and behaviors of the eigenvalues using the techniques outlined here.

The connection between random matrix theory and one and two-dimensional electrostatics is well-known, and indeed understanding the electrostatics provides insight into the eigenvalues of random matrices. This perspective was introduced by Dyson in a series of papers [**7**] and explored in detail in the archimedean (real and complex) setting by Forrester in [**10**].

The current work is also connected to potential theory on non-archimedean spaces (See [**1**] and the reference therein). Expressions similar to the potential energy of our particle system appear in that domain, where much work is done to investigate low-energy configurations especially as connected to problems in number theory [**8, 9**]. For low temperatures we expect our particles to 'jostle' around these low-energy configurations, and it would be worthwhile to explore the implications of our results to fluctuations about the ground state as it arises in potential theory.

For a broader survey of p-adic mathematical physics, see [**6**].

We provide a brief introduction to both non-archimedean fields and statistical physics as we go. A more complete (and well-written) introduction to p-adic numbers can be found here [**12**]. Likewise an approachable introduction to statistical physics can be found in David Tong's lecture notes on the subject [**19**].

2. Non-Archimedean fields

2.1. Absolute values. An absolute value $|\cdot|$ on $\mathbb{Q}$ satisfies the axioms:

(1) $|x| \in [0, \infty)$ with $|x| = 0$ iff $x = 0$;
(2) $|xy| = |x||y|$;
(3) $|x + y| \leq |x| + |y|$.

The absolute value with $|0| = 0$ and $|x| = 1$ for all other x is called the *trivial* absolute value and we will exclude it from all consideration. The usual absolute value $|x|_\infty := \operatorname{sign}(x)x$ is, of course, an absolute value. Given a prime integer p, we may factor $x \in \mathbb{Q}$ as

$$x = p^b \frac{m}{n}$$

where b, m, n are integers with m and n relatively prime to p. The *p-adic absolute value* is then specified by

$$\left| p^b \frac{m}{n} \right|_p = p^{-b}.$$

It is easily verified that $|\cdot|_p$ is an absolute value, which satisfies a stronger version of 3, called the *strong triangle inequality*

$$|x + y|_p \leq \max\{|x|_p, |y|_p\}. \tag{2.1}$$

Absolute values which satisfy the strong triangle inequality are called *non-archimedean* absolute values.

Two absolute values are equivalent if one is a power of the other, and an equivalence class of absolute values is called a *place* of $\mathbb{Q}$. A celebrated theorem of Ostrowski [**17**] shows that any non-trivial absolute value on $\mathbb{Q}$ is equivalent to either the usual absolute value $|\cdot|_\infty$ or to $|\cdot|_p$ for some prime p.

2.2. Completions. The real numbers are constructed from the rational numbers by completing $\mathbb{Q}$ with respect to $|\cdot|_\infty$. Recall the construction: two Cauchy sequences of rational numbers (x_n) and (y_n) are equivalent if $(x_n - y_n)$ converges to 0. The real numbers are then defined to be the set of equivalence classes of Cauchy sequences, and the algebraic operations of addition and multiplication are given by term-wise addition and multiplication of equivalence class representatives. The rational numbers can be represented by constant sequences, and these are dense in the completion, and using this fact we may extend the absolute value $|\cdot|_\infty$ to the usual absolute value on $\mathbb{R}$.

The p-adic numbers $\mathbb{Q}_p$ are constructed in the same way, except that the notions of convergence are with respect to the p-adic absolute value. That is, (x_n) is Cauchy if given $\epsilon > 0$ there exists M such

$$\sup_{n,m>M} |x_n - x_m|_p < \epsilon$$

and we say (x_n) is equivalent to (y_n) if there exists M such that

$$\sup_{m>M} |x_m - y_m|_p < \epsilon.$$

As with the real numbers, $\mathbb{Q}$ is dense in $\mathbb{Q}_p$ and the absolute value $|\cdot|_p$ extends to an absolute value on $\mathbb{Q}_p$.

2.3. Differences and similarities between $\mathbb{R}$ and $\mathbb{Q}_p$. Despite the similarity in construction, $\mathbb{Q}_p$ has important structural differences from $\mathbb{R}$. Here is a brief compendium of facts about $\mathbb{Q}_p$ which illustrate such differences:

(1) $|\cdot|_p$ takes values in the discrete set $\{p^n : n \in \mathbb{Z}\}$.
(2) If $n \in \mathbb{Z}$, then $|n|_p \leq 1$.
(3) The completion of $\mathbb{Z}$ in $\mathbb{Q}_p$ is given by $\mathbb{Z}_p := \{x \in \mathbb{Q}_p : |x| \leq 1\}$ and called the p-adic integers. $\mathbb{Z}$ is dense in $\mathbb{Z}_p$.
(4) $\mathbb{Z}_p$ is a ring with a unique maximal ideal $\mathfrak{m}_p := \{x \in \mathbb{Z}_p : |x| < 1\}$. Moreover $\mathfrak{m}_p$ is a principal ideal generated by p (i.e. $\mathfrak{m}_\mathfrak{p} = p\mathbb{Z}_p$).
(5) $\mathbb{Z}_p/\mathfrak{m}_p \cong \mathbb{Z}/p\mathbb{Z}$.

In spite of these differences, there are also similarities which we will exploit. Perhaps most important is that both $\mathbb{R}$ and $\mathbb{Q}_p$ are locally compact abelian groups under addition, and $\mathbb{R}^\times$ and $\mathbb{Q}_p^\times$ are locally compact abelian groups under multiplication. This is useful because it means that $\mathbb{Q}_p$, like $\mathbb{R}$, has a Haar measure.

2.4. Haar measure on $\mathbb{Q}_p$. A Haar measure on a locally compact abelian group is a Borel measure which is invariant under the action of the group on itself. For instance, Lebesgue measure is a Haar measure on $\mathbb{R}$, since the Lebesgue measure of an interval (and hence any Lebesgue measurable set) is invariant under translation. Haar measures are not unique, though once one specifies the measure of a compact set containing an open set, the measure is completely specified. Thus we may contruct a unique Borel measure μ_p on $\mathbb{Q}_p$ with the following properties: $\mu_p(\mathbb{Z}_p) = 1$ and for any $x \in \mathbb{Q}_p$ and Borel subset B, $\mu_p(x + B) = \mu_p(B)$. This measure also behaves nicely with respect to multiplication $\mu_p(xB) = |x|_p\mu_p(B)$. In particular, $\mu_p(\mathfrak{m}_p) = \mu_p(p\mathbb{Z}_p) = 1/p$. We remark that $\mathbb{Z}_p$ is a compact abelian group, and μ_p restricted to $\mathbb{Z}_p$ is the unique Haar probability measure on this group.

We will have limited need for a Haar measure on $\mathbb{Q}_p^\times$, but for the record, it is absolutely continuous with respect to μ_p on $\mathbb{Q}_p^\times$, and a natural Haar measure is given by $\mu_p^\times(dx) = \mu_p(dx)/|x|_p$.

2.5. Non-Archimedean completions of number fields. We may generalize the previous discussion somewhat by letting K be a number field with ring of integers $\mathfrak{o}$ and a chosen prime ideal $\mathfrak{p} \subset \mathfrak{o}$. $\mathfrak{o}/\mathfrak{p}$ is a finite field, say $\mathbb{F}_q$ where q is a power of a rational prime. Each element x of K lies in $\mathfrak{p}^n/\mathfrak{p}^{n+1}$ for some rational integer n, and we define $|x|_\mathfrak{p} = 1/q^n$. Completing K with respect to $|\cdot|_\mathfrak{p}$ produces the field $K_\mathfrak{p}$. As before, we define the ring of integers of $K_\mathfrak{p}$ and its unique maximal ideal by

$$\mathfrak{o}_\mathfrak{p} = \{x \in K_\mathfrak{p} : |x|_\mathfrak{p} \leq 1\} \qquad \text{and} \qquad \mathfrak{m}_\mathfrak{p} = \{x \in \mathfrak{o}_\mathfrak{p} : |x|_\mathfrak{p} < 1\}.$$

The units in $\mathfrak{o}_\mathfrak{p}$ are given by $U_\mathfrak{p} = \{x \in \mathfrak{o}_\mathfrak{p} : |x|_\mathfrak{p} = 1\} = \mathfrak{o}_\mathfrak{p} \setminus \mathfrak{m}_\mathfrak{p}$.

By general principles, $\mathfrak{o}_\mathfrak{p}/\mathfrak{m}_\mathfrak{p}$ is a finite field which can be shown to be isomorphic to $\mathbb{F}_q$. That is, $\mathfrak{m}_\mathfrak{p}$ has q cosets, and we denote these by $\mathfrak{m}_\mathfrak{p}, 1+\mathfrak{m}_\mathfrak{p}, \ldots, q-1+\mathfrak{m}_\mathfrak{p}$. It can be shown that $\mathfrak{m}_\mathfrak{p}$ is a principal ideal, and if π is a generator (or *uniformizer*) for $\mathfrak{m}_\mathfrak{p}$ then $|\pi|_\mathfrak{p} = 1/q$.

There is a unique Haar measure $\mu_\mathfrak{p}$ on $\mathfrak{o}_\mathfrak{p}$ satisfying $\mu_\mathfrak{p}(\mathfrak{o}_\mathfrak{p}) = 1$. Since $\mathfrak{o}_\mathfrak{p}$ is the disjoint union of q cosets of $\mathfrak{m}_\mathfrak{p}$ each of which is a translation of $\mathfrak{m}_\mathfrak{p}$, we have that $\mu_\mathfrak{p}(\mathfrak{m}_\mathfrak{p}) = 1/q$.

2.6. Notation. We will be working in a single non-archimedean field of characteristic 0. We may take this to be the completion of a number field with respect to an absolute value induced by a prime ideal, but we need not burden our notation with explicit dependence on the particular number field or the prime ideal.

Thus, we set $\mathbb{K}$ to be a field complete with respect to non-archimedean absolute value $|\cdot|$, with ring of integers $\mathfrak{o}$, maximal ideal $\mathfrak{m}$ with q (a prime power) cosets generated by uniformizer π, and Haar measure μ normalized so that $\mu(\mathfrak{m}) = 1/q$. We will often be integrating over the cartesian product of a number of copies of a subset of $\mathfrak{o}$. We will denote the product measure on $\mathfrak{o}^N$ by μ^N (this is the unique Haar probability measure on $\mathfrak{o}^N$). There is ambiguity interpreting $\mathfrak{m}^N$ and we will interpret it as the N-fold copy of the maximal ideal $\mathfrak{m}$. If we need to denote the *ideal* given by the nth power of $\mathfrak{m}$–all such ideals in $\mathfrak{o}$ are of this form–we will write $\pi^n\mathfrak{o}$.

3. Electrostatics

Imagine two like charged particles identified with points α and α' in $\mathfrak{o}$. We define the *interaction energy* of this simple system by

$$E(\alpha, \alpha') := -\log|\alpha - \alpha'|.$$

Note that $E(\alpha, \alpha') \in [0, \infty]$, with $E(\alpha) = \infty$ if and only if $\alpha = \alpha'$. Notice also that $E(\alpha, \alpha')$ takes its minimal value 0, exactly when $(\alpha - \alpha')$ is a unit—that is, when α and α' are in different cosets of $\mathfrak{m}$.

Given a system of N such particles, its potential energy is the sum of interaction energies over all $\binom{N}{2}$ pairs of particles. That is, if we identify the *state* of a system with N charged particles by $\boldsymbol{\alpha} \in \mathfrak{o}^N$, the potential energy of that state is given by

$$E(\boldsymbol{\alpha}) = -\sum_{m<n} \log|\alpha_n - \alpha_m|. \tag{3.1}$$

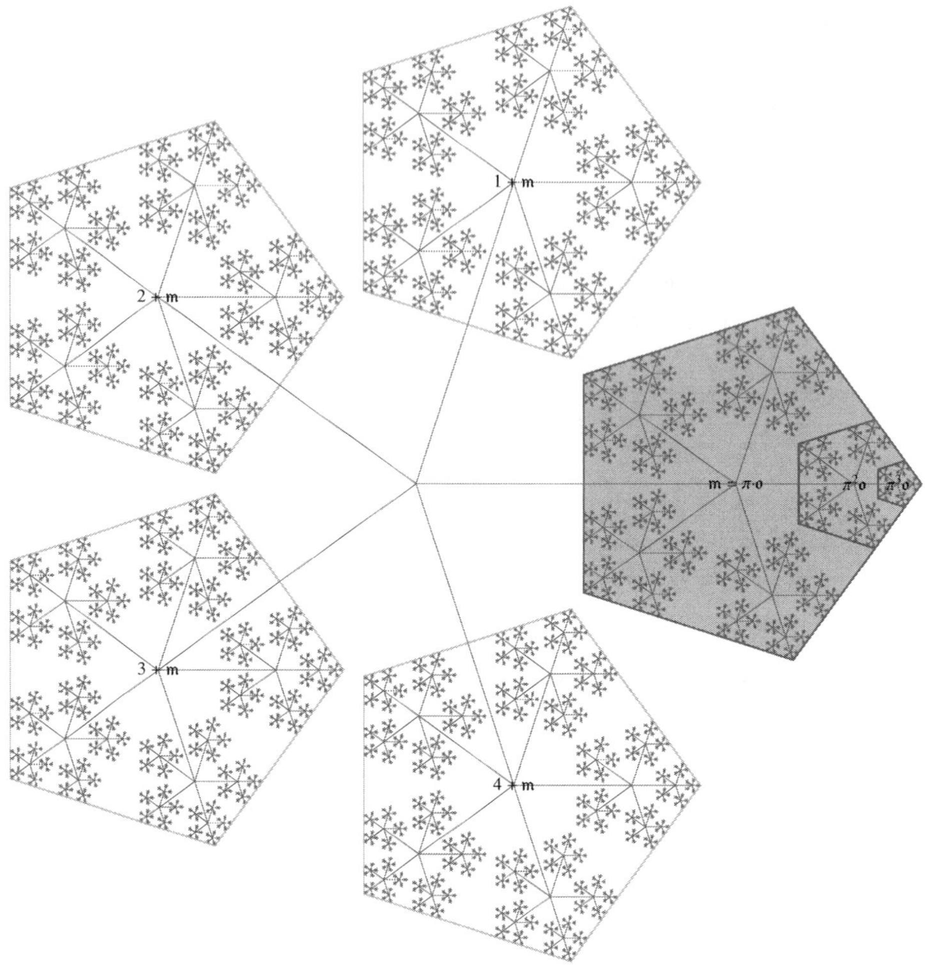

FIGURE 1. A schematic diagram of $\mathfrak{o} = \mathbb{Z}_5$. Cosets and powers of the maximal ideal $\mathfrak{m}$ are represented. The 5-adic numbers are the fractal boundary of the tree, *not* the interstitial edges and vertices. Every ball in $\mathbb{Z}_5$ is the fractal boundary of one of the naturally appearing 'pentagons' in the tree (at all scales, assuming it had infinite resolution), and each is naturally in bijective equivalence with $\mathbb{Z}_5$ itself. Addition by a fixed α in $\mathbb{Z}_5$ can be thought of as an epicyclic rotation where the cosets of $\mathfrak{m}$ are rotated, as are their neighborhoods of the next smallest radius, and so on down the tree. Addition by a fixed rational integer can be distinguished from addition by an arbitrary α since in the former situation there is some radius below which the epicyclic rotations are trivial. Multiplication by a unit (any element not in $\mathfrak{m}$) stabilizes $\mathfrak{m}$ and permutes in some manner the non-trivial cosets of $\mathfrak{m}$. Multiplication by an element of $\mathfrak{m}$ sends $\mathfrak{o}$ onto some smaller ideal $\pi^r\mathfrak{o}$; a contraction of sorts onto a smaller ball. There are many ways to embed $\mathbb{Z}$ into this diagram, but one simple model has 0 as the right-most point on the diagram, and given any rational integer there is some 'pentagon' for which that integer is its right-most point. This provides a heuristic verification of the density of the rational integers in $\mathbb{Z}_5$. Haar measure corresponds to areal measure: the largest 'pentagon' (all of $\mathbb{Z}_5$) is assumed to have measure/area 1, and then the measure of any smaller 'pentagon' is its area.

As defined, each physical state is overcounted by a factor of $N!$ since permuting the coordinates of $\boldsymbol{\alpha}$ does not alter the identity of the system.[1] This overcounting will be adjusted for later.

3.1. The microcanonical ensemble. An *ensemble* is a probability measure on a set of states of a physical system.

The *microcanonical* ensemble is that given by our system conditioned so that the total energy of the system is some fixed value E_*. Since our absolute value is discrete, not all values are allowed for E_*, and we will assume that E_* is an attainable value of the energy as specified by (3.1). The set of attainable states of the microcanonical ensemble is then

$$\Omega_N(E_*) := \{\boldsymbol{\alpha} \in \mathfrak{o}^N : E(\boldsymbol{\alpha}) = E_*\}.$$

A first obvious question is what is the volume of the set of attainable states? That is, what is $\mu^N(\Omega_N(E_*))$?

To answer this question, it will be convenient to define

$$F_N(\xi) := \mu^N\{\boldsymbol{\alpha} \in \mathfrak{o}^N : E(\boldsymbol{\alpha}) \leq \log \xi\} \qquad \xi \geq 0. \tag{3.2}$$

It follows then that the volume of accessible states for the energy E_* is

$$\mu^N(\Omega_N(E_*)) = F_N(e^{E_*}) - \lim_{E \to E_*-} F_N(e^E).$$

This volume is 0 if E_* is not an allowable value of the energy.

The introduction of the $\log \xi$ term on the right-hand-side of (3.2) will be convenient in the sequel, since we can rewrite $F_N(\xi)$ as

$$F_N(\xi) = \mu^N\left\{\boldsymbol{\alpha} \in \mathfrak{o}^N : |\Delta_N(\boldsymbol{\alpha})| < \xi\right\},$$

where Δ_N is the Vandermonde determinant

$$\Delta_N(\boldsymbol{\alpha}) := \prod_{m<n} (\alpha_n - \alpha_m).$$

Written in this way, we see that F_N is the cumulative distribution function for the random variable $|\Delta_N(\boldsymbol{\alpha})|$ where the $\alpha_1, \ldots, \alpha_N$ are independent uniform random variables in $\mathfrak{o}$.

4. The canonical ensemble

The canonical ensemble differs from the microcanonical ensemble in that we now allow the energy to vary. Since energy is a conserved quantity, we do this by placing our system of particles in contact with another, typically much larger system, so that the energy of the aggregate system is constant, but energy is allowed to flow between our system of N particles and the larger system. The larger system is called a *heat reservoir* and we will view it as being at a fixed temperature T. It is usually more convenient to introduce the inverse temperature parameter $\beta = (kT)^{-1}$. Here k is Boltzmann's constant and β is a unitless quantity.

---

[1] We are implicitly deciding that the particles are indistinguishable, and the probability that two particles are co-located is zero—the first of these assumptions is definitional and the second will be justified in later sections.

The relative density of states is given by the *Boltzmann factor* $e^{-\beta E(\boldsymbol{\alpha})}$. That is, the probability density of finding the system in state $\boldsymbol{\alpha}$ is given by

$$\frac{1}{Z}e^{-\beta E(\boldsymbol{\alpha})} = \frac{1}{Z}\prod_{m<n} |\alpha_n - \alpha_m|^\beta \tag{4.1}$$

where

$$Z = \int_{\mathfrak{o}^N} \prod_{m<n} |\alpha_n - \alpha_m|^\beta \, d\mu^N(\boldsymbol{\alpha}).$$

Z is called the *partition* function of the *canonical ensemble* of particles, and it is more than just a normalization constant necessary to make a probability measure.

It is sometimes useful to make explicit the variables on which Z depends, and so we write

$$Z(N, V, \beta) = \int_{V^N} |\Delta_N(\boldsymbol{\alpha})|^\beta \, d\mu^N(\boldsymbol{\alpha}).$$

Here V is a finite measure subset of $\mathbb{K}$, whose measure plays the role of volume in traditional statistical physics. For our purposes, V will usually be either $\mathfrak{o}$ or $\mathfrak{m}$. We define $Z(0, V, \beta) := 1$ and, with the interpretation that an empty product is equal to 1, we see that $Z(1, V, \beta) = \mu(V)$. In particular, $Z(0, \mathfrak{o}, \beta) = Z(1, \mathfrak{o}, \beta) = 1$.

For those uncomfortable with the sudden introduction of temperature, or those unfamiliar with the derivation of the Boltzmann factor, we can take (4.1) as an axiomatic relationship between the energy and density of states, and the temperature as represented by β. Let us see, however that for the value $\beta = 0$, the density of states satisfies our intuition: When $\beta = 0$ the temperature is infinite. It is reasonable to suppose that in such a situation thermal fluctuations of particle positions will overwelm any repulsion stemming from electrical charge. That is, when $\beta = 0$ the particles are independent and uniform over $\mathfrak{o}$. That is the relative density of states should be constant on $\mathfrak{o}^N$. The intuition in this case agrees with the result given by (4.1).

Similarly, as $\beta \to \infty$, the temperature is tending toward 0, and we expect the repulsion from the charge to overwhelm the thermal fluctuations. From a physical perspective, we therefore expect that the system will find itself in a state with minimal energy. The minimal energy configurations correspond to states $\boldsymbol{\alpha}$ where $|\Delta(\boldsymbol{\alpha})|$ is maximal (note that the maximum is attained because $|\Delta(\boldsymbol{\alpha})| \leq 1$ and the absolute value is discrete). Looking at the integrand in $Z(N, V, \beta)$ as $\beta \to \infty$, the contributions to the integral from low energy configurations (exponentially!) overwhelm higher energy configurations, and the resulting density of states becomes localized around the states with minimal energy.

4.1. Relating the microcanonical and canonical ensembles. Our goal in the study of the microcanonical ensemble is to determine $\mu^N(\Omega_N(E_*))$. This information is encoded into the distribution function F_N and we will attempt to derive useful information about F_N by considering its Mellin transform[2]

$$\widehat{F_N}(s) := \int_0^\infty \xi^{-s} F_N(\xi) \, \frac{d\mu(\xi)}{\xi}.$$

The following lemma relates the Mellin transform of F_N to $Z(N, \mathfrak{o}, \beta)$.

[2]For those unfamiliar with the Mellin transform, it is a multiplicative version of the Fourier transform and has the same utility. In particular, there is an inversion formula which links a function and its Mellin transform.

LEMMA 4.1. *For* $\mathrm{Re}(s) > 0$, $\widehat{F_N}(s) = -\frac{1}{s}Z(N, \mathfrak{o}, s)$.

PROOF. Using Lebesgue-Stieltjes integration by parts,

$$\int_0^\infty \xi^s F_N(\xi)\,\frac{d\mu(\xi)}{\xi} = \frac{\xi^s}{s}F_N(\xi)\bigg|_0^\infty - \frac{1}{s}\int_0^\infty \xi^s dF_N(\xi).$$

Since $F_N(0) = 0$ and $F_N(\xi) \le 1$, the first term vanishes, and we find

$$\widehat{F_N}(s) = -\frac{1}{s}\int_0^\infty \xi^s dF_N(\xi) = -\frac{1}{s}\int_{\mathfrak{o}^N} |\Delta(\boldsymbol{\alpha})|^s\, d\mu^N(\boldsymbol{\alpha})$$

as claimed. □

4.2. The additivity of energy over cosets of $\mathfrak{m}$. By our previous remarks, if α and α' are particles in different cosets of $\mathfrak{m}$, then this pair of particles does not contribute to the energy of the configuration. This is a primary observation: *Particles in different cosets can't 'sense' each other.*

LEMMA 4.2 (Additivity of energy over cosets.). *Suppose* $\boldsymbol{\alpha}$ *is a state with* n^0 *particles in* $\mathfrak{m}$, n^1 *particles in the coset* $1 + \mathfrak{m}$, *etc. If we represent the state in the* r*th coset by* $\boldsymbol{\alpha}^r$, *then by reordering the coordinates if necessary, we can write* $\boldsymbol{\alpha} = (\boldsymbol{\alpha}^0, \boldsymbol{\alpha}^1, \ldots, \boldsymbol{\alpha}^{q-1})$. *Then,*

$$E(\boldsymbol{\alpha}) = \sum_{r=0}^{q-1} E(\boldsymbol{\alpha}^r).$$

Alternatively,

$$|\Delta(\boldsymbol{\alpha})| = \prod_{r=0}^{q-1} |\Delta(\boldsymbol{\alpha}^r)|.$$

We will call $(\boldsymbol{\alpha}^0, \boldsymbol{\alpha}^1, \ldots, \boldsymbol{\alpha}^{q-1})$ a *factored state* of $\boldsymbol{\alpha}$ with occupancy vector $\mathbf{n} = (n^0, \ldots, n^{q-1})$ which sums to N. The set of factored states with occupancy vector $\mathbf{n}$ is given by

$$U_{\mathbf{n}} = \mathfrak{m}^{n^0} \times (1 + \mathfrak{m})^{n^1} \times \cdots \times (q - 1 + \mathfrak{m})^{n^{q-1}}$$

4.3. The partition function in the canonical ensemble. Here we derive a way of expressing $Z(N, \mathfrak{o}, \beta)$ in terms of $Z(n, \mathfrak{m}, \beta)$ for $n < N$. This will provide a recursive way to determine $Z(N, \mathfrak{o}, \beta)$.

THEOREM 4.3. *For* $N > 0$,

$$Z(N, \mathfrak{o}, \beta) = N! \sum_{\mathbf{n}} \left\{ \prod_{r=0}^{q-1} \frac{Z(n^r, \mathfrak{m}, \beta)}{n^r!} \right\},$$

where the sum is over all occupancy vectors $\mathbf{n}$ *with* $n^0 + \cdots + n^{q-1} = N$.

PROOF. If $\alpha = \alpha'$ for some two coordinates of $\boldsymbol{\alpha}$, then $\Delta(\boldsymbol{\alpha}) = 0$. This means that such states make no contribution to $Z(N, \beta, \mathfrak{o})$ and we can safely ignore such inadmissable states. Each admissible state with occupancy vector $\mathbf{n}$ corresponds to $n_0! \cdots n_{q-1}!$ factored states which differ by permuting the particles in each coset seperately. Moreover, each admissible factored state corresponds to $N!$ admissible states formed by permuting the coordinates indiscriminantly. Thus,

$$Z(N, \mathfrak{o}, \beta) = N! \sum_{\mathbf{n}} \left\{ \prod_{r=0}^{q-1} \frac{1}{n_r!} \right\} \int_{U_{\mathbf{n}}} |\Delta(\boldsymbol{\alpha}_0, \ldots, \boldsymbol{\alpha}_{q-1})|^\beta\, d\mu^{n_0}(\boldsymbol{\alpha}_0) \cdots d\mu^{n_{q-1}}(\boldsymbol{\alpha}_{q-1}).$$

The additivity of energy over cosets implies that the integrand factors, and Fubini's theorem implies then that

$$Z(N,\mathfrak{o},\beta) = N!\sum_{\mathbf{n}}\left\{\prod_{r=0}^{q-1}\frac{1}{n_r!}\int_{(r+\mathfrak{m})^{n_r}}|\Delta(\boldsymbol{\alpha}_r)|^{\beta}\,d\mu^{n_r}(\boldsymbol{\alpha}_r)\right\}.$$

As a final maneuver, we note that both the integrand and the measure are invariant under the change of variables $\boldsymbol{\alpha}_r \mapsto r + \boldsymbol{\alpha}_r$. That is, the physics can't distinguish the identity of cosets so we can replace the integrals over individual cosets with integrals over independent copies of $\mathfrak{m}$. In any event,

$$Z(N,\mathfrak{o},\beta) = N!\sum_{\mathbf{n}}\left\{\prod_{r=0}^{q-1}\frac{1}{n_r!}\int_{\mathfrak{m}^{n_r}}|\Delta(\boldsymbol{\alpha})|^{\beta}\,d\mu^{n_r}(\boldsymbol{\alpha})\right\} = N!\sum_{\mathbf{n}}\left\{\prod_{r=0}^{q-1}\frac{Z(n_r,\mathfrak{m},\beta)}{n_r!}\right\}.$$

□

By rescaling the domain we can write $Z(n,\mathfrak{m},\beta)$ in term of $Z(n,\mathfrak{o},\beta)$ which leads to a recursive formula for $Z(N,\mathfrak{o},\beta)$.

LEMMA 4.4. *If $B = \zeta + \pi^r\mathfrak{o}$ then,* $Z(n,B,\beta) = q^{-r\beta\binom{n}{2}-rn}Z(n,\mathfrak{o},\beta)$.

PROOF. This is a special case of Lemma 4.9 below. □

This lemma leads immediately to the following theorem.

THEOREM 4.5. *For $N > 0$,*

$$Z(N,\mathfrak{o},\beta) = N!\sum_{\mathbf{n}}\left\{\prod_{r=0}^{q-1}\frac{q^{-\beta\binom{n_r}{2}-n_r}}{n_r!}Z(n_r,\mathfrak{o},\beta)\right\}. \tag{4.2}$$

Solving for $Z(N,\mathfrak{o},\beta)$ (which appears on both sides of 4.2),

$$Z(N,\mathfrak{o},\beta) = \left(q^N - q^{1-\beta\binom{N}{2}}\right)^{-1} N!\sum_{\mathbf{n}}{}'\left\{\prod_{j=0}^{q-1}\frac{q^{-\beta\binom{n_r}{2}}}{n_r!}Z(n_j,\mathfrak{o},\beta)\right\},$$

where $\sum'_{\mathbf{n}}$ is over all occupancy vectors except those of the form $(0,\ldots,N,\ldots 0)$ (that is, except those that correspond to all particles being in the same coset of $\mathfrak{m}$).

THEOREM 4.6. *The $\{Z(n,\mathfrak{o},\beta) : n = 0,1,\ldots,N-1\}$ satisfy*

$$\sum_{n=0}^{N}\frac{(N-(q+1)n)}{n!(N-n)!}q^{-\binom{n}{2}\beta}q^{-n}Z(n,\mathfrak{o},\beta)Z(N-n,\mathfrak{o},\beta) = 0.$$

See the remark after Theorem 5.2 for the proof.

Theorem 4.6 gives us an easy way to compute $Z(N,\mathfrak{o},\beta)$ for small values of N. These are increasingly complicated rational functions in $q^{-\beta}$.

$$\begin{aligned}
Z(0,\mathfrak{o},\beta) &= 1\\
Z(1,\mathfrak{o},\beta) &= 1\\
Z(2,\mathfrak{o},\beta) &= \frac{(q-1)q^{\beta}}{q^{\beta+1}-1}\\
Z(3,\mathfrak{o},\beta) &= \frac{(q-1)q^{3\beta}\left(-2q^{\beta+1}+q^{\beta+2}+2q-1\right)}{\left(q^{\beta+1}-1\right)\left(q^{3\beta+2}-1\right)}\\
Z(4,\mathfrak{o},\beta) &= \frac{A+B+C}{\frac{1}{24}(4-4(q+1))q^{-6\beta-4}+\frac{1}{6}},
\end{aligned}$$

where

$$A = -\frac{(q-1)^2(4-2(q+1))q^{\beta-2}}{4\left(q^{\beta+1}-1\right)^2}$$

$$B = -\frac{(3-q)(q-1)\left(-2q^{\beta+1}+q^{\beta+2}+2q-1\right)q^{3\beta-1}}{6\left(q^{\beta+1}-1\right)\left(q^{3\beta+2}-1\right)}$$

$$C = -\frac{(q-1)(4-3(q+1))\left(-2q^{\beta+1}+q^{\beta+2}+2q-1\right)}{6q^3\left(q^{\beta+1}-1\right)\left(q^{3\beta+2}-1\right)}$$

4.4. The distribution of energies in the microcanonical ensemble. That $Z(N,\mathfrak{o},\beta)$ is expressible in terms of the Mellin transform of F_N (Lemma 4.1) means that analytic information, viewing $Z(N,\mathfrak{o},\beta)$ as a function of a complex variable β, can provide information about the nature of the distribution of allowable energies in the microcanonical ensemble. In order to distinguish β the complex variable in this expression from the inverse temperature (which must necessarily be real and positive) we will write $s = \sigma + it$ for a complex variable and write $Z(N,\mathfrak{o},s)$ for the partition function as a function of a complex variable.

A first observation is that $Z(N,\mathfrak{o},s)$ is analytic in the right half plane $\{\sigma + it : \sigma > 0\}$. To see this suppose $\sigma > 0$ and with the usual absolute value on $\mathbb{C}$ denoted by $|\cdot|_\infty$,

$$\int_{\mathfrak{o}^N} \left||\Delta(\boldsymbol{\alpha})|^{\sigma+it}\right|_\infty d\mu^N(\boldsymbol{\alpha}) = \int_{\mathfrak{o}^N} |\Delta(\boldsymbol{\alpha})|^{\sigma}\, d\mu^N(\boldsymbol{\alpha}) \leq \int_{\mathfrak{o}^N} d\mu^N(\boldsymbol{\alpha}) = 1. \tag{4.3}$$

It follows that, if T is an oriented triangle in $\{\sigma > 0\}$ then

$$\int_T Z(N,\mathfrak{o},s)\,ds = \int_{\mathfrak{o}^N} \left\{ \int_T |\Delta(\boldsymbol{\alpha})|^s ds \right\} d\mu^N(\boldsymbol{\alpha}),$$

where we used (4.3) to justify the use of Fubini's Theorem, and the ds integrals are (complex) line integrals around T. But, for every fixed $\boldsymbol{\alpha}$, the function $s \mapsto |\Delta(\boldsymbol{\alpha})|^s$ is analytic and hence

$$\int_T |\Delta(\boldsymbol{\alpha})|^s ds = 0 \qquad \text{and thus} \qquad \int_T Z(N,\mathfrak{o},s)\,ds = 0.$$

Since this is true for all triangles in the right half-plane, Morera's Theorem implies $Z(N,\mathfrak{o},s)$ is analytic there. In fact, we will see below that the domain of convergence of $Z(N,\mathfrak{o},s)$ is the half-plane $\sigma > -2/N$. This was previously established in [**24**].

THEOREM 4.7. *The integral defining $Z(N,\mathfrak{o},s)$ converges to an analytic function of $s = \sigma + it$ in the half-plane $\{\sigma > -2/N\}$ and is absolutely divergent on $\{\sigma \leq -2/N\}$.*

Another proof of this theorem will come after the development of a handful of lemmas.

One importance of recognizing the Mellin transform of $F_N(\xi)$ in terms of the function $Z(N,\mathfrak{o},s)$ is that information from any analytic/meromorphic continuation of $Z(N,\mathfrak{o},s)$ beyond the initial domain of convergence gives us new information about $F_N(\xi)$.

Our first observation is that $Z(N,\mathfrak{o},s)$ analytically continues to a meromorphic function. In fact it is a rational function in q^{-s}. We will provide a proof of this fact, but it is also the consequence of a much deeper theorem of Igusa on the continuation of certain types of L-functions, of which $Z(N,\mathfrak{o},s)$ is a particular example.

LEMMA 4.8. *For $N \geq 2$, there exists a non-constant rational function R_N with rational coefficients so that $Z(N, \mathfrak{o}, s) = R_N(q^{-s})$ for all $s = \sigma + it$ with $\sigma > 0$.*

PROOF. The proof is an easy consequence of Theorem 4.5. By definition $Z(0, \mathfrak{o}, s) = 1$ and a trivial calculation shows $Z(1, \mathfrak{o}, s) = 1$ as well. The second equation in Theorem 4.5 then implies that $Z(N, \mathfrak{o}, s)$ can be expressed as a ratio where the numerator is a rational linear combination of products of $Z(n, \mathfrak{o}, s)$ with $n < N$, and the denominator is a polynomial in q^{-s} with rational coefficients. The strong inductive hypothesis is that $Z(n, \mathfrak{o}, s)$ is a rational function in q^{-s} with rational coefficients for all $2 \leq n < N$, and the result follows. □

4.5. The σ-algebra of symmetrized sets. An ensemble is merely a probability space, and we here we set up the formal machinery to compute probabilities of events of physical interest.

We set $\mathcal{B}$ and $\mathcal{B}^N$ to be the Borel σ-algebras on $\mathfrak{o}$ and $\mathfrak{o}^N$ as usual. We think of $\mathfrak{o}$ as the one-particle space and $\mathfrak{o}^N$ as the state space of our system of N particles. The σ-algebra $\mathcal{B}^N$, however, is too large for our purposes in the sense that it contains events that would be out of reach of an observer of the system. Consider, for instance the set $A \times \mathfrak{o} \times \cdots \times \mathfrak{o} \subseteq \mathfrak{o}^N$ for some $A \in \mathcal{B}$. This event is equivalent to knowing whether or not the particle in the first coordinate of the vector $\boldsymbol{\alpha}$ is in A. However, since our particles are indistinguishable, we can't discern whether or not the *first* particle is in A; the closest we can come is to discern whether or not *one* of the particles is in A.

This reasoning suggests we should only consider events in $\mathfrak{o}^N$ which are stabilized by the natural action of the symmetric group S_N. We denote the σ-algebra generated by all such *symmetrized* Borel subsets by $\mathcal{S}_N$. We explicitly define our probability space by $(\mathfrak{o}^N, \mathcal{S}_N, \mathbb{P}_N)$ where,

$$\mathbb{P}_N(B) = \frac{1}{Z(N, \beta, \mathfrak{o})} \int_B \prod_{m<n} |\alpha_n - \alpha_m|^\beta \, d\mu^N(\boldsymbol{\alpha}).$$

$\mathbb{P}_N(B)$ depends implicitly on β. If we need to make this dependence explicit, we will write $\mathbb{P}(N, B, \beta)$.

To see what events in $\mathcal{S}_N$ look like, consider a rectangle $A = A_1 \times A_2 \times \cdots \times A_N$ in $\mathcal{B}^N$. Given a permutation $\tau \in S_N$ we define $\tau \cdot A = A_{\tau(1)} \times A_{\tau(2)} \times \cdots \times A_{\tau(N)}$. Using this, we define

$$S_N \cdot A = \bigcup_{\tau \in S_N} \tau \cdot A.$$

$S_N \cdot A$ is in $\mathcal{S}_N$ and is the *symmetrized* rectangle formed from A. $\mathcal{S}_N$ is generated by all symmetrized rectangles—in fact, since $\mathbb{P}_N$ is a Borel measure, we may restrict our attention to symmetrized rectangles A where the A_m are all balls.

LEMMA 4.9. *Given $E \subseteq \mathcal{S}_n$, $\zeta \in \mathfrak{o}$ and $r \in \mathbb{N}$, define the translated contraction of E by*

$$\zeta + \pi^r E = \{(\zeta + \pi^r \alpha_1, \ldots, \zeta + \pi^r \alpha_n) : \boldsymbol{\alpha} \in E\}.$$

Then,

$$\mathbb{P}_n(\zeta + \pi^r E) = q^{-r\beta\binom{n}{2} - rn} \mathbb{P}_n(E)$$

PROOF. Each point in $\zeta + \pi^r E$ can be written in the form $\zeta + \pi^r \boldsymbol{\alpha}$ where $\boldsymbol{\alpha} \in E$. Moreover this map is a bijection. We may thus write

$$\mathbb{P}_n(\zeta + \pi^r E) = \int_E |\Delta_n(\zeta + \pi^r \boldsymbol{\alpha})|^\beta \, d\mu^n(\zeta + \pi \boldsymbol{\alpha}),$$

where

$$|\Delta_n(\zeta+\pi^r \boldsymbol{\alpha})| = \prod_{\ell<m} |\zeta+\pi^r\alpha_m - (\zeta+\pi^r\alpha_\ell)| = |\pi|^{r\binom{n}{2}} \prod_{\ell<m} |\alpha_m - \alpha_\ell| = q^{-r\binom{n}{2}} |\Delta_n(\boldsymbol{\alpha})|.$$

The translation invariance of μ implies that

$$d\mu^n(\zeta + \pi^r \boldsymbol{\alpha}) = |\pi|^{rn} d\mu^n(\boldsymbol{\alpha}) = q^{-rn} d\mu^n(\boldsymbol{\alpha}).$$

Putting this all together, we see that $\mathbb{P}_n(\zeta + \pi^r E) = q^{-r\beta\binom{n}{2} - rn}\mathbb{P}_n(E)$. □

4.6. The σ-algebra of cylinder sets. Given Borel $B \subseteq \mathfrak{o}$ we define $N_B : \mathfrak{o}^N \to \mathbb{Z}_{\geq 0}$ by $N_B(\boldsymbol{\alpha}) = \#\{\alpha_1, \ldots, \alpha_N\} \cap B$. That is $N_B(\boldsymbol{\alpha})$ is the number of coordinates of $\boldsymbol{\alpha}$ in B. Put another way, N_B is a random variable counting the number of particles in B. We call $N_B^{-1}(n) \subseteq \mathfrak{o}^N$ a *simple* cylinder set. Clearly $N_B^{-1}(n)$ is in $\mathcal{S}_N$. The σ-algebra generated by all simple cylinder sets is called the *cylinder* σ-algebra and denoted $\mathcal{C}_N$. More specifically,

$$\mathcal{C}_N = \sigma\{N_B : B \in \mathcal{B}(\mathfrak{o})\} \subseteq \mathcal{S}_N.$$

We also define $\mathcal{C}_N(B) = \sigma(N_B) \subseteq \mathcal{C}_N$ which contains information (only) about the number of particles in B. Note that $\mathcal{C}_N(B) = \mathcal{C}_N(B^c)$ is a finite σ-algebra.

To get a feeling for what cylinder sets look like, fix $B \subseteq \mathfrak{o}$ and note that

$$\{N_B = n\} = S_N \cdot \underbrace{B \times \cdots \times B}_{n} \times \underbrace{B^c \times \cdots \times B^c}_{N-n}$$

Likewise, if B and E are disjoint sets and n and m non-negative integers with $n + m \leq N$, then

$$\{N_B = n, N_E = m\} = S_N \cdot \underbrace{B \times \cdots \times B}_{n} \times \underbrace{E \times \cdots \times E}_{m} \times \underbrace{(B \cup E)^c \times \cdots \times (B \cup E)^c}_{N-n-m},$$

and this pattern continues. Suppose $\mathbf{B} = (B_1, \ldots, B_M)$ is a finite open cover of $\mathfrak{o}$, and $\mathbf{n} = (n_1, \ldots, n_M)$ is an *occupation vector* such that $n_1 + \cdots + n_M = N$. We define the event

$$\{N_{\mathbf{B}} = \mathbf{n}\} = \{N_{B_1} = n_1, \ldots, N_{B_M} = n_M\} = S_N \cdot B_1^{n_1} \times \cdots \times B_M^{n_M}.$$

All cylinder sets can be described as unions and intersections of such sets.

We will be especially interested in such cylinder sets where each of the B_m is a ball. Each B_m can then be written as $\zeta_m + \pi^{r_m}\mathfrak{o}$ where $\zeta_m \in B_m$ and r_m is a non-negative integer. The strong triangle inequality implies that any ζ in B_m has equal claim to being its center, and we use this fact to our advantage in the proof of the next theorem.

THEOREM 4.10. *Let $(B_m = \zeta_m + \pi^{r_m}\mathfrak{o} : m = 1, \ldots, M)$ be a collection of disjoint balls, and let $\mathbf{n}$ be an occupation vector which sums to N. Then,*

$$\begin{aligned}\mathbb{P}_N\{N_{\mathbf{B}} = \mathbf{n}\} &= \frac{N!}{Z(N, \mathfrak{o}, \beta)}\left\{\prod_{k<\ell}^{M} |\zeta_\ell - \zeta_k|^{\beta n_\ell n_k}\right\} \prod_{m=1}^{M} \frac{Z(n_m, \pi^{r_m}\mathfrak{o}, \beta)}{n_m!} \\ &= \frac{N!}{Z(N, \mathfrak{o}, \beta)}\left\{\prod_{k<\ell}^{M} |\zeta_\ell - \zeta_k|^{\beta n_\ell n_k}\right\} \prod_{m=1}^{M} q^{-r_m\beta\binom{n_m}{2} - r_m n_m} \frac{Z(n_m, \mathfrak{o}, \beta)}{n_m!}.\end{aligned}$$

Before the proof, a couple of remarks are in order. First, this gives an exact method of computing the probabilities of these special cylinder sets in finite time. In fact, since any finitely-described cylinder set is a disjoint union of these special cylinder sets we can now compute exactly, and in finite time, the probability for any cylinder set we may care about. Finally we remark that the $\prod_{k<\ell}^{M} |\zeta_\ell - \zeta_k|^{\beta n_m n_k}$ term that appears is the Boltzmann factor for a system at inverse temperature β and a particle at each of the ζ_m with integer charge n_m. This connects to the *multi-component ensemble* where we allow particles to have different integer multiple charges, and is considered in Section 6.

PROOF. There are $\binom{N}{n_1, \ldots, n_M}$ images of $B_1^{n_1} \times \cdots \times B_M^{n_M}$ under the action of S_N. Thus,

$$\begin{aligned}\mathbb{P}_N\{N_{\mathbf{B}} = \mathbf{n}\} = \frac{1}{Z(N, \mathfrak{o}, \beta)} \frac{N!}{n_1! \cdots n_M!} \int_{B_1^{n_1}} \cdots \int_{B_M^{n_M}} &\left\{\prod_{k<\ell}^{M} \prod_{i=1}^{n_\ell} \prod_{j=1}^{n_k} |\alpha_i^\ell - \alpha_j^k|^\beta\right\} \\ &\times \prod_{m=1}^{M} |\Delta_{n_m}(\boldsymbol{\alpha}^m)|^\beta d\mu^{n_1}(\boldsymbol{\alpha}^1) \cdots d\mu^{n_M}(\boldsymbol{\alpha}^M).\end{aligned}$$

But now, if $\alpha_\ell \in B_\ell$ and $\alpha_k \in B_k$ then,

$$|\alpha_\ell - \alpha_k| = |\alpha_\ell - \zeta_\ell + \zeta_\ell - \zeta_k + \zeta_k - \alpha_k| \leq \max\{|\alpha_\ell - \zeta_\ell|, |\zeta_\ell - \zeta_k|, |\zeta_k - \alpha_k|\}.$$

Notice also that since $\zeta_k \notin B_\ell$ we must have

$$|\zeta_\ell - \zeta_k| > |\alpha_\ell - \zeta_\ell| \qquad \text{and similarly} \qquad |\zeta_\ell - \zeta_k| > |\alpha_k - \zeta_k|.$$

The strong triangle inequality is thus an *equality*, and $|\alpha_\ell - \alpha_k| = |\zeta_\ell - \zeta_k|$. It follows that

$$\prod_{k<\ell}^{M} \prod_{i=1}^{n_\ell} \prod_{j=1}^{n_k} |\alpha_i^\ell - \alpha_j^k|^\beta = \prod_{k<\ell}^{M} |\zeta_\ell - \zeta_k|^{\beta n_\ell n_k},$$

and

$$\mathbb{P}_N\{N_{\mathbf{B}} = \mathbf{n}\} = \frac{N!}{Z(N, \mathfrak{o}, \beta)}\left\{\prod_{k<\ell}^{M} |\zeta_\ell - \zeta_k|^{\beta n_m n_k}\right\} \prod_{m=1}^{M} \frac{Z(n_m, B_m, \beta)}{n_m!}.$$

To arrive at the expressions given in the statement of the theorem, we first notice that translating the ball $B_m = \zeta_m + \pi^{r_m}\mathfrak{o}$ to $\pi^{r_m}\mathfrak{o}$ does not change the partition function. That is $Z(n_m, B_m, \beta) = Z(n_m, \pi^{r_m}\mathfrak{o}, \beta)$. The second expression follows from the first by using the fact that $\pi^{r_m}\mathfrak{o}$ is a contraction of $\mathfrak{o}$ and the integral defining $Z(n_m, \pi^{r_m}\mathfrak{o}, \beta)$ can be expressed in terms of $Z(n_m, \mathfrak{o}, \beta)$ accordingly. $\square$

To derive probabilities of more general cylinder sets $\{N_{\mathbf{B}} = \mathbf{n}\}$ where $n_1 + \cdots + n_M = N' < N$, we note that B being a finite union of balls implies that B^c too can be expressed as the disjoint union of finitely many balls in $\mathfrak{o}$. That is there exist balls $C_1, \ldots, C_K$ so that $B^c = C_1 \sqcup \cdots \sqcup C_K$. There is more than one way of doing this, but there is a unique set of balls (up to reordering) with minimal K. We then write

$$\{N_{\mathbf{B}} = \mathbf{n}\} = \bigsqcup_{\Sigma \ell_k = N - N'} \{N_{\mathbf{B}} = \mathbf{n}, N_{\mathbf{C}} = \boldsymbol{\ell}\},$$

where the union is over all occupation vectors $\boldsymbol{\ell} = (\ell_1, \ldots, \ell_K)$ summing to $N - N'$. Events like $\{N_{\mathbf{B}} = \mathbf{n}, N_{\mathbf{C}} = \boldsymbol{\ell}\}$ are now computable by Theorem 4.10.

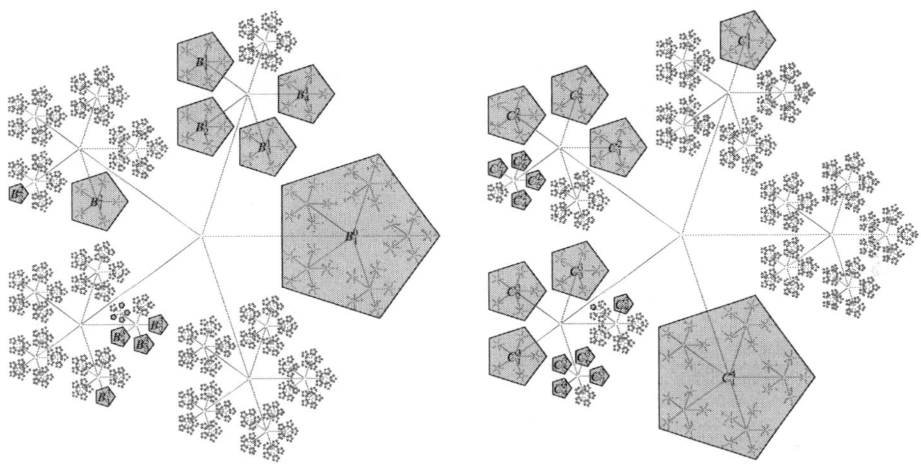

FIGURE 2. An example of a finite disjoint union of balls $\mathbf{B}$ in $\mathfrak{o} = \mathbb{Z}_5$, and its complementary disjoint union $\mathbf{C}$. These are labeled as in Corollary 4.12, with labels suppressed for smaller diameter balls.

COROLLARY 4.11. *Suppose* $B = B_1 \sqcup \cdots \sqcup B_M$ *is a proper subset of* $\mathfrak{o}$ *with each* $B_m = \zeta_m + \pi^{r_m}\mathfrak{o}$, *and* $B^c = C_1 \sqcup \cdots \sqcup C_K$ *where each* $C_k = \xi_k + \pi^{t_k}\mathfrak{o}$. *Then, if* $\mathbf{n} = (n_1, \ldots, n_M)$ *is an occupation vector with* $n_1 + \cdots + n_M = N' < N$. *Then,*

$$\mathbb{P}_N\{N_{\mathbf{B}} = \mathbf{n}\} = \frac{N!}{Z(N, \mathfrak{o}, \beta)} \left[\frac{Z(N', \mathfrak{o}, \beta)}{N'!} \mathbb{P}_{N'}\{N_{\mathbf{B}} = \mathbf{n}\} \right.$$

$$\left. \times \sum_{\boldsymbol{\ell} : \Sigma \ell_k = N - N'} \left\{ \prod_{m=1}^{M} \prod_{k=1}^{K} |\zeta_m - \xi_k|^{\beta n_m \ell_k} \right\} \frac{Z(N - N', \mathfrak{o}, \beta)}{(N - N')!} \mathbb{P}_{N-N'}\{N_{\mathbf{C}} = \boldsymbol{\ell}\} \right],$$

where the sum is over all occupation vectors $\boldsymbol{\ell} = (\ell_1, \ldots, \ell_K)$ *summing to* $N - N'$.

In many situations we may simplify Corollary 4.11 by using the additivity of energy over cosets: If $\zeta + \pi^r\mathfrak{o}$ and $\xi + \pi^s\mathfrak{o}$ are in different cosets of $\mathfrak{m}$, then $|\zeta - \xi| = 1$. This is the basis for the following formidable-looking simplification.

COROLLARY 4.12. *Suppose for each* $j = 0, \ldots, q-1$, $\mathbf{B}^j = (B_1^j, \ldots, B_{M_j}^j)$ *is a set of disjoint balls in* $j + \mathfrak{m}$ *and* $\mathbf{C}^j = (C_1^j, \ldots, C_{K_j}^j)$ *is a family of disjoint balls whose union* C^j *is the complement of* $B^j = B_1^j \sqcup \cdots \sqcup B_{M_j}^j$ *in* $j + \mathfrak{m}$. *Suppose*

$B^j_m = \zeta^j_m + \pi^{r^j_m}\mathfrak{o}$ *and* $C^j_\ell = \xi^j_\ell + \pi^{t^j_\ell}\mathfrak{o}$. *Set* $N^j = n^j_1 + \cdots + n^j_{M_j}$ *and suppose* $N' = N^0 + \cdots + N^{q-1} \le N$, *and* $L = N - N'$.

$$\mathbb{P}_N\{N_{\mathbf{B}^0} = \mathbf{n}^0, \ldots, N_{\mathbf{B}^{q-1}} = \mathbf{n}^{q-1}\} = \frac{N!}{Z(N,\mathfrak{o},\beta)}\left\{\prod_{j=0}^{q-1} \frac{Z(N_j,\mathfrak{o},\beta)}{N_j!}\mathbb{P}_{N_j}\{N_{\mathbf{B}^j} = \mathbf{n}^j\}\right\}$$

$$\sum_{\Sigma L^j = L}\sum_{\Sigma l^0_k = L^0}\cdots\sum_{\Sigma l^{q-1}_k = L^{q-1}}\left\{\prod_{j=0}^{q-1}\left\{\prod_{m=1}^{N_j}\prod_{k=1}^{L_j}|\zeta^j_m - \xi^j_k|^{\beta n^j_m \ell^j_k}\frac{Z(L_j,\mathfrak{o},\beta)}{L_j!}\right\}\mathbb{P}_{L_j}\{N_{\mathbf{C}^j} = \boldsymbol{\ell}^j\}\right\},$$

where the sums are over $L^0 + \cdots + L^{q-1} = L$ *and all* $\boldsymbol{\ell}^j = (\ell^j_1, \ldots, \ell^j_{K_j})$ *where* $\ell^j_1 + \cdots + \ell^j_{K_j} = L^j$ *for* $j = 0, \ldots, q-1$.

Note that any cylinder set given as in Corollary 4.11 can be written as $\{N_{\mathbf{B}^0} = \mathbf{n}^0, \ldots, N_{\mathbf{B}^{q-1}} = \mathbf{n}^{q-1}\}$ for an appropriate choice of the $\mathbf{B}^j$ and $\mathbf{n}^j$.

4.7. Conditioning on $\mathcal{C}_N(B)$. Now that we understand the σ-algebra of cylinder sets, at least in principle, we turn to the conditional distribution of particles given events like $\{N_B = n\}$. Specifically, we consider events of the form

$$S_N \cdot A_1 \times \cdots \times A_n \times \underbrace{B^c \times \cdots \times B^c}_{N-n},$$

where $A_1 \times \cdots \times A_n$ is a measurable rectangle in B^n. We define $\mathcal{L}_N(B)$ to be the *local* σ-algebra generated by all such sets (over all possible $0 \le n \le N$). We also denote the set $\mathcal{L}^n_N(B)$ by

$$\mathcal{L}^n_N(B) = \{E \cap \{N_B = n\} : E \in \mathcal{L}_{N,B}\}$$

Events in $\mathcal{L}^n_N(B)$ are events on which $N_B = n$ and which contain information about those n particles in B. We remark that $\mathcal{L}^n_N(B)$ is itself *not* a σ-algebra. However, it *is* the image of the Borel σ-algebra on B^n under the map

$$A \mapsto S_N \cdot A \times (B^c)^{N-n}$$

In fact, we could take $A \subseteq B^n$ to be symmetrized with respect to S_n and we get a bijective correspondence between $\mathcal{L}^n_N(B)$ and the σ-algebra of symmetrized Borel subsets of B^n denoted $\mathcal{S}_n(B)$. The complementary information, about the particles in B^c, is given by the σ-algebras $\mathcal{L}_{N,B^c}$ and $\mathcal{S}_{N-n}(B^c)$ (which is in correspondence with $\mathcal{L}^{N-n}_N(B^c)$).

Our main result for this section is the following.

THEOREM 4.13. *Let* $B \subseteq \mathfrak{o}$ *be a ball. Then* $\mathcal{L}_N(B)$ *and* $\mathcal{L}_N(B^c)$ *are conditionally independent given* $\mathcal{C}_N(B)$. *That is,* $\mathcal{L}_N(B)]\mathcal{C}_N(B)[\mathcal{L}_N(B^c)$. *Moreover, the conditional distribution on* $\mathcal{L}_N(B)$ *given the event* $\{N_B = n\}$ *is*

$$\mathbb{P}_N\{S_N \cdot A_1 \times \cdots A_n \times (B^c)^{N-n} | N_B = n\} = \mathbb{P}_n(S_n \cdot A_1 \times \cdots \times A_n).$$

PROOF. On the event $\{N_B = n\}$, a generic set $\mathcal{L}^n_N(B)$ is a union of events which look like $A = S_N \cdot A_1 \times \cdots \times A_n \times (B^c)^{N-n}$. Similarly, a generic event in $\mathcal{L}^{N-n}_N(B^c)$ looks like $E = S_N \cdot B^n \times E_1 \times \cdots \times E_{N-n}$. The intersection of A and E is then given by $S_N \cdot A_1 \times \cdots \times A_n \times E_1 \times \cdots \times E_{N-n}$.

It follows that

$$\mathbb{E}[\mathbf{1}_{\{N_B=n\}}\mathbf{1}_A\mathbf{1}_E] = \frac{N!}{Z(N,\mathfrak{o},\beta)}\int_{A_1}\cdots\int_{A_n}\int_{E_1}\cdots\int_{E_{N-n}}$$

$$|\Delta_n(\boldsymbol{\alpha})|^\beta|\Delta_{N-n}(\boldsymbol{\gamma})|^\beta\prod_{j=1}^{n}\prod_{\ell=1}^{N-n}|\alpha_j-\gamma_\ell|^\beta\,d\mu^n(\boldsymbol{\alpha})d\mu^{N-n}(\boldsymbol{\gamma}).$$

Suppose now $\alpha \in B$ and $\gamma \in B^c$. Since B is assumed to be a ball, there is some radius on which $B = \{\alpha' : |\alpha'-\alpha| \leq r\}$. Since $\gamma \notin B$ we must also have $|\alpha-\gamma| > r$. It follows then that if α' is any other element of B,

$$|\alpha'-\gamma| = |\alpha'-\alpha+\alpha-\gamma| \leq \max\{|\alpha-\alpha'|, |\alpha-\gamma|\} = |\alpha-\gamma|$$

by the strong triangle inequality. But since $|\alpha-\alpha'| < |\alpha-\gamma|$, in fact the inequality becomes an equality, and we conclude $|\alpha'-\gamma| = |\alpha-\gamma|$. It follows that, as a function of $\boldsymbol{\alpha}$,

$$\prod_{j=1}^{n}\prod_{\ell=1}^{N-n}|\alpha_j-\gamma_\ell|$$

is constant on B^n, and we write $R_B(\boldsymbol{\gamma})$ for the resulting function on $(B^c)^{N-n}$. Consequently,

$$\mathbb{E}[\mathbf{1}_{\{N_B=n\}}\mathbf{1}_A\mathbf{1}_E] = \frac{N!}{Z(N,\mathfrak{o},\beta)}\int_{A_1}\cdots\int_{A_n}|\Delta_n(\boldsymbol{\alpha})|^\beta d\mu^n(\boldsymbol{\alpha})\int_{E_1}\cdots\int_{E_{N-n}}$$

$$R_B(\boldsymbol{\gamma})^\beta|\Delta_{N-n}(\boldsymbol{\gamma})|^\beta d\mu^{N-n}(\boldsymbol{\gamma}).$$

Similar reasoning shows

$$\mathbb{E}[\mathbf{1}_{\{N_B=n\}}\mathbf{1}_A] = \frac{N!}{(N-n)!Z(N,\mathfrak{o},\beta)}\int_{A_1}\cdots\int_{A_n}|\Delta_n(\boldsymbol{\alpha})|^\beta d\mu^n(\boldsymbol{\alpha})\int_{(B^c)^{N-n}}$$

$$R_B(\boldsymbol{\gamma})^\beta|\Delta_{N-n}(\boldsymbol{\gamma})|^\beta d\mu^{N-n}(\boldsymbol{\gamma}),$$

$$\mathbb{E}[\mathbf{1}_{\{N_B=n\}}\mathbf{1}_E] = \frac{N!}{n!Z(N,\mathfrak{o},\beta)}\int_{B^n}|\Delta_n(\boldsymbol{\alpha})|^\beta d\mu^n(\boldsymbol{\alpha})\int_{E_1}\cdots\int_{E_{N-n}}$$

$$R_B(\boldsymbol{\gamma})^\beta|\Delta_{N-n}(\boldsymbol{\gamma})|^\beta d\mu^{N-n}(\boldsymbol{\gamma}),$$

and

$$\mathbb{E}[\mathbf{1}_{\{N_B=n\}}] = \frac{N!}{n!(N-n)!Z(N,\mathfrak{o},\beta)}\int_{B^n}|\Delta_n(\boldsymbol{\alpha})|^\beta d\mu^n(\boldsymbol{\alpha})$$

$$\int_{(B^c)^{N-n}}R_B(\boldsymbol{\gamma})^\beta|\Delta_{N-n}(\boldsymbol{\gamma})|^\beta d\mu^{N-n}(\boldsymbol{\gamma}).$$

It follows that

$$\frac{\mathbb{E}[\mathbf{1}_{\{N_B=n\}}\mathbf{1}_A]}{\mathbb{E}[\mathbf{1}_{\{N_B=n\}}]} = \frac{n!}{Z(n,B,\beta)}\int_{A_1}\cdots\int_{A_n}|\Delta_n(\boldsymbol{\alpha})|^\beta\,d\mu^n(\boldsymbol{\alpha}).$$

and

$$\frac{\mathbb{E}[\mathbf{1}_{\{N_B=n\}}\mathbf{1}_A\mathbf{1}_E]}{\mathbb{E}[\mathbf{1}_{\{N_B=n\}}]} = \frac{\mathbb{E}[\mathbf{1}_{\{N_B=n\}}\mathbf{1}_E]}{\mathbb{E}[\mathbf{1}_{\{N_B=n\}}]}\frac{\mathbb{E}[\mathbf{1}_{\{N_B=n\}}\mathbf{1}_A]}{\mathbb{E}[\mathbf{1}_{\{N_B=n\}}]}$$

That is, $\mathbf{1}_A$ and $\mathbf{1}_E$ are independent given the event $\{N_B = n\}$. Since $\mathcal{C}_N(B)$ consists of disjoint unions of the sets $\{N_B = n\}$ we conclude that $\mathcal{L}_N(B)$ and $\mathcal{L}_N(B^c)$ are conditionally independent given $\mathcal{C}_N(B)$. □

That is, the conditional distribution on $L_N(B)$ is given by

$$d\mathbb{P}_N(\boldsymbol{\alpha}|N_B = n) = \frac{|\Delta_n(\boldsymbol{\alpha})|^\beta}{Z(n, B, \beta)} d\mu^n(\boldsymbol{\alpha}).$$

Note that if $C \subseteq B^c$, then $\mathcal{L}_N(C) \subseteq \mathcal{L}_N(B^c)$. It follows that if we condition on the event $\{N_B = n, N_C = m\}$ the distribution of the n particles in B is independent of the m particles in C. In particular, if C is also a ball, then the conditional density of $\boldsymbol{\alpha} \in B^n$ and $\boldsymbol{\gamma} \in C^m$ is proportional to $|\Delta_n(\boldsymbol{\alpha})|^\beta |\Delta_m(\boldsymbol{\gamma})|^\beta$. We may view this as giving a probability distribution on $B^n \times C^m$ with the σ-algebra $\mathcal{S}(B^n) \otimes \mathcal{S}(C^m)$.

This is the basis for the following corollary.

COROLLARY 4.14. *Suppose* $\mathbf{B} = (B_1, \ldots, B_M)$ *is a vector of pairwise disjoint balls in* $\mathfrak{o}$, *and let* $\mathbf{n} = (n_1, \ldots, n_M)$ *be a vector of non-negative integers with* $n_1 + \cdots + n_M \leq N$. *Write*

$$\{N_{\mathbf{B}} = \mathbf{n}\} = \{N_{B_1} = n_1, \ldots, N_{B_M} = n_M\}.$$

Then the conditional distribution given $\{N_{\mathbf{B}} = \mathbf{n}\}$ *on* $\mathcal{S}(B_1^{n_1}) \otimes \cdots \otimes \mathcal{S}(B_M^{n_M})$ *is given by*

$$d\mathbb{P}_N(\boldsymbol{\alpha}_1, \ldots, \boldsymbol{\alpha}_M|N_{\mathbf{B}} = \mathbf{n}) = \frac{|\Delta_{n_1}(\boldsymbol{\alpha}_1)|^\beta}{Z(n_1, B, \beta)} d\mu^{n_1}(\boldsymbol{\alpha}_1) \cdots \frac{|\Delta_{n_M}(\boldsymbol{\alpha}_M)|^\beta}{Z(n_M, B, \beta)} d\mu^{n_M}(\boldsymbol{\alpha}_M).$$

5. The grand canonical ensemble

We now allow the system to exchange not only energy with the reservoir but also particles. That is N is no longer fixed. In this *grand canonical ensemble*, there is a parameter χ called the *chemical potential* which represents the energy cost/reward per particle. In this setting, the energy of a system with N particles in $\mathfrak{o}$ is given by

$$E(N, \boldsymbol{\alpha}) = \chi N - \sum_{n<m} \log |\alpha_n - \alpha_m|.$$

The density of states is then given by

$$\frac{1}{Z} e^{-\beta E(N, \boldsymbol{\alpha})}$$

where, the grand canonical partition function is given by

$$Z = Z(\chi, V, \beta) = \sum_{N=0}^{\infty} \frac{1}{N!} Z(N, V, \beta) e^{\chi \beta N}.$$

Notice the introduction of the $N!$ term in the summand. This term is now necessary since we are assuming that our particles are indistinguishable.[3] Sometimes, instead of the chemical potential, the energy cost per particle is encoded in the *fugacity* parameter $t := e^{-\beta\chi}$ and we write

$$Z(t, V, \beta) = \sum_{N=0}^{\infty} \frac{1}{N!} Z(N, V, \beta) t^N.$$

for this version of the grand canonical potential function. We will view t as independent of β, so that we may also view $Z(t, V, \beta)$ as the exponential generating function

[3] Even though the particles are indistinguishable in the canonical ensemble, there the $N!$ term is unnecessary since it appears in both the Boltzmann factor and the partition function and hence cancels in the probability density of states.

for the $Z(N,V,\beta)$. The physical partition function is then given by $Z(e^{-\beta\chi},V,\beta)$, and when deriving physical quantities, the latter should be used.

Note that, if $\boldsymbol{\alpha}\in\mathfrak{o}^N$, then the strong triangle inequality implies that $|\Delta(\boldsymbol{\alpha})|\leq 1$. Thus, for fixed $\beta\geq 0$,

$$\begin{aligned} Z(t,V,\beta) &= \sum_{N=0}^{\infty}\frac{t^N}{N!}\int_{V^N}|\Delta(\boldsymbol{\alpha})|^{\beta}\,d\mu^N(\boldsymbol{\alpha}) \\ &\leq \sum_{N=0}^{\infty}\frac{t^N}{N!}=e^t, \end{aligned}$$

with equality when $\beta=0$. That is, $Z(t,V,\beta)$ is an entire function of t for all $\beta\geq 0$.

5.1. The partition function in the grand canonical ensemble. First we consider the zero temperature ($\beta=+\infty$) and infinite temperature ($\beta=0$) regimes. In these cases we can compute the partition function exactly.

THEOREM 5.1. *At infinite temperature,*

$$Z(t,\mathfrak{o},0)=e^t \qquad \text{and} \qquad Z(t,\pi^{\ell}\mathfrak{o},0)=e^{t/q^{\ell}}$$

At zero temperature,

$$Z(t,\mathfrak{o},+\infty)=\left(1+\frac{t}{q}\right)^q \qquad \text{and} \qquad Z(t,\pi^{\ell}\mathfrak{o},+\infty)=1+\frac{t}{q^{\ell}}$$

PROOF. The $\beta=0$ situation is obvious, since $Z(N,\mathfrak{o},0)=1$ and $Z(N,\pi^{\ell}\mathfrak{o},0)=q^{-\ell N}$ for all N. When $\beta=+\infty$ only states with zero energy contribute to the partition function. Zero energy states can only occur when each of the particles is in a different coset of $\mathfrak{m}$. Since there are q cosets we must have $0\leq N\leq q$. For such an N there are $\binom{q}{N}$ ways of choosing which cosets get the particles, and $N!$ ways of distributing the particles amoungst these cosets. This collection of occupied cosets has volume $1/q^N$ and hence,

$$Z(t,\mathfrak{o},+\infty)=\sum_{N=0}^{q}\binom{q}{N}\left(\frac{t}{q}\right)^N=\left(1+\frac{t}{q}\right)^q.$$

For the neighborhood $\pi^{\ell}\mathfrak{o}$, the energy is only zero if there are 0 or 1 particles. Since $Z(0,V,+\infty)=1$ and $Z(1,V,+\infty)=\mu(V)$ the result follows. □

From Theorems 4.3 and 4.5 we quickly find the following

THEOREM 5.2.

$$Z(t,\mathfrak{o},\beta)=Z(t,\mathfrak{m},\beta)^q.$$

PROOF.

$$\begin{aligned} Z(t,\mathfrak{o},\beta) &= \sum_{N=0}^{\infty}Z(N,\mathfrak{o},\beta)\frac{t^N}{N!} \\ &= \sum_{N=0}^{\infty}\sum_{\mathbf{n}}\left\{\prod_{r=0}^{q-1}Z(n_r,\mathfrak{m},\beta)\frac{t^{n_r}}{n_r!}\right\}. \end{aligned}$$

The inner sum over $\mathbf{n}$ depends on N since we require that $n_0 + \cdots + n_{q-1} = N$. However, since we are summing over all N we may replace the double sum over N and $\mathbf{n}$ with a sum over all q-tuples of non-negative integers. That is,

$$Z(t, \mathfrak{o}, \beta) = \sum_{n_0} \cdots \sum_{n_{q-1}} \left\{ \prod_{r=0}^{q-1} Z(n_r, \mathfrak{m}, \beta) \frac{t^{n_r}}{n_r!} \right\}.$$

But then, Fubini's theorem (factoring the product over the sum) implies that

$$Z(t, \mathfrak{o}, \beta) = \sum_{n_0} Z(n_0, \mathfrak{m}, \beta) \frac{t^{n_0}}{n_0!} \cdots \sum_{n_{q-1}} Z(n_{q-1}, \mathfrak{m}, \beta) \frac{t^{n_{q-1}}}{n_{q-1}!} = Z(t, \mathfrak{m}, \beta)^q. \qquad \square$$

There is a physical explanation for this formula. The quantity

$$\Phi(t, \mathfrak{o}, \beta) = -\frac{1}{\beta} \log Z(t, \mathfrak{o}, \beta)$$

is called the grand canonical *potential* and plays the role of free energy in the canonical ensemble. Since it is an energy, we expect that if we have two non-interacting systems that we formally join into a single system, then their grand canonical potentials add. In our situation, the system with a variable number of particles in $\mathfrak{o}$ can be expressed as the union of q independent systems each with a variable number of particles in a coset of $\mathfrak{m}$. But the interaction between particles within a coset is independent of the identity of the coset, so we may as well assume that we have q independent copies of an ensemble with a variable number of particles in $\mathfrak{m}$. The grand canonical potentials add, and thus we have

$$\Phi(t, \mathfrak{o}, \beta) = q \cdot \Phi(t, \mathfrak{m}, \beta)$$

which leads to a heuristic proof of Theorem 5.2.

PROOF OF THEOREM 4.6. This proof follows from an observation of Cauchy about the relationship between two power series, one of which is an integer power of the other. $\square$

5.2. Algebraic formalism for grand canonical partition functions. We have seen that many important probabilities and physical quantities of our system reduce to expressions involving $Z(t, \mathfrak{m}, \beta)$. In particular, our main result $Z(t, \mathfrak{o}, \beta) = Z(t, \mathfrak{m}, \beta)^q$ demonstrates the primacy of the quantity $Z(t, \mathfrak{m}, \beta)$. Since $\mathfrak{m}$ and $\mathfrak{o}$ are homeomorphic and isomorphic, we might be tempted to think that $Z(t, \mathfrak{m}, \beta)$ is itself a q power of another function ($Z(t, \pi^2\mathfrak{o}, \beta)$ springs to mind). This is not the case, at least not in the traditional way.

To see why, consider a system with all particles in $\mathfrak{m}$. By counting the number of particles in each coset of $\pi^2\mathfrak{o}$ we get an occupation vector $\mathbf{n}$. Two particles in different cosets of $\pi^2\mathfrak{o}$ will be distance $1/q$ from each other, and each such pair will contribute $\log q$ to the energy of the system. That is, if $(\boldsymbol{\alpha}_0, \ldots, \boldsymbol{\alpha}_{q-1})$ is a factored state of the system so that $\boldsymbol{\alpha}_r$ is an n_r vector whose coordinates are the location of particles in $\pi(r + \mathfrak{m})$, then

$$E(\boldsymbol{\alpha}_0, \ldots, \boldsymbol{\alpha}_{q-1}) = \sum_{0 \le r < s < q} n_r n_s \log q + \sum_{r=0}^{q-1} E(\boldsymbol{\alpha}_r) \tag{5.1}$$

In contrast, if two particles are in different cosets of $\mathfrak{m}$ then they are distance 1 from each other, and there is no interaction energy, $\log q$, introduced by such pairs.

We introduce some algebraic formalism to deal with the terms like $q^{-\beta n_r n_s}$ which arise for the factors introduced into $Z(N, \mathfrak{m}, \beta)$ from the interaction energy between pairs in different cosets of $\pi^2\mathfrak{o}$.

Let R be a commutative ring and let $(R[[t]], +)$ be the group of formal power series in the indeterminant t with the usual definition of $+$. Given two series $\sum a_n t^n$ and $\sum b_m t^m$ and non-zero constant $C \in R$, we define their $\star$-product to be

$$\sum a_n t^n \star \sum b_m t^m = \sum_{m,n} C^{nm} a_n b_m t^{n+m}.$$

LEMMA 5.3. *$\star$ is an associative operation on $R[[t]]$.*

PROOF. Consider

$$\Big(\sum a_n t^n \star \sum b_m t^m\Big) \star \sum c_\ell t^\ell = \Big(\sum_{m,n} C^{nm} a_n b_m t^{n+m}\Big) \star \sum c_\ell t^\ell.$$

Simplify

$$\sum_{m,n} C^{nm} a_n b_m t^{n+m} = \sum_k \Big[\sum_{m+n=k} C^{nm} a_n b_m\Big] t^k$$

Hence,

$$\begin{aligned}\Big(\sum a_n t^n \star \sum b_m t^m\Big) \star \sum c_\ell t^\ell &= \sum_{\ell,k} C^{k\ell} c_\ell \Big[\sum_{m+n=k} C^{nm} a_n b_m\Big] t^{k+\ell} \\ &= \sum_{\ell,m,n} C^{(nm+n\ell+m\ell)} a_n b_m c_\ell t^{n+m+\ell} \\ &= \sum a_n t^n \star \Big(\sum b_m t^m \star \sum c_\ell t^\ell\Big).\end{aligned}$$

where the conclusion follows from symmetry in the penultimate equation. □

THEOREM 5.4. *$R[[t]]$ is a commutative ring under the operations $+$ and $\star$ with unity $1 = 1 + 0t + 0t^2 + \cdots$. Moreover, if R is an integral domain, then so too is $(R[[t]], +, \star)$.*

PROOF. Commutativity and that 1 is the multiplicative identity are obvious from the definition of $\star$. To verify the distributive property of $\star$ over $+$, consider

$$\begin{aligned}\sum a_n t^n \star \Big(\sum b_m t^m + \sum b'_m t^m\Big) &= \sum a_n t^n \star \sum (b_m + b'_m) t^m \\ &= \sum_{n,m} C^{nm} a_n (b_m + b'_m) t^{n+m} \\ &= \sum_{n,m} C^{nm} a_n b_m t^{n+m} + \sum_{n,m} C^{nm} a_n b'_m t^{n+m} \\ &= \sum a_n t^n \star \sum b_m t^m + \sum a_n t^n \star \sum b'_m t^m.\end{aligned}$$

Next suppose R is an integral domain and $\sum a_n t^n$ and $\sum b_m t^m$ are nonzero. Then there exist n_0 and m_0 so that a_{n_0} and b_{m_0} are non-zero. It follows that the coefficient of $t^{n_0+m_0}$ in $\sum a_n t^n \star \sum b_m t^m$ is $a_{n_0} b_{m_0} C^{n_0 m_0} \neq 0$. Hence, in this situation $(R[[t]], +, \star)$ has no zero divisors. □

LEMMA 5.5. *In this ring, the Jth power of a power series satisfies*

$$\Big(\sum a_n t^n\Big)^{\star J} = \sum_{n_1,\dots,n_J} \Big\{\prod_{j<k}^{J} C^{n_j n_k}\Big\} \prod_{j=1}^{J} a_{n_j} t^{n_j}.$$

PROOF. We induct on J. The base case is trivial. Then,

$$\begin{aligned}
\Big(\sum a_n t^n\Big)^{\star J} &= \Big(\sum a_n t^n\Big)^{\star(J-1)} \star \sum a_n t^n \\
&= \Big(\sum_{n_1,\dots,n_{J-1}} \Big\{\prod_{j<k}^{J-1} C^{n_j n_k}\Big\} \prod_{j=1}^{J-1} a_{n_j} t^{n_j}\Big) \star \sum a_n t^n \\
&= \Big(\sum_m \sum_{n_1+\cdots+n_{J-1}=m} t^m \Big\{\prod_{j<k}^{J-1} C^{n_j n_k}\Big\} \prod_{j=1}^{J-1} a_{n_j}\Big) \star \sum a_n t^n \\
&= \sum_m \sum_n C^{nm} \Big(\sum_{n_1+\cdots+n_{J-1}=m} \Big\{\prod_{j<k}^{J-1} C^{n_j n_k}\Big\} \prod_{j=1}^{J-1} a_{n_j}\Big) a_n t^{n+m},
\end{aligned}$$

and this is equal to

$$\sum_m \sum_n \Big(\sum_{n_1+\cdots+n_{J-1}=m} \Big\{C^{n(n_1+\cdots+n_{J-1})} \prod_{j<k}^{J-1} C^{n_j n_k}\Big\} \prod_{j=1}^{J-1} a_{n_j}\Big) a_n t^{n+m}.$$

Renaming the index variable n to n_J,

$$\Big(\sum a_n t^n\Big)^{\star J} = \sum_m \Big(\sum_{n_J} \sum_{n_1+\cdots+n_{J-1}=m} \Big\{\prod_{j<k}^{J} C^{n_j n_k}\Big\} \prod_{j=1}^{J} a_{n_j} t^{n_j}\Big).$$

The sum over m is superfluous if we remove the restriction on the n_j, and we arrive at the formulation in the statement of the lemma. □

5.3. $\star$ as a convolution operator. We define a transform on $(R[[t]], +, \star)$ by

$$\overline{\sum a_n t^n} := \sum C^{\binom{n}{2}} a_n t^n,$$

where $C^{\binom{0}{2}} = C^{\binom{1}{2}} := 1$. If C is invertible in R, then this transform has an inverse given by

$$\underline{\sum a_n t^n} := \sum C^{-\binom{n}{2}} a_n t^n.$$

Clearly then, using this notation,

$$\underline{\overline{\sum a_n t^n}} = \sum a_n t^n.$$

LEMMA 5.6.

$$\overline{\sum a_n t^n} \star \overline{\sum b_m t^n} = \overline{\sum a_n t^n \sum b_m t^m},$$

where the multiplication on the right hand side is the usual multiplication of power series.

PROOF.

$$\begin{aligned}\overline{\sum a_n t^n} \star \overline{\sum b_m t^n} &= \sum C^{\binom{n}{2}} a_n t^n \star \sum C^{\binom{m}{2}} b_m t^m \\ &= \sum_{n,m} C^{\binom{n}{2}} C^{nm} C^{\binom{m}{2}} a_n b_m t^{n+m} \\ &= \sum_N \left(\sum_{n+m=N} C^{\binom{n}{2}} C^{nm} C^{\binom{m}{2}} a_n b_m \right) t^N.\end{aligned}$$

Finally, since

$$\binom{n}{2} + \binom{m}{2} + nm = \binom{n+m}{2},$$

we have

$$\overline{\sum a_n t^n} \star \overline{\sum b_m t^n} = \sum_N \left(\sum_{n+m=N} a_n b_m \right) C^{\binom{N}{2}} t^N = \overline{\sum a_n t^n \sum b_m t^m}. \qquad \square$$

It will be useful to iterate the transform and its inverse. We will denote the ℓth iteration of the transform and its inverse (when it exists) by, respectively

$$\overline{\sum a_n t^n}^{\,\ell} \qquad \text{and} \qquad \underline{\sum a_n t^n}_{\,\ell}$$

Note that these are just the transforms formed by replacing C in the original definition with C^ℓ. As such, there is a product $\star^\ell$, formed by replacing C in the definition of $\star$ with C^ℓ so that, for instance

$$\overline{\sum a_n t^n}^{\,\ell} \star^\ell \overline{\sum b_m t^n}^{\,\ell} = \overline{\sum a_n t^n \sum b_m t^m}^{\,\ell}.$$

To mirror our notation with transforms and inverses, it makes sense to set $\star_\ell$ for $\star^{(-\ell)}$. That is $\star_\ell$ is the convolution operator formed by replacing C with $C^{-\ell}$. Of course, this only makes sense if C is invertible in R. In this situation,

$$\underline{\sum a_n t^n}_{\,\ell} \star_\ell \underline{\sum b_m t^n}_{\,\ell} = \underline{\sum a_n t^n \sum b_m t^m}_{\,\ell}.$$

Note when $\ell = 0$ the transform (and its inverse) are the identity transform and $\star_0 = \star^0$ is simply the usual multiplication of power series.

5.4. Partition functions as $\star q$-powers. In this section, and throughout whenever we are discussing the grand canonical partition function, we specify that $C = q^{-\beta}$. The utility of the algebraic constructions we have introduced becomes apparent in the next two results.

THEOREM 5.7. *For any Borel set V and $\ell \geq 0$,*

$$Z(t, \pi^\ell V, \beta) = \overline{Z}^{\ell}(t/q^\ell, V, \beta).$$

In particular, $Z(t, \mathfrak{m}, \beta) = \overline{Z}(t/q, \mathfrak{o}, \beta)$.

This puts Theorem 5.2 into a new light, as we can now express the q-power relationship between $Z(t, \mathfrak{m}, \beta)$ and $Z(t, \mathfrak{o}, \beta)$ as a functional equation that $Z(t, \mathfrak{o}, \beta)$ must satisfy.

COROLLARY 5.8. $Z(t, \mathfrak{o}, \beta) = \overline{Z}(t/q, \mathfrak{o}, \beta)^q$. *That is* $Z(t, \mathfrak{o}, \beta)$ *satisfies* $\alpha(t) - (\overline{\alpha}(t/q))^q = 0$ *in* $\mathbb{R}[[t]]$.

THEOREM 5.9.

$$Z(t, \mathfrak{m}, \beta) = Z(t, \pi^2\mathfrak{o}, \beta)^{\star^1 q}.$$

More generally, for any integer ℓ,

$$Z(t, \pi^\ell\mathfrak{o}, \beta) = Z(t, \pi^{\ell+1}\mathfrak{o}, \beta)^{\star^\ell q}.$$

COROLLARY 5.10.

$$Z(t, \mathfrak{m}, \beta) = \overline{Z}(t/q, \mathfrak{m}, \beta)^{\star^1 q}$$

Some remarks:

(1) This says that the grand canonical partition function for any ideal is the qth power, using the appropriate operator $\star^\ell$, of the grand canonical partition function for its unique maximal ideal;
(2) Theorem 4.5 corresponds to $\ell = 0$;
(3) This formula is valid for any integer ℓ, including negative integers. This explains how to extend to fractional ideals outside of $\mathfrak{o}$;
(4) We may iterate so that, for instance

$$Z(t, \mathfrak{o}, \beta) = \left(Z(t, \pi^2\mathfrak{o}, \beta)^{\star^1 q}\right)^{\star^0 q}.$$

PROOF OF THEOREM 5.9. In the most general setting, this follows by replacing $\log q$ in (5.1) with $\ell \log q$. Then, using familiar maneuvers,

$$Z(N, \pi^\ell\mathfrak{o}, \beta) = \sum_{n_0+\cdots+n_{q-1}=N} N!\left\{\prod_{r<s} q^{-\beta\ell n_r n_s}\right\}\prod_{r=0}^{q-1}\frac{1}{n_r!}Z(n_r, \pi^{\ell+1}\mathfrak{o}, \beta),$$

and,

$$Z(t, \pi^\ell\mathfrak{o}, \beta) = \sum_{n_0,\ldots,n_{q-1}}\left\{\prod_{r<s} q^{-\beta\ell n_r n_s}\right\}\prod_{r=0}^{q-1}\frac{t^{n_r}}{n_r!}Z(n_r, \pi^{\ell+1}\mathfrak{o}, \beta).$$

The theorem now follows from Lemma 5.5. □

5.5. Probabilities of cylinder sets. We turn to the explicit construction and analysis of the underlying probability space induced by physical considerations. This will be constructed from the probability spaces from the canonical ensemble.

Our probability space

$$\Omega := \bigsqcup_N \mathfrak{o}^N$$

with σ-algebra

$$\mathcal{S} = \sigma\{A_1 \sqcup \cdots \sqcup A_M : M \in \mathbb{N}, A_n \in \mathcal{S}_n\}.$$

A generic element in $\mathcal{S}$ looks like $A = A_1 \sqcup A_2 \sqcup \cdots$ where $A_n \in \mathcal{S}_n$. The probability measure induced by the Boltzmann factor is then

$$\mathbb{P}(A) := \frac{1}{Z(t, \mathfrak{o}, \beta)}\sum_{N=0}^{\infty} Z(N, A_N, \beta)\frac{t^N}{N!},$$

or equivalently,

$$\mathbb{P}(A|N_\mathfrak{o} = N) = \frac{Z(N, A_N, \beta)}{Z(N, \mathfrak{o}, \beta)} \qquad \text{and} \qquad \mathbb{P}\{N_\mathfrak{o} = N\} = \frac{Z(N, \mathfrak{o}, \beta)}{Z(t, \mathfrak{o}, \beta)}\frac{t^N}{N!}.$$

$\mathbb{P}(A)$ depends implicitly on β and t. If we need to make this dependence explicit, we will write $\mathbb{P}(t, A, \beta)$. Theorem 5.1 implies that when $\beta = 0$, $\mathbb{P}\{N_\mathfrak{o} = N\} =$

$e^{-t}t^N/N!$, and hence $N_{\mathfrak{o}}$ is a Poisson random variable with parameter t. When $\beta = +\infty$, $N_{\mathfrak{o}}$ is a binomial random variable with q trials each with probability of success $t/(q+t)$.

The numerator of $\mathbb{P}(A)$ is an important generating series, and we define

$$Z(A) = Z(t, A, \beta) := \sum_{N=0}^{\infty} Z(N, A_N, \beta)\frac{t^N}{N!}.$$

Given A as above, $\zeta \in \mathfrak{o}$ and $r \in \mathbb{N}$, define

$$\zeta + \pi^\ell A = \bigsqcup_N (\zeta + \pi^\ell A_N). \tag{5.2}$$

Like in the canonical ensemble, there is a simple relationship between $Z(t, \zeta + \pi^\ell A, \beta)$ and $\overline{Z}^\ell(t/q^\ell, A, \beta)$. This is recorded in the following lemma, which follows immediately from Lemma 4.9 and the definition of $\overline{\sum a_n t^n}^\ell$.

LEMMA 5.11.

$$Z(t, \zeta + \pi^\ell \mathfrak{o}, \beta) = \overline{Z}^\ell(t/q^\ell, \mathfrak{o}, \beta).$$

Given Borel $B \subseteq \mathfrak{o}$ we define $\mathcal{C}(B) = \sigma(N_B)$ as before, and we define the σ-algebra of cylinder sets $\mathcal{C} \subseteq \mathcal{S}$ to be that generated by $\{N_B : B \in \mathcal{B}\}$. Each C in $\mathcal{C}$ can be written as $C = C_1 \sqcup C_2 \sqcup \cdots$ where $C_N = C \cap \{N_{\mathfrak{o}} = N\} \in \mathcal{C}_N$. To get a feel for cylinder sets in the grand canonical setting, consider the event $\{N_B = n\}$,

$$\{N_B = n\} = \bigsqcup_{N=n}^{\infty} S_N \cdot B^n \times (B^c)^{N-n}$$

The event that *all* particles are in B is given by $\bigsqcup_N B^N$.

As in the canonical ensemble, the cylinder sets whose probabilities are most easy to describe are those of the form $\{N_{\mathbf{B}} = \mathbf{n}\}$ where $\mathbf{B} = (B_1, \dots, B_M)$ is a family of disjoint balls. Here we do not necessarily assume that the union of the balls is all of $\mathfrak{o}$, but we do relabel the balls so that $\mathbf{B}^j = (B_1^j, \dots, B_{M_j}^j)$ are the family of balls contained in $j + \mathfrak{m}$. By likewise reorganizing the coordinates of $\mathbf{n}$ we can write

$$\{N_{\mathbf{B}} = \mathbf{n}\} = \{N_{\mathbf{B}_0} = \mathbf{n}_0, \dots, N_{\mathbf{B}_{q-1}} = \mathbf{n}_{q-1}\}.$$

We use the fact that $j + \mathfrak{m}$ is in bijection with $\mathfrak{o}$ under the map $\alpha \mapsto j + \pi\alpha$, and write $\underline{B}_k^j$ for the pre-image of B_k^j under this map. If B_k^j has radius $q^{-r_k^j}$ then $\underline{B}_k^j$ has radius $q^{-r_k^j+1}$. The $\underline{\mathbf{B}}^j = (\underline{B}_1^j, \dots, \underline{B}_{M_j}^j)$ are now disjoint balls in $\mathfrak{o}$. If we can write $Z(t, \{N_{\mathbf{B}} = \mathbf{n}\}, \beta)$ in terms of the $Z(t, \{N_{\underline{\mathbf{B}}^j} = \mathbf{n}^j\}, \beta)$ and provide a formula for $Z(t, \{N_{\mathfrak{o}} = n\}, \beta)$ we will have an inductive formula for $Z(t, \{N_{\mathbf{B}} = \mathbf{n}\}, \beta)$.

As in the canonical case we take $\mathbf{C}^j = (C_1^j, \dots, C_{K_j}^j)$ to be a collection of disjoint balls complementary to the $\mathbf{B}^j$ in $j + \mathfrak{m}$. Clearly then $\underline{\mathbf{C}}^j$ is complementary to $\underline{\mathbf{B}}^j$ in $\mathfrak{o}$. We remark that it is possible that either $\mathbf{B}^j$ or $\mathbf{C}^j$ are empty. If, for instance $\mathbf{B}^j$ is empty, then necessarily $\mathbf{n}^j$ is empty and $\mathbf{C}^j = j + \mathfrak{m}$. We need to then decipher what the event $\{N_{\mathbf{B}^j} = \mathbf{n}^j\}$ means when $\mathbf{B}^j$ and $\mathbf{n}^j$ are empty. In this case, we are putting no restriction on the number of particles in $j + \mathfrak{m}$, and this event is equal to

$$E := \bigsqcup_N (j + \mathfrak{m})^N.$$

Note that $\underline{E}$ is the probability that all particles are in $\mathfrak{o}$ with no restrictions on their number or location—that is with no restriction whatsoever. It follows that $\underline{E} = \Omega$ and $Z(t, \underline{E}, \beta) = Z(t, \mathfrak{o}, \beta)$. In essense, this says that if the event $\{\mathbf{N_B} = \mathbf{n}\}$ makes no specifications on the number of particles in a coset, then that coset contributes a factor of $Z(t, \mathfrak{m}, \beta)$ to $Z(t, \{N_{\mathbf{B}} = \mathbf{n}\}, \beta)$. Put another way, this provides one base case necessary for an inductive formula for $Z(t, \{N_{\mathbf{B}} = \mathbf{n}\}, \beta)$ in terms of the $Z(t, \{N_{\underline{\mathbf{B}}^j} = \mathbf{n}^j\}, \beta)$. The other base case is given by

$$Z(t, \{N_{\mathfrak{o}} = n\}, \beta) = \frac{t^N}{N!} Z(n, \mathfrak{o}, \beta).$$

THEOREM 5.12.

$$Z(t, \{N_{\mathbf{B}} = \mathbf{n}\}, \beta) = \prod_{j=0}^{q-1} \overline{Z}(t/q, \{N_{\underline{\mathbf{B}}^j} = \mathbf{n}^j\}, \beta).$$

PROOF. For each J, $\mathbf{n}^j = (n_1^j, \ldots, n_{M_j}^j)$, define $N^j := n_1^j + \cdots + n_{M_j}^j$. On the event $\{N_{\mathbf{B}} = \mathbf{n}\}$, N_j the minimum number of particles in $j + \mathfrak{m}$.
and using Corollary 4.12,

$$Z(t, \{N_{\mathbf{B}} = \mathbf{n}\}, \beta) = \sum_{N=N'}^{\infty} t^N \left\{ \prod_{j=0}^{q-1} \frac{Z(N^j, \{N_{\mathbf{B}^j} = \mathbf{n}^j\}, \beta)}{N^j!} \right\}$$

$$\sum_{\Sigma L^j = N - N'} \sum_{\Sigma l_k^0 = L^0} \cdots \sum_{\Sigma l_k^{q-1} = L^{q-1}} \left\{ \prod_{j=0}^{q-1} \left\{ \prod_{m=1}^{N^j} \prod_{k=1}^{L^j} |\zeta_m^j - \xi_k^j|^{\beta n_m^j \ell_k^j} \frac{Z(L^j, \{N_{\mathbf{C}^j} = \boldsymbol{\ell}^j\}, \beta)}{L^j!} \right\} \right\},$$

where the sums are over $L^0 + \cdots + L^{q-1} = N - N'$ and all $\boldsymbol{\ell}^j = (\ell_1^j, \ldots, \ell_{K_j}^j)$ where $\ell_1^j + \cdots + \ell_{K_j}^j = L^j$ for $j = 0, \ldots, q-1$.

The sum over N and frees the constraints on $L^0 + \cdots + L^{q-1} = N - N'$, and allows us to sum over all non-negative $L^0, \ldots, L^{q-1}$. That is,

$$Z(t, \{N_{\mathbf{B}} = \mathbf{n}\}, \beta) = t^{N'} \left\{ \prod_{j=0}^{q-1} \frac{Z(N^j, \{N_{\mathbf{B}^j} = \mathbf{n}^j\}, \beta)}{N^j!} \right\}$$

$$\sum_{L^0, \ldots, L^{q-1}} \sum_{\Sigma l_k^0 = L^0} \cdots \sum_{\Sigma l_k^{q-1} = L^{q-1}} \left\{ \prod_{j=0}^{q-1} t^{L^j} \frac{Z(L^j, \{N_{\mathbf{C}^j} = \boldsymbol{\ell}^j\}, \beta)}{L^j!} \left\{ \prod_{m=1}^{M^j} \prod_{k=1}^{K^j} |\zeta_m^j - \xi_k^j|^{\beta n_m^j \ell_k^j} \right\} \right\}.$$

This allows us to exchange the (inside) product over $j = 0, \ldots, q-1$ and the sums over the L^j. That is,

$$Z(t, \{N_{\mathbf{B}} = \mathbf{n}\}, \beta) = t^{N'} \left\{ \prod_{j=0}^{q-1} \frac{Z(N^j, \{N_{\mathbf{B}^j} = \mathbf{n}^j\}, \beta)}{N^j!} \right.$$

$$\left. \times \sum_{L} \sum_{\Sigma l_k = L} t^L \left\{ \prod_{m=1}^{M^j} \prod_{k=1}^{K^j} |\zeta_m^j - \xi_k^j|^{\beta n_m^j \ell_k^j} \right\} \frac{Z(L, \{N_{\mathbf{C}^j} = \boldsymbol{\ell}\}, \beta)}{L!} \right\}$$

$$= \left\{ \prod_{j=0}^{q-1} \sum_{L} t^{N^j + L} \frac{Z(N^j, \{N_{\mathbf{B}^j} = \mathbf{n}^j\}, \beta)}{N^j!} \right.$$

$$\left. \times \sum_{\Sigma l_k = L} \left\{ \prod_{m=1}^{M^j} \prod_{k=1}^{K^j} |\zeta_m^j - \xi_k^j|^{\beta n_m^j \ell_k^j} \right\} \frac{Z(L, \{N_{\mathbf{C}^j} = \boldsymbol{\ell}\}, \beta)}{L!} \right\}.$$

Note that $\{N_{\mathbf{B}^j} = \mathbf{n}^j\}$ accounts for the whereabouts of N^j particles in $j + \mathfrak{m}$, and hence

$$Z(N^j, \{N_{\mathbf{B}^j} = \mathbf{n}^j\}, \beta)$$
$$= Z(N^j, \{N_{j+\pi\underline{\mathbf{B}}^j} = \mathbf{n}^j\}, \beta) = q^{-\beta\binom{N^j}{2} - N^j} Z(N^j, \{N_{\underline{\mathbf{B}}^j} = \mathbf{n}^j\}, \beta).$$

Similarly,

$$Z(L, \{N_{\mathbf{C}^j} = \boldsymbol{\ell}\}, \beta) = q^{-\beta\binom{L}{2} - L} Z(L, \{N_{\underline{\mathbf{C}}^j} = \boldsymbol{\ell}\}, \beta).$$

And, since

$$\binom{N^j}{2} + \binom{L}{2} = \binom{N^j + L}{2} - LN^j,$$

we have

$$t^{N^j+L} Z(L, \{N_{\mathbf{C}^j} = \boldsymbol{\ell}\}, \beta) Z(N^j, \{N_{\mathbf{B}^j} = \mathbf{n}^j\}, \beta) \left\{ \prod_{m=1}^{M^j} \prod_{k=1}^{K^j} |\zeta_m^j - \xi_k^j|^{\beta n_m^j \ell_k^j} \right\}$$
$$= q^{-\beta\binom{N^j+L}{2}} \left(\frac{t}{q}\right)^{N^j+L} Z(N^j, \{N_{\underline{\mathbf{B}}^j} = \mathbf{n}^j\}, \beta) Z(L, \{N_{\underline{\mathbf{C}}^j} = \boldsymbol{\ell}\}, \beta)$$
$$\left\{ \prod_{m=1}^{M^j} \prod_{k=1}^{K^j} q^{\beta} |\zeta_m^j - \xi_k^j|^{\beta n_m^j \ell_k^j} \right\}.$$

Now note that, if we denote the centers of $\underline{B}_m^j$ and $\underline{C}_k^j$ by $\underline{\zeta}_m^j$ and $\underline{\xi}_k^j$, then $|\underline{\zeta}_m^j - \underline{\xi}_k^j| = q|\zeta_m^j - \xi_k^j|$. It follows that

$$t^{N^j+L} Z(L, \{N_{\mathbf{C}^j} = \boldsymbol{\ell}\}, \beta) Z(N^j, \{N_{\mathbf{B}^j} = \mathbf{n}^j\}, \beta) \left\{ \prod_{m=1}^{M^j} \prod_{k=1}^{K^j} |\zeta_m^j - \xi_k^j|^{\beta n_m^j \ell_k^j} \right\}$$
$$= q^{-\beta\binom{N^j+L}{2}} \left(\frac{t}{q}\right)^{N^j+L} Z(N^j, \{N_{\underline{\mathbf{B}}^j} = \mathbf{n}^j\}, \beta) Z(L, \{N_{\underline{\mathbf{C}}^j} = \boldsymbol{\ell}\}, \beta)$$
$$\left\{ \prod_{m=1}^{M^j} \prod_{k=1}^{K^j} |\underline{\zeta}_m^j - \underline{\xi}_k^j|^{\beta n_m^j \ell_k^j} \right\},$$

and hence

$$Z(t, \{N_{\mathbf{B}} = \mathbf{n}\}, \beta) = \left\{ \prod_{j=0}^{q-1} \sum_{L} q^{-\beta\binom{N^j+L}{2}} \left(\frac{t}{q}\right)^{N^j+L} \frac{Z(N^j, \{N_{\underline{\mathbf{B}}^j} = \mathbf{n}^j\}, \beta)}{N^j!} \right.$$
$$\left. \times \sum_{\Sigma l_k = L} \left\{ \prod_{m=1}^{M^j} \prod_{k=1}^{K^j} |\underline{\zeta}_m^j - \underline{\xi}_k^j|^{\beta n_m^j \ell_k^j} \right\} \frac{Z(L, \{N_{\underline{\mathbf{C}}^j} = \boldsymbol{\ell}\}, \beta)}{L!} \right\}.$$

Since the $\underline{\mathbf{B}}^j$ and $\underline{\mathbf{C}}^j$ union to all of $\mathfrak{o}$, we can use Corollary 4.12 again to find

$$Z(t, \{N_{\mathbf{B}} = \mathbf{n}\}, \beta) = \left\{ \prod_{j=0}^{q-1} \sum_{L} \frac{(t/q)^{N^j+L}}{(N^j+L)!} q^{-\beta\binom{N^j+L}{2}} Z(N^j + L, \{N_{\underline{\mathbf{B}}^j} = \mathbf{n}^j\}, \beta) \right\}$$

In fact, since $Z(K, \{N_{\mathbf{B}^j} = \mathbf{n}^j\}, \beta) = 0$ if $K < N^j$,

$$Z(t, \{N_{\mathbf{B}} = \mathbf{n}\}, \beta) = \left\{ \prod_{j=0}^{q-1} \sum_{K} \frac{(t/q)^K}{K!} q^{-\beta\binom{K}{2}} Z(K, \{N_{\underline{\mathbf{B}}^j} = \mathbf{n}^j\}, \beta) \right\}$$

where the last equality follows from Theorem 5.7. That is,

$$Z(t, \{N_{\mathbf{B}} = \mathbf{n}\}, \beta) = \prod_{j=0}^{q-1} \overline{Z}(t/q, \{N_{\underline{\mathbf{B}}^j} = \mathbf{n}^j\}, \beta). \qquad \square$$

As an example, consider the cylinder set $\{N_{\mathbf{B}} = \mathbf{n}\}$ in the 5-adics given diagramatically by

$$\{N_{\mathbf{B}} = \mathbf{n}\} =$$ 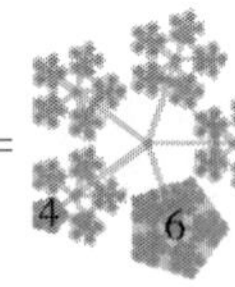

.

In words, $\{N_{\mathbf{B}} = \mathbf{n}\}$ is the set of states for which there are 6 particles in a fixed coset of $\mathfrak{m} = 5\mathbb{Z}_5$, and 4 particles in a distinct coset of $\pi^2\mathfrak{o} = 25\mathbb{Z}_5$. In our case, there are three cosets of $\mathfrak{m}$ whose occupation number is unspecified. Each of these three will contribute a factor of $\overline{Z}(t/5, \mathfrak{o}, \beta)$. The coset containing 6 particles will contribute a factor of

$$\overline{Z}(t/5, \{N_{\mathfrak{o}} = 6\}, \beta) = Z(6, \mathfrak{o}, \beta) 5^{-\beta\binom{6}{2}} \frac{(t/5)^6}{6!}.$$

The remaining coset contributes a factor of

$$\begin{aligned}\overline{\overline{Z}(t/25, \{\mathbb{N}_{\mathfrak{m}} = 4\}, \beta)\overline{Z}(t/25, \mathfrak{o}, \beta)^4} &= \overline{Z}^2(t/25, \{\mathbb{N}_{\mathfrak{o}} = 4\}, \beta) \star \overline{Z}^2(t/25, \mathfrak{o}, \beta)^{\star 4} \\ &= Z(4, \mathfrak{o}, \beta) 5^{-2\beta\binom{4}{2}} \frac{(t/25)^4}{4!} \star \overline{Z}^2(t/25, \mathfrak{o}, \beta)^{\star 4}.\end{aligned}$$

Putting it all together, we have

$$\begin{aligned}Z(t, \{N_{\mathbf{B}} = \mathbf{n}\}, \beta) = \overline{Z}(t/5, \mathfrak{o}, \beta)^3 Z(6, \mathfrak{o}, \beta) 5^{-\beta\binom{6}{2}} \frac{(t/5)^6}{6!} \\ \left(Z(4, \mathfrak{o}, \beta) 5^{-2\beta\binom{4}{2}} \frac{(t/25)^4}{4!} \star \overline{Z}^2(t/25, \mathfrak{o}, \beta)^{\star 4} \right).\end{aligned}$$

6. Multi-component ensembles

Here we allow particles to have different integer charges. Here we suppose $\mathbf{Q} = (Q_1, \ldots, Q_M)$ be distinct positive integers representing allowable charges. To eliminate notational complexity, we will view $\mathbf{Q}$ as fixed once and for all. Let $\mathbf{N} = (N_1, \ldots, N_M)$ be positive integers representing the number of each species of particles.

If we suppose that the particles of charge Q_m are identified with the coordinates of $\boldsymbol{\alpha}_m = (\alpha_m^1, \ldots, \alpha_m^{N_m}) \in \mathfrak{o}^{N_m}$ then the state of the system is specified by $(\boldsymbol{\alpha}_1, \ldots, \boldsymbol{\alpha}_M)$ and the energy of such a system is given by

(6.1)
$$E(\boldsymbol{\alpha}_1, \ldots, \boldsymbol{\alpha}_M) = -\sum_{m=1}^{M} Q_m^2 \sum_{j<k}^{N_m} \log|\alpha_m^j - \alpha_m^k| - \sum_{\ell<m}^{M} Q_\ell Q_m \sum_{j=1}^{N_\ell} \sum_{k=1}^{N_m} \log|\alpha_\ell^j - \alpha_m^k|.$$

This is the sum of the interaction energies between all pairs of particles.

The partition function in the canonical ensemble is thus

$$Z(\mathbf{N}, \mathfrak{o}, \beta) = \int_{\boldsymbol{\alpha}_1 \in \mathfrak{o}^{N_1}} \cdots \int_{\boldsymbol{\alpha}_M \in \mathfrak{o}^{N_M}} \left\{ \prod_{m=1}^{M} \prod_{j<k}^{N_m} |\alpha_m^j - \alpha_m^k|^{\beta Q_m^2} \right\}$$
$$\times \left\{ \prod_{\ell<m}^{M} \prod_{j=1}^{N_\ell} \prod_{k=1}^{N_m} |\alpha_\ell^j - \alpha_m^k|^{\beta Q_m Q_\ell} \right\} d\mu^{N_1}(\boldsymbol{\alpha}^1) \cdots d\mu^{N_M}(\boldsymbol{\alpha}^M),$$

and the partition function for the grand canonical ensemble (in the fugacity variables $\mathbf{t} = (t_1, \ldots, t_M)$) is given by

$$Z(\mathbf{t}, \mathfrak{o}, \beta) = \sum_{\mathbf{N}} Z(\mathbf{N}, \mathfrak{o}, \beta) \frac{\mathbf{t}^{\mathbf{N}}}{\mathbf{N}!} \qquad \text{where} \qquad \frac{\mathbf{t}^{\mathbf{N}}}{\mathbf{N}!} = \prod_{m=1}^{M} \frac{t_m^{N_m}}{N_m!}$$

THEOREM 6.1.

$$Z(\mathbf{t}, \mathfrak{o}, \beta) = Z(\mathbf{t}, \mathfrak{m}, \beta)^q$$

PROOF. A bit of strategy is in order. As before we will partition particles according to which coset they reside in. The allowance of different charges doesn't change the fact that the interaction energy between particles in different cosets is always zero. That is, the energy is still additive over cosets. This implies the integrand in $Z(\mathbf{N}, \mathfrak{o}, \beta)$ factors over cosets and Fubini's Theorem allows us to move the product, which is indexed by the cosets of $\mathfrak{m}$, outside the integral. The translation invariance of the remaining integrands allows us to translate each coset to $\mathfrak{m}$. Some combinatorial reorganization of the sums appearing in the partition function will then produce the result.

Given a state $(\boldsymbol{\alpha}_1, \ldots, \boldsymbol{\alpha}_M)$, for each $1 \leq m \leq M$ we introduce a factored state for $\boldsymbol{\alpha}_m$ given by $(\boldsymbol{\alpha}_m^0, \ldots \boldsymbol{\alpha}_m^{q-1})$. We will denote the jth entry of $\boldsymbol{\alpha}_m^r$ be $\alpha_m^r(j)$. We then define $\mathbf{N}_m := [N_m^0, \ldots, N_m^{q-1}]$ for the vector of non-negative integers counting how many of the particles with charge Q_m are in each of the cosets. We view $\mathbf{N}$ as an $M \times q$ matrix of non-negative integers the m, r entry of which N_m^r is the number of particles with charge Q_m in coset $r + \mathfrak{m}$. Note that $N_m^0 + \cdots + N_m^{q-1} = N_m$. In summary N_m is the number of particles with charge Q_m. These particles are located at the coordinates of $\boldsymbol{\alpha}_m$. The vector $\mathbf{N}_m$ gives the number of particles in each coset and $(\boldsymbol{\alpha}_m^0, \ldots, \boldsymbol{\alpha}_m^{q-1})$ is the factored state with $\boldsymbol{\alpha}_m^r \in \mathfrak{o}^{N_m^r}$ representing the location of all charge Q_m particles lying in the coset $r + \mathfrak{m}$.

Then,

(6.2)

$$E(\boldsymbol{\alpha}_1, \ldots, \boldsymbol{\alpha}_M) =$$
$$-\sum_{r=0}^{q-1} \left[\sum_{m=1}^{M} Q_m^2 \sum_{j<k}^{N_m^r} \log |\alpha_m^r(j) - \alpha_m^r(k)| - \sum_{\ell<m}^{M} Q_\ell Q_m \sum_{j=1}^{N_\ell^r} \sum_{k=1}^{N_m^r} \log |\alpha_\ell^r(j) - \alpha_m^r(k)| \right].$$

We will reorganize the integral defining the grand canonical partition function replacing the integral over all state vectors with an integral over factored state vectors. In order to do this correctly we need to account for the number of state vectors corresponding to a factored state. By permuting the corrdinates of $(\boldsymbol{\alpha}_m^0, \ldots, \boldsymbol{\alpha}_m^{q-1})$ we arrive at generic state vector specifying the positions of the particles with charge Q_m. Thus a choice of $\boldsymbol{\alpha}_m^r$ for all $m = 1, \ldots, M$ and $r = 0, \ldots, q-1$

leads to $N_1! \cdots N_M!$ different states of the system. The set of factored states still overcounts unique states, since permuting the coordinates of any one $\boldsymbol{\alpha}_m^r$ does not actually change the state of the system. Thus, in order to compensate for the over-counting within factored states, we need to introduce terms like $N_m^0! \cdots N_m^{q-1}!$ into the denominator. That is, if we integrate over factored states instead of all state vectors we need to compensate with the combinatorial term

$$\frac{N_1!}{N_1^0! \cdots N_1^{q-1}!} \cdot \cdots \cdot \frac{N_M!}{N_M^0! \cdots N_M^{q-1}!}$$

Now,

$$Z(\mathbf{t}, \mathfrak{o}, \beta) = \sum_{N_1=0}^{\infty} \frac{t_1^{N_1}}{N_1!} \cdots \sum_{N_M=0}^{\infty} \frac{t_M^{N_M}}{N_M!} \int_{\boldsymbol{\alpha}_1 \in \mathfrak{o}^{N_1}} \cdots \int_{\boldsymbol{\alpha}_M \in \mathfrak{o}^{N_M}} \left\{ \prod_{m=1}^{M} \prod_{j<k}^{N_m} |\alpha_j^m - \alpha_k^m|^{\beta Q_m^2} \right\}$$
$$\times \left\{ \prod_{\ell<m}^{M} \prod_{j=1}^{N_\ell} \prod_{k=1}^{N_m} |\alpha_j^\ell - \alpha_k^m|^{\beta Q_m Q_\ell} \right\} d\mu^{N_1}(\boldsymbol{\alpha}_1) \cdots d\mu^{N_M}(\boldsymbol{\alpha}_M).$$

First we note that by integrating over factored states we replace integrals such as

$$\int_{\boldsymbol{\alpha}_m \in \mathfrak{o}^{N_m}} \qquad \text{with} \qquad \left\{ \int_{\boldsymbol{\alpha}_m^0 \in \mathfrak{m}^{N_m^0}} \cdots \int_{\boldsymbol{\alpha}_m^{q-1} \in (q-1+\mathfrak{m})^{N_m^{q-1}}} \right\}$$

The combinatorial factor together with the fact that $N_m^0 + \cdots + N_m^{q-1} = N_m$ allows us to replace the sums

$$\sum_{N_m} \frac{t_m^{N_m}}{N_m!} = \sum_{N_m} \sum_{\{\mathbf{N}_m : N_m^0 + \cdots + N_m^{q-1} = N_m\}} \frac{t_m^{N_m^0} \cdots t_m^{N_m^{q-1}}}{N_m^0! \cdots N_m^{q-1}!},$$

The sum over N_m however, makes the condition that $N_m^0 + \cdots + N_m^{q-1} = N_m$ superfluous, and since the summand depends in no other way on N_m, we can in fact write

$$\sum_{N_m} \frac{t_m^{N_m}}{N_m!} = \sum_{\mathbf{N}_m} \frac{t_m^{N_m^0} \cdots t_m^{N_m^{q-1}}}{N_m^0! \cdots N_m^{q-1}!},$$

where the sum over the $\mathbf{N}_m$ is no longer constrained.

It follows then that

$$Z(\mathbf{t}, \mathfrak{o}, \beta) = \sum_{\mathbf{N}_1} \frac{t_1^{N_1^0} \cdots t_1^{N_1^{q-1}}}{N_1^0! \cdots N_1^{q-1}!} \cdots \sum_{\mathbf{N}_M} \frac{t_M^{N_M^0} \cdots t_M^{N_M^{q-1}}}{N_M^0! \cdots N_M^{q-1}!}$$
$$\times \left\{ \int_{\boldsymbol{\alpha}_1^0 \in \mathfrak{m}^{N_1^0}} \cdots \int_{\boldsymbol{\alpha}_1^{q-1} \in (q-1+\mathfrak{m})^{N_1^{q-1}}} \right\} \cdots \left\{ \int_{\boldsymbol{\alpha}_M^0 \in \mathfrak{m}^{N_M^0}} \cdots \int_{\boldsymbol{\alpha}_M^{q-1} \in (q-1+\mathfrak{m})^{N_M^{q-1}}} \right\}$$
$$\times \prod_{r=0}^{q-1} \left\{ \prod_{m=1}^{M} \prod_{j<k}^{N_m^r} |\alpha_m^r(j) - \alpha_m^r(k)|^{\beta Q_m^2} \times \prod_{\ell<m}^{M} \prod_{j=1}^{N_\ell^r} \prod_{k=1}^{N_m^r} |\alpha_\ell^r(j) - \alpha_m^r(k)|^{\beta Q_\ell Q_m} \right\}$$
$$\times\, d\mu^{N_1^0}(\boldsymbol{\alpha}_1^0) \cdots d\mu^{N_1^{q-1}}(\boldsymbol{\alpha}_1^{q-1}) \cdots d\mu^{N_M^0}(\boldsymbol{\alpha}_M^0) \cdots d\mu^{N_M^{q-1}}(\boldsymbol{\alpha}_M^{q-1}).$$

The factorization of the integrand over cosets is the multiplicative version of the additivity of energy over cosets, 6.1.

Next, we reorder the integrals so that, instead of being grouped by the charge Q_m, they are instead grouped by cosets. That is,

$$\left\{\int_{\boldsymbol{\alpha}_1^0\in\mathfrak{m}^{N_1^0}}\cdots\int_{\boldsymbol{\alpha}_1^{q-1}\in(q-1+\mathfrak{m})^{N_1^{q-1}}}\right\}\cdots\left\{\int_{\boldsymbol{\alpha}_M^0\in\mathfrak{m}^{N_M^0}}\cdots\int_{\boldsymbol{\alpha}_M^{q-1}\in(q-1+\mathfrak{m})^{N_M^{q-1}}}\right\}$$
$$=\left\{\int_{\boldsymbol{\alpha}_1^0\in\mathfrak{m}^{N_1^0}}\cdots\int_{\boldsymbol{\alpha}_M^0\in\mathfrak{m}^{N_M^0}}\right\}\cdots\left\{\int_{\boldsymbol{\alpha}_1^{q-1}\in(q-1+\mathfrak{m})^{N_1^{q-1}}}\cdots\int_{\boldsymbol{\alpha}_M^{q-1}\in(q-1+\mathfrak{m})^{N_M^{q-1}}}\right\}$$

We are now in position to use Fubini's Theorem,

$$\begin{aligned}Z(\mathbf{t},\mathfrak{o},\beta)&=\sum_{\mathbf{N}_1}\frac{t_1^{N_1^0}\cdots t_1^{N_1^{q-1}}}{N_1^0!\cdots N_1^{q-1}!}\cdots\sum_{\mathbf{N}_M}\frac{t_M^{N_M^0}\cdots t_M^{N_M^{q-1}}}{N_M^0!\cdots N_M^{q-1}!}\\&\quad\times\prod_{r=0}^{q-1}\int_{\boldsymbol{\alpha}_1\in(r+\mathfrak{m})^{N_1^r}}\cdots\int_{\boldsymbol{\alpha}_M\in(r+\mathfrak{m})^{N_M^r}}\prod_{m=1}^{M}\prod_{j<k}^{N_m}|\alpha_{m,j}-\alpha_{m,k}|^{\beta Q_m^2}\\&\quad\times\prod_{\ell<m}^{M}\prod_{j=1}^{N_\ell^r}\prod_{k=1}^{N_m^r}|\alpha_{\ell,j}-\alpha_{m,k}|^{\beta Q_\ell Q_m}\\&\quad\times d\mu^{N_1^r}(\boldsymbol{\alpha}_1)\cdots d\mu^{N_M^r}(\boldsymbol{\alpha}_M).\end{aligned}$$

Note that each of the integrands appearing in the inner most product is invariant under translation $r+\mathfrak{m}\mapsto\mathfrak{m}$, and we find our first major simplification (by appropriate definition, but still!),

$$Z(\mathbf{t},\mathfrak{o},\beta)=\sum_{\mathbf{N}_1}\frac{t_1^{N_1^0}\cdots t_1^{N_1^{q-1}}}{N_1^0!\cdots N_1^{q-1}!}\cdots\sum_{\mathbf{N}_M}\frac{t_M^{N_M^0}\cdots t_M^{N_M^{q-1}}}{N_M^0!\cdots N_M^{q-1}!}\prod_{r=0}^{q-1}Z(\mathbf{N}^r,\mathfrak{m},\beta)$$

Now, the sum over $\mathbf{N}_1,\ldots,\mathbf{N}_M$ can be thought of as a sum over the rows of all $M\times q$ matrices with positive integer coefficients. We may reindex this sum so that instead of summing over rows, we sum over the columns $\mathbf{N}^0,\ldots,\mathbf{N}^{q-1}$. When we do this, we regroup the monomials accordingly

$$\begin{aligned}&\frac{t_1^{N_1^0}\cdots t_1^{N_1^{q-1}}}{N_1^0!\cdots N_1^{q-1}!}\cdots\frac{t_M^{N_M^0}\cdots t_M^{N_M^{q-1}}}{N_M^0!\cdots N_M^{q-1}!}=\frac{t_1^{N_1^0}\cdots t_M^{N_M^0}}{N_1^0!\cdots N_M^0!}\cdots\frac{t_1^{N_1^{q-1}}\cdots t_M^{N_M^{q-1}}}{N_1^{q-1}!\cdots N_M^{q-1}!}=\\&=\frac{\mathbf{t}^{\mathbf{N}^0}}{\mathbf{N}^0!}\cdots\frac{\mathbf{t}^{\mathbf{N}^{q-1}}}{\mathbf{N}^{q-1}!}.\end{aligned}$$

It follows that

$$\begin{aligned}Z(\mathbf{t},\mathfrak{o},\beta)&=\sum_{\mathbf{N}^0}\cdots\sum_{\mathbf{N}^{q-1}}\frac{\mathbf{t}^{\mathbf{N}^0}}{\mathbf{N}^0!}\cdots\frac{\mathbf{t}^{\mathbf{N}^{q-1}}}{\mathbf{N}^{q-1}!}Z(\mathbf{N}^0,\mathfrak{m},\beta)\cdots Z(\mathbf{N}^{q-1},\mathfrak{m},\beta)\\&=\sum_{\mathbf{N}^0}\frac{\mathbf{t}^{\mathbf{N}^0}}{\mathbf{N}^0!}Z(\mathbf{N}^0,\mathfrak{m},\beta)\cdots\sum_{\mathbf{N}^{q-1}}\frac{\mathbf{t}^{\mathbf{N}^{q-1}}}{\mathbf{N}^{q-1}!}Z(\mathbf{N}^{q-1},\mathfrak{m},\beta)\\&=\left[\sum_{\mathbf{N}}\frac{\mathbf{t}^{\mathbf{N}}}{\mathbf{N}!}Z(\mathbf{N},\mathfrak{m},\beta)\right]^q\\&=Z(\mathbf{t},\mathfrak{m},\beta)^q.\end{aligned}$$

□

Acknowledgements

I have had many conversations about the 'p-adic Selberg integral' and 'repelling p-adic random variables' over the last decade that ultimately lead to the current work. In particular, Oregon colleagues Matt Grimes, Jonathan Wells, Joe Webster and Ben Young have shared their insights on the topic, and a subset of them are working on continutions of the work presented here. Much of the groundwork was laid at a series of small workshops (SQuaRE) hosted by the American Institute of Mathematics where the author were joined by Igor Pritsker, Jeff Vaaler and Maxim Yattselev, who encouraged our fascination in the current work.

I am in debt to the organizers of the XX School of Mathematics Lluís Santaló 2019: p-Adic Analysis, Arithmetic and Singularities. It was here where I had the opportunity to learn the connections between p-adic physics, algebraic geometry, number theory and place my work in a larger framework.

I also appreciate the comments of the reviewer(s) who provided many of the connections between the present work and existing results in p-adic physics including and beyond the electrostatic model. I am indebted to the generosity of the organizations and individuals who contributed.

References

[1] Matthew Baker, *An introduction to Berkovich analytic spaces and non-Archimedean potential theory on curves*, p-adic geometry, Univ. Lecture Ser., vol. 45, Amer. Math. Soc., Providence, RI, 2008, pp. 123–174, DOI 10.1090/ulect/045/04. MR2482347

[2] Smain Balaska and Toufik Sahabi, *Boundary states and correlation functions of tricritical Ising model from Coulomb-gas formalism*, Commun. Theor. Phys. (Beijing) **51** (2009), no. 1, 115–122, DOI 10.1088/0253-6102/51/1/22. MR2548809

[3] M. Bocardo-Gaspar, Willem Veys, and W. A. Zúñiga-Galindo, *Meromorphic continuation of Koba-Nielsen string amplitudes*, J. High Energy Phys. **9** (2020), 138, 43, DOI 10.1007/jhep09(2020)138. MR4190286

[4] Miriam Bocardo-Gaspar, H. García-Compeán, and W. A. Zúñiga-Galindo, *Regularization of p-adic string amplitudes, and multivariate local zeta functions*, Lett. Math. Phys. **109** (2019), no. 5, 1167–1204, DOI 10.1007/s11005-018-1137-1. MR3946489

[5] Joe Buhler, Daniel Goldstein, David Moews, and Joel Rosenberg, *The probability that a random monic p-adic polynomial splits*, Experiment. Math. **15** (2006), no. 1, 21–32. MR2229382

[6] B. Dragovich, A. Yu. Khrennikov, S. V. Kozyrev, I. V. Volovich, and E. I. Zelenov, *p-adic mathematical physics: the first 30 years*, p-Adic Numbers Ultrametric Anal. Appl. **9** (2017), no. 2, 87–121, DOI 10.1134/S2070046617020017. MR3652559

[7] Freeman J. Dyson, *Statistical theory of the energy levels of complex systems. III*, J. Mathematical Phys. **3** (1962), 166–175, DOI 10.1063/1.1703775. MR143558

[8] Paul Fili and Zachary Miner, *Equidistribution and the heights of totally real and totally p-adic numbers*, Acta Arith. **170** (2015), no. 1, 15–25, DOI 10.4064/aa170-1-2. MR3373827

[9] Paul Fili, Clayton Petsche, and Igor Pritsker, *Energy integrals and small points for the Arakelov height*, Arch. Math. (Basel) **109** (2017), no. 5, 441–454, DOI 10.1007/s00013-017-1080-x. MR3710764

[10] P. J. Forrester, *Log-gases and random matrices*, London Mathematical Society Monographs Series, vol. 34, Princeton University Press, Princeton, NJ, 2010, DOI 10.1515/9781400835416. MR2641363

[11] Peter J. Forrester and S. Ole Warnaar, *The importance of the Selberg integral*, Bull. Amer. Math. Soc. (N.S.) **45** (2008), no. 4, 489–534, DOI 10.1090/S0273-0979-08-01221-4. MR2434345

[12] Fernando Q. Gouvêa, *p-adic numbers*, 2nd ed., Universitext, Springer-Verlag, Berlin, 1997. An introduction, DOI 10.1007/978-3-642-59058-0. MR1488696

[13] Jun-ichi Igusa, *An introduction to the theory of local zeta functions*, AMS/IP Studies in Advanced Mathematics, vol. 14, American Mathematical Society, Providence, RI; International Press, Cambridge, MA, 2000, DOI 10.1090/amsip/014. MR1743467

[14] Shinsuke Kawai, *Coulomb-gas approach for boundary conformal field theory*, Nuclear Phys. B **630** (2002), no. 1-2, 203–221, DOI 10.1016/S0550-3213(02)00150-5. MR1903806

[15] Shinsuke Kawai, *Free-field realization of boundary states and boundary correlation functions of minimal models*, J. Phys. A **36** (2003), no. 24, 6875–6893, DOI 10.1088/0305-4470/36/24/321. MR2004318

[16] Madan Lal Mehta, *Random matrices*, 3rd ed., Pure and Applied Mathematics (Amsterdam), vol. 142, Elsevier/Academic Press, Amsterdam, 2004. MR2129906

[17] Alexander Ostrowski, *Über einige Lösungen der Funktionalgleichung* $\psi(x) \cdot \psi(x) = \psi(xy)$ (German), Acta Math. **41** (1916), no. 1, 271–284, DOI 10.1007/BF02422947. MR1555153

[18] Atle Selberg, *Remarks on a multiple integral* (Norwegian), Norsk Mat. Tidsskr. **26** (1944), 71–78. MR18287

[19] David Tong, *Lectures on statistical physics*, 2011.

[20] V. S. Vladimirov, I. V. Volovich, and E. I. Zelenov, *p-adic analysis and mathematical physics*, Series on Soviet and East European Mathematics, vol. 1, World Scientific Publishing Co., Inc., River Edge, NJ, 1994, DOI 10.1142/1581. MR1288093

[21] Igor V. Volovich, *Number theory as the ultimate physical theory*, p-Adic Numbers Ultrametric Anal. Appl. **2** (2010), no. 1, 77–87, DOI 10.1134/S2070046610010061. MR2594440

[22] Joe Webster, log-*Coulomb gas with norm-density in p-fields*, p-Adic Numbers Ultrametric Anal. Appl. **13** (2021), no. 1, 1–43, DOI 10.1134/s2070046621010015. MR4218379

[23] W. A. Zúñiga-Galindo and Sergii M. Torba, *Non-Archimedean Coulomb gases*, J. Math. Phys. **61** (2020), no. 1, 013504, 16, DOI 10.1063/1.5127191. MR4048290

[24] W. A. Zúñiga Galindo, B. A. Zambrano-Luna, and Edwin León-Cardenal, *Graphs, local zeta functions, log-coulomb gases, and phase transitions at finite temperature*, 2020.

University of Oregon, Eugene Oregon, 97403
Email address: csinclai@uoregon.edu

Published Titles in This Subseries